Holger Preuschoft and
David J. Chivers (eds.)

Hands of Primates

Springer-Verlag Wien New York

Dr. Holger Preuschoft
Abteilung für Funktionelle Morphologie
Ruhr-Universität Bochum
Bochum, Federal Republic of Germany

Dr. David J. Chivers
Department of Anatomy
University of Cambridge
Cambridge, United Kingdom

Softcover reprint of the hardcover 1st edition 1993

Printed on acid-free paper

With 224 Figures

ISBN-13: 978-3-7091-7434-0 e-ISBN-13: 978-3-7091-6914-8
DOI:10.1007/ 978-3-7091-6914-8

CONTENTS

LIST OF CONTRIBUTORS

Prof. Dr. R. McNeill Alexander
Dept. of Pure and Applied Zoology, University of Leeds, Leeds LS2 9JT, U.K.

Hans-Friedrich Bär
Chirurgische Klinik und Poliklinik, «Bergmannsheil Bochum» Universitätsklinik, Gilsingstrasse 14, D-4630 Bochum 1, FRG

Dr. Christopher Beard
Dept. of Cell Biology and Anatomy, Johns Hopkins University School of Medicine, Baltimore, M.D.
New address: Section of Vertebrate Fossils, Carnegie Museum of Natural History, 4400 Forbes Ave., Pittsburgh, PA 15213, USA

Dr. Christophe Boesch
Ethologie und Wildforschung, Zoologisches Institut der Universität CH-8057 Zürich, Zürich
New address: Zoologisches Institut der Universität Basel, Rheinsprung 9, CH-4051 Basel, Switzerland

Dr. Hedwige Boesch
Ethologie und Wildforschung, Zoologisches Institut der Universität CH-8057 Zürich, Zürich
New address: Zoologisches Institut der Universität Basel, Rheinsprung 9, CH-4051 Basel, Switzerland

Dr. Beate Brand-Saberi
Institut für Anatomie, Ruhr-Universität Bochum Gebäude MA 5/148, D-4630 Bochum 1, FRG
New address: Abteilung Anatomie II, Anatomisches Institut der Albert-Ludwigs-Universität Freiburg, Albertstrasse 17, D-7800 Freiburg i.Br., FRG

Dr. Philippe Brenot
Laboratoire d'Anthropologie, Université de Bordeaux I, F-33400 Talence, France

Dr. Christoph Buck
Abteilung Funktionelle Morphologie, Ruhr-Universität Bochum, D-4630 Bochum 1, FRG
New address: Gyrenkampstrasse 5, D-4300 Essen-Werden, FRG

Prof. Dr. Bodo Christ
Abteilung Anatomie II, Anatomisches Institut der Albert-Ludwigs-Universität Freiburg, Albertstrasse 17, D-W-7800 Freiburg i.Br., FRG

Marianne Christel
Abteilung Funktionelle Morphologie, Ruhr-Universität Bochum, D-4630 Bochum 1, FRG
New address: Hochkirchstrasse 5, D-1000 Berlin 62, FRG

Dr. Brigitte Demes
Dept. of Anatomical Sciences, Health Sciences Center, State University of New York at Stony Brook, Stony Brook, N.Y. 11794-8081, USA

Dr. Erika Doden
Klinik für Mund-, Kiefer- und Gesichtschirurgie, Klinikum der Christian-Albrechts-Universität, Arnold-Heller-Strasse 16, D-W-2300 Kiel 1, FRG

Dr. Jens Lorenz Franzen
Forschungsinstitut Senckenberg, Senckenberganlage 25, D-W-6000 Frankfurt a.M. 1, FRG

Dr. Marc Godinot
Laboratoire de Paléontologie, Institute des Sciences de l'Evolution, Université Montpellier II, Place Eugène Bataillon, F-34060 Montpellier, France

Dr. Milos Grim
Abteilung Anatomie II, Anatomisches Institut der Albert-Ludwigs-Universität Freiburg, Albertstrasse 17, D-7800 Freiburg i.Br., FRG

Dr. Michael M. Günther
Abteilung Funktionelle Morphologie, Ruhr-Universität Bochum, D-4630 Bochum 1, FRG
New address: Dept. of Human Anatomy & Cell Biology, University of Liverpool, P.O. Box 147, Liverpool L69 3BX, U.K.

Prof. Dr. Wolfgang F. Gutmann
Forschungsinstitut Senckenberg, Senckenberganlage 25, D-6000 Frankfurt a.M. 1, FRG

Prof. Dr. Akira Ikeda
Dept. of Anatomy, Kawasaki Medical School, Kurashiki City, Okayama 70101, Japan

Dr. Heinz Jürgen Jacob
Abteilung Anatomie und Embryologie, Ruhr-Universität Bochum, Gebäude MA 6/30, D-4630 Bochum 1, FRG

Dr. Françoise K. Jouffroy
Unité de Recherche Associée 1137, Laboratoire d'Anatomie Comparé, M.N.H.N. Paris VII, 55, rue Buffon, F-75005 Paris, France

Dr. Robert F. Ker
Dept. of Pure and Applied Biology, University of Leeds, Leeds LS2 9JT, U.K.

Prof. Dr. Jürgen Koebke
Anatomisches Institut der Universität Köln, Joseph-Stelzmann-Strasse 9, D-5000 Köln 41, FRG

Prof. Dr. J.M.F. Landsmeer
Vakgroep Anatomie, Faculteit de Geneeskunde Rijksuniversiteit te Leiden
Wassenaarseweg 62, NL-2300 RC Leiden, Netherlands

Prof. Dr. Wolfgang Maier
Lehrstuhl für Spezielle Zoologie, Institut für Biologie III, Universität Tübingen, Auf der Morgenstelle 28, D-7400 Tübingen, FRG

Dr. Shin Matsumoto
Dept. of Anatomy, Kawasaki Medical School, Kuraschiki City, Okayama, 70101, Japan

Dr. Yoshihiko Nakano
Japan Monkey Centre, Institute and Museum of Primatology, Inuyama, Aichi 484, Japan

Dr. Ute Nieschalk
Abteilung Funktionelle Morphologie, Ruhr-Universität Bochum, D-4630 Bochum 1, FRG
New address: Silberbachstrasse 5a, D-6239 Eppstein/Ts. 6, FRG

Prof. Dr. Holger Preuschoft
Abteilung Funktionelle Morphologie, Ruhr-Universität Bochum, Gebäude MA 0/44, D-4630 Bochum 1, FRG

Dr. Richard G. Rawlins
Dept. of Obstetrics and Gynecology, Rush-Presbyterian, St. Luke's Medical Center, 1653 West Congress Parkway, Chicago, IL 61612, USA

Dr. M.E. Redshaw
Dept. of Child Health, Institute of Child Health, University of Bristol and Royal Hospital for Sick Children, St. Michel's Hill, Bristol 52 U.K.
New address: 4, Northcote Road, Clifton, Bristol BS8 3HB, U.K.

Ir. G. Pauline Rijken-Rauwerdink
Vakgroep Funktionele Anatomie, Rijksuniversiteit Utrecht, Catharinjesingel 59, NL-3511 GG Utrecht, Netherlands

Dr. Sabine Schöneich
Abteilung für Verhaltensforschung, Zoologisches Institut der Westfälischen Wilhelms-Universität Münster, Badestrasse 9, D-4400 Münster, FRG
New address: Deller Heide 15, D-W-4220 Dinslaken, FRG

Dr. Mark F. Teaford
Dept. of Cell Biology and Anatomy, Johns Hopkins University School of Medicine, 725 North Wolfe Street, Baltimore, MD 21205, USA

Dr. Naoto Umeda
Dept. of Anatomy, Kawasaki Medical School, Kurashiki City, Okayama 70101, Japan

Prof. Dr. Alan Walker
Dept. of Cell Biology and Anatomy, Johns Hopkins University School of Medicine, 725 North Wolfe Street, Baltimore, MD 21205, USA

Introduction

The hand commonly is considered to have exerted great influence on the evolution of typically human characteristics, like upright posture, stereoscopic vision, «manipulative» handling of parts of the environment. The German term «Begreifen», which is commonly used for the understanding of complex relationships in a generalised, abstract sense, always implies the original meaning of seizing objects with the aid of the hands. The hands are also of greatest importance for the survival of the other, non-human primates. Hands are absolutely essential for locomotion in an arboreal habitat, and the intake of food is dependant on the use of the hands as well: primates very rarely take in food directly with their mouths, in the wast majority of cases they seize food items with their hands. Even drinking is often performed by dipping the hand into the water and licking the drops from hand and forearm.

An organ of such importance will very probably be «adapted» to its «function». This statement is made so often, that any further considerations seem superfluous. Nobody doubts, that the hands of primates are highly adaptive organs, the general form and internal structure of which are closely related to the necessities of life.

However, if one tries to go beyond this general statement, he finds himself confronted with several problems:

First of all, a point which W. GUTMANN has emphasized repeatedly: according to the results obtained by genetics, the first thing to appear is the mutated character. It follows that the animal who possesses this mutation has to arrange its life in such a way to make optimal use of the character: that means to find a «niche». Most of the discussions about adaptation, however, start off from the silent precondition that the niche exists first, and the only thing the animal has to do is to adjust, or to "adapt" its body structure to the requirements of that niche.

Second, if one asks to which «biological roles» and to which «functions» exactly the hand in a given species is adapted, one finds oneself soon surrounded by numerous neither very detailed, nor clear statements, and only few precise analyses. Many considerations end up with NAPIER's discrimination and definition of power grip and precision grip — which has turned out to be very useful in surgery — and the restating of man's distinctiveness.

The use primates make of their hands seems to be incredibly variable. In a close-up view, however, much of the variability is confined to situations in which the fingers of the hand do not exert high forces, or, to put it in a positive way, in which the hands exert forces below the maximal forces that can be produced. The same force which is exerted by the hand against a part of the primate's environment evokes of course a «reaction force» of the same size, but opposite direction from outside on the hand. It seems easy to understand that a primate's hand must be able to sustain even the highest of the «external» or reaction forces, to which it is normally exposed. «Normally» means, in events which occur daily, or once a week in the animal's life. The strength of the tissues probably exceeds these necessities by a certain safety factor, which allows exposure in cases of emergency to still higher forces than those occuring «normally». The resistence to the external forces evokes inside the hand (as well as other parts of the locomotor apparatus) «internal» forces, or mechanical «stresses».

The strength of body segments (and of the hand) reaches its limits by the weight of the bony, ligamentous, tendinous, or muscular material necessary in view of limitations of the strength properties inherent to the materials.

To sustain the highest internal forces with a minimum of material means being adapted to a «normal function» (in a strict, mechanical sense). Evidently, locomotion , or exerting a «power grip», evokes higher external forces against a hand than most sorts of «precision handling». Therefore, we have to consider — in order to understand the morphological adaptations of primate hands — first of all the way in which high forces are applied to the hand in locomotion and power-involving behaviors.

Detailed, close-up analyses made earlier (PREUSCHOFT, 1968, 1969, 1971, 1973,a,b, 1987) investigated the relation between reaction forces, internal stress distribution and shape. They also showed the shapes of primate hands to be easier understood as adaptations to their specific locomotor behavior than to their precision handling. In short, primates seem to acquire their hand shape with respect to locomotion and to make the best of it when in the need of performing precise manipulations. This also explains the variability of precision handling among species: a given shape dictates the way in which the hand can be used.

As a whole, studies on the hand and upper extremity are much rarer than on the foot and lower limb. This may be due to the lesser degree of specialisation of the human hand in comparison to the foot. Another reason may be the tremendous variability of hand use, which makes it difficult to identify the most characteristic functions, or ways in which the hand is stressed. The characteristic features of the human hand are to a large extent shared by the hands of other primates, and therefore alone it seems logical to approach the human hand by looking into the details of hand function and hand morphology in non-human primates. In addition: evolution is a process, in which one functioning organism is transformed into another. Therefore the result depends from the starting point of the process. A comparison with the forelimbs of other mammals, or of lower vertebrates also seems promising, since a well-known hypothesis interpretes the human hand as «primitive». All these aspects need a thourough reconsideration in view of our present knowledge.

One prerequisite for making progress on this field is an intimate knowledge of hand use in a variety of animals. Even basic informations about hand use in primates, which could serve as a starting point for a better approach are rare. Only recently, some knowledge is being accumulated, largely through the great number of behavioral or field studies on primates.

The other essential for approaching the problems mentioned is the detailed knowledge of the anatomical structure «hand» and of the evolutionary pathway which led to its present form. Available is a bulk of descriptions of the anatomy, while the survey of general metrics is incomplete. Whereas the anatomy of the hand is largely known (though not completely), our knowledge of the variations (recent and phylogenetic) is still severely limited.

In view of this not very satisfactory situation some of the authors have met at the Reisenburg in 1985 (sponsored by the Deutsche Forschungsgemeinschaft) to collate and to discuss the available information. It became evident, that there are two ways for further improvement of our understanding the primate hand: Either by taking the morphological structure as a starting point and by raising the question how it functions, and which biological role it plays for the individual. Or by starting from the requirements of actual life of the organism, and searching for possibilities which improve the chances of survival. To make use of these possibilities, the animal needs an organ, the form of which is optimised during evolution within the limits set by the Bauplan. With the increasing consideration of biomechanic principles, a new understanding of morphology becomes possible. In other words, the necessities of hand shape can sometimes be identified with physical exactness. The ontogenetic origin of the hand gives an example of the origin of shape in general.

The discussions between the authors have been continued in the years after the symposium. Other colleagues have produced additional information and developed new ideas, and a number of new investigations have been carried out in persuit of the discussions.

Now we can present a well-integrated series of articles which are products of the extended exchange. All of them are approaches to the above outlined problem: What are the biological roles of the various primate hands, what is their function, which morphological changes did the terminal segments of the primate forelimb undergo during evolution? What characterises their present forms and which ontogenetic processes have lead to their definite shapes?

All these aspects are delt with in several papers, starting off with divergent viewpoints. Some of them are aimed at more than one aspect. The arrangement of the articles therefore is to some extent arbitrary and cannot claim systematic correctness. All contributors have done considerable empirical work, in most cases on hands. We, the editors, have tried to make them present their results and to evaluate them with regard to the above-mentioned ideas. In a few cases, contributions are not directly based on new empirical information, but these authors open new ways of interpreting known data and thus help us to understand better the available data or see clearer the weakness of common ideas.

A first group of articles is focused on the use made of the hands by non-human primates. Admittedly, only some species are considered. But new evidence as to the manipulative abilities is collected, and the importance of non-locomotory hand-use is investigated down to the selective advantage contained in using the hands for elaborate tasks (tool-using) in terms of energy-expenditure.

A second group of contributions contains new information about how primate hands are shaped and about the «mechanical functions» of morphological traits and the «biological roles» the hand plays in the life of some primates. Entirely new approaches to hand function are made and known ones are reviewed. Some articles fill long-standing gaps in our knowledge.

The last section of the book is devoted to the phylogenetic and ontogenetic evolution of the primate hand. Whereas an article on ontogeny aims at the processes which lead to the development of the characteristic form, its variability through evolutionary time is the subject of others. The papers add new information about the earliest evolutionary stages of primates as a whole, or to the infra-order Hominoidea, to which humans belong. A comprehensive survey of the available paleontological evidence is also included.

The editors:
Holger Preuschoft
David Chivers

I

HAND USE

S. Schöneich

Hand usage in the ring-tailed lemur (*Lemur catta* Linnaeus 1758) when solving manipulative tasks *

Two ringtailed lemurs (1,1) learned to open an increasing number of different fastenings (e.g. to press a lever, to open a hook, to throw back the bar of a padlock loop) of a problem box containing food reward. They both opened the fastenings normally with their hands, very seldom with their snout or teeth. The female subject preferred manipulating with her right hand, whereas the male preferred his left hand. Until the lemurs solved and mastered a newly-presented manipulative task they tested various movements, all which they were able to perform with their hands. They developed individual solving strategies in opening unknown fastenings and opening fastenings presented as reflected in a mirror. The results show that lemurs do not use just one prehensive pattern in every situation. They use their hands for fine manipulations in very different and individual grasping motions, although they are only capable of whole-hand control.

Key words: action chain, fine manipulation, hand preference, hand use, problem box.

Introduction

Since the first studies of Thorndike (1898), who introduced the problem-box, the removing of obstacles and releasing of fastenings of such a box offer the possibility to gain some knowledge on one hand about the manipulative ability of animals and on the other hand about how to solve problems and understanding the task. The behaviour of various species, e.g. procyonids (Cole, 1907; Rensch & Dücker, 1969; Löhmer, 1975), sciurids (Yoakum, 1909; Dücker *et al.*, 1982), tree shrews (Schöneich, 1986 and 1988), primates (Thorndike, 1901; Kinnaman, 1902; Kohts, 1928; Rensch & Dücker, 1966; Simons & Holtkötter, 1986) and birds (Dücker *et al.*, 1986) has been examined by this method. In order to determine individual limits, the difficulty of the necessary manipulation (by using different obstacles), as well as the number of obstacles to be removed have to be increased.

This paper presents some observations on how prosimians, which are said to be only capable of whole-hand control, are able to use their hands when solving these tasks.

In the early research of Bierens de Haan in 1930, as well as in the observations of Petter in 1962 and Jolly 1964, lemurs are said to prefer using their snouts and teeth for fine manipulations instead of their hands. Bishop wrote in 1964 about the use of hands, that the hook-like prehension of the lemur is well adapted to pulling a branch and holding food. Lemuriforms grasp with parallel fingers, touching the branch or food with the finger tips first. They use just one prehensive pattern in every situation, no matter whether they are reaching for a raisin or a branch. However, the shape of the object determines the final, variable grasping motion.

* *Bernhard Rensch in memoriam.*

Field studies, on the other hand (e.g. SUSSMANN, 1977; SUSSMANN & TATTERSALL, 1976), describe varied manipulative abilities of the lemurs during foraging and ingestion.

The ability of lemurs to open a locked problem box was first investigated by VÖSSING (1984 and 1987). As his investigation was focussed on work division within a social group, his research does not contain a description of the manipulative abilities of the animals.

Method

Subjects and setting

Two adult ring-tailed lemurs, male and female, took part in the experiments. The ring-tailed lemur is a diurnal herbivorous species. HILL (1953) described it as the most terrestrial species in the family Lemuridae. Its terrestrial habit and behavioural thermoregulation seem to be factors that enable the ring-tailed lemur to exploit different habitats (SUSSMAN, 1974 and GANZHORN, 1985).

The two lemurs lived in a cage (2x4m base, 2m height), which was divided into two equal parts (2x2m each) during experiments, so that the lemurs operated in a familiar environment. Their diet consisted of fruit, vegetables, nuts and special dumplings, made of cottage cheese, eggs, mash for babies, boiled rice, rolled oats, honey, dextrose, vitamin preparation and a mineral preparation. The lemurs were not deprived of food, although before the experiments started in the morning all food remnants were removed from their cage.

Problem box

Figure 1 shows the problem box (48 cm high, 45 cm wide, 32 cm deep) with the first 4 successively-presented fasteners on the outer door, a snap door. The problem box was designed in such a way that the fasteners could be fixed in different positions (4 positions at each side of the outer door). During the first experiment, the snap door was hinged on the right-hand side, in the second experiment on the left-hand side. In the inner part of the box was a second door. This was a sliding door which could be opened likewise either to the left or to the right (distance to the snap door: 12 cm). At this door the fasteners could be fixed in 3 positions: at both sides and at the upper part of the door (*Figure 2*).

Figure 1 - Problem-box with fasteners at the snap door: D1 = snap door, L = lever, H = hook, P = peg, B = bolt.

Figure 2 - Problem bow with fasteners at the sliding door: D2 = sliding door, H = hook, P = peg, BP = bar of a padlock, CH = chain.

The fasteners and the doors had been chosen in such a manner that occasionally other movements, like pressing, pulling, pushing, and so forth, were necessary for opening. In the second experiment, the animals had to transfer some movements into several directions. At both sides of the problem box a window was put between snap- and sliding-door in order to observe the manipulations done by the subject at the sliding door.

During the first experiment all fasteners were presented at the left-hand side of the box, while in the second they were presented as reflected in a mirror to the right side. In the first experiment the lemurs had to master 10 tasks to be successful. At the end of the first experiment, they had to manipulate the doors, which were arranged in the following sequence (the number in parenthesis shows the sequence, in which the individual task was introduced):

Experiment 1:

Snap door

1. press a lever (2.)
2. open a hook (3.)
3. push a wooden pin (fixed with a thread) out of a padlock loop (4.)
4. remove a bolt from the right- to the left-hand side (5.)
5. open a snap door (1.).

The opening sequence of the fasteners could be chosen freely. After pressing the lever the snap door flew open a narrow slit (approximately 1cm).

Sliding door

6. open a hook into the direction of the door (7.)
7. push a wooden pin out of a padlock loop (8.)
8. open the bar of a padlock from left- to the right-hand side (9.)
9. take off a chain out of a loop (10.)
10. open a sliding door from left- to right-hand side (6.)

Experiment 2:

Like experiment 1, but all fasteners and doors were presented as reflected in a mirror.

Procedure

While the experiments were carried out (between 1100 and 1400 h) each lemur was occasionally kept separately in one part of the cage. The course of the experiments was filmed (for later analysis) with a video camera.

The problem box was baited with a delicacy, normally a piece of a fig or raisin.

The problem box was placed within the cage in such a way, that the non-active subject was not able to observe the manipulations done by the other subject at the box.

Normally, a maximum of five trials took place each day. Sometimes only one experiment or even none took place, because the animals were not interested in opening the box.

After habituating the animals to the experimental situation and training them to find bait in the box, the experiments began. For this experiment the box was closed successively.

The first task was to open the snap door. Having mastered opening this door, an additional fastener was introduced. The sequence of different fastenings was the following: lever, hook, padlock with pin and bolt.

Learning criterion was to open the box 5 times in quick succession (and 10 times respectively, if the first trial had been successful). The learning criterion which had to be reached before a new fastener was fixed is a so-called «straight solution». This means the animals were only confronted with a new task once they had solved the previous one or, the other way round, when they failed, this task was eliminated from the experiment.

Each lemur had ten minutes in which to open the box completely. If one failed several times, he or she lost all interest in opening the box after the first few minutes of each trial.

If an animal failed to open a newly-introduced fastening frequently, two indirect learning aids were given:

a) to renew motivation: the unknown fastening was opened or removed so that the subject only had to open fastenings already known (these trials have not been valued);

b) to focus attention on the new fastening: already known fastenings had been removed and the subject was only confronted with the new fastening.

The snap door was the only door opened a little at the beginning of experiment 1. All the other fastenings and the sliding door were introduced already locked.

Results

Experiment 1

At the beginning, those parts of the animal's body will be described which were used while learning to solve the task. This will be followed by a description of the lemur's behaviour while opening the box.

WHICH PARTS OF THE BODY ARE USED FOR THE MANIPULATIONS AT THE PROBLEM BOX?

Until the lemurs solved and mastered a new manipulative task they tested various movements, all of which they were able to perform with their hands, snout or teeth. *Figure 3* summarizes the male's and female's preferences for using either hands, snout or teeth when manipulating at the box in the first experiment (to learn to open the snap door with all fastenings and also the sliding door with all fastenings). This preference is expressed in percent frequency (of the right hand, RH, left hand, LH, both hand together, BH, of the snout, S and of the teeth, T).

Both subjects preferred manipulating with their hands. The female preferred manipulating with her right hand whereas the male preferred his left hand. Besides making use of just one hand, each subject manipulated relatively often with both hands. Snout and teeth were hardly used.

First, both animals manipulated indiscriminately at the left door slit and the fastenings of the snap door and likewise at the sliding door. This is typical for trial-and-error learning.

Figure 4 shows which parts of the body were used for opening a snap door and its different fastenings.

The use of different parts of the body to open different fastenings at the snap door is shown in percent frequency, for the male on the left side and for the female on the right side.

The male exclusively released the fastenings with his hands, whereas the female at least occasionally used her snout for removing the hook.

The results show that the subjects made use of their snouts and their teeth while learning but definitely preferred their hands once they learned the correct opening movement. Whether they employed their right or left hand was determined by their individual preference.

The female failed on the bolt. This fastening appeared to be an unsolvable problem for her.

Figure 5 again presents the use of different parts of the body among the animals for opening the given fastenings at the sliding door expressed in percent frequency. Whereas tasks 6 and 7 correspond to tasks 2 and 3 at the snap door, the tasks 8, 9 and 10 are newly introduced. The subjects transferred at once the successful movements of tasks 2 and 3 upon tasks 6 and 7. This was clearly recognizable in the nearly exclusive use of just one part of the body, whereas the use of different parts of the body in order to open the unknown fastenings or doors varied more in the three new tasks. Unequivocally, they prefer their hands.

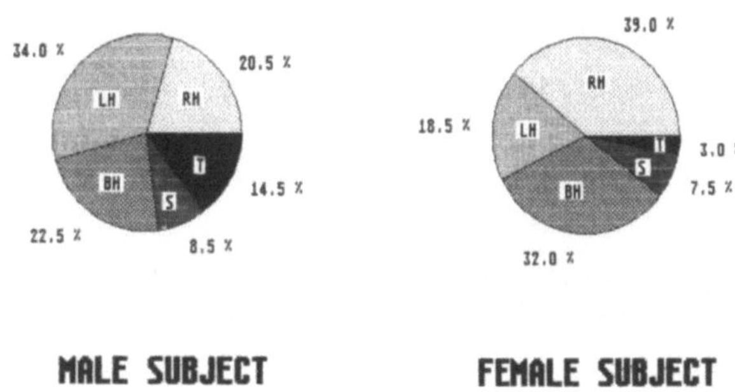

Figure 3 - Preferences for using different parts of the body when manipulating at the box (percent frequency): RH = right hand, LH = left hand, BH = both hands together, S = snout, T = teeth.

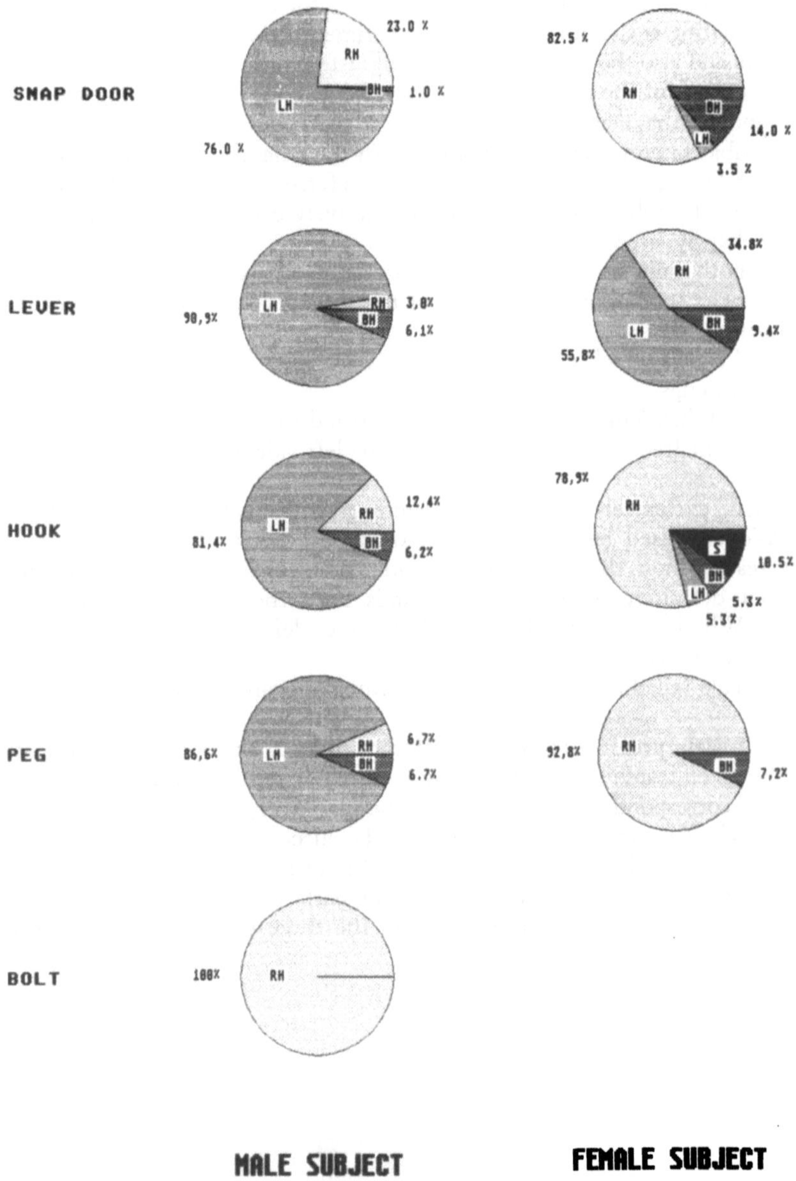

Figure 4 - Preferences for using different parts of the body when opening the snap door and its different fastenings (percent frequency): RH = right hand, LH = left hand, BH = both hands together, S = snout.

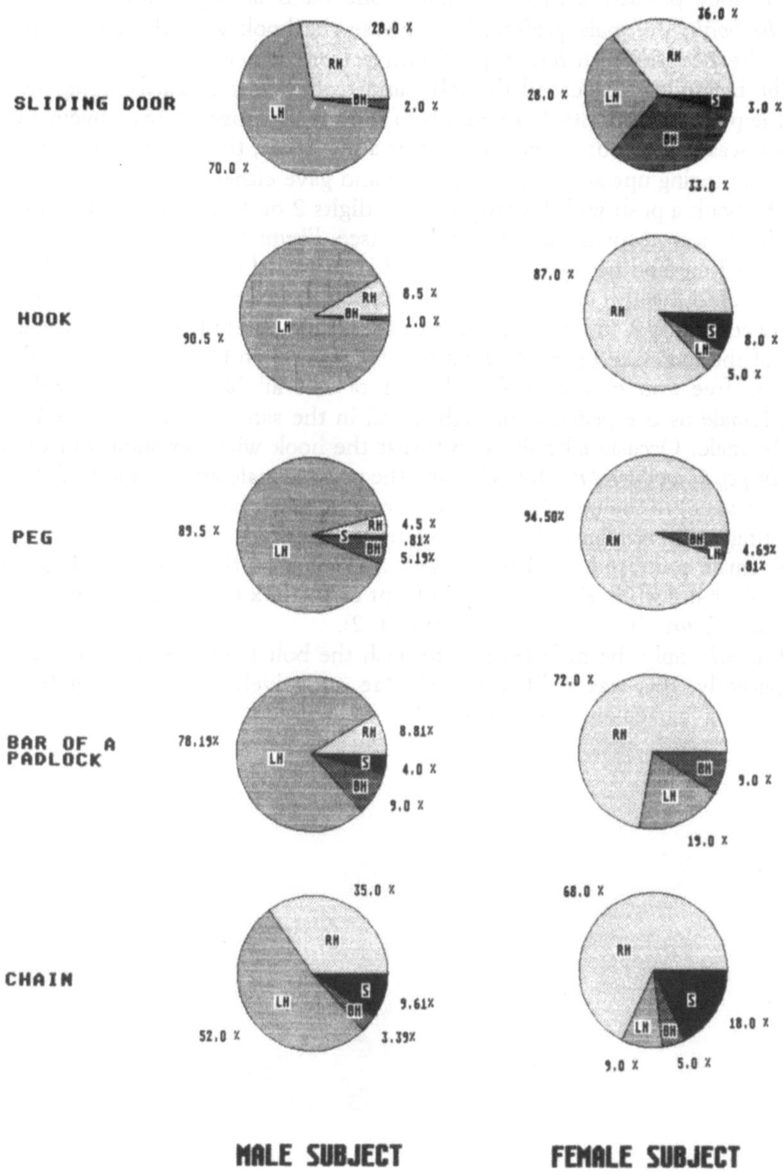

Figure 5 - Preferences for using different parts of the body when opening the sliding door and its different fastenings (percent frequency): RH = right hand, LH = left hand, BH = both hands together, S = snout.

How do they release the fastenings and open the doors?

1. *The lever*: both subjects pressed the lever with the front part of their palms down, seldom with the tips of their digits 2 to 4 only, at which in both cases the digits were crooked. If they pressed with both hands, one hand lay upon the other.

2. *The hook*: the male preferred to release the hook with the left hand. For this movement he had three different possibilities from which to choose.

In the beginning he pushed the left hand from below upwards along the door slit pressing the pads of his digits 2, 3 and 4 into the slit. Because of this movement, the arm of the hook was moved out of the loop. After a few trials, the lemur moved the left hand with the palm facing upwards quickly upward and gave either the arm of the hook or the point of the hook a push with his finger tip of digits 2 or 3. This was done in such a way that the hook was thrown out of the loop (see *Figure 6*).

The third method he used was to place his left hand upon the arm of the hook, his fingers parallel to the arm and his finger tips facing the loop. Usually the arm of the hook rested between digits 2 and 3. A rotation of the whole hand upwards tore the arm of the hook out of the loop (see *Figure 7*). As a rule the male then tried at once to pull the door open at the free arm of the hook, whether or not all fastenings were released.

The female used especially the right hand in the same way as described in method one for the male. Occasionally she also thrust the hook with her snout out of the loop.

3. *The peg (a wooden pin)*: for releasing the peg the male applied method 2 of opening the hook (*Figure 6*) to the peg (*Figure 8*). As a rule he gave the point of the peg such a push with his finger tip of digits 2 or 3 that it came out of the loop.

The female preferred another method: She released the peg by pulling the thread with her right hand while either sitting in front of the box (see *Figure 9*), or sometimes on the box (see *Figure 10*, made in experiment 2).

4. *The bolt*: only the male learned to push the bolt from the right- to the left-hand side. Apparently, this was a difficult task. He exclusively used his right hand for the correct opening movement (see *Figure 11*).

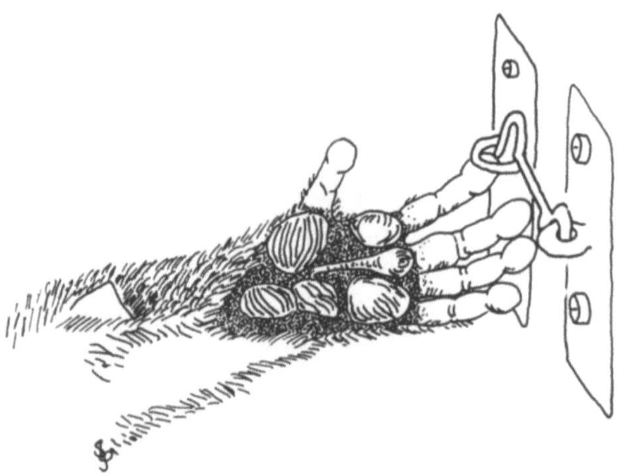

Figure 6 - Male subject opens the hook with his left hand.

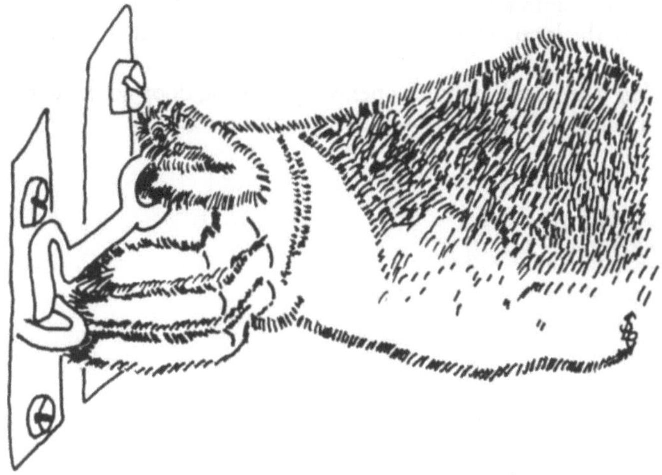

Figure 7 - Male subject opening the hook with his left hand.

5. *The snap door*: the male preferred to open the snap door with his left hand at the door edge. Later he opened the door by the free arm of the hook. The female used her right hand for opening the door by the door edge. In the course of this procedure she sometimes sat on the box.

6. and 7. *The hook and the peg at the sliding door*: both subjects used their hands in the same way as at the snap door.

8. *The bar of a padlock*: the animals put their hand from above onto the bar and took it off with the tips of digits 2 or 3. The male preferred using his left hand, whereas the female preferred her right hand.

Figure 8 - Male subject removes the peg.

9. *The chain*: The animals reached for the chain and dragged it upwardly, until the chain removed from the loop. The female sometimes used her snout as if she were opening the hook.

10. *The sliding door*: Both animals opened the sliding door with their hands touching and pulling at a doorknob (see *Figure 12*) or a fastener.

Figure 9 - Female subject removes the peg.

Figure 10 - Female subject removes the peg.

Figure 11 - Male subject removes the bolt with his right hand from the right to the left-hand side.

Experiment 2

The subjects needed far less trials for solving the new tasks (fasteners and doors presented as reflected in a mirror) than in experiment 1: only 3-5 trials compared to 15-25 trials in experiment 1. The subjects applied the old successful movements to the fasteners in the new position. *Figure 13*, for instance, shows the male opening the hook (compare with *Figure 6*).

Figure 12 - Male subject opens the sliding door with his left hand.

Figure 13 - Male subject opens the hook with his left hand.

Figure 14 - Male subject removes the bolt with his left hand.

The bolt was an exception. The male needed 32 trials to perform a new successful opening movement. First he tried to remove the bolt with his right hand from the right- to the left-hand side. He was not able to change the direction of the movement. After changing the hand he performed the successful opening movement. He now removed the bolt with his left hand from the left- to the right-hand side (see *Figure 14*).

In the course of his first straight solutions he spontaneously changed his hand for opening the hook: he opened the hook with his right hand in the way shown in *Figure 7* for the left hand.

Discussion

Two ringtailed lemurs (1,1) learned to open an increasing number of different fasteners of a problem box containing a food reward. The male subject mastered in experiment 1 all 10 tasks, whereas the female failed on the bolt. This fastener requires pulling or pushing from the right- to the left-hand side. The female showed this movement with her hands when manipulating the box, but not the bolt. The differences in the performance seem to be connected with a different motivation. The male had more perseverance when manipulating the box. Action chains in which they had to undo 10 (9) fastenings cannot be regarded as their limit. The aim of our experiments was not to test their limitations, but to show that the different manipulations seem to be present in their natural behaviour. The performance can be compared with that of procyonids (e.g. LÖHMER, 1975), sciurids (e.g. DÜCKER *et al.*, 1982) and tree shrews (SCHÖNEICH, 1986 and 1988).

First the lemurs opened fasteners just by chance, but already after a small number of successful solutions they developed individual and variable correct opening movements.

L. catta use their hands for fine manipulations, despite the fact that they are only capable of whole hand control. BIERENS DE HAAN (1930), PETTER (1962) and JOLLY (1964) observed that lemurs seem to prefer their snouts for fine manipulations (e.g. to pick up a raisin from a flat ground; to open a drawer first with their snouts and then, having opened it a narrow slit, they used their hand as if pulling a branch). The results of TORIGOE (1985) showed that lemurs manipulate with simple objects (a nylon rope and a wooden cube) in a «primate-common way»: they prefer touching the object with the mouth, biting or licking it. These results differ from the results obtained at the problem box in this research and at the Hasp Puzzle (JOLLY, 1964). In more difficult tasks lemurs used their hands for fine manipulations.

In the research of MILLIKEN *et al.* (1989), in which 13 animals (7,6) were observed, it is described that male ringtailed lemurs prefer their left hand when reaching an object. Female ringtailed lemurs are ambipreferent or prefer using their right hand. The authors connect the observed lateralization with sex differences and early experience. The research of WARD *et al.* (1990), in which a population of 194 lemurs (*Lemur* spp.) was studied, shows that there is a population-level bias in males toward use of the left hand in food reaching. My own results seem to confirm a left-hand preference in the male gender, if this conclusion is allowed after all, because only one male took part in the experiments.

In experiment 2 the animals learned to open fasteners and doors presented as reflected in a mirror. The results show that they applied the old successful movements to the fasteners in the new position. They seem to associate the opening movement with the fastener. If fasteners or doors needed the same grasping motion for opening, the subjects were at once successful. This is a so-called «positive transfer». First, the male was not able to change the direction of the movement for opening the bolt. After changing the hand he performed the successful opening movement. This is a so-called «negative transfer».

It could not be observed whether the subjects control their manipulation visually. Manipulating the fasteners, they often looked around the cage.

The animal's perseverance in solving a problem, and the rapid and rational performance of the manipulation chain, indicate that the behavioural chains are probably steered by a dominant concept of a goal. Opening the fasteners of a problem box does not necessarily give insight into how to open a fastener (see the negative transfer), but the animal which can not see or smell the bait must have a concept of a goal when manipulating at the box.

ACKNOWLEDGEMENTS — My thanks go to Prof. Dr. G. Dücker for her generous provision of facilities and assistance throughout this study.

I also wish to express my gratitude to the «Allwetterzoo Münster», Dr. G. Ruempler, Zoo Director, and Dr. K. Kaiser, scientific assistant, for letting me work with their animals for this project.

I should also like to give special thanks to my colleague, H. Illy, who offered to correct the translation of the manuscript for this article.

References

COLE L.W., 1907. *Concerning the intelligence of raccons.* J. comp. Neurol. Psychol., 17: 211-261.

BIERENS DE HAAN J.A., 1930. *Über das Suchen nach unsichtbarem Futter bei Affen und Halbaffen. Zugleich ein Beitrag zur Frage nach dem Verständnis dieser Tiere.* Z. vergl. Physiol., 11: 630-655.

BISHOP A., 1964. *Use of the hand in lower primates.* Evolutionary and genetic biology of primates. Ed. J. Buettner-Janusch, Academic Press, New York, London: 132-225.

DÜCKER G., HORTER M. & LUSCHER C., 1982. *Manipulierfähigkeit und Erlernen von Handlungsketten. Untersuchungen an Sciuriden.* Z. Tierpsychol, 60: 146-165.

DÜCKER G., LUSCHER C. & SCHULZ P., 1986. *Problemlöseverhalten von Stieglitzen* (Carduelis carduelis) *bei «manipulativen» Aufgaben.* Zool. Beitr. N. F., 29: 377-412.

GANZHORN J. U., 1985. *Habitat separation of semifree-ranging* Lemur catta *and* Lemur fulvus. Folia Primatol., 45: 76-88.

HILL W.C.O., 1953. *Primates: comparative anatomy: Vol. 1.* Strepsirhini. Edinburgh, Edinburgh University Press.

JOLLY A., 1964. *«Prosimians» manipulation of simple object problems.* Anim. Behav., 12: 560-570.

KINNAMAN A.J., 1902. *Mental life of two* Macacus rhesus *monkeys in captivity.* Am. J. Psychol., 13: 98-148.

KOHTS N., 1928. *Adaptive motor habits of the* Macacus rhesus *under experimental conditions: a contribution to the problem of «labor processes» of monkeys.* Zoopsychologisches Laboratorium des Museums Darwinianum, Moskau.

LÖHMER R., 1975. *Vergleichende Untersuchungen über die Manipulier- und Lernfähigkeit von Waschbären* (Procyon lotor) *und Krabbenwaschbären* (Procyon cancrivorus cancrivorus). Z. Säugetierk., 40: 36-49.

MILLIKEN G.W., FORSYTHE C. & WARD J.P., 1989. *Multiple measures of hand-use lateralization in the ring-tailed lemur* (Lemur catta). J. comp. Psychol., 103 (3): 262-268.

PETTER J.J., 1962. *Recherches sur l'ecologie et l'ethologie des lemuriens malgaches.* Mem. Mus. Natl. Hist. Nat., Paris, ser. A27(1): 1-147.

RENSCH B. & DÜCKER G., 1966. *Manipulierfähigkeit eines jungen Orang Utans und eines jungen Gorillas.* Z. Tierpsychol., 23: 874-892.

RENSCH B. & DÜCKER G., 1969. *Manipulierfähigkeit eines Wickelbären bei längeren Handlungsketten.* Z. Tierpsychol., 26: 104-112.

SCHÖNEICH S., *Problem solving behaviour in tree-shrews* (Tupaia belangeri). Primate Report 14. Abstracts and Papers presented at the XIth. Congress of the International Primatological Society 20-25 July, 1986, Göttingen, FRG: 258.

SCHÖNEICH S., 1988. *Manipulative Fähigkeiten des Spitzhörnchens* (Tupaia belangeri THOMAS 1914) *an der Problembox: Bewältigung längerer Handlungsketten, Lösen verschiedener Transpositionsaufgaben und Überprüfung der Gedächtnisleistung.* Sonderheft zum Band 53 der Z. Säugetierk; 24-25.

SIMONS D. & HOLTKÖTTER M., 1986. *Cognitive processes in Cebus monkeys* (Cebus apella) *when solving problem-box tasks.* Folia primatol., 46: 149-163.

SUSSMAN R.W., 1974. *Ecological distinctions in sympatric species of* Lemur. In: Prosimian Biology. Eds. Martin, R. D., Doyle, G. A., Walker, A. C. G. Duckworth, London: 75-108.

SUSSMAN R.W., 1977. *Feeding behavior of* Lemur catta *and* Lemur fulvus. In: Primate Ecology. T. H. Clutton-Brock (ed.) London: 1-36.

SUSSMAN R.W., TATTERSALL I., 1976. *Cycles of activity, group composition and diet of* Lemur mongoz mongoz *Linnaeus 1766 in Madagascar.* Folia primatol., 26: 270-283.

THORNDIKE E.L., 1898. *Animal intelligence: an experimental study of the associative processes in animals.* Psych. Rev. Mon., Suppl., 2, 8: 1-109.

THORNDIKE E.L., 1901. *The mental life of the monkeys.* Psych. Rev. Mon., Suppl. 3, 5.

TORIGOE T., 1985. *Comparison of object manipulation among 74 species of non-human primates.* Primates, 26 (2): 182-194.

VÖSSING A., 1984. *Problemlöseverhalten in Primatengruppen: Untersuchungen an Weißkopfmakis* (Lemur fulvus albifrons). Verh. Dtsch. Zool. Ges., 77: 213.

VÖSSING A., 1987. *Arbeitsteiliges und kooperatives Problemlöseverhalten in Primatengruppen.* Zool. Beitr., N. F. 31: 305-340.

WARD J.P., MILIKEN G.W., DODSON D.L., STAFFORD D.K. & WALLACE M., 1990. *Handedness as a function of sex and age in a large population of Lemur.* J. comp. Psychol., 104 (2): 167-173.

YOAKUM C.S., 1909. *Some experiments upon the behavior of the squirrels.* J. Comp. Neurol. Psychol., 19: 541-568.

R.G. Rawlins

Locomotive and manipulative use of the hand in the Cayo Santiago macaques (*Macaca mulatta*)

Primates retain a generalized hand morphology which functions to transmit body weight to the substrate and acts as a contact interface for the animal in manipulating its physical environment via eye-hand coordination. Routine hand use is directly related to habitual behavior and habitat exploration, inviting speculation that neurological and morphological correlates of propulsive and manipulative functions can be found. Frequency of hand use and preference was recorded for 45 members of a single free-ranging social group (Troop L) of rhesus monkeys on Cayo Santiago over a six month period. Of these, 20 animals showed a definite right hand preference, 17 were left hand dominant, and eight were ambidextrous. No significant difference ($p > 0,05$) in manual asymmetry was found for the troop as a whole. The animals were not predominantly dextrous as in human populations. Preferred hand use cut across task performance. Animals with hand preferences in terrestrial foraging and locomotion retained those preferences in the arboreal substrate. When partitioned by genealogical affiliation, homogeneity of hand use was found in one of four matrilines ($p < 0,001$). The members of this genealogy were right hand dominant. Wounds and injuries inhibited use of the preferred hand, but dominance was reasserted after periods of healing lasting up to four months. Hand preference emerged within the first three to eight months of life for infants born into the troop. The results are consistent with available neurological evidence which suggests an interaction of both genetic and ontogenetic components leads to manual asymmetry in macaques.

Key words: gait change, handedness, hand posture, injuries, weight support.

Introduction

Animals interact with objects, events and other individuals within the environment through integrated action of the neuro-musculo-skeletal system, therefore behavior is bounded by anatomical limits. Conversely, behavioral constraints influence patterns of positional behavior, regulating the degree and type of mechanical demands placed on that anatomy. Thus, anatomical potential for locomotion need not be fully exploited in daily activity and observations of positional behavior in free-ranging primate populations are necessary for accurate biomechanical interpretation of morphology.

Two previous studies in this series on positional behavior in rhesus monkeys, at Cayo Santiago (Rawlins, 1976, 1977) have shown the interaction of locomotor morphology and habitual behavior can be studied by documentation of the ontogeny of locomotor and postural activities in a single population. Sociometric and morphometric observations on a free-ranging social group at Cayo Santiago have revealed significant differences in the frequencies with which adults, juveniles and infants employ locomotor and postural modes and that morphology of the shoulder girdle and forelimb is functionally associated with observed shifts in actual positional frequencies. Keeping in mind that location of the

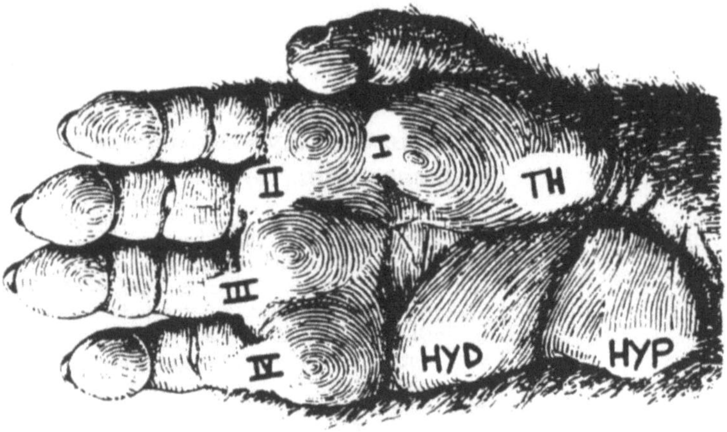

Figure 1 - Areas of maximum dermatoglyphic complexity as found by Meier in Cayo Santiago rhesus.

animal in its life cycle affects both locomotor morphology and habitual positional behavior, congruence of morphology and behavior in a single primate troop supports use of anatomical data for inference of possible locomotor activity.

Mammals have evolved striking morphological adaptations to provide for maximum locomotor efficiency, culminating in the highly specialized digitigrade and unguligrade cursors. As an order, primates retain a generalized hand morphology, which functions not only to transmit body weight to the substrate, but also as a qualitative contact interface for the animal in manipulating its physical environment via eye-hand coordination. Routine hand employment is directly related to habitual behavior and habitat exploitation, inviting speculation that morphological correlates of its manipulative and propulsive functions can be found.

In 1972, ROBERT MEIER examined variability of palmar dermatoglyphics in the Cayo Santiago rhesus (44 males, 53 females), finding epidermal pattern complexity greatest at interdigital areas II, III, IV, while complexity of the thenar and hypothenar distal regions was significantly lower (*Figure 1*). Apical digital areas were invariant, all having complex whorl patterns. MEIER hypothesized that either the epidermal patterns are functionally related to hand use or that pattern complexity reflects the degree of volar pad development, in turn relating to hand employment. Beyond SCHULTZ's (1969) statement that rhesus monkeys are plantigrade animals, no qualitative or quantitative data were available to test for functional relations of dermatoglyphs to locomotor and manipulative behavior. This study clarifies some aspects of the relation of epidermal patterning to hand use in the rhesus monkey.

Hand use in locomotor and postural behavior

Topography of Cayo Santiago is extremely varied and the heavy forest canopy provides an ample range of locomotor substrates. From July 1974 to July 1975, the ontogeny of positional behavior in a group of 77 free-ranging macaques (Group L) was studied on Cayo Santiago. A total of 1753 hours of direct observation and 2,400 meters of

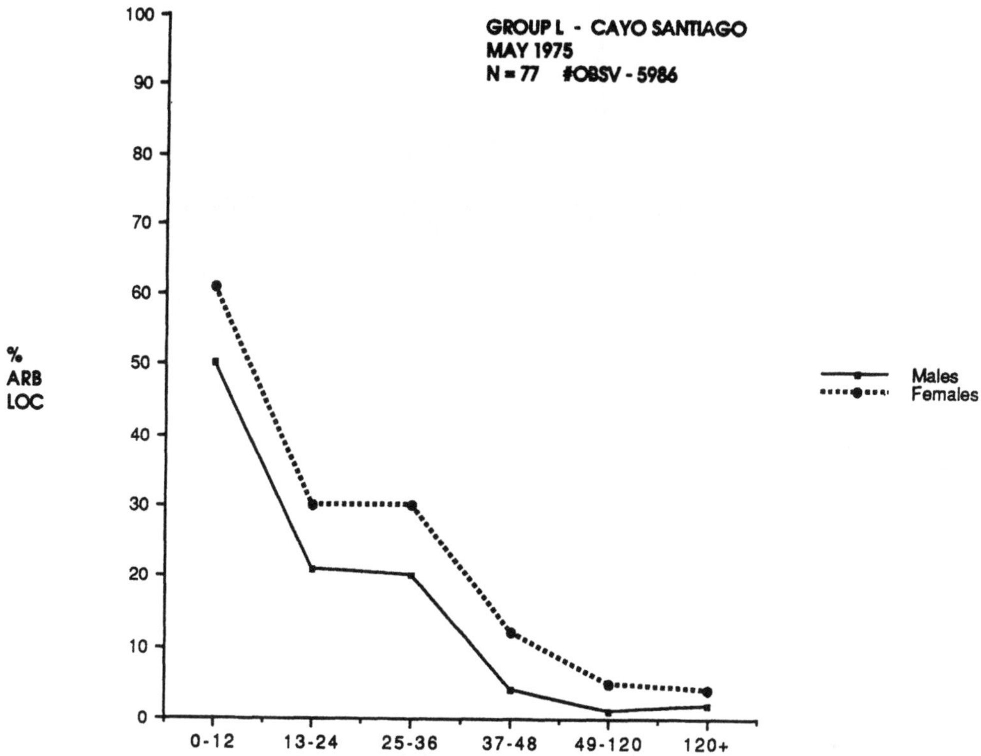

Figure 2 - Mean arboreal activity of all locomotor behavior by age-sex.

8 mm film were obtained to document the locomotor and postural repertoire. Forty positional modes were described and hand use in each recorded. Animals were of known age, sex, maternal genealogy and dominance rank and the effects of these factors on positional frequencies were assessed. Qualitative aspects were noted in diary form and drawn from analysis of film taken at 54 f/s. Quantitative data were obtained by taking randomized, multiple three minute ad libidum samples of all observed positional behavior in the group. Only the first observed mode was recorded each hour. The resulting frequency data was cast into monthly contingency tables and forms the data base for this analysis.

Considerable variation exists in hand use for Cayo Santiago rhesus monkeys. Arboreal climbing, leaping, and dropping is predominantly palmigrade. Though the relative amount of arboreal activity (*Figure* 2) is greater for females than males and significantly decreases with age for both sexes, hand use is consistent. When adults, juveniles and infants traverse a horizontal branch in diagonal couplet or sequence climbing, the substrate passes across the palm in a power grip, running between interdigitals I and II and the thenar and hypothenar distal regions, Axis 1 (*Figure 3*). Branch ascents also dictates use of this axis, but on head first descents, an alternative to the power grip may be used if branch diameter is great enough. Axis 4, the entire palmar surface, with interdigitals II, III, IV and hypothenar distal and thenar areas bearing the weight is

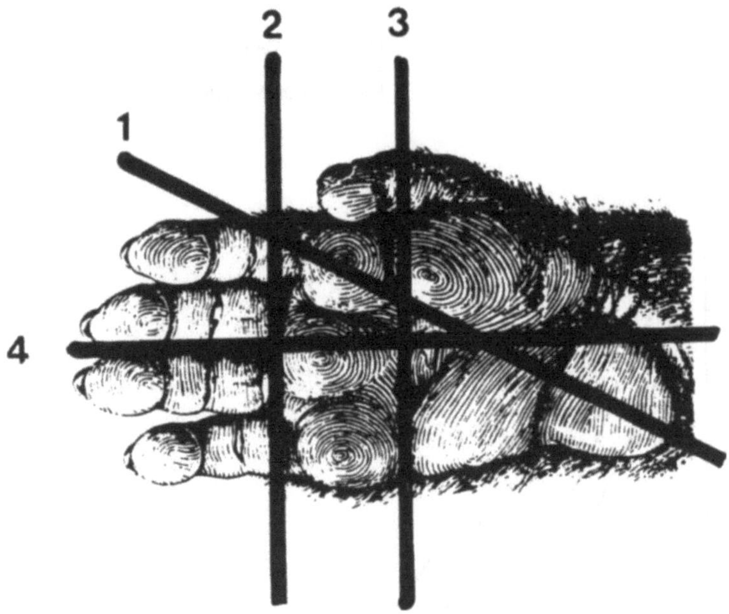

Figure 3 - Major axes of hand use in Cayo rhesus.

employed, resulting in a more rapid, less cautious descent. The extent to which Axis 4 is used varies with body size, as infants easily descend branches where an adult would require the power grip of Axis 1. In climbing vertical vines and small branches Axis 1, Axis 2 (the junction of the proximal and intermediate phalanges) or Axis 3, the metacarpophalangeal junction, may be used in a modified power grip for ascent or for a tail first descent. In each case, the substrate is pressured against interdigitals II, III, IV by the apical digits. Vertical ascent on large diameter tree trunks, by paired limb hop-climbing, involves powerful hindleg thrusts, propelling the body upwards to be recovered by full palmigrade contact around the tree until the next impulse is delivered. Vertical tail first descents are braked by palmigrade contact as well, with full use of thenar, hypothenar distal and hypothenar proximal regions, as a frictional surface in a form of «rappelling».

Leaps and drops use palmigrade contact, relying on the power grip of Axis 1 and 2 when the target is forest canopy or palmigrade contact with wrist and elbow flexion, for shock absorption, employing Axis 4 when the target is level with ground or water. In landing, first contact is made by finger tips and interdigitals II, III, IV, followed by dorsiflexion of the wrists and elbow, at which time the thenar and hypothenar regions contact. Axis 2 is also used when the animal is suspended from the arboreal substrate, either hanging arms above head with body axis normal to ground, or with complete quadrupedal inversion and the body parallel to ground. Here the finger tips bear against the clasped substrate and are opposed on the opposite side by pressure from the volar pads of interdigitals II, III, IV. For both low and high speed arboreal locomotion, the interdigital areas II, III, IV are the main points of substrate contact. The hypothenar and thenar regions play a reduced role in carrying body weight, but resist shearing forces and slippage in head first descents.

Rhesus monkeys walk in diagonal couplets, a symmetrical gait bending the axial skeleton in the horizontal plane and using evenly spaced foot-falls and support periods. Each movement is counterbalanced by antagonistic action of the corresponding limb, such that the hindfeet pass to the same side of their respective forefeet, except when turning, and duration of anterior contact is about equal to posterior. HILDEBRAND (1976, 1977) has shown that as the hindfoot passes lateral to the fore, the spine orients at an angle to the line of travel, such that one shoulder advances ahead of its opposite and the leading limb takes a longer stride. This gait is unique among walking mammals as non-primates employ the lateral couplet. GRAY (1944) has shown the diagonal couplet is the least efficient limb sequencing pattern of six possible walking gaits with respect to providing stable support for the moving body. However, PROST'S (1969) later kinematic study indicates that for arboreal climbing the diagonal couplet is efficient, maintaining forward movement and compensating for lack of stable triangular support by counterthrusting the diagonal leg to control rolling torques which could cause a lateral fall. Retention of this gait during terrestrial locomotion, rather than a shift to the lateral couplet, reflects anatomical constraints of joint placement which direct walking footfalls immediately below the body midline, ruling out formation of support triangles even on ground level.

In terrestrial walking, Cayo Santiago macaques are facultatively digitigrade or palmigrade. In a slow diagonal couplet, either the digitigrade or palmigrade mode is used when moving in relaxed fashion on open even ground, but shifts to digitigrade when picking a path through brush or detritus such as fallen tree limbs or areas of widely scattered rocks. The digitigrade mode is also used when ascending evenly surfaced slopes. Full palmigrady reappears when animals traverse uneven rocky ground.

A G statistic test for heterogeneity on 6 months of hand use during walking showed no significant differences ($p > 0{,}05$) in employment of digitigrade or palmigrade modes by sex, but palmigrade contact was used with twice the frequency, except for three animals. The first, second and third ranking adult males were significantly different, employing digitigrade contact in a ratio of two to one over palmigrady, introducing the notion that behavioral roles rather than substrate can also dictate patterns of habitual locomotor behavior. In walking, body weight is carried primarily on interdigitals II, III, IV with little or no involvement of the hypothenar or thenar regions, a fact supported by examining plaster casts of impressions made by footprints of monkeys walking in wet sand at Cayo Santiago.

Speed of body propulsion depends on body weight, leg length, stride frequency and duration of contact with the substrate. Since the rhesus hindlimb is longer than the forelimb, with an intermembral index of 87.4 -87.6 (WASHBURN, 1942; SCHULTZ, 1969), a full hindleg stride cannot be equalled by the foreleg. At a slow walk, no problem is encountered in maintaining balance, but at higher speeds HOWELL (1944) has shown that, although front feet contact ground symmetrically with the hind, they must leave before them to compensate for the difference in stride length. As stride frequency increases, the effect can be reduced in part by extending forelimb length during the symmetrical gait. Presumably this is the reason that, as Cayo Santiago rhesus grade from a walk into a trot (a diagonal couplet gait at about double the walking stride frequency), they shift from palmigrade to digitigrade mode. Again, interdigitals II, III, IV and the finger tips strike first and bear body weight. Eventually, increased stride length gained by the shift to digitigrady can no longer compensate for timing and balance problems introduced by limb length disparity.

In reptiles, HOWELL (1944) has shown that when a symmetrical gait is retained as the high velocity mode, functional accentuation in the hindlimb is seen, leading to bipedalism

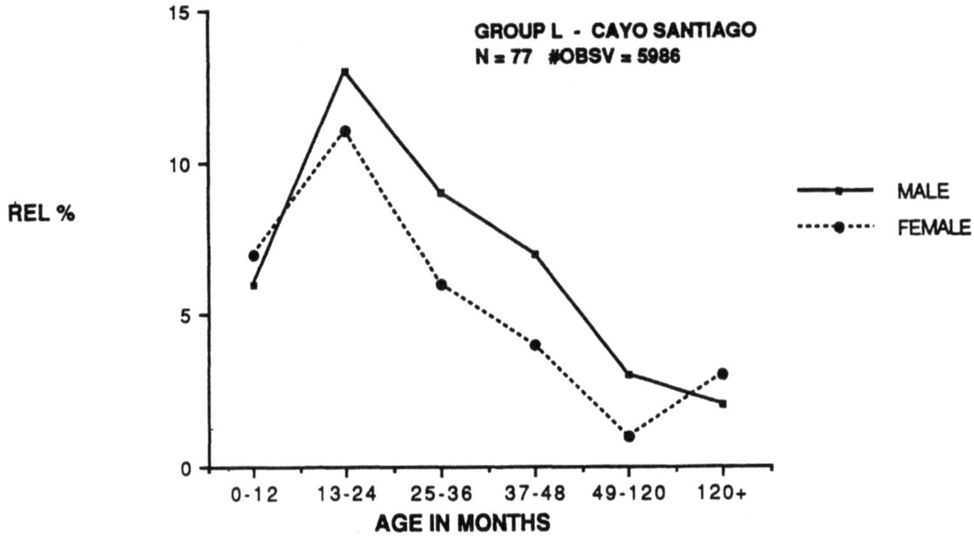

Figure 4 - Relative frequency of gallop as % of total positional behavior.

(such as that of the extinct Saurischians) or facultative bipedalism as in the Jesus Christ lizard, *Basiliscus sp.* Alternatively, high-speed balance problems can be solved and effective stride length increased by adopting an asymmetrical gait such as the gallop. Here, footfall intervals are unevenly spaced, pronounced bending occurs in the midsagittal plane and asymmetry of support patterns and elimination of damping counteraction of corresponding limbs occurs. One or two periods of total body suspension for each complete limb cycle are possible in these gaits and primary impulse comes from the hindfeet, followed by a long recovery interval. The distance the body moves forward while unsupported adds to length of the stride.

Cayo Santiago rhesus monkeys shift from the walk or trot into a rotary gallop with palmigrade contact on all substrates. Frequency of use (*Figure 4*) declines with age, yet for all it is the mode of maximum velocity. Partial synchronization of the two fore and hind limbs is distinctive but complete synchrony does not occur (PROST, 1969). The left forelimb moves before the right and right hindlimb before the left. In two of eight phases (gathered and fully extended) all four feet clear the ground and stride is lengthened by radical flexion and extension of the spine so that the hind feet contact ground ahead of forefoot liftoff.

During a gallop, a macaque's hand hits ground at increasingly acute angles, causing the body to rise as the shoulders pass over the forelimb and to fall when the limb extends to the front or rear. HILDEBRAND (1960) has shown that aclavicular cursors offset this bobbing motion by displacing the scapula dorsally and by dorsiflexing (over extending) wrists and ankles, thus shortening effective leg length the moment the body passes across the vertical limb. Though some scapular displacement does occur, the claviculate rhesus monkey must smooth forward motion by returning to the palmigrade mode during the gallop, striking first with the digits and palmar II, III, IV regions, followed by the thenar and hypothenar distal contact. The hand again becomes digitigrade as the body's center of gravity moves forward relative to the shoulder. Body weight passes posterior to the wrist and maximum pressure hits interdigitals II, III, and IV, then moves to the digits.

Once body weight rests on the digits, stress in the extensor compartment from dorsiflexion of the wrist is reduced and the joint raised from the ground as the elbow extends, bringing the limb forward onto and then in front of the metacarpal heads.

TUTTLE (1975) suggests predominance of the flexor digitorum profundus muscle amongst the deep extrinsic flexors of the hand in *Macaca mulatta* reflects mechanical demands on the distal phalanges to produce a strong propulsive toe snap as the forelimb enters swing phase which smooths out the stride. However, this idea has not been confirmed by force plate recordings for Japanese macaques and other quadrapeds (KIMURA *et al.*, 1979). In addition, RAUWERDINK (this volume) finds that among the primates, this muscle is uniformly the heaviest muscle of the forearm, regardless of the species' primary locomotor mode.

Though an analysis has not yet been done on Cayo Santiago data, it is reasonable to speculate that the point of transition from the walk or trot to the gallop is scaled by body size and that gait changes occur at lower speeds and higher stride frequencies in juveniles than for fully grown adults.

Thus, hand use during terrestrial locomotion is determined by quality of the terrain at slow speeds and by mechanical constraints on the locomotor system at higher velocities. Social interactions also plays a determining role in that high speed gaits are used primarily in context of play chasing, aggressive pursuit, and escape from other troop members.

Hand use in arboreal locomotion maximizes palmar friction to prevent sliding on the substrate. Shearing forces encountered on inclined surfaces may be behaviorally offset by selecting substrates with high coefficients of friction. Alternatively, the integument may serve as a friction enhancing system. The palmar dermatoglyphics and volar pads supporting them may function in this capacity (see BUCK & BÄR, this volume). For both terrestrial and arboreal locomotion, interdigitals II, III, and IV are the areas of greatest complexity most frequently subjected to striking shock in contacting the substrate and sustain maximum loads in bearing body weight.

Postural employment of the hand involves two functions 1) bracing the animal, either on the ground or in the trees, to maintain resting or foraging positions, and 2) active manipulation of the environment. For Group L of the Cayo Santiago population, postural behavior accounts for 30% of all positional activity in juveniles and 55-60% for adults (females exceed males). A significant increase in postural behavior is seen with age.

On the ground, animals in relaxed sitting postures employ a single or double hand brace to the ground or nearby roots and branches. Hand use can be palmigrade, digitigrade or a combination of both and depends on the type of surface touched. Alert sitting postures are normally digitigrade so that the shoulders are elevated and the spine erect. Sleeping in a sitting position, the animal is palmigrade and slumped forward. Alternatively, elbows rest on flexed knees and no hand contact is made. Sitting or sleeping in trees, a single brace is used routinely. Often both hands contact branches for support in addition to foot propping. Axis 1 and Axis 3 are most frequently used for bracing contact with a passive digit grip.

Both arboreal and terrestrial foraging involve use of precision and power grips. Axes 1,2 and 3 are used to clasp the object and a precision grip, opposing the terminal phalanx of the thumb against either the tip of the flexed first finger or the radial surface of its second phalanx is used to pick. If the arboreal posture is secure, both hands may be used together or in alternating fashion to forage, but as the animal leans forward, moving its center of gravity anterior to the pelvis, a bracing hand is selected, using Axis 1 or 3 to stabilize the body as the opposite hand works. The animal may continue to brace and forage with the trunk extended, or return to a neutral sitting posture to eat, later bracing again to resume collecting.

Only infants and juveniles use supensory foraging postures, hanging tail first, head first or quadrapedally from the trees and employing Axis 2 in the terminal branches or Axes 1, 2 or 3 when suspended below limbs of relatively large diameter with respect ot the hand. Again the precision grip described above is used to acquire food items. Though trees at Cayo Santiago would support supensory activity of adults (males reach 16-18 kgs body weight), the animals seem very uncomfortable in these postures and rarely use them, instead preferring the upright sitting mode.

All age classes use the tripedal foraging position, where three limbs of the standing animal remain in contact with the substrate, tree or ground, and one hand is extended, with a precision grip clasp, to reach for food items. Substrate dictates whether the bracing forelimb is digitigrade of palmigrade, following the same criteria outlined for walking. The precision grip is also employed during auto- and allogrooming, to extract hair, detritus and ectoparasites from the body. One hand adopts the precision mode and the other slides palmigrade counter to the lay of the fur, exposing the skin below.

Distribution of handedness

Distribution of handedness in the rhesus monkey has been studied anatomically and behaviorally in the laboratory with inconclusive results. DHALL & SINGH (1977) looked at forelimb muscle weights in only 12 rhesus, but concluded handed preference is dextrous since mean weights were significantly greater on the right side of nine animals (see also chapters by Alexander and Rauwerdink, this volume). BLUM (1950) and co-workers report rhesus handedness about equal between right, left and ambidexterity and that unilateral resections of the parietal cortex contralateral to the preferred hand shifts dominance to the other. BROOKSHIRE & WARREN (1962) found a left hand bias in manipulative responses to bait recovery, as did BECK & BARTON (1972), but the latter also observed a right hand preference for voluntary non-incentive manipulations. Holding responses were also labile; when food was retrieved from an object that had to be manually steadied, a right hand preference appeared, casting doubt on findings of left hand predominance (BROOKSHIRE & WARREN, 1962; ETTLINGER, 1961; ETTLINGER & MOFFETT, 1964; MILNER, 1969). More recently, BRENOT (1982) has reported no lateral trend in food catching in 14 captive rhesus monkeys and notes a high proportion of ambidextrous individuals (see also BRENOT, this volume). To my knowledge, no observations of spontaneous manual asymmetry have been reported on free-ranging rhesus monkeys. The frequency with which a given hand is used for support of precision manipulation and therefore the areas of palmar contact employed, varies by individual at Cayo Santiago.

Frequency of hand preference for each animal in Group L, was tabulated for 6 consecutive months of observation for both terrestrial and arboreal bracing and foraging activity. Grooming was excluded since allo- and auto-grooming were combined during sampling and choice of hand in auto-grooming was often simply that closest to the irritation. Of 77 animals in the group, sufficient data were obtained to test for handedness in 45. Of these 20 showed right hand preference, 17 left hand and 8 were ambidextrous. Test made using Yates corrected Chi square, with goodness of fit set for a 50:50 ratio and p of exclusion $p < 0{,}05$ showed no significant manual asymmetry in this sample. Animals with definite hand preferences in ground foraging retained those preferences when foraging in the trees, using the nonpreferred hand to brace. Preferred hand use clearly cut across task performance. If an animal used the left hand to pull leaves from a tree and braced with the right, it would raise chow hopper lids with the right and forage with the left. Of particular interest, an r * c interaction test based on the G statistic showed significant ($p < 0.001$) homogeneity of hand use amongst one of four genealogies in Group L. Female S' genealogy was right hand dominant.

Four animals in the study were wounded or suffered breaks of the preferred hand. Lacerations did not inhibit use unless severe mechanical damage had been done. Where trauma was severe, or bones broken, use of the preferred hand was relinquished and the opposite hand functioned perfectly well. Upon healing (as in the case of female 238), the preferred hand reasserted dominance, sometimes after periods of as long as four months (female 431) quiescence.

Behavioral and neurological evidence indicates acquisition of rhesus handedness results from an interaction of learning and phylogenetic status (BROOKSHIRE & WARREN, 1962), however, no studies have determined exact or relative contributions of genetic and ontogenetic components. ETTLINGER 1961) noted enhancement of handedness over a six to ten month testing period attributing the reinforcement to learning. However, JENSEN (1961) has found ontogenetic development of asymmetry.

Observations on Cayo Santiago, though not conclusive, show hand preference emerges slowly when it occurs, usually within the first three to eight months of life. Though preference may be learned, it is interesting to recall KUYPERS (1962) findings that the bulk of direct cortico-motoneuronal connections in rhesus monkeys are established postnatally, also during the first eight months. Perhaps latent expression of the genome, through neural differentiation of fine forelimb control via pyramidal cortex plays as significant a role in the generating handedness as reinforcement during the learning process.

Data on hand use from the Cayo Santiago population do not resolve the question of dermatoglyphic primary function or its relation to volar pad development, but they strongly support MEIER's hypothesis. Areas of greatest pattern complexity are those that bear body weight most frequently.

Mechanical reinforcement from complex whorls probably does maximize resistance to shear and sliding of the epidermis over connective tissue and fat during locomotion (see also BUCK & BÄR, this volume). In all positional behavior a high coefficient of friction is provided for secure substrate contact.

ACKNOWLEDGEMENTS — The assistance of C.S. Evans in preparation of data for this manuscript is gratefully acknowledged. I would also like to thank A.M. Dawson for her secretarial support and the staff of the Caribbean Primate Research Center, University of Puerto Rico School of Medicine, for their assistance in carrying out this study. Special thanks are due Dr. Matt J. Kessler.

References

BECK C.H.M. & BARTON R.L., 1972. *Deviation and laterality of hand preference in monkeys.* Cortex, 8: 339-363.

BLUM J.S., CHOW K.L. & PRIBRAM K.H., 1950. *A behavioral analysis of the organization of the parieto-temporo-occipital cortex.* J. Comp. Neurol., 93: 53-100.

BRENOT P., 1982. *Evolution functionelle du cortex des primates.* Bull Soc. Anthrop. s-o, 17: 1-33.
BROOKSHIRE K.H. & WARREN J.M., 1962. *The generality and consistency of handedness in monkeys.* Animal Behavior, 10: 222-227.
DHALL V. & SINGH I., 1977. *Anatomical evidence of one sided forelimb dominance in the rhesus monkey.* Anat. Anz., 141: 420-425.
ETTLINGER G., 1961. *Lateral preferences in monkeys,* Behavior. 17: 275-287.
ETTLINGER G. & MOFFETT A., 1964. *Lateral preferences in monkeys.* Nature, 204: 406.
GRAY J., 1944. *Studies in the mechanics of the tetrapod skeleton.* J. Exp. Biol., 19: 88-116.
HILDEBRAND M., 1967. *Symmetrical gaits of primates.* Amer. J. Phys. Anthrop., 26: 119-130.
HILDEBRAND M., 1977. *Analysis of asymmetrical gaits.* J. Mammalogy, 58: 131-155.
HILDEBRAND M., 1960. *How animals run.* Scientific American, 202: 148-157.
HOWELL A.B., 1944. *Speed in Animals*; reprinted in 1965, Hafner Publishing Co., NY.
JENSEN G.S., 1961. *The development of prehension in a macaque.* J. Comp. Physiol., Psychol., 54: 11-12.
KIMURA T., OKADA M. & ISHIDA H., 1979. *Kinesiological characteristics of primate walking: Its significance in human walking.* In: M.E. Morbeck, H. Preuschoft and H. Gomberg (eds.) Environment, Behavior and Morphology: Dynamic interactions in Primates, G. Fischer, New York, Stugat, 297-311.
KUYPERS H.G.J.M., 1962, *Corticospinal connections: postnatal development in the rhesus monkey.* Science, 138: 678-680.
MEIER R.J., 1973. *Considerations of function in macaque dermatoglyphics.* Folia primatol., 20: 112-124.
MILNER A.D., 1969. *Distribution of hand preference in monkeys.* Neuropsychologia, 7: 375-377.
PROST J.H., 1969. *A replication study on monkey gaits.* Amer. J. Phys. Anthrop., 30: 203-208.
RAWLINS R.G., 1976. *Locomotor ontogeny in Macaca mulatta: I. Behavioral strategies and tactics.* Amer. J. Phys. Anthrop., 44: 201.
RAWLINS R.G., 1977. *Locomotor ontogeny in Macaca mulatta: II. Locomotor dimensions of the shoulder girdle and forelimb.* Amer. J. Phys. Anthrop., 47: 156.
SCHULTZ A., 1969. *The Life of Primates.* Universe Books, N.Y.
TUTTLE R.H., 1972. *Cheiridial muscles in catarrhine primates.* In: The Functional and Evolutionary Biology of Primates, ed. R.H. Tuttle. AldineAtherton, Chicago
WASHBURN S.L., 1942. *Skeletal proportions of adult langurs and macaques.* Human Biol., 14: 444-472

C. Boesch
H. Boesch

Different hand postures for pounding nuts with natural hammers by wild chimpanzees

Despite the fact that chimpanzees use mainly their hands for locomotion, they have developed great manipulatory skills. Wild chimpanzees of the Taï National Park, Côte d'Ivoire, use a very large variety of hand postures when cracking nuts with a hammer. To crack nuts directly in the producing tree, they may hold their hammer with 6 different grips and may hold, at the same time, some collected nuts with the other hand in 5 different ways. The dexterity of the chimpanzee's hands allows complete individuality of finger movements and the possibility to perform multiple functions with both hands, simultaneously if necessary. It is proposed that the hand of the chimpanzee is already morphologically able to perform most tasks attributed to *Homo habilis.*

Key words: grip variation, manipulation, nut cracking, precision grip, tool use.

Introduction

Although the chimpanzee is known to use tools often (Boesch & Boesch, 1984a; McGrew *et al.*, 1979; van Lawick-Goodall, 1970; Nishida & Hirawa, 1982; Teleki, 1974), its hands are used mainly for locomotion. Chimpanzees possess morphological adaptations in their hands for arboreal foraging and climbing distinct from those of hylobatids. In addition to primary adaptations for climbing (Preuschoft, 1965, 1973), they became secondarily adapted for terrestrial knuckle walking (Tuttle, 1969 a, b). This represents a compromise between manipulative and locomotor functions.

Anthropologists tend to associate the evolution of humans with bipedalism, tool use and making. More recently, following the discovery of the fossils of *Australopithecus afarensis*, the appearance of bipedalism in humans was considered as preceeding the evolution of tool use and tool making (Johansen & Edey, 1981; Leakey, 1981; Pilbeam, 1980), emphasizing that the freeing of the hands from locomotion would have facilitated the evolution of tool use in human ancestors. However, since then the proportion of arboreality attributed to *A. afarensis* has increased, and some of its bipedal abilities have been put in doubt (Susman *et al.*, 1984). Recent work on body weight support by the four limbs in primates (Kimura *et al.*, 1979; Reynolds, 1985) shows a clear trend in the reduction of the weight supported by the forelimbs compared to other quadrupedal mammals. Reynolds (1985) proposes that a similar tendency to reduce loads on forelimbs, combined with changes in the habitat, might be sufficient for the evolution of habitual bipedalism. An intriguing example of the latter is seen in the occasional bipedalism of spider monkeys and gibbons when on the ground or on large branches. It may accordingly be proposed that the tendency to manipulate with the hands increases proportionally to the decrease in body weight supported by the forelimbs. Man would thus be the extreme example supporting no body weight with the hands and having the best manipulative hands among primates. Some primates like the gibbons, however, may specialize so much in arboreality that this may be at the cost of some manipulative skills. Our observations on wild chimpanzees may fit in with these ideas, since they show elaborate manipulative skills.

Nut-cracking behaviour

We have studied a community of chimpanzees living in the tropical rain-forest of the Taï National Park, Côte d'Ivoire, since 1979 (BOESCH & BOESCH, 1983, 1984 a, b). This population is, to our knowledge, the first animal population for which the importance of tool use for their survival could be proved. During 4 months of the year, the main Coula-nut season, each independent individual will use wooden clubs or stones as a hammer to crack nuts for a mean duration of 2½ hours each day. This represents an energetic intake of about 18.000 kJ and 64g of proteins (new estimate based on our observations of undisturbed animals during the 1984-85 nut season, whereas BOESCH & BOESCH, 1984a, present an evaluation with less habituated individuals). An individual may even crack nuts without interruption for up to 5 hours. The chimpanzees of the Taï forest crack 5 different species of nuts, two of them very frequently, *Coula edulis* and *Panda oleosa* (see BOESCH & BOESCH, 1983, for more details). The Coula nuts are opened directly in the tree or on the ground; the much harder *Panda* nuts always on the ground and always with stone hammers (see BOESCH & BOESCH, 1981, for more details).

The usual procedure for cracking nuts is as follows: The chimpanzees will collect as many nuts as they can carry in their hands, mouth and even feet, carry them to a hard surface, commonly a root, used as an anvil. They will crack and eat them one by one, collect more nuts and crack them either on the same anvil or on another one closer to the collected nuts. They may either choose an anvil with a hammer lying close to it, or carry the hammer along while collecting the nuts and moving to other anvils. These tool transports are frequent and over distances up to 600 m. An individual may crack nuts and transport the same hammer over such distances for more than 3 hours. Normally, chimpanzees walk on their four limbs when carrying the hammer, the nuts or both. But according to the amount of nuts or the weight of the hammer, they may change their locomotor posture: In 14% of the 287 transports observed, they hold one hand against the chest walking tripedally and in 8% of these transports, they walk bipedally.

Use of the hands for nut cracking

Cracking nuts in the tropical rain-forest can be more difficult than one might suspect. First, some nut species are very hard and stones, which are rare in the forest, are necessary to open them; a *Panda* nut requires a 10 kg stone to fall from a height of 1.20 m to be opened. Second, the surface roots (*Figure 1*) that represent the great majority of the anvils do not have a regular surface and oblige the chimpanzees to adapt in each case its posture and the hitting movements to these irregularities. Third, and most important, the optimal hammers are hard to find and, if an individual does not want to loose much time and energy in searching for such a perfect tool, it must adapt itself to the particular physical problems each hammer presents (*Figure 2*). We have to underline the fact that in the forest less than 1% of the available fallen branches may be used as a hammer. Most of them are

Figure 1 - Coula anvil: surface root of a Coula tree (*Coula edulis*) used by the chimpanzees as anvil. Note the shell remains of the nuts eaten by them. The inclination and deformation of the root's shape are the major difficulties associated with such roots, forcing the nut-cracker to adopt a specific angle in hitting the nuts.

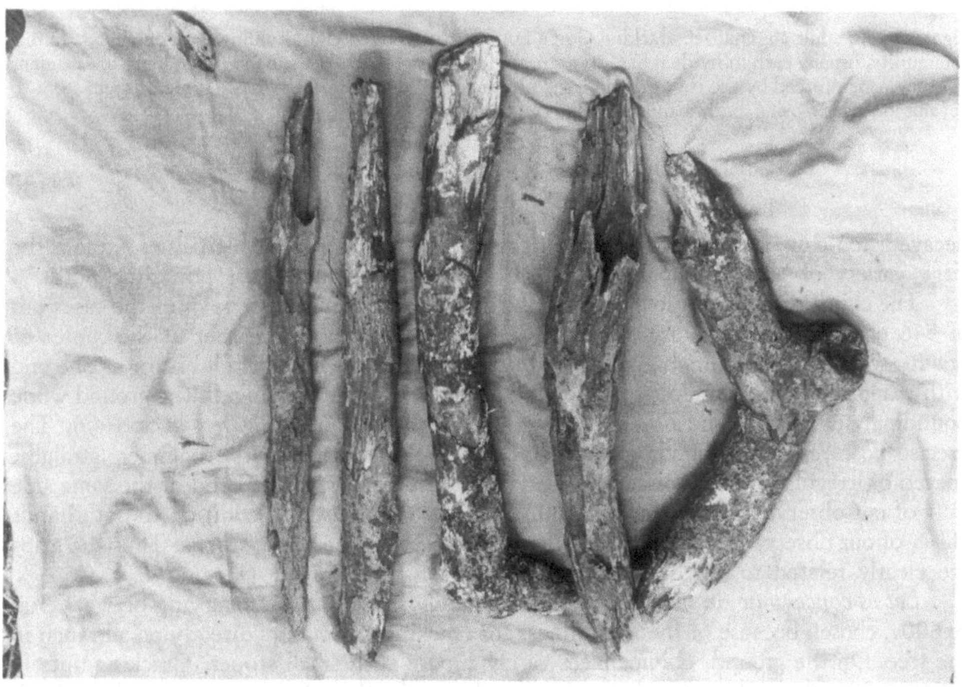

Figure 2 - Some wooden clubs used by chimpanzees to crack Coula nuts (*Coula edulis*). All these different shapes have to be taken into account by the nut cracker. In addition, the wear apparent in most clubs force them to be very precise in their hitting movements so as to hit the nut and not the root.

Figure 3 - An adult chimpanzee cracking Coula nuts with a wooden club, holding it in one hand. In our community, usually each individual has a preference for the right or the left hand for holding the hammer but may, when forced by the root morphology, use the other one efficiently. Only some of them seem to be ambidextrials, using both hands equally frequently.

decayed, too thin, too big or too light in density. These physical difficulties explain the large variety of grips chimpanzees use to hold their hammers for pounding nuts.

The chimpanzees can use up to 3 limbs to hold the hammer: In 51% of the cases out of 546 observations, they use one hand (*Figure 3*). They hold it either at the center of gravity to prevent the nuts bouncing out of the hole on the anvil when hit, or at one end with hammers over 80 cm long, the other end of the hammer resting on the ground while pounding. In 34% of the observations, they use both hands (*Figures 4a* and *4b*). The precise postures of the hands change according to the shape of the hammer (straight, curved or irregular). One foot can be used complementary with the hand of the same side (3% of our observations) for stone hammers. One foot is used in addition to both hands (11% of our observations) when using long and big clubs or heavy stones. Thus the grips are clearly related to the size and the weight of the hammer.

Let us concentrate on the small hammers, mainly stone hammers varying between 300 to 600g, chosen because of their handiness for cracking *Coula* nuts directly on a branch in the tree. On the ground, chimpanzees rarely choose such small stones. Cracking nuts in the tree, 20m above the ground, requires good coordination of movements by the animal, as the hammer and all the nuts must constantly be held while pounding eating and placing a new nut on the anvil.

Figure 4 - Two adult chimpanzees using both hands to open a) Coula nuts with a big wooden club and b) Panda nuts with a 5,2 kg granite stone.

We have observed 6 different grips for holding such small hammers. Due to the very difficult observational conditions in the tropical forest it was impossible for us to take close pictures of the chimpanzees' hands cracking nuts at such a height. But we could observe it perfectly with our binoculars and we had drawings made of these grips in order to describe them. They present many variations in the respective position of each finger according to the morphology of the hammer; the drawings represent the most often observed variation for each grip.

— Grip 1 (*Figure 5a*): This is a typical power grip with the thumb opposed to the other fingers when using a small club or a stone of about 500g (77% of 144 observations). If the size of the stone is smaller in such a power grip, the fingers may be exposed to the hits against the anvil and the nut. So, chimpanzees will adapt their grip to the size of the hammer as follows:

— Grip 2 (*Figure 5b*): With stones of about 400g, the chimpanzees conserve the power grip, but turn the hand in order to hit the nut with the hand held more vertically (4% of the observations).

— Grip 3 (*Figure 5c*): With smaller stones of about 300g, the chimpanzees may keep the stone low against the palm of the hand by holding it with the thumb inside the grip (3% of observations). This adaptation of grip 2 was observed only with juveniles, as this grip may be difficult to practice with a larger hand. Adults with such small hammers will use the two following alternatives:

— Grip 4 (*Figure 5d*): The stone is held in a modified power grip using only the thumb, the second and the third fingers. The two other fingers are kept stretched out (6% of observations).

— Grip 5 (*Figure 5e*): In this case the hammer is held between the thumb and the fourth and the fifth fingers. The two other fingers are kept in a relaxed position (7% of observations).

— Grip 6 (*Figure 5f*): For thin wooden clubs with less than 5cm of diameter, the chimpanzees change the grip 1, normally used for small clubs, in order to prevent the fingers being exposed to the hits. The club is held only by the long fingers and the stability of the grip is increased by large round hitting movement (2% of observations).

When cracking nuts directly in the tree, the chimpanzees must hold the hammer and the collected nuts simultaneously. These are usually kept in the mouth and one hand, while the other hand pounds the nut with the hammer (*Figure 6a*). However, in 69% of the 150 observations, they not only held the collected nuts in one hand, but with this same hand they also hold the nut which is being pounded on its anvil, to prevent it falling on the ground after a hit when using a thin or an inclined branch as anvil (*Figure 6b*).

We have observed 5 different postures of the hand to fulfil these two functions on 103 nut-cracking occasions. As for the grips, no close pictures could be made of the hand postures and the drawings represent the most often observed variations for each posture.

— Posture 1 (*Figure 7a*): In 54% of observations, the pounded nut is held by a precision grip between the thumb and the index. The other fingers hold some nuts against the palm of the hand. In order not to hit its fingers (especially the thumb) when opening the nuts, however, the space between the hammer and the anvil should remain large enough. Imagine yourself hammering a small nail into a wall. If this space is too small, the chimpanzee has to adapt its hand posture by using fingers thinner than the thumb to hold the nut.

— Posture 2 (*Figure 7b*): In 26% of observations, when using a stable and horizontal anvil, the chimpanzees may just control the pounded nut with their index finger while pounding. The index finger may touch the nut from different directions according to each

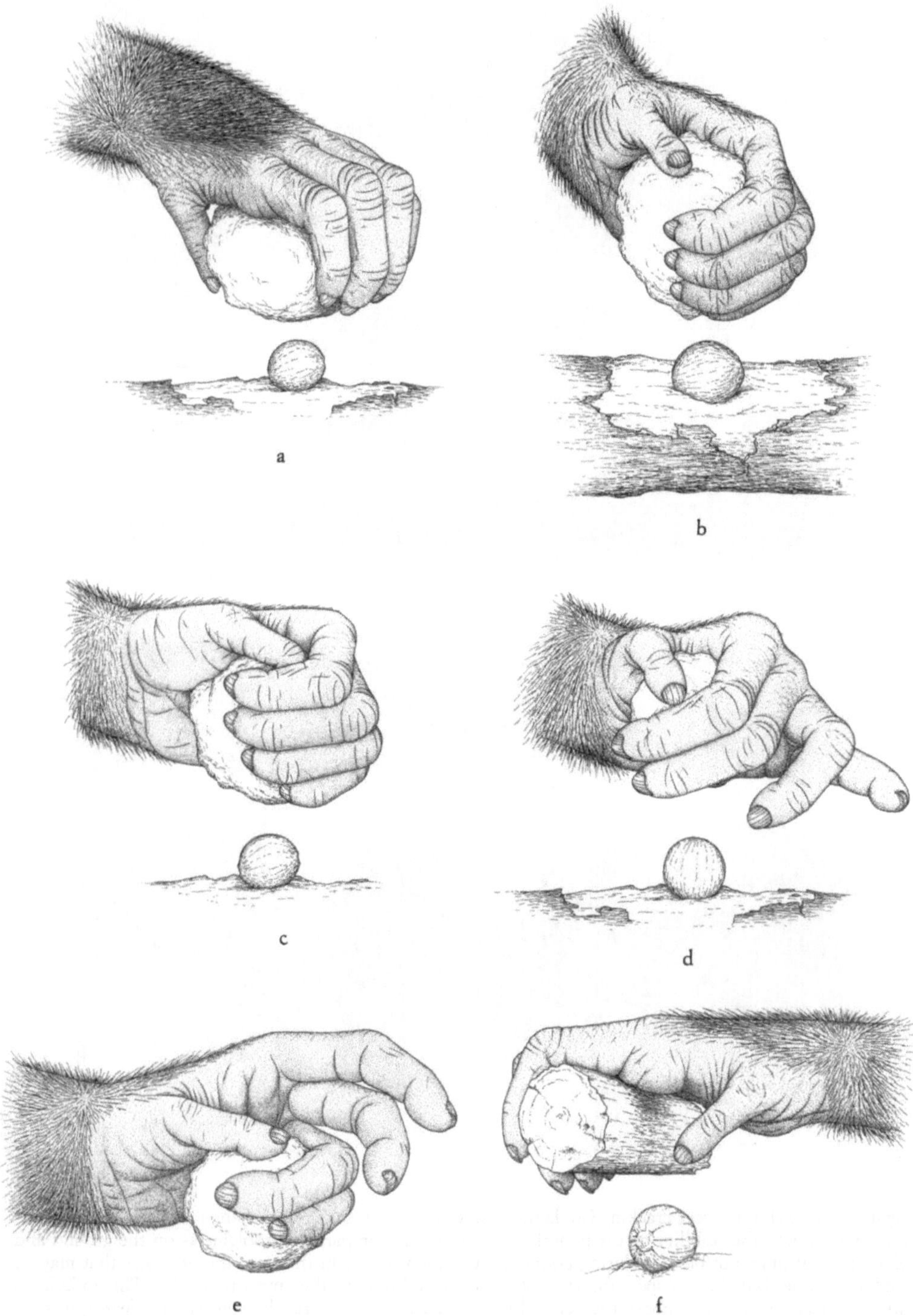

Figure 5 - Drawings illustrating the grips 1 to 6. See text for further explanations.

Figure 6 - Two chimpanzees cracking Coula nuts directly in the tree. a) Here using a low and large branch as an anvil where the stability of the pounded nuts presents comparable problems as on the ground and where the nut must not be held during pounding. b) Much higher in the tree, the branches that may be used are thinner but the transport distance of the collected nuts to the anvils is smaller. This individual sits on a small branch and choose the vertical trunk near a crossing branch as an anvil. Clearly, in such situation the pounded nut has to be held permanently, otherwise it will immediately fall on the ground 20 m below.

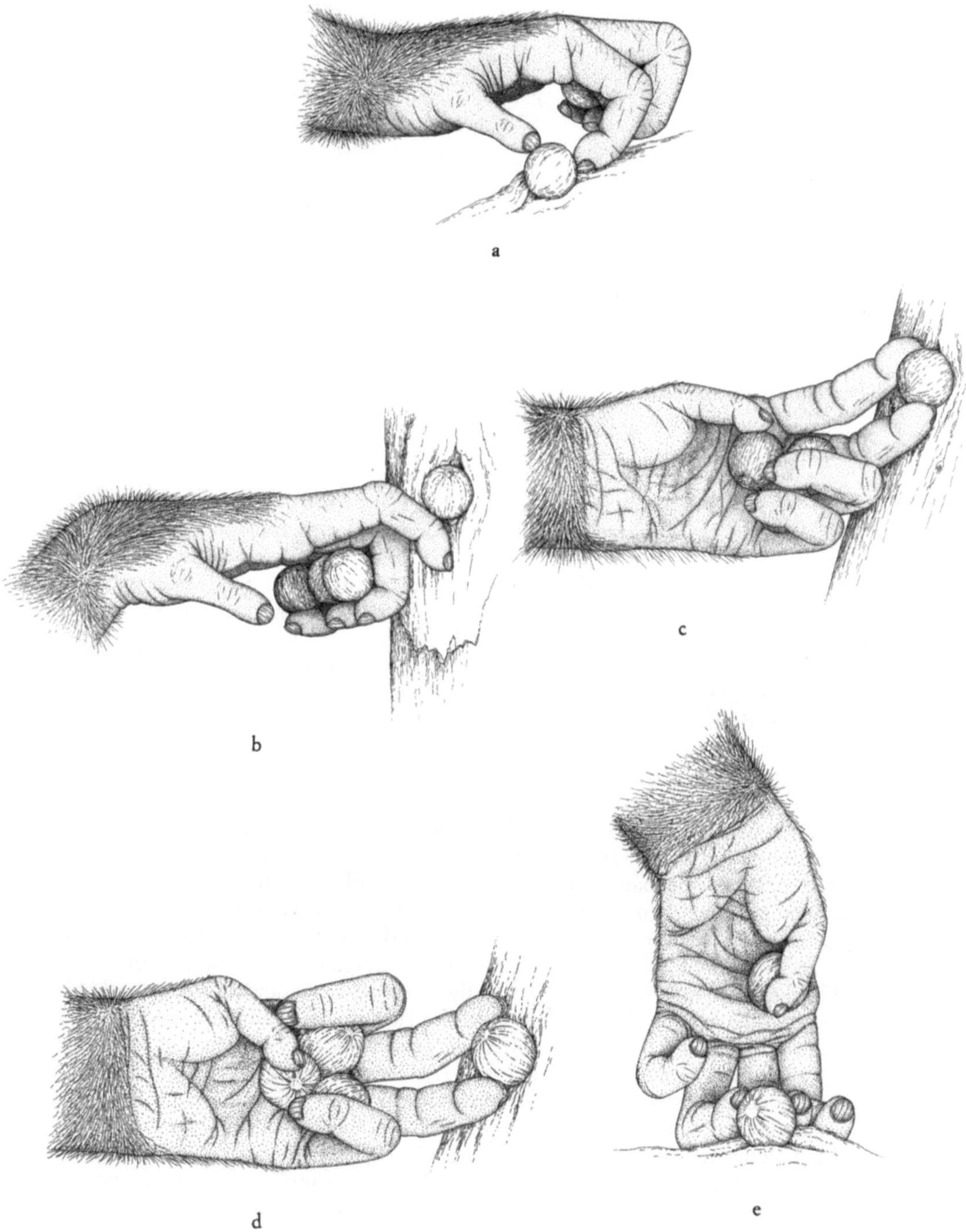

Figure 7 - Drawings illustrating the hand postures 1 to 5. See text for further explanations.

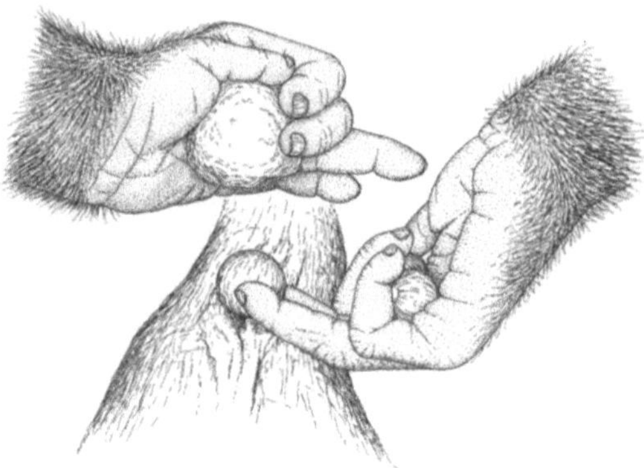

Figure 8 - Drawing illustrating the hand postures of figure 6, combining a grip 4 with a hand posture 4. See text for further explanations.

situation. If the anvil is in a more precarious position (as for example in figure 6a), the chimpanzees will use a firmer grip to control the pounded nut (as below).

— Posture 3 (*Figure 7c*): In 10% of observations, the pounded nut is held between the index and the third finger, the palm upwards,. Some collected nuts are held in the palm with the thumb and the fourth and fifth finger.

— Posture 4 (*Figure 7d*): In 5% of observations, the nut is held between the third and the fourth fingers, the palm upwards. Some more nuts are held in the palm with the thumb, second and fifth fingers.

— Posture 5 (*Figure 7e*): In 5% of the observations, the chimpanzees, besides controlling the pounded nut and holding some collected nuts, need some body weight support for their balance. In such situations, they will hold their hand on the anvil in the knuckle-walking position, the palm facing the pounded nut, controlled by the back of the third phalanges bent upwards. One or two more nuts are held by the thumb against the palm. In this posture, the hand fulfils 3 functions simultaneously.

These hand postures (1 to 5) are used in two different ways in order to avoid hitting the fingers: The chimpanzees may release the pounded nut and withdraw the hand sideways at each hit, then hold the nut again with the previous posture before lifting the hammer. Or they can rotate their hand slightly outwards and downwards at each hit without releasing the grip on the pounded nut. The first way is commonly used by most subadults and some adults, but is less efficient as it does not prevent the nut from bouncing away during the hit. The second way is typical for experienced nut-crackers.

Besides the dexterity shown by the chimpanzees in these various hand postures, we wish to emphasize the multiplicity of functions fulfilled by both hands *simultaneously*: One hand is holding the hammer and pounding with it, the other hand is holding the collected nuts as well as the nut that is being pounded on the anvil. In posture 5, we saw that some body weight can be also supported. It may even be more complicated: *Figures 8* and *9* illustrate a situation in which a chimpanzee cracks nuts by using grip 4 for the hammer and hand posture 4 for the nuts. With these hand postures, right at the moment of hitting the nut, the palm of both hands form a bowl-like shape around the pounded nut

Figure 9 - An adult female with a newborn infant clinging on her belly is cracking Coula nuts directly in the tree with a small stone hammer of about 300 g. She holds the stone with a grip 4 in her right hand; with the left one, using a hand posture 4, she holds some nuts in the palm and the pounded nut with the index and the major fingers. This picture illustrates the difficulty to crack nuts in the tree, which explains the adaptive advantage to use such elaborate hand postures.

on the anvil, thus being able to catch the nut when it bounces off the anvil. Here 5 functions are fulfilled by both hands at the same time.

Conclusion

Our observations of hand use by wild chimpanzees practised in the nut-cracking context, as well as the observations made on termite-fishing and other uses of sticks (Boesch & Boesch, pers. obs.; McGrew, 1977; Van Lawick- Goodall, 1975), emphasize the very large variety of manipulative skills in this primate. Some modifications of their gait allow the chimpanzees to reduce the loads on their forelimbs at increasing speeds more than seen in most other primates (Kimura *et al.*, 1979, Reynolds, 1985). Accordingly, this may have allowed them more mobility of their hands than would be expected, if adopting the normal quadrumanous climbing solution, the basic locomotor adaptation shared with many species.

When looking at the frequent, complex and asymmetric use of tools on which chimpanzees rely in the wild, one may ask if the tool use attributed to *Homo habilis*, such as pounding bones or stones to produce flakes, and butchering large dead mammals

(ISAAC, 1978, 1980; TOTH, 1985) is not within the manipulative abilities of the chimpanzee's hand.

ACKNOWLEDGEMENT — This study was supported by grant No. 3.420.83 from the Swiss National Science Foundation to Prof. Hans Kummer. We are greatly indebted to the «Ministère de la Recherche Scientifique» and the «Ministère de l'Agriculture et des Eaux et Forêts» of Ivory Coast for their support and facilities for working in the Taï National Park. This project is integrated in the UNESCO project TAI-MAB and we wish to thank the general supervisor, Dr Henri Dosso, for his support.

We are very grateful to the following persons: Prof. F. Boulière who initially proposed this project to us; Prof. A. Aeschlimann for his constant support; Dr F. Naef, director of the «Centre Suisse de Recherches Scientifiques en Côte d'Ivoire» and T. Zoroa Tiepkan for their efficient logistic help; Prof. H. Kummer and Prof. H. Preuschoft for their kind help in having the drawings made; and last but not least Helga Schulze for drawing the hands in different postures.

Finally, we wish to express particular thanks to Prof. H. Kummer for his valuable guidance, critical discussion and constant encouragement throughout the study.

References

BOESCH C. & BOESCH H., 1981. *Sex differences in the use of natural hammers by wild chimpanzees: a preliminary report.* Journal of Human Evolution, 10: 583-593.

BOESCH C. & BOESCH H., 1983. *Optimization of nut cracking with natural hammers by wild chimpanzees.* Behaviour, 3/4: 265-286.

BOESCH C. & BOESCH H., 1984a. *Possible causes of sex differences in the use of natural hammers by wild chimpanzees.* Journal of Human Evolution, 13: 414-440.

BOESCH C. & BOESCH H., 1984b. *Mental map in wild chimpanzees: An analysis of hammer transports for nut cracking.* Primates, 25 (2): 160-170.

ISAAC G., 1978. *The food-sharing behaviour of protohuman hominids.* Scientific American, 238: 90-108.

ISAAC G., 1980. *Casting the net wide: A review of archeological evidence for early hominid land use and ecological relations.* In: Current Argument on Early Man (ed. Lars-Koenig Koenigsson). Pergamon Press, Oxford: 226-251.

JOHANSON D. & EDEY M. 1981. *Lucy: the beginning of humankind.* Simon and Schuster. New York.

KIMURA T., OKADA M. & ISHIDA H., 1979. *Kinesiological characteristics of primate walking: its significance in human walking.* In: Environment, Behavior and Morphology; Dynamic Interactions in Primates, M.E. Morbeck, H. Preuschoft & N. Gomberg, eds. pp. 297-311. G. Fischer. New York and Stuttgart.

LEAKEY R., 1981. *The making of mankind.* Dutton, New York.

MCGREW W., 1977. *Socialization and object manipulation of wild chimpanzees.* In: Primate Bio-Social Development: Biological, Social and Ecological Determinants (eds. Chevalier-Skolnikoff, S. and Poirier, F.). Garland, New York: 261-288.

MCGREW W., TUTIN C. & BALDWIN P., 1979. *Chimpanzees, tools and termites: Cross-cultural comparisons of Senegal, Tanzania and Rio Muni.* Man, 14: 185-214.

NISHIDA T. & HIRAWA M. 1982. *Natural history of a tool-using behavior by wild chimpanzees in feeding upon wood-boring ants.* Journal of Human Evolution, 11: 73-99.

PILBEAM D., 1980. *Majors trends in human evolution.* In: Current Argument on Early Man. Lars-Koenig Koenigsson, ed. Pergamon Press, Oxford: 261-285.

PREUSCHOFT H., 1965. *Muskeln und Gelenke der Vorderextremität des Gorilla.* Morph. Jb., 107: 99-183.

PREUSCHOFT H., 1973. *Functional anatomy of the upper extremity.* In: G.H. Bourne (ed). The Chimpanzee, Vol. 6, Karger-Verlag, Basel, München und New York: 34-120.

REYNOLDS T. 1985. *Stresses on the limbs of quadrupedal primates.* American Journal of Physical Anthropology, 67: 351-362.

SUSMAN R., STERN J. & JUNGERS W., 1984. *Arboreality and bipedality in the Hadar hominids.* Folia Primatologica, 43: 113-156.

TELEKI G., 1974. *Chimpanzee subsistence technology: material and skills.* Journal of Human Evolution, 3: 575-594.

TOTH N., 1985. *The Oldowan reassessed: a close look at early stone artefacts.* Journal of Archaeological Science, 12: 101-120.
TUTTLE R., 1969. *Quantitative and functional studies on the hands of the Anthropoidea. I. The Homonidea.* Journal of Morphology, 128 (3): 309-363.
TUTTLE R., 1969. *Knuckle walking and the problem of human origins.* Science, 166: 953-961.
VAN LAWICK-GOODALL J., 1970. *Tool using in primates and other vertebrates.* In: The Study of Behavior (eds. Lehrman D., Hinde, R.A. and Shaw, E.). vol 3: 195-249.
VAN LAWICK-GOODALL J., 1973. *Cultural elements in a chimpanzee community.* Proceedings of the 4th International Congress of Primatology, vol. 1. Karger, Basel: 144-184.

P.H. Brenot

Lateralised handedness, bipedalism and cortical specialisation

Key words: cortex, bipedalism, right-left handedness, laterality index, hemispheric specialisation.

Handedness and bipedalism are two human features which are not diagnostic because they occur in many species. Humans are unique, however, in their combination of bipedalism and habitual orthogradism with a marked manual dexterity; the former seems to be connected with the latter through lateral specialisations of the cortex.

Bipedalism occurs in many animal groups (modern reptiles, saurians, birds, mammals) in various modes (KUMMER, 1965) and with extremely varied side-effects on the skeleton. In *Homo s. sapiens*, bipedalism is permanent, not facultative (COPPENS, 1981), and involved anatomical modelling of skull, spine, pelvis and limbs; the lower limbs used solely for locomotion became specialised by lengthening; the upper limbs, used solely for prehension, were shortened. The discharge of the upper limbs from their locomotor functions allows hand specialisation. These characters seem to have been present 3.7 million years ago (according to the first human footprints found in Laetoli, Tanzania; WHITE, 1980) and in *Australopithecus* (according to fossil remains, e.g. ROBINSON, 1972; PREUSCHOFT, 1971; SUSMAN & STERN, 1983) and probably much earlier in that phylum (*Dryopithecus ?*) (O'CONNOR, 1976) owing to a progressive acquisition over about fifteen million years.

Handedness exists occasionally in arthropods, rodents, (TSAI & MAURER, 1930; COLLINS, 1968; PETERSON, 1934), bovids (UHRBOCK, 1969; WAGNON & ROLLINS, 1972), felines (COLE, 1955), and primates. It is permanent in *Homo s. sapiens*, incomplete in *Pongo* and frequent in *Pan* and *Gorilla*. This upper limb laterality testifies an incipient cortical specialisation. It has been found, though not constantly, in other mammals, in bovids, rats, cats (WARREN, 1958), cetaceans (LILLY, 1962) and passerine birds (NOTTEBOHM, 1971), as well as in primates: monkeys (BECK, 1972; COLE, 1957; LEHMAN, 1978; WARREN, 1958) and apes (FINCH, 1941). In macaques, neither WARREN (1958), nor LEHMAN (1978), nor BRENOT (1982b) found any homogeneous trend of handedness; but we did find a lack of manual preference in the youngest individuals. One can even talk of some «competition» between the two upper limbs. In apes, beyond isolated observations, we only know of three studies of handedness, those of FINCH in 1941, CHIVERS in 1974 and ours in 1982. Out of 31 free living chimpanzees FINCH found a clear laterality in 25 — on the right as well as on the left — which could be translated into two-fifths right-handed, as many left-handed, and one-fifth undetermined (see also RAWLINS, REDSHAW & CHRISTEL, all in this volume).

My own work was done on nine captive young apes (BRENOT, 1982b): 5 *Pan troglodytes*, 1 *Gorilla gorilla*, 3 *Pongo pygmaeus*. Each animal's handedness was repeatedly studied during ten periods of «food-catching» (choice of a strategy to get hold of the food they are presented with: use of right or left hand) and periods of free behavior, each lasting 5 minutes, according to an evaluation grid (0 = right-handed, 1 = left-handed

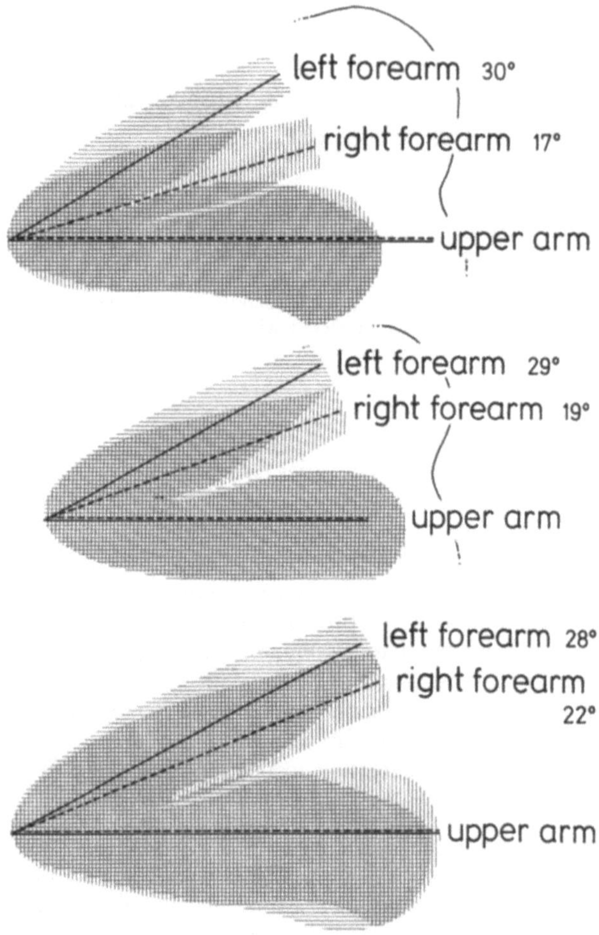

Figure 1 — Flexibility of the elbow joint in apes (BRENOT, 1982). Schematic drawings of *P.troglodytes* (Charlotte, 4 1/2 years, top), *G.gorilla* (Aline, 4 years, middle), and *P.pygmaeus* (Borg, 4 1/2 years, bottom), in side view with strongly flexed elbow joints. The angles of flexion are sharper on the right side in left handed subjects, and vice versa — probably because of the better development of the muscles in the dominant arm. The three apes shown here are left handed subjects, the drawing is in accordance with *Figure* 2.

catching). All those sequences were video recorded and analysed using the still-frame facility.

Results were described in the form of a laterality index (LI), as the ratio of the left acts (L) over the total number (T): LI = L/T, with the ratio varying from 0 (full rigth-handedness) to 1 (full left-handedness). We compared this estimate with that of «extensibility» (STAMBAK, 1958; differential bending of the forearm on the upper arm as related to the muscular development: *Figure* 1) which strengthens and confirms the LI measurement: This simple test, tried and tested on large human population sample (HECAEN & AJURIAGUERRA, 1963), makes it possible to determine laterality in a child while it is not yet established. It is commonly admitted that in adults the left arm closes at a sharper angle in right handed subjects and conversely the right arm closes at a sharper angle in left handed subjects.

Concerning the laterality index, *Figure* 2 (LI according to its Standard Deviation) shows a clear separation between high LI an low SD individuals and those with low or indifferent LI (around 0.50) and high SD.

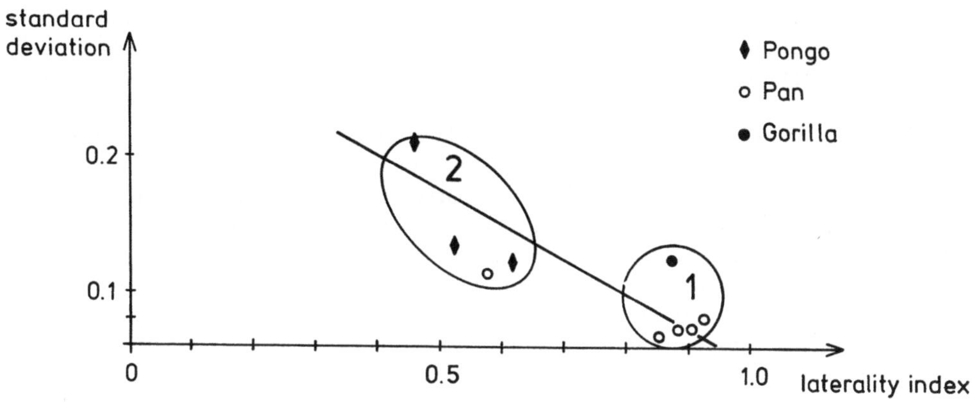

Figure 2 — Laterality index with standard deviation and regression (explanations in the text).

Laterality indices appear higher in chimpanzee and gorilla (0.8 to 0.9) as opposed to those of the orang-utan (o.45/0.53/0.64). This strong laterality (left-handed in this case) of the whole population of both ape species is not confirmed by the behaviour of free-ranging adults (Finch, 1941, and ours).

The sharp separation according to manual laterality between chimpanzee/gorilla and orang-utan, that is to say between the purely arboreal species and those which prefer living on the ground, and have a tendency towards bipedalism, is worth noticing. This point is further proven by the presence in the orang-utan group of a young female chimpanzee (*Figure* 2) with an unsteady median laterality (0.58 ± 0.11). Her individual behaviour is very different from that of other chimpanzees: while the others spent most of their time on the ground, she was very arboreal, rarely bipedal. In *Figure* 2, the two sub-groups do not seem to be characterised by taxonomy, but by the mode of living:

1 - Strongly lateralized ground-living, occasionally bipedal animals.

2 - Hardly lateralized arboreal four-handed animals.

Among modern humans, many researchers (Annett, 1972; Marshall, 1975; Hardyck, 1977) found a constant proportion of about 80% right-handed individuals in a variety of populations.

Human lateral handedness leaves its print in the flaking off of stones and this probably already in *Homo habilis* (Lewin, 1986). Nicolas Toth (1985) seems to have found indications of right-handedness at seven sites in Northern Kenya, dated between 1.5 and 2 Million years back. Observations pointing to similar conclusions have also been made following the occasional discovery of tools, such as a bauxite knife from Petralona (about 690.000 years old: Poulianos, 1977), which could only be held by a right-handed individual and bore signs of such use. Those isolated observations nevertheless do not tell much of the general right- or left-handedness of those populations. The first evidence regarding frequency of hand preference can be found in the upper Paleolithic with the negative handprints of Gargas (France), a large number of which were left hands (159 left as against 23 right), which probably documents right-handedness of their bearers, and come close to the present proportion.

In most mammals the early signs of *cortical specialisation* (listed on first page) are connected to the evolution of communicative systems, particular those based on sound and, therefore, connected to the social evolution of the species. And with the development of communication (birds, cetaceans, primates), likewise, those early signs of specialisation are predominantly linked with upper limb activity (rodents, felines, cetaceans, primates).

With hominisation there appeared a set of cerebral phenomena, which are already present in primitive representatives of the phylum (BAUCHOT, 1972; BAUCHOT and STEPHAN, 1969; STEPHAN, 1972). They are markedly developed in modern humans: a gradual increase in brain — and above all cortical — volume; a gradual increase in brain — and above all cortical — complexity; the growth and growing intricacy of the cortex, especially in pre-frontal regions (telencephalisation); finally, the secondary specialisation of this new cortex and of homologous regions in each hemisphere.

Beyond the much-described change in quantity, the evolution in quality lends itself far less observations, yet it is that which constitutes the originality and specificity of present man and we may observe its landmarks in the course of hominisation. These landmarks are twofold: direct evidence (cerebral and cortical indices), and indirect evidence (remaining signs of lateralised behaviour). The latter becomes obvious in the right-handedness of every human group so far investigated. The behavioural right-handedness reflects the functional asymmetry of the cortex, a most varied sensory, motor and cognitive mosaic in present-day man (BRENOT, 1982b). Cortex specialisation and functional complementarity of the two hemispheres entails an increase of potential abilities of the brain and is a great adaptative advantage over strict functional symmetry, which is to be found in most mammalian species.

This specialisation of the cortex seems to be linked to language, to its appearance in humans and to the following structural development of the cortex through language. A first fundamental question is raised: Does the functional cortex asymmetry occur before language or not?

The extremely rich specialisation of the cortex of present-day adult man testifies the fundamental role played by man's communicative system — language — in the functional structural modelling of the cortex. The functional patterns which can now be visualised, thanks to computer graphics, particularly of cerebral blood flow (INGVAR, 1977), show this mosaic-like specialisation which makes it possible to go beyond the old-fashioned notions of «centres» and «asymmetry» and to consider the more global notion of «cortical specialisation». There are variations especially in the use of hand and touching, for instance between the Japanese and the British (HATTA, 1984), also in the location of motor and sensory language structures in homogeneous right-handed subjects, for instance in the native American Navajo (MC KEEVER, 1984). Those variations in adult functional patterns are cultural ones and they let us understand how far «the language» can structure that very plastic biological substrate which present-day man's cortex constitutes. Very early left-sided cortical lesions in the child (before language is acquired) by no means prevent its subsequent setting in (JEANNEROD, 1984) and document that the child, unlike the adult, recovers very quickly from aphasia. This structuring of the biological cortex is effected according to language features (BRENOT, 1984, 1986; HINES, 1984), it obeys very roughly the following rules:

— the left hemisphere is more efficient in verbal coding and decoding tasks.
— the right hemisphere is more efficient in time-and-space recognition tasks.

One might as well say that we are faced with a verbal hemisphere and a non-verbal one, *or rather* with one hemisphere which is competent in «time-and-space» and one which is not. We have come to the crossroads of primate and human evolution with social changes, and lifestyles that shift from space/time control — of environment — to the psychic dimension which language grants.

Some structural asymmetries in humans correspond to this functional asymmetry, particular in language areas, with the greater development of the «planum temporale» on the left, twice as large as on the right (BONIN, 1955; GESCHWIND & LEVITSKY, 1968). This change in the left temporal region can be seen from the outside in the asymmetry of the sylvian fissure, shorter and reaching higher on the right (GESCHWIND & LEVITSKY, 1968). On a macrospopic level we notice in modern *Homo sapiens* the advent of new cortical — frontal and pre-frontal — areas, which are known to be related with cognition and the temporal region asymmetry, which is known to be linked with sensor and motional language functions (*Figure* 3).

Those elements already appear in non-human primates: increasing cerebral volume, increasing intricacy, early signs of the frontal regions growing forward, but also various attempts of cortical specialisations to grow laterally. Anatomical asymmetries are many in every cortical region (LEMAY, 1976). One may suggest that the most varied and marked human cortex lateralisation is but the final stage of only one of the numerous evolutionary attempts to this aim in the primate group. The asymmetric «planum temporale» is already established, although not significantly, in macaques, and, very significantly, in chimpanzees (YENI-KOMSHIAN & BENSON, 1976). In primate cortices (*Macaca mulatta, Pan troglodytes*) can already be found electro-cortical asymmetries (BRENOT, 1982, a, b) and functional specialisations (JASON, 1984). It seems that biological structures have already appeared in monkeys and particularly in apes, which will lead to the extremely varied human cortical mosaic.

Both direct and indirect elements found in the process of hominisation bring into relief the functional use of this lateral specialisation trend. However, those elements need some readjustment and sometimes must be viewed critically so that their limits should be underlined. The growing complexity of tools and the increasingly diverse ordering of the hominoid environment make it possible to establish a link between industry and the development of the related cortical region, particularly those of manual control, sensory discrimination as well as of language motor areas that are nearest, and which transmitting technology requires (*Figure* 3). Can it not be said that, on a cortical level, language motor areas are a downward extension in the Rolandian fissure of manual areas?

We know of fossil four-handed arboreal primates (I assume that they were not lateralized) then terrestrial bipedal hominoids (*Australopithecus*): I put forward the hypothesis that they were already lateralized, but that indirect documents of manual laterality were too slight to confirm this. Later, in *Homo habilis*, it seems possible to find clear evidence of handedness (TOTH, 1985).

Brain endocasts do not show clearly any structural asymmetry yet. They do show the forward development (to frontal regions) of the new structures and the remodelling of the sylvian region (SABAN, 1980). A few samples of a temporal region (planum temporale) more developed on the left side suggest a diagnosis of «right-handedness». Reservations should be made as to the validity of such an interpretation inasmuch as cross-asymmetries do exist and that a more developed left region can also be found in left-handed individuals. However, the state of conservation and reconstruction of bones does not always permit such conclusions. One must state that, on certain fossils like the neanderthaler from La Chapelle aux Saints, cortical asymmetry is obvious, but no positive identification of lateral handedness can be made.

All these facts point to the existence of an increasing cortical asymmetry during hominisation with the earlier occurence of some behavioural and manual laterality. Verbal

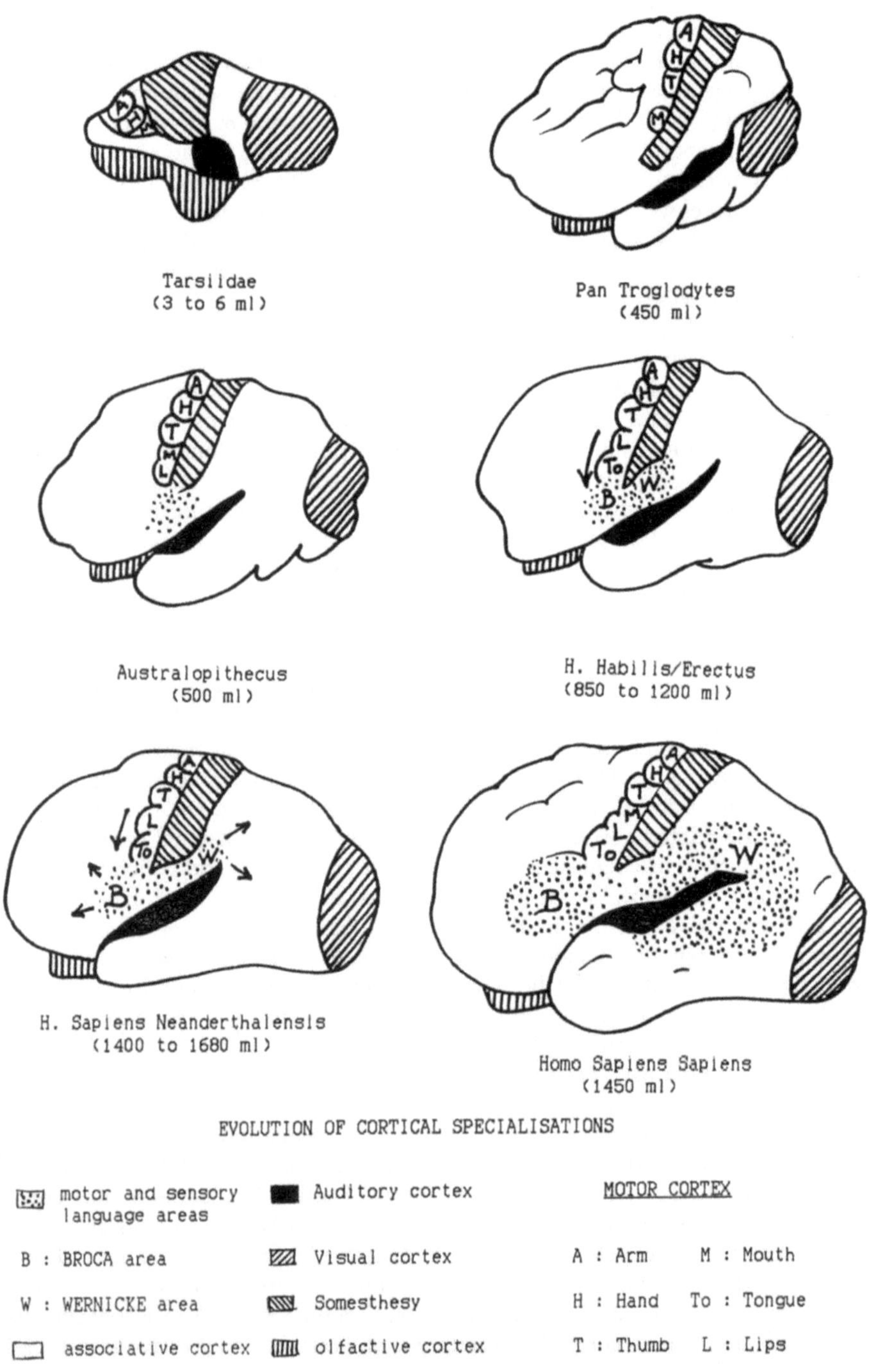

Figure 3 — Evolution of cortical specialisation according to my personal hypothesis. (further explanations in the text).

language, which is in many points linked to the manual refining evolution and to the specialisation of this cortical region, followed an identical modelling most probably later because of the position of its motor areas.

So, the functional evolution of the human cortex actually started long ago in vertebrate evolution and, as has been noted above, its attempts to reach its present form were many and occurred in different animal groups. With the primary symmetry of cerebral structures, which appeared as early as the prochordates, a functional symmetry, likely to have been linked to the development of the visual system (YOUNG, 1962), made it possible to dominate space better, through the combination and coordination of both right and left eyesight fields, which constitute an adaptation advantage over separate cerebral structures. The secondary specialisation of those symmetrical stuctures now appears as an additional advantage over strict functional symmetry (BRENOT, 1982, a, b). The mechanisms of such changes have not been clearly understood yet, but it can be argued that there has been a selection of some hominoids, already more lateralised than others, in which these evolutionary dynamics have permitted the colonisation by language of cortical regions already specialised in hand activity. The hypothesis of a gestural or semi-gestural language as a transitory stage seems plausible. Finally, the great functional stimulation of this region in the temporo-parieto-frontal crossroad must have helped in permitting great synaptic diversity, of which these incipient functions could then make use (CHANGEUX, 1983). Cortex hominisation is a recent phenomenon: it all started with the growing intricacy and specialisation of the existing structures; it has not come to an end, owing to the fundamental subsequent modelling of those new structures by language which still seems nowadays the truly human dimension.

The foregoing seems to suggest a relation between bipedalism and handedness, both depending on cortical specialisations that have developed in the beginning of hominisation.

References

ANNETT M., 1972. *The distribution of manual asymmetry.* Brit. J. Psychol., 63: 343-358.

BAUCHOT R., 1972. *Encéphalisation et phylogénie.* Comptes rendus Acad. Sciences, Paris, 275 (3): 441-443.

BAUCHOT R. & STEPHAN H., 1969. *Encéphalisation et niveaux evolutifs chez les simiens.* Mammalia, 33: 225-275.

BECK C.H.M. & BARTONN R.L., 1972. *Deviation and laterality of hand preference in monkeys.* Cortex, 4: 339-363.

BONIN G. VON, 1955. *Essai sur le Cortex Cerebral.* Masson, Paris, 160 pp.

BRENOT PH., 1982a. *Evolution fonctionelle du cortex des primates.* Soc. Estudios Vascos, 261-265, and Bull. Soc. Anthrop. S-O, 33.

BRENOT PH., 1982b. *Evolution de l'asymetrie électro-corticale chez les primates.* These Doc. ès Sciences, Bordeaux, No. 728, 325 pp.

BRENOT PH., 1984. *Evolution et symétrie des structures cerebrales.* Seminaire du Lab. de Paleontologie humaine de l'E.P.H.E., Bordeaux, (not published).

BRENOT PH., 1984. *Langage et hominisation.* Revue de phonétique appliqué, 71-72: 217-226.

BRENOT PH., 1986. *Bipédie, lateralité et langage.* Seminaire Coll. de France, Paris, (not published).

CHANGEUX J.P., 1983. *L'Homme neuronal.* Fayard, Paris, 419 pp.

CHIVERS D.J., 1974. *The siamang in Malaya. A field study of a primate in tropical rain forest.* Contributions Primatol., 4, 335 pp.

CHRISTEL, *this volume.*

COLE J., 1955. *Paw preference in cats related to hand preference in animals and man.* J. Compl. Physiol. Psychol. 48: 137-140.

COLE J., 1957. *Laterality in the use of the hand, foot and eye in monkeys.* J. compl. Physiol. Psychol., 50: 296-299.

COLLINS R.L., 1968. *On the inheritance of handedness; laterality in inbred mice.* J. Heredity, 59: 9-12.

COPPENS Y., 1981. *L'origine du genre Homo.* In: Les Processus de l'Hominisation, CNRS, Paris, 55-60.

FINCH G., 1941. *Chimpanzee handedness.* Science, 94: 117-118.

GESCHWIND N. & LEVITZKY W., 1968. *Human brain: left-right asymmetries in the temporal speech region.* Science, 161: 186-187.

HARDYCK C. & PETRINOVICH L., 1977. *Left-handedness.* Psychol. Bull., 84(3): 385-404.

HATTA T. & DIMOND S.J., 1984. *Difference in face touching by Japanese and Britain people.* Neuropsychologia, 22(4): 427-433.

HECAEN H. & AJURIAGUERRA, J. DE, 1963. *Les gauchers, prévalence manuelle et dominance cérébrale.* P.U.F., Paris, 172 p.

HINES D. *et al.*, 1984. *Hemispheric asymmetry in use of semantic category information.* Neuropsychologia, 22(4): 531-534.

INGVAR D.H., 1977. *Functional responses of the human brain studied by regional cerebral blood flow techniques.* Act. Clin. Belgica, 32(2): 68-83.

JEANNEROD M., 1984. *Neurobiology of language in children.* Neuropsychiatr. Enfance, Adolesc., Paris, 32: 10-11, 529-532.

KUMMER B., 1965. *Das Mechanische Problem der Aufrichtung auf die Hinterextremität im Hinblick auf die Evolution der Bipedie des Menschen.* In: G. Heberer (ed.), Menschliche Abstammungslehre, G. Fischer, Stuttgart, pp. 225-248.

LEHMAN R.A.W., 1978. *The handedness of Rhesus monkeys: concurrent reaching.* Cortex, 14: 190-196.

LEMAY M., 1976. *Morphological cerebral asymmetries of modern man, fossil man, and non-human primate.* In: Origins and Evolution of Language and Speech. Ann. N.Y. Acad. Sci., 280: 349-366.

LEWIN R., 1986. *When stones can be deceptive.* Science, 23: 113-115.

LILLY J.C., 1962. *Discussion paper.* In: Mountcastle (ed.), Interhemispheric Relations and Cerebral Dominance. Baltimore, 112-114.

MARSHALL J.C., CAPLAN D. & HOLMES. J.M., 1975. *The measure of laterality.* Neuropsychol., 13: 315-321.

MCKEEVER W.F. & HUNT L.J., 1984. *Failure to replicate the Scott et al. finding of reserved ear dominance in the native American navajo.* Neuropsychologia, 22(4): 539-541.

NOTTEBOHM F., 1971. *Neural lateralization of vocal control in a passerine bird.* J. Exp. Zool., 177: 229-261.

O'CONNOR B.L., 1976. *Dryopthecus (Proconsul) africanus: quadruped or non-quadruped?* J. Hum. Evol., 5(3): 279-283.

PETERSON G.M., 1934. *Mechanisms of handedness in the rat.* Comp. Psychol. Monogr., 9: 1-67.

POULIANOS A.N., 1977. *Petralonian archanthropus was right-handed. — Anthropos, Athens,* 4(1-2): 147-151.

PREUSCHOFT H., 1971. *Body posture and locomotion in Early Pleistocene hominids.* Folia primat., 14: 209-240.

RAWLINS, *this volume.*

REDSHAW, *this volume.*

ROBINSON 1972. *Early Hominid Posture and Locomotion.* University Chicago Press, Chicago, London. 362 p.

SABAN R., 1980. *Le systeme des veines méningées moyennes chez Homo erectus, d'après le moulage endocranien.* 105 ème Congrés Nat. Soc. Savantes, Caen, 4: 8-12.

STAMBAK M. & AJURIAGUERRA J. DE, 1958. *Evolution de l'extensibilité musculaire depuis la naissance jusqu' a l'âge de 2 ans.* Presse Med., Paris, 2: 24-27.

STEPHAN H., 1972. *Evolution of primate brains: a comparative anatomical investigation* In: R. Tuttle (ed.) The Functional and Evolutionary Biology of Primates. Aldine- Atherton, Chicago, 155-174.

STERN J.R. & SUSSMAN R.L., 1983. *The locomotor anatomy of Australopithecus afarensis.* Amer. J. Phys. Antrop., 60: 279-317.

TOTH N., 1985. *Archeological evidence for preferential right-handedness in the lower and middle Pleistocene, and its possible implications.* J. Hum. Evol., 14: 607.

TSAI L. & MAURER S., 1930. *Right-handedness in white rats.* Science, 72: 436-438.

UHRBOCK R.S., 1969. *Bovine laterality.* J. Genetic Psychol., 115: 77-79.

WAGNON K.A. & ROLLINS W.C., 1972. *Bovine laterality* . J. Animal. Sci., 35: 468-488.

WARREN J.M., 1958. *The development of paw preference in cats and monkeys.* J. genet. Psychol., 93: 229-236.

WHITE T.D., 1980. *Evolutionary implications of Pliocene hominid footprints.* Science, 208: 175-176.

YENI-KOMSHIAN G. & BENSON D., 1976. *Anatomical study of cerebral asymmetry in the temporal lobe of humans, chimpanzees and rhesus monkeys.* Science, 192: 387-389.

YOUNG J.Z., 1962. *Why do we have two brains?* In: Mountcastle (ed.), Interhemispheric Relations and Cerebral Dominance. Baltimore, 7-25.

M.E. Redshaw

The development of prehension in human and gorilla infants

Experimental observational methods were used to follow the development of prehension skills in human and gorilla infancy and to compare this early behaviour with that observed in human adults. Responses to a wide range of different objects of varying sizes were recorded on videotape and using still-frame and slow-motion techniques the fine detail of hand movements and postures were examined.
Developmental changes were found at all stages in the prehension sequence. from approach to object capture. Over time both species of infant showed a progression in which this manual skill became more smoothly co-ordinated and more dextrous. They became more direct in approaching objects, more appropriate in matching the configuration of the hand to the characteristics of the objects grasped, and increasingly used the radial portion of the hand in grasping.
Development in the two species was in general similar, though some differences were evident: the young gorillas used more flexion at the inter-phalangeal joints, were less able to hyperextend the forefinger or to differentiate this digit and the adjacent finger jointly, and over all they used significantly fewer precision grips than the human infants. Such differences, which contribute to the more robust appearance of gorilla prehension, can be linked with the dual function of the gorilla hand in locomotor and manipulative activity.
Radical changes take place in the acquisition of prehension skills in infancy and yet it is evident from comparisons with human adults on the same task that there are further developments which have yet to take place. Faster, more precise, clearly lateralised movements in which, anticipating, the hand closely matches the shape of the object or else relates to aspects of its function, are the keynote of human adult prehensive behaviour.

Key words: dexterity, hand movements, hand postures, manipulation, manual skill, precision grips, prehensive act.

Introduction

Non-human primates provide much material that can be used in considering the evolution of the hand and its functions, but the primates of today must not be thought of as merely part of a step-like sequence leading upwards to man. A range and variety of manipulative hand functions have evolved and can be seen in use among a wide range of primate groups living in many different habitats. Though these provide useful contextual background material in studying hand use in man, the data on some species of primate are more relevant than that on others.

The prosimians possess a fairly fixed pattern of whole hand control and species that are insectivorous are considerably more adept at prehension of small objects than those that are frugivorous (Bishop, 1962a). Neither type differentiates power and precision movement, and the prosimian level of prehension appears to be a conservative one involving a single prehensive pattern without much fine control. In contrast, some of the New World monkeys have precision patterns, but these do not involve divergence of the thumb from the other digits, rather a scissor-like action with two fingers is used or the forefinger alone. These grips are used for fine manipulation and are clearly different from the power grips employed in locomotion (Bishop, 1962b; see also Hofer, 1960).

The hand of most Old World monkeys can be distinguished from that of New World forms by the independence of the thumb, but the functional effectiveness of the opposability achieved with this digit is variable (Napier, 1960, 1961). The ability to place the terminal volar pads of the index finger and thumb in pulp-to-pulp contact depends on the relative proportions of these two digits (Napier & Napier, 1967). Unlike the arboreal cercopithecoid monkeys, species such as the patas (*Erythrocebus*) have completely separated precision from power grip as one of the adaptations to terrestrial life, walking with digits 2-5 flat on the ground and the thumb not in contact with the substrate at all, while small objects are prehended with the tip of the thumb and the side of the forefinger. Hall & Mayer (1966) linked this ability to prehend extremely small objects to the diet of small plants, seeds, grass stems and insects available in the savanna environment which the patas inhabit. A similar situation exists in *Theropithecus*, as described by Maier in this volume. The mangabeys, macaques and baboons all have separate control of the forefinger and thumb and use a precision grip in which the ball of the thumb is used against the ball of d2. In contrast *Colobus* has a thumb which is markedly reduced.

Rollinson (1975), using a large number of Old World monkey species as experimental subjects confirmed that there is a «ground-to-tree» gradient in terms of manipulative skill, with the terrestrial forms showing a greater degree of fine control in handling small objects than the arboreal species.

The fine control that seems to have evolved first in the separation of power and precision patterns and then in separate control of the radial digits, has progressed further in the apes, into individual control of all the digits.

The structure of the hand in the gibbon (Hylobatidae) assists in the highly specialised form of locomotion observed in these primates, namely brachiation. During such locomotion all the fingers, including the thumb are parallel, forming a hook which allows the hand to rotate around a branch during the rapid grasping and releasing sequence that is required (Lorenz, 1971). However, during climbing and feeding, grips are favoured in which the relatively long thumb (with a ball and socket joint) is opposed to the other digits. Thus, despite a highly specialised locomotor pattern, a large strong and mobile thumb is available for coping with situations requiring a degree of manual power or skill. According to Lorenz, gibbons also use the thumb as a «probe finger» in investigating and exploring novel objects, a function which in the great apes is performed by the finger tips or the forefinger alone.

Unlike the Hylobatidae, the great apes have relatively short thumbs. However, their long and powerful fingers can be used similarly with a hook-like grip during suspensory locomotion. Despite the discrepancy in length between the thumb and the index finger these digits can be placed in opposition by acutely flexing the forefinger at its interphalangeal joints and by widely abducting the thumb (Napier, 1961). A tip-to-tip contact results which, though precise, may lack the sensitivity and control that is possible with full contact of the volar pads. Napier (1960) described the thumb as having little or no contribution to the hook or fist-like grips which constitute the power grip complex. A variety of precision movements were also described ranging from the double-locking mechanism in which the digits can grasp a slender object with little or no use of the palmar surface, to those in which the side or pulp of the thumb is directed towards the middle or terminal phalanx, or dorsum of the index finger and even one in which there is bimanual use of the forefingers jointly (CHRISTEL, in this volume, has a more detailed discussion of the grips used by apes in fine prehension).

Though there generally appears to be greater control of the digits than in other primate groups, with the exception of the gelada baboon, (see MAIER, this volume), the hands represent a very obvious compromise between the requirements of weight-bearing in climbing, arm suspension and knuckle-walking and the need for manipulative skill in feeding, food preparation and activities like grooming.

Special features of the human hand, that is the hand seen in modern man today, include relatively broad terminal phalanges that ensure a good thumb-to-finger grip, a thumb that is long in proportion to the index finger, set at a wide angle and able to rotate to meet all the other digits in good opposition, a wrist joint with enhanced mobility, and a highly sensitive palmar surface (NAPIER, 1962. See also LANDSMEER in this vol.).

The evolution of the hand and the relationship between the morphology and various types of movement and forms of locomotion, including brachiation and knuckle walking have been studied using many techniques. These include dissection, radiographic and biomechanical studies, multiple measurements and multivariate statistical analyses, lesion and micro-electrode studies and observation. It is with the last of these that this chapter is largely concerned. Morphological and neurological studies can provide information about the mechanical constraints and limits to movement, but they do not tell us how an ape or a human actually uses his hands in a fairly continuous way during the locomotion, feeding, social contact and object manipulation that is an integral part of daily life. It was with the aim of investigating normal hand use that a comparative observational study of the development of manual prehension was undertaken.

The experiment; Background

The aim was to make a detailed study of the development of prehension in one of the great apes and to compare this with that observed in man, with a view to elucidating some of the similarities and differences seen in this aspect of hominoid behaviour. It was decided to examine the behaviour of infants prior to, in the case of the humans, language development and in the case of both species, the beginnings of tool-use. Many researchers on motor skill have directed their attention to the first year, but have largely centred their efforts on the development of visually guided reaching and related responses. In contrast this study sought to examine the development and consolidation of early prehensive behaviour, once visually guided reaching had begun. The changing pattern of grasping, its increasing appropriateness and complexity and the development of hand preference are all part of the substrate on which many later skills, including the use and manufacture of tools, are built.

The first eighteen months of life, that is «the sensorimotor period» (PIAGET, 1952), is the most relevant time at which to attempt comparisons of ape and human development, for at this stage human infants are without the language skills that are known to contribute a great deal to later learning and development. As part of a broader, long-term study on the development of cognitive, social and manipulative skills in human and gorilla infants (HUGHES & REDSHAW, 1974; REDSHAW, 1975, 1978), a detailed longitudinal study was made on the development of prehension from twenty-two to sixty-two weeks of age. Additional data were also collected on adult humans. This was in order to gain some appreciation of the end point and goal of the observed infant development, at least in man, giving the study a wider and longer term perspective.

In view of the relatively close phylogenetic relationship between man and the gorilla, the considerable similarity of limb structure and neonatal reflexes (CARTER, 1973; REDSHAW, 1989) and the fact that of the great apes, the gorilla has the most similar hand proportions (NAPIER & NAPIER, 1967; NAPIER, 1980) it was predicted that the human and gorilla infants would follow much the same course in developing prehensive skills. However, it was also thought that as locomotor skills developed and the knuckle-walking posture was increasingly used by the gorilla infants, this would gradually exert an influence on prehension. It was thus predicted that any species differences that were found could be linked with the knuckle-walking posture becoming habitual. A greater difference was expected between the behaviour of the infants and that of the adults, who, with so much experience of object handling, might yet show more refinements, even in this basic skill.

Method

Subjects

The humans were three normal full-term infants (one male and two female; SM, NA, SR), brought up by their own parents, but who visited the laboratory at regular intervals for experimental sessions, and eight adults (four males and four females) who came for a single experimental session.

The gorilla infants (A and M), both males, were resident at the Jersey Wildlife Preservation Trust. They had been removed soon after birth for hand rearing. Neither had suffered maltreatment from its mother. both of whom were primiparous, one infant being entirely ignored after cleaning and the other being cared for but not allowed to suckle. These normal, healthy infants assessed by a paediatrician at one and two days of age (CARTER, 1973), were reared together and throughout the period of study were looked after by a conscientious and enthusiastic caretaker in a stimulating environment. They could not be said to be «deprived» relative to the human infants reared in their own homes, though of course their experience was different from that of mother-reared infants.

Procedure

The human infants were placed in a high chair with an extended table-top and presented with the task, while the gorilla infants were seated on their caretaker's lap in front of a table top, having strenuously resisted and then become distressed at any attempt to position them similarly. Both species of infant needed little encouragement to manipulate the objects presented.

The human adults were seated in front of the same kind of table top as the infants, however, unlike them they would not spontaneously grasp and manipulate the objects without verbal instruction. They were thus requested to pick up each object after it had been presented and give it back to the experimenter. This may have reduced the variety of prehensive patterns observed, but the responses recorded show the very minimum range of behaviours of which a human adult's hand is capable on this simple task.

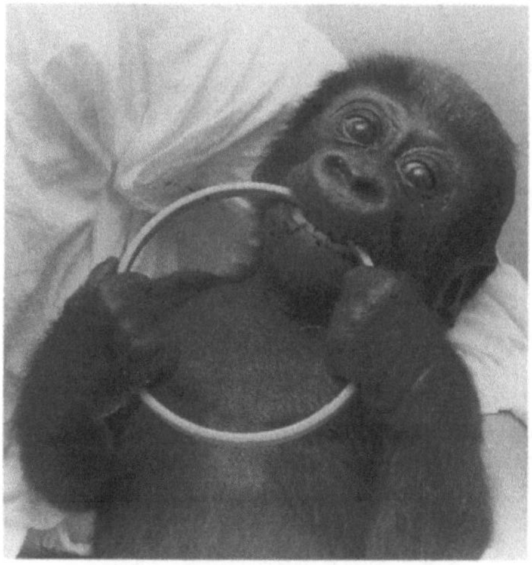

a

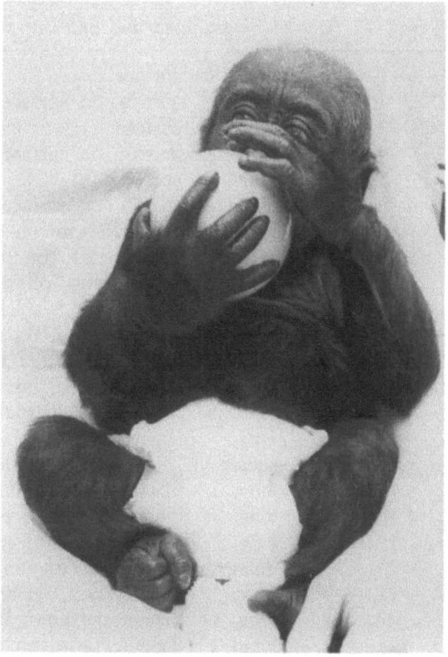

b

c

d

Plate 1 - (a-d) Gorilla infants aged between 18 and 30 weeks of age, holding four of the experimental objects. (Photograph P. Coffey).

Fifteen different objects were used in this experiment, four of which are shown in *Plate 1.* Some were presented in different orientations, and so for each subject at each session, a total of twenty-four presentations was made. All were presented in the midline. Dimensions of the different objects and the method and order of presentation are as listed in *Table 1.* They were chosen for incorporating a variety of properties, in addition to

TABLE 1 - *List of objects used and order of presentation.*

1.	1½″	CUBE, on surface
2.	1½″	diameter SPHERE, on surface
3.	2″	CUBE, on surface
4.	6″	diameter SPHERE, on surface
5.	1″	CUBE, on surface
6.	3½″	diameter SPHERE, on surface
7.	5″	long, ¼″ diameter ROD, on surface, left to right
8.	6″	square, ¼″ thick TILE, on surface
9.	10″	long, ″ diameter ROD, on surface, upright
10.	6″	diameter, ½″ wide, ¼″ thick RING, in air, vertical, left to right
11.	5″	long, ¼″ diameter ROD, on surface, left to right
12.	6″	square, ¼″ thick TILE, horizontal in air above table-top
13.	5″	long, ¼″ diameter ROD, on surface, upright
14.	6″	diameter, ½″ wide, ¼″ thick RING, horizontal on surface
15.	10″	long, ″ diameter ROD, horizontal in air, pointing at subject
16.	6″	square, ¼″ thick TILE, vertical, in air, pointing at subject
17.	5″	long, 1″ diameter ROD, on surface, left to right
18.	14″	long, ¼″ diameter STRING, on surface, left to right
19.	10″	long, ″ diameter ROD, on surface, left to right
20.	½″	diameter convex PELLET, on surface
21.	5″	long, ″ diameter ROD, on surface, left to right
22.	14″	long, ¼″ diameter STRING, in air, vertically
23.	5″	long, 1″ diameter ROD, on surface, upright
24.	¼″	diameter, convex PELLET, on surface

including items used in other prehension studies. Some had flat surfaces, others were curved, some were rigid and others were not. Wide, thick, thin, large and small objects were included, some of which projected considerably above the surface of the table-top. With this range of fairly simple objects it was hoped that a variety of approaches, prehensive patterns and grips would be provoked. The colour of the objects was held constant and since human infants seem attracted to the red-yellow part of the colour spectrum, these colours were used. For reasons of contrast red objects were used with the humans and yellow with the young gorillas. All of them were light enough for the infants to pick up easily. The same order was used for all the subjects.

Data collection

The aim of the experiment was to describe many features of the prehensive act and to compare the responses of the different subjects in a detailed way. For this video-recording was used as the first stage of data collection. Using a wide-angle lens, fitted to a video-camera supported on a tripod which was positioned in front of the table-top, the movements of the hands, arms, shoulders, and in the case of the infants the head, were recorded.

Details of the behaviour were transcribed using a viewing monitor and a video-recorder with still-frame and variable slow-motion facilities. Each prehensive act was observed several times at normal speed, after which particular segments of the act were viewed using the slow-motion and still-frame controls. Particular attention was paid to the approach, the configuration and orientation of the hand just prior to contact, grasping behaviour and object capture.

Results

The longitudinal and comparative data are largely displayed graphically. Unless otherwise stated, mean frequency data for each of the subject groups are shown in the figures. Where comparisons were made between the subject groups most usually a Chi Squared Test was used, though occasionally where the lowest expected frequency was less than five, the Fisher exact probability test was used.

Approaches to objects

The way a hand is projected in space during an approach can be described with reference to movement in the horizontal and vertical planes, detailed descriptions of which are given by HALVERSON (1931). A simplified diagram illustrating the possibilities is shown in *Figure 1.* With regard to the horizontal aspect of the hand's movement (seen from above), the approach is categorised as backhand, circuitous or straight, and for the

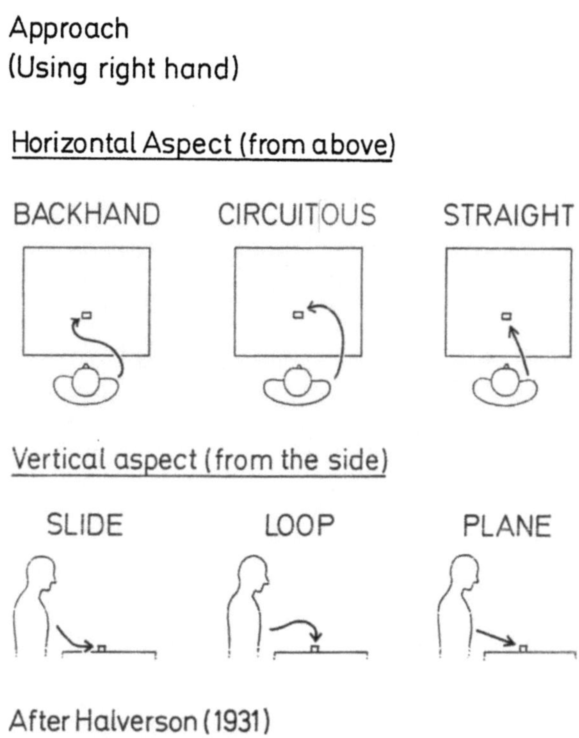

Figure 1 - Modes of approach prior to object prehension.

vertical aspect (seen from the side) as a slide, loop or a plane. Examples of approaches would be a «circuitous loop» or a «backhand slide». In the first of these there is a good deal of lateral movement of the hand as it moves forward while, at the same time, it is lifted well above the height of the object and then lowered onto it. In the second, the hand moves laterally across the midline to the opposite side of the table-top, sweeping the surface so that the back of the hand contacts the object first.

Horizontal movement

The behaviour of the subjects is illustrated in *Figure 2.* Both species of infant showed a decline in the use of the circuitous approach, compensated for by an increase in more direct movement. Neither used the backhand route to any great extent and even at twenty-two weeks of age, when the study began, a third or more of each subject's approaches were already straight, and this increased over subsequent sessions. However, it

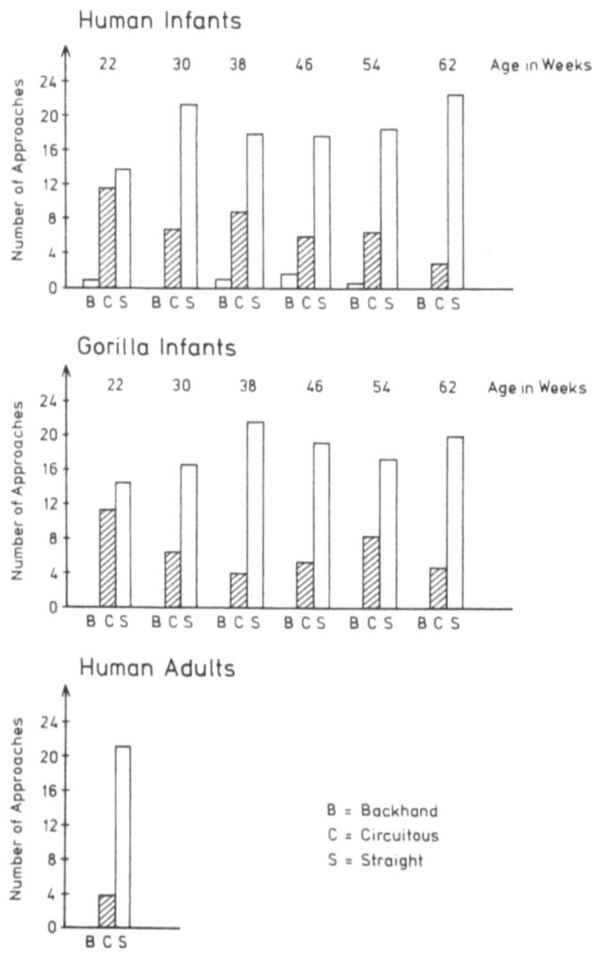

Figure 2 - Type of approach used: Horizontal aspect.

must be emphasised that the straight approach, generally thought of as the most preferable, is not always used, even by mature adult humans, for as *Figure 2* shows, approximately a fifth of their approaches were observed to be circuitous.

Vertical movement

At twenty-two weeks the human and gorilla infants' predominant technique was to approach the objects with a looping movement, that is, a kind of vertical over-shooting in which the hand gains height as it moves forward, rather than being lifted initially and sweeping down more directly. By one year of age the less direct looping approach was only rarely observed by Halverson (1931), whereas in this study at well over twelve months, between a quarter and one half of the approaches made by each infant involved this kind of movement. The data from the human adults present the same picture in which a looping vertical movement is used more frequently than might have been expected. As *Figure 3* shows, the human adults on average used this movement more than twice as often as the direct planing action.

Such looping movements may allow a more precise location of the hand in relation to the object that is about to be grasped than is possible with a more direct approach in which all or part of the object may be obscured from view as the hand moves forward. A slightly less direct approach may also allow a variety of prehension possibilities to be assessed and this may be more important when the object to be grasped is larger or of a more complex shape than the one-inch cubes or small pellets used in prehension studies (Halverson, 1931; Castner, 1932). Both, infants and adults, were at times observed to move the hand towards one surface of an object, only to divert it to another aspect. Though in fact this kind of behaviour involves some delay, such an adjustment may contribute to a more appropriate grip on the object, or possibly assist in leaving a specific portion of it available for further manipulation.

In the early sessions there was some indication of the infants using the surface of the table-top as a guide, as evidenced in the number of sliding approaches made. The horizontal surface on which a small object is presented, besides acting as a guide for the movement, also serves to steady an approaching hand. During later sessions the human and the gorilla infants appeared to benefit from this strategy in approaching the pellets, though for bigger objects like the large ball or the tall rod, this behaviour was never observed. The human adults appeared to need no such assistance with any object, for only in one instance out of all their prehensive episodes was a sliding movement used.

A three-dimensional picture of the approximate way the hands are projected in space is gained by combining the descriptions of the vertical and horizontal movements. These data are shown in *Tables 2a, 2b, 2c*. At the youngest age the most frequent method of approach used by the human and gorilla infants was a «straight loop», though a «circuitous loop» was used almost as often. The second of these had declined by thirty weeks of age, while the former had increased in frequency of use. The most economical approach in terms of expenditure of time and energy is the «straight plane», and though this was used by all the infants, even at twenty-two weeks (at a low level), the infants behaved like the adults and it was still not the most frequent approach forty weeks later. Thus, though there was an overall reduction in the number of different approaches used and an increase in those that were more direct, these were fewer than had been expected. This was perhaps not a surprising finding in view of the fact that, in comparing the responses of the adults with the infants at sixty-two weeks, the former were found to use significantly

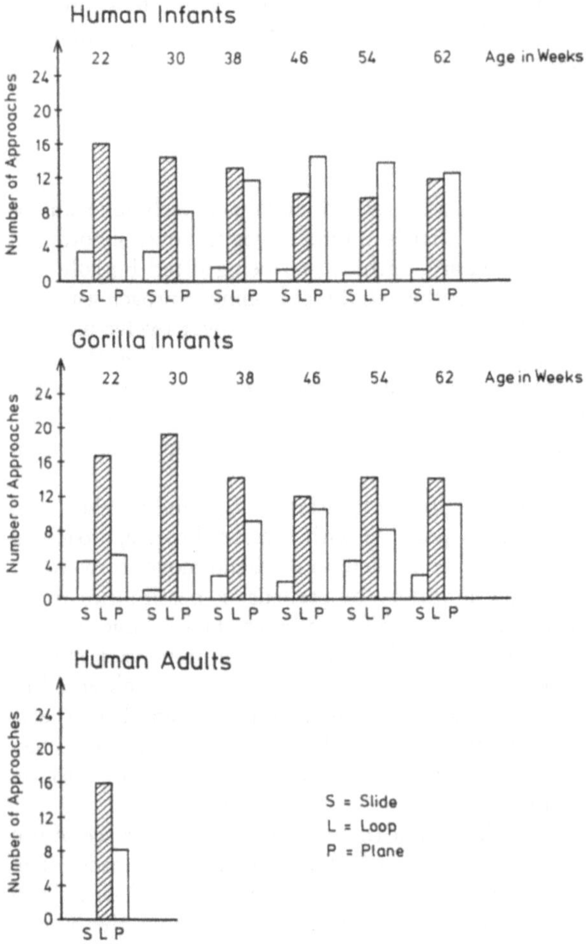

Figure 3 - Type of approach used: Vertical aspect.

more looping approaches than the infants of either species ($p = < 0.01$), probably for the reasons mentioned previously.

The configuration of the hand before object contact and prehension

Data on this aspect of behaviour were taken from the videotape recording at the same point in each prehensive act, 0.04 seconds before contact.

For all the subjects, humans and gorillas, a horizontal orientation of the hand appeared to predominate, as *Figure 4* shows, followed by an oblique and then vertical orientation. To what extent the method of presentation, using a flat table-top, may have contributed to this is uncertain. Fewer objects were presented with the main axis vertical, but this was not to the extent that the subjects' responses would indicate.

TABLE 2 a - *Frequency of different types of approach used by human infants during prehension.*

		Human Infants								
		NA			SR			SM		
Age in Weeks		SL	LO	PL	SL	LO	PL	SL	LO	PL
22	BA	0	0	0	0	0	0	0	2	0
	CI	3	6	4	0	6	2	1	9	2
	ST	1	9	2	7	7	4	0	9	1
30	BA	0	0	0	0	0	0	0	0	0
	CI	2	3	2	1	3	5	0	2	1
	ST	5	11	4	3	10	7	0	17	5
38	BA	0	1	0	0	0	0	0	1	0
	CI	0	4	8	1	5	3	2	2	0
	ST	0	5	10	1	12	5	1	9	9
46	BA	0	0	3	0	0	0	0	1	0
	CI	1	2	3	1	3	4	1	0	2
	ST	1	7	7	0	11	5	1	6	13
54	BA	0	0	1	0	0	0	0	0	0
	CI	0	3	9	0	1	3	1	0	1
	ST	0	4	7	0	16	5	2	5	15
62	BA	0	0	0	0	0	0	0	0	0
	CI	0	0	3	1	0	1	0	1	2
	ST	2	12	8	3	13	8	0	6	15

BA Backhand — SL Slide
CI Circuitous — LO Loop
ST Straight — PL Plane

The wrist posture at this stage of the prehensive act was described as being clearly flexed or extended or neither of these things (WOOD JONES, 1944). Flexion consists in bending the hand towards its palmar surface, while extension (or dorsiflexion) is a movement in the opposite direction.

As *Figure 5* shows, for the gorilla infants the wrist posture in which the joint is neither flexed nor markedly extended (dorsiflexed) predominated at all the ages tested, and the use of flexion and extension movements was generally low, though these did increase slightly with age. This may be related to the increasingly used «stiff» knuckle-walking posture adopted by the young gorillas at sitting, at rest, standing and walking. The considerable variation found in this aspect of the human infants' behaviour, perhaps reflects the instability and immaturity of wrist control at such a young age.

It had been thought that the human adults would use a relatively higher proportion of flexion and large extension movements of the wrist than that observed. However, though flexed wrist postures were observed five times more often than dorsiflexed ones, postures in which the hand and lower arm were in the same plane were most common. It may be

TABLE 2 b - *Frequency of different types of approach used by gorilla infants during prehension.*

		Gorilla Infants					
		A			M		
Age in Weeks		SL	LO	PL	SL	LO	PL
	BA	0	0	0	0	0	0
22	CI	2	7	6	3	3	2
	ST	2	9	1	1	14	2
	BA	0	0	0	0	0	0
30	CI	0	7	1	0	3	2
	ST	1	13	2	1	15	3
	BA	0	0	0	0	0	0
38	CI	0	6	1	1	0	0
	ST	1	11	7	3	11	10
	BA	0	0	0	0	0	0
46	CI	0	0	6	0	2	3
	ST	4	8	7	0	14	5
	BA	0	0	0	0	0	0
54	CI	1	3	3	1	3	6
	ST	6	11	3	1	10	4
	BA	0	0	0	0	0	1
62	CI	1	2	6	0	0	1
	ST	6	11	3	1	10	4

BA Backhand SL Slide
CI Circuitous LO Loop
ST Straight PL Plane

TABLE 2 c - *Frequency of different types of approach used by a group of human adults during prehension.*

	SL	LO	PL
BA	0	0	0
CI	0	14	10
ST	1	121	46

that on contact with an object, the posture changes and increased movement takes place at this joint during grasping, but it was not possible to check as the prehended objects often obscured this aspect of the hand. NAPIER (1980) pointed out that the optimal position in gripping is with the wrist in 30-40 degrees of extension, for with flexion there is a large reduction in grasping power. A possible explanation for the low incidence of this amount

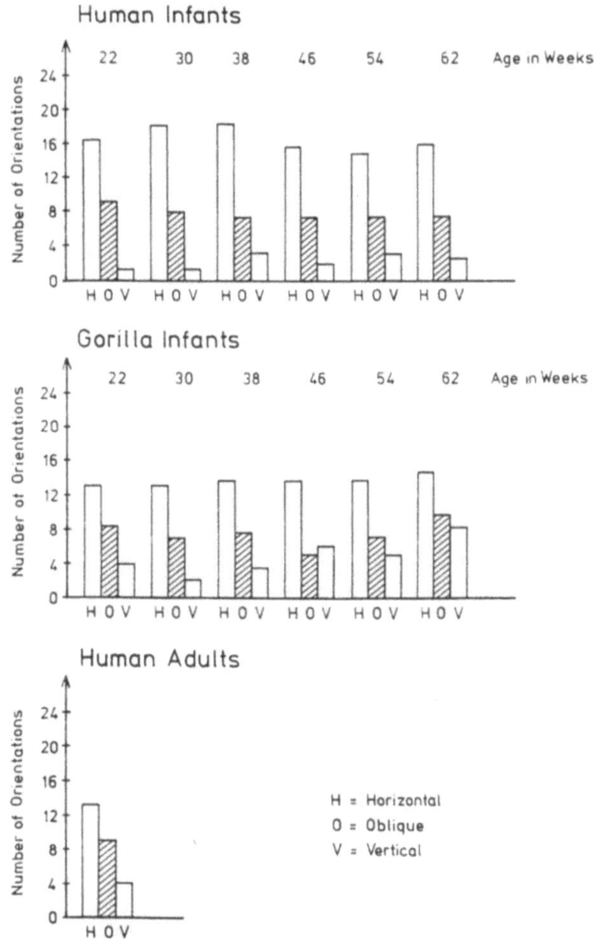

Figure 4 - Orientation of the hand in space prior to object contact.

of extension may lie in the task requirements, for the objects were light and simply had to be grasped and returned to the experimenter.

In addition to flexion-extension movements of the wrist joint, at the same time a lateral movement in either an ulnar or radial direction could also occur (Wood Jones, 1944). Two human infants and one gorilla used this kind of movement occasionally from thirty-eight weeks of age onwards. It appears that such behaviour may be a relatively late development in the acquisition of manual skill and control as it was more frequently observed in the human adults (in approximately 12% of prehensive acts). Among them, deviation of the hand in an ulnar or radial direction was almost equally distributed, though in the human infants the nine instances recorded were all ulnar movements and in the one young gorilla six were radial and three ulnar. It had been thought that the human infants, in the course of developing prehension skills, would use this kind of wrist movement increasingly and that in view of the relative rigidity required at the wrist

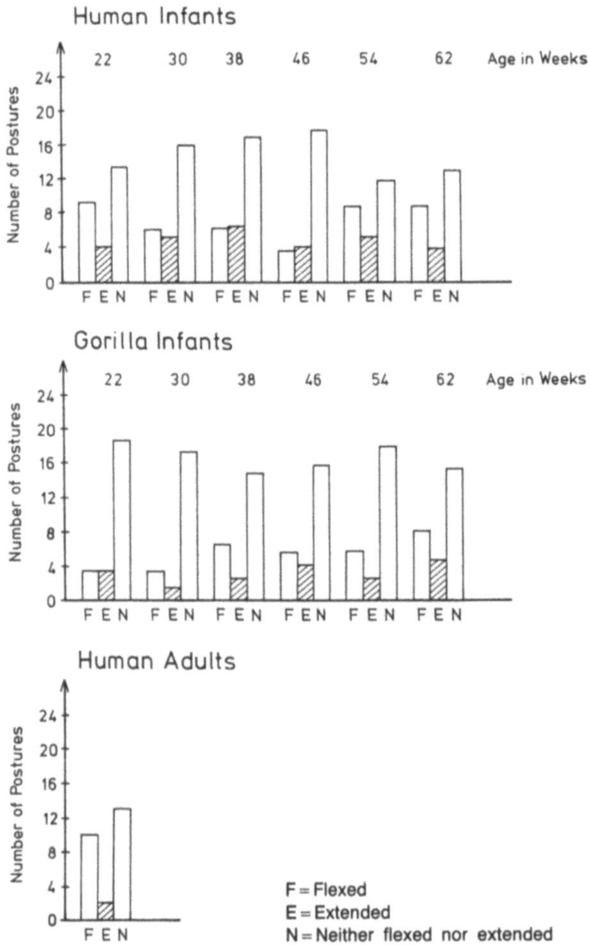

Figure 5 - Wrist posture prior to object contact.

during knuckle-walking, there would be a species difference in its use. There was some evidence of a developmental progression, but with one infant of each species showing no such deviation at all, there was no evidence of a difference between them.

During prehension, prior to contacting an object, many other anticipatory adjustments and positioning movements are made with the hand. These include closure, differentiation of the digits, particularly the forefinger, and opposition of the thumb. It was expected that the behaviour of the infants, even in the last experimental session, would contrast with the mature behaviour of the adult human subjects in the extent of the adjustment made.

Closure involves an inward movement of the digits leading to capture of an object between them and the palmar or other volar surfaces of the hand. It was categorised as absent, in which case the fingers were extended and the thumb adducted, or slight,

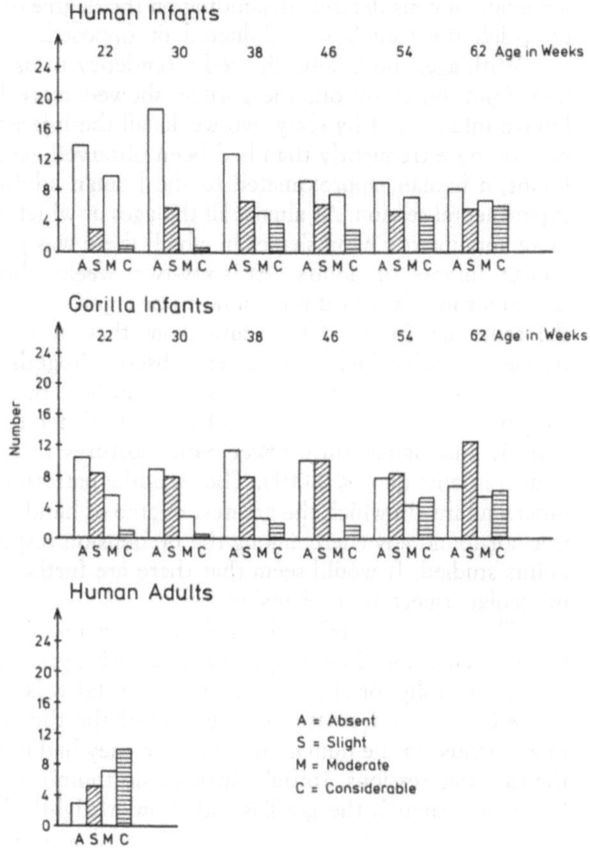

Figure 6 - Amount of hand closure prior to object contact.

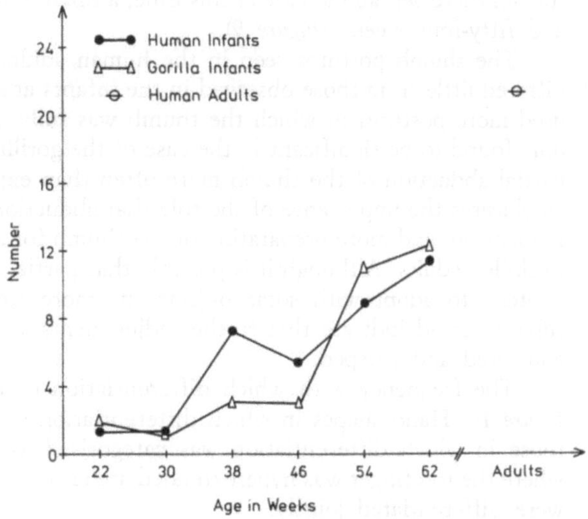

Figure 7 - Frequency with which a considerable degree of hand closure was used, prior to object contact.

moderate or considerable, depending on the degree of flexion of the fingers and the extent to which the thumb was abducted or opposed.

With age, the infants showed a tendency to use increasing amounts of hand closure (see *Figure 6*). Early on, the gorillas showed more hand closure of some sort than the human infants, but by sixty-two weeks all the infants used a considerable degree of hand closure more frequently than had been observed earlier (*Figure 7*). Nevertheless only one infant, a human, approximated to the human adults' level of performance by the final experimental session. At almost all the ages at which they were observed, the gorillas used more anticipatory hand shapes in which there was just slight closure of the hand than the human infants or adults. At sixty-two weeks this was quite pronounced and their behaviour in this regard was found to be significantly different from the human infants at the same age ($p = <0.05$) and from that of the human adults ($p = <0.01$). The frequency with which closure was absent altogether or was present to a considerable degree was also found to differ significantly. Thus the human adults used the greatest amounts of hand closure more often than either the human or the gorilla infants ($p= <0.01$) and significantly fewer hand postures in which there was no closure than the human infants ($p= <0.001$). The step-like pattern of the results on hand closure for the human adults, in which the greatest degree of hand closure was observed most frequently, was not found anywhere among the prehension responses of the infants at any of the age points studied. It would seem that there are further developments that take place in this particular aspect of prehension.

The thumb, though little used by very young infants, gradually acquires a more active role in prehension. In anticipation, and during grasping, it may be adducted, or abducted, partially or fully, or else opposed to the distal volar pad of the forefinger or other digits. *Figure 8* shows a decrease in adduction of the thumb by all the infants, though this was more marked in the human infants since they had used this movement more frequently in the first two sessions. Initially little or no thumb opposition was observed in the infants' behaviour, though the gorillas did show a slightly higher level of full abduction of the thumb. Later, the behaviour could not be distinguished. In the human infants the most sudden increase in preparation of the thumb for an active role, by its abduction and opposition, took place between thirty and thirty-eight weeks of age, but for the gorillas, though there was an increase at this time, a further sharp rise took place between forty-six and fifty-four weeks (*Figure 9*).

The thumb postures seen in the human adults at this point in the prehensive act differed little from those observed in the infants at sixty-two weeks. However, the latter used more postures in which the thumb was fully abducted, though the difference was only found to be significant in the case of the gorillas ($p = <0.01$), and the adults used partial abduction of the thumb more often than expected. The hand function literature emphasises the importance of the role that abduction and opposition of this digit play in prehension, and more preparation of the thumb for active use might have been predicted in skilled adults. Although it is possible that partial abduction may be a more favourable strategy to adopt with some objects, its more frequent occurrence among the adult subjects could indicate that further adjustments are likely to be made after objects are contacted and grasped.

The frequency with which differentiation of the forefinger occurred is shown in *Figure 10*. Hand shapes in which differentiation was slight or absent are shown, as are those in which differentiation was categorised as moderate or considerable. Postures where the forefinger was hyperextended are also shown, and those in which digits 2 and 3 were differentiated jointly.

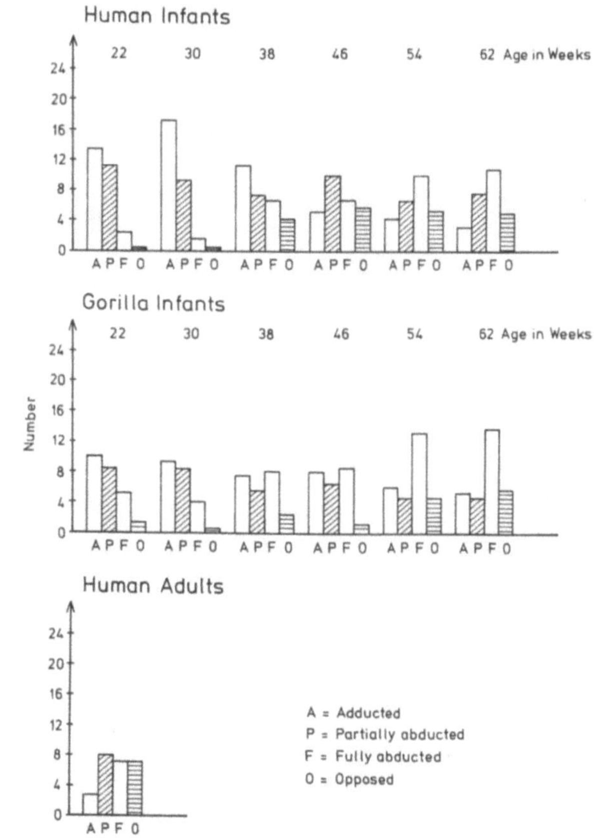

Figure 8 - Posture of the thumb prior to object contact.

Hyperextension of the forefinger may be an immature way of differentiating it from the other digits. The data on the human infants generally endorse this view, though even at sixty-two weeks all of them, and occasionally adults, used this posture at some time during the session. In contrast with these human subjects, one gorilla, on only one occasion, was recorded to have hyperextended this digit. The otherwise total absence of hyperextension of the forefinger by the gorillas stands out as a clear species difference in manual behaviour. The finding that, when compared with the human subjects, they also showed a low frequency of differentiating digits 2 and 3 jointly may be associated, for both may be related to the adaptations necessary for climbing and knuckle-walking. Separation of these digits is neither necessary nor desirable during such locomotion. What was surprising was the low frequency of differentiation of the forefinger at this point in the prehensive act for both the infant and adult humans, though of course this may take place in the grasping part of the act itself.

The effect of object characteristics on the configuration of the hand prior to prehension

The development of prehension, like the acquisition of other motor skills, is linked with the growth of cognitive competence. This part of the study was concerned with

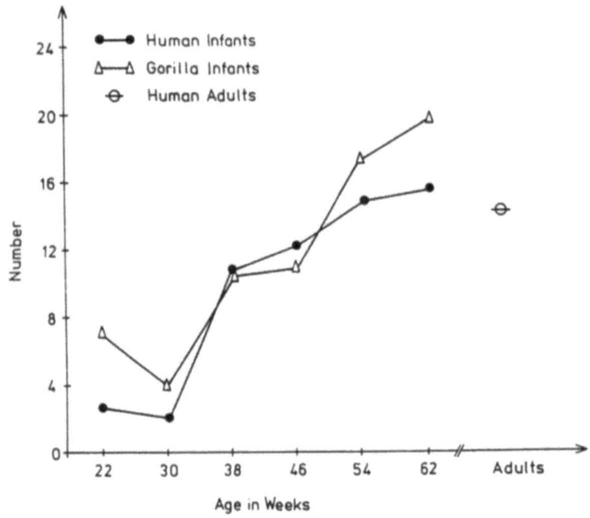

Figure 9 - Frequency of full abduction and opposition of the thumb prior to object contact.

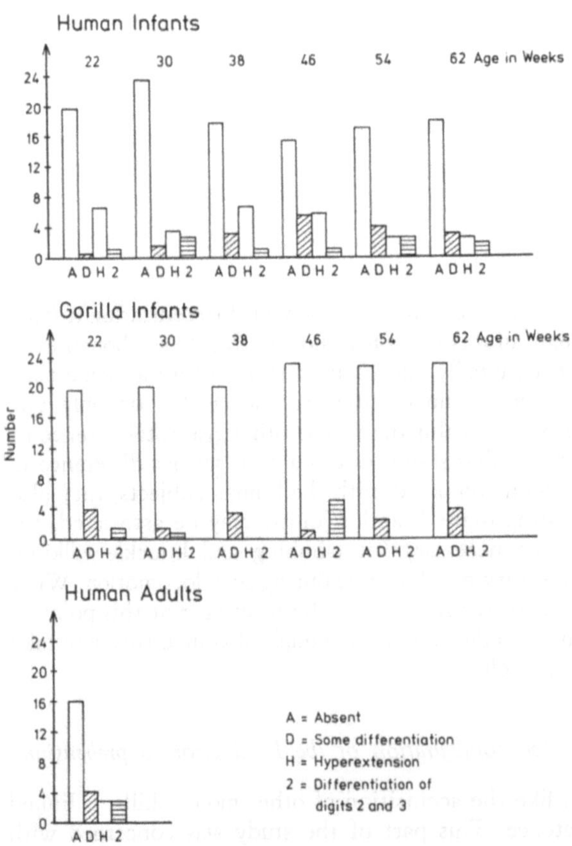

Figure 10 - Differentiation of the forefinger prior to object contact.

examining the way in which the size, shape and orientation of an object may affect the hand before grasping actually takes place. Information about this preparatory prehensive behaviour was transcribed from the videotape 0.04 seconds before contact.

Based on informal observations it was predicted that, for the adult subjects, large and long objects would encourage use of two hands and less hand closure, that small objects would provoke thumb opposition and differentiation of the forefinger and that the orientation of an approaching hand would match the main orientation of the object that was to be grasped. For the infants, a developmental progression was predicted during which they would gradually take account of object properties at this phase of prehension. Data on this aspect of the infants' behaviour were analysed with reference to the experimental sessions at twenty-two, thirty-eight, forty-six and sixty-two weeks of age. Numerous comparisons were made between the subjects' responses to the many objects used in this task (REDSHAW, 1985) some of which are discussed here.

The effect of size

Anticipatory behaviour towards the large sphere, the small sphere and the small pellet was compared in an attempt to examine the effect of size on manual response.

In the human adults use of two hands was limited to the large sphere. The infants behaved similarly except at twenty-two weeks when two tried to pick up one of the smaller objects using two hands. Wrist posture and hand orientation were quite varied and gave little clue to the size of object being approached. However, most of the adults in approaching the pellet which projected least above the surface did not use flexed wrist postures as they had with the more markedly projecting spheres. Flexion at the wrist appears to allow a hand to go round a larger object better, a strategy that is obviously unnecessary when the object is very small.

Closure was found to be a good discriminator of object size. For all the adults it was absent before grasping the large sphere, except for one who used slight closure. This contrasts with the moderate amount of hand closure most often recorded with the much smaller sphere, and the greatest degree of closure observed with the pellet. The infants of both species, at all the ages examined, behaved in much the same way as the adults with the large sphere, but with the small pellet there was a developmental change, such that by the last session a considerable degree of closure was present in all but one instance, while at twenty-two weeks this had been absent or slight.

Differentiation of the forefinger also seemed to discriminate large from small objects. In all but two instances it was absent from the preparatory behaviour of the adults towards the large sphere. The reverse situation occurred with the small sphere for all the adults showed differentiation as they did with the pellet, though with this last object digits 2 and 3 were sometimes differentiated jointly. With the smaller objects the infants were more variable. They too sometimes differentiated digits 2 and 3 together and, for example, using the thumb too, made a three pronged attack on the small unstable sphere.

Based on CASTNER'S (1932) observations of human infants, it had been expected that at the beginning of the second year pellet prehension would be preceded by clearance of the other fingers, making only the distal volar surfaces of the thumb and forefinger available for controlled and efficient object capture. However only one infant of each species behaved in this way and examination of the other postures adopted showed that these were typically ones in which the heel of the palm, tips of the fingers and side of the thumb rested on the table-top almost surrounding the pellet which was then closed in upon from all directions. This would seem to be a less mature behaviour, which though successful in terms of object capture, leaves little room for further manipulation.

Checks on the effect of size on the hand shape observed before prehension were also made in a broader comparison using more objects. Thus in addition to the uniformly shaped spheres, objects that were thin, wide, short and long were included. When the responses to these were compared with those described above similar results were obtained. All the subject groups used two hands significantly more often with the larger and wider objects ($p = <0.05$) and overall the human infants were more likely than the adults to use two hands ($p = <0.05$). The ages at which the infants first used two hands with the small and large objects are listed in *Table 3*. From this it can be seen that even at twenty-two weeks, one infant of each species differentiated their behaviour in this way.

Differences were also found in relation to the amount of hand closure observed just prior to prehension of the large and small objects. All three subject groups used moderate or considerable degrees of hand closure more often with the small objects. This was a highly significant finding in the case of the human adults and infants ($p = <0.001$) and only slightly less so for the gorilla infants ($p = <0.01$). However, the young gorillas, as

TABLE 3 - *A comparison of the ages (in weeks) at which two hands were first used with a number of large objects and the ages at which two hands were used with a number of small objects, if at all.*

	Human Infants			Gorilla Infants	
	SM	NA	SR	A	M
Large Objects					
Large Ball	22	22	62	22	62
Medium Ball	22	22	—	—	—
Surface Tile	22	38	—	22	46
Vertical Tile	22	22	—	38	62
Long Rod Surface	22	46	46	22	—
Small Objects					
Small Pellet	—	22	—	—	—
Small Cube	22	—	—	—	—
Small Thin Rod	22	38 46	—	—	—
Small Ball	22	—	—	—	—
Vertical String	62	38 62	—	62	—

has been discussed previously, used less closure with the small objects than the infant humans ($p = < 0.05$).

With a wide range of objects, size was again found to affect the frequency of differentiation of the radial digits. All the subjects differentiated these significantly more in responding to the smaller objects ($p = < 0.01$). In spite of this, the gorilla infants also used fewer postures of this type in prehension of the small objects than did the human infants ($p = < 0.001$). Though overall the infants' behaviour in this regard appeared to discriminate between large and small objects, there was also a developmental progression in which differentiation was increasingly used with small objects, but dropped out in the course of prehending the large items.

The effect of orientation

In the course of the experiment the same four objects were presented both horizontally and vertically. *Table 4* lists these and the ages at which the human and the gorilla infants first used horizontal hand positions with the horizontal objects and vertical or oblique positions with the vertical objects. At twenty-two weeks, with all but one object, the infants of both species were able to approximate to the orientation given. When the horizontal object data from all sessions were combined the different subjects were found to behave similarly, positioning their hands horizontally or obliquely 90% of the time. Data on the vertically presented objects showed that with these, all three groups used

TABLE 4 - *A comparison of the ages (in weeks) at which horizontal and postures were first used with horizontal objects and vertical or oblique hand postures with vertically presented objects.*

	Human Infants			Gorilla Infants	
	SM	NA	SR	A	M
Horizontal Objects					
String	22	22	22	22	22
Short Rod	22	46	22	22	22
Long Rod	22	38	22	22	22
Tile Surface	22	22	22	46	22
Vertical Objects					
String	22	22	22	22	22
Short Rod	22	38	22	22	38
Long Rod	22	22	22	22	22
Tile Air	22	46	46	22	38

vertical or oblique postures 75% of the time or more. No significant differences were found between them.

That the matching of hand orientation to the vertical objects was less pronounced than to those presented horizontally seems due in part to the human infants and adults, and less frequently the gorillas, approaching from above. This was most obvious with the vertical rods which were sometimes grasped from above, rather than the more easily available side. In some cases the human infants even approached as if to take hold of a rod from above, but at the last moment applied their hands to the vertical surface using an inappropriate orientation, making adjustments in the grip after contact.

In general among the adults there was good matching to orientation, but where this did not occur it was due to directing the hand to features like the corner or edge of the tile or to the top or end of the string or rod. The infants, particularly the gorillas, seemed to take less account of corners, ends and edges. The more mixed behaviour of the human adults should perhaps have been expected, in view of their concern about the next step, for they were not simply interested in prehension for its own sake, but were thinking about what the next task would be. Thus at times they used an unexpected orientation and then grasped a smaller, more peripheral part of an object, perhaps the better to hand it back.

The effect of shape

A judgement was made about each hand posture and the extent to which it was suited to the dimensions and shape of the object about to be prehended. Appropriateness was considered to involve making suitable surfaces of the hand available for prehension and in the hand to some extent mimicking some property or properties of the object itself. Only in this way could the behaviour of the subjects across a wide range of objects be compared. The four categories ranged from quite inappropriate to highly appropriate.

The extent to which the human and gorilla infants used moderately or highly appropriate hand shapes in object prehension is illustrated in *Figure 11*. From this a similar developmental pattern can be seen, in which at least 40% of configurations of the hand at twenty-two weeks were judged to be appropriate. Thereafter, though an improvement was shown such that by sixty-two weeks this figure had reached 70%, the performance of the infants was still found to be significantly different from that of the adults ($p = < 0.01$), who used moderately or highly appropriate hand shapes in 94% of the instances observed.

Table 5 gives the ages at which appropriate hand shapes were first recorded for the human and gorilla infants during prehension of twelve of the experimental objects. From this it can be seen that all the infants responded appropriately to the large sphere in the earliest session, but that the horizontal and vertical presentations of the tile shape were more difficult to match in terms of hand shape. Appropriate behaviour with the pellet occurred sooner in the human infants, though the gorilla infants seemed more able with the small sphere, horizontal ring and string.

In summary, it would seem that the prediction about the human adults being able to match the configuration of their hands to the shape of the objects they were about to grasp was confirmed. It was also confirmed that with age the human and gorillas infants showed an improvement in their matching ability, but the data also indicate that the infants were more able in some respects initially than had been expected. This aspect of the prehension experiment does show that objects of different size provoked differences in the behaviour of all three subject groups, as did objects presented in differing

TABLE 5 - *A comparison of the ages (in weeks) at which the first moderately or highly appropriate hand shape was observed, if at all, with twelve differently shaped objects.*

Objects	Human Infants			Gorilla Infants	
	SM	NA	SR	A	M
Small Pellet	38	38	38	46	46
Small Sphere	38	22	38	22	22
Large Sphere	22	22	22	22	22
Large Cube	22	38	38	38	62
Small Cube	38	22	62	38	—
Tall Rod	22	38	38	46	22
Short Rod	22	38	22	—	22
Long Rod	22	22	38	22	46
Tile Surface	62	46	—	62	—
Tile Vertical	22	—	62	—	—
String Surface	38	62	—	22	22
Ring Surface	22	38	38	22	22

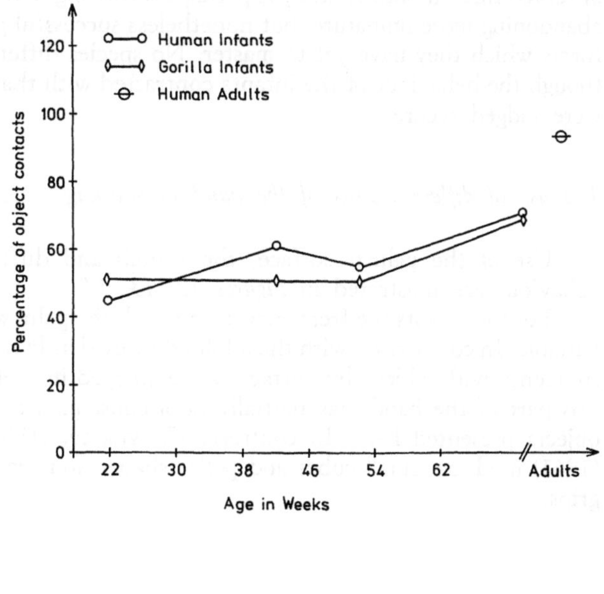

Figure 11 - Proportion of moderately or highly appropriate hand shapes observed before object contact.

orientations. That during visually guided reaching the human and the gorilla infants were able, in some cases, to match their hand shape to that of an object, prior to any tactile feed-back and from as early as twenty-two weeks of age, is of considerable interest. However the skill evident in the achievements of the human adults should not be underestimated. For much of the time it seemed, simply from the configuration of the hand, that one could work out the approximate size, shape, and orientation of the object being approached for prehension.

Object capture

Successful capture and manual control of objects depends on the use of accurate and efficient grips. The development of prehension proceeds such that initially the goal is simply to take hold of an object, but later it is grasped with the aim of further manipulation and use.

At the youngest age the infants were less able and less motivated to grasp the smallest and largest objects. The former were often reached for, contacted and then ignored, while with the latter the tendency was to rest a hand on them or to push them around. At sixty-two weeks of age the human infants successfully grasped all the objects presented while the gorilla infants did not. Though these young gorillas could have grasped the complete set of objects as indicated by their manual behaviour at other times, apparently they preferred to hit one or two with the flat of the hand or to slide them around the surface of the table in a playful manner. This kind of response may be related to their usual social play behaviour in which chest-beating was beginning to emerge. They also showed a marked enthusiasm for hitting surfaces with their open hands, during solitary locomotor and exploratory play (Redshaw & Locke, 1976; Redshaw, 1985). Success in grasping improved with age, but not all the grips used were secure ones in which the captured object was held firmly. At times the objects were grasped and then, due to an inefficient or inappropriate grip, often involving an excess of force, were subsequently lost. In general insecure grips occurred more frequently at the younger ages and by sixty-two weeks these were infrequent or absent. That, at a low level, insecure grips continued to be observed throughout the study is perhaps surprising, but these may be the result of infants abandoning more immature, but nonetheless successful grips, in favour of more advanced forms which they have yet to master. No species differences were found in this regard, though the behaviour of the infants contrasted with that of the adults for whom all grips were judged secure.

The use of different parts of the hand in grasping

Use of the palmar surface, the thumb and differentiated forefinger in grasping behaviour are illustrated in *Figures 12, 13, 14.*

For the infants the frequency with which the palm was involved in grasping was quite variable. In comparison with the adults they evidently had some way to go in reducing the frequency with which this surface was employed in prehension. The high level of use of this part of the hand may partially be a consequence of the variety of large and small objects presented here. In contrast, Halverson (1931); Castner (1932) and Jensen (1961) used just small cubes and pellet shapes and recorded lower incidences of palmar grips.

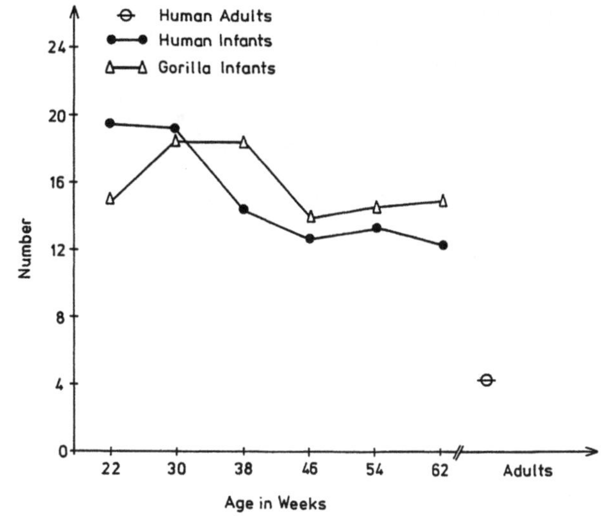

Figure 12 - Use of the palmar surface in grasping.

Differentiation of the forefinger among the infants increased only slightly with age, but like use of the palmar surface, this also may have been related to the mixture of large and small objects presented, for as was discussed above, only the latter continued to provoke separate use of this digit.

The thumb was found to play quite an active role in prehension, even from early on, and this increased with age such that by sixty-two weeks the frequency of use in grasping was comparable with that recorded for the human adults.

Power and precision grips

The human and gorilla subjects' grasping movements were categorised according to NAPIERS's (1956) classification of grips and the resulting picture of the developing use of precision movements is shown in *Figure 15*.

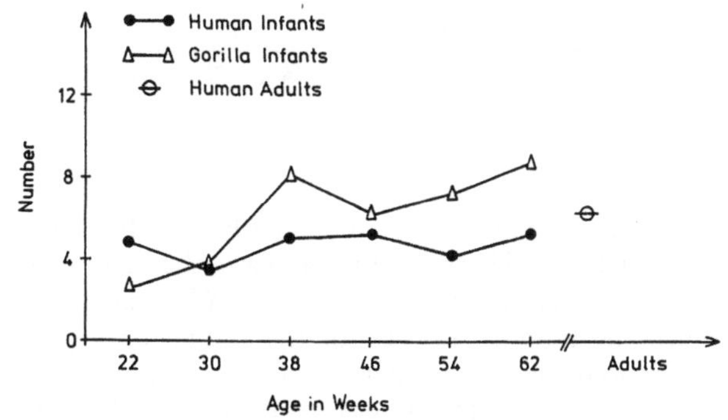

Figure 13 - Use of the differentiated forefinger in grasping.

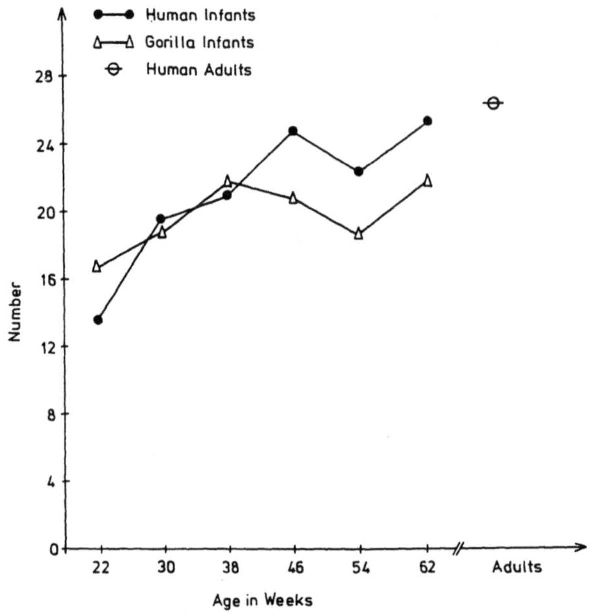

Figure 14 - Active use of the thumb in grasping.

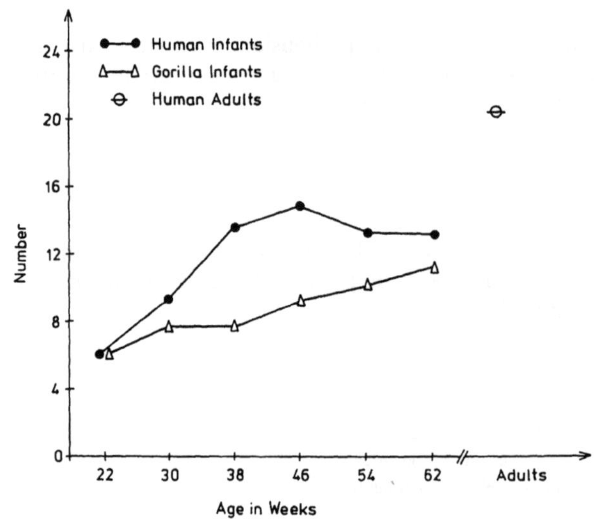

Figure 15 - Use of the precision grip in grasping.

All the infants started off similarly, mostly using power grips in which the prehended objects were held between the palm and the inner surface of the digits with a considerable degree of force and very little control. With age, they increasingly used precision grips in which the terminal portion of the thumb provided a counter-balancing force and the objects were held between its distal volar pad and the terminal pads of the opposing digits. However, at sixty-two weeks of age only 50% or less of prehensive episodes involved precision movements of this type, contrasting with the 80% level or more recorded for the adults.

When data on the different sessions were combined, the gorilla infants were found to have used more power and fewer precision grips than the human infants ($p = < 0.02$), though in the last session their behaviour did not stand out markedly.

Use of the volar pads in grasping

The frequencies with which the individual subjects used the different terminal volar pads are shown in *Figure 16.*

For all the infants use of the thumb had increased by sixty-two weeks, though

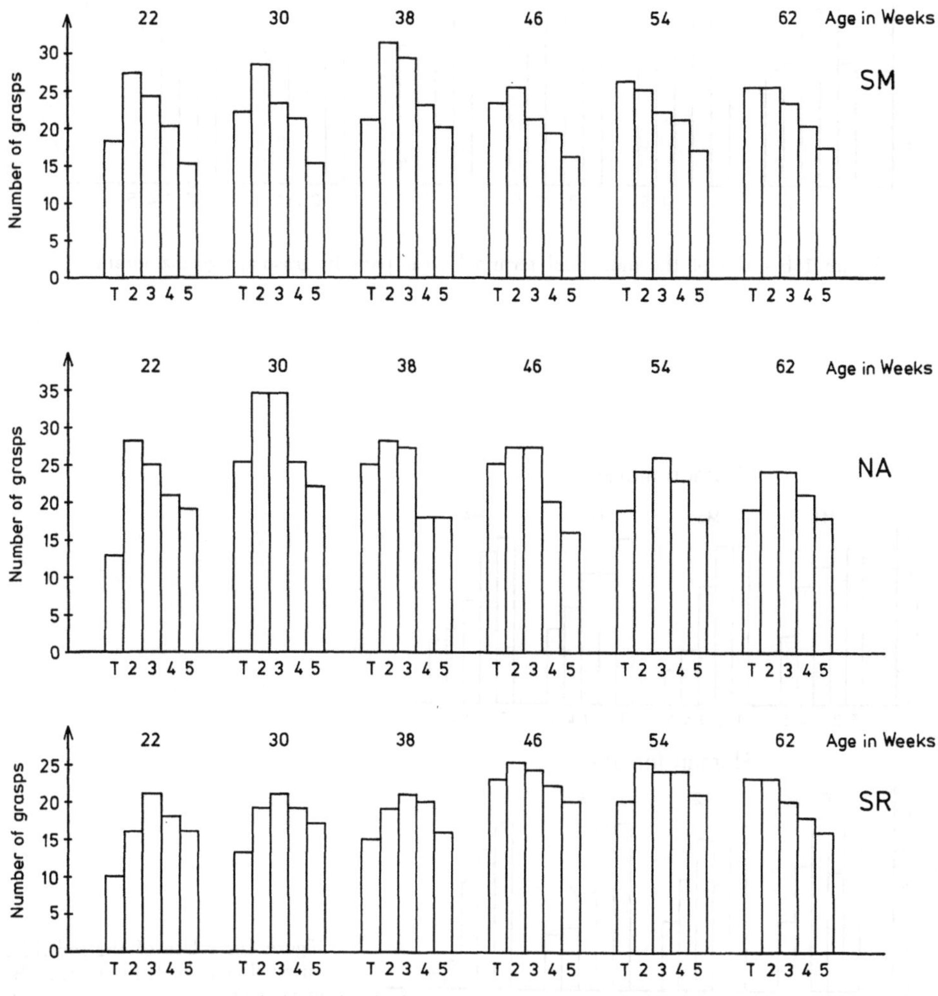

Figure 16a - Use of the individual terminal volar pads in grasping: human infants.

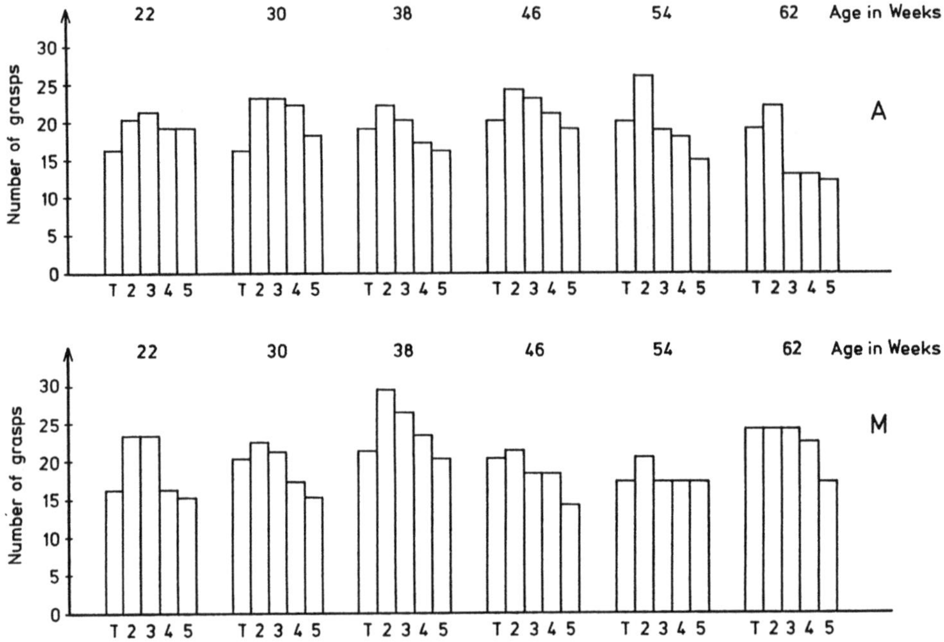

Figure 16b - Use of the individual terminal volar pads in grasping: gorilla infants.

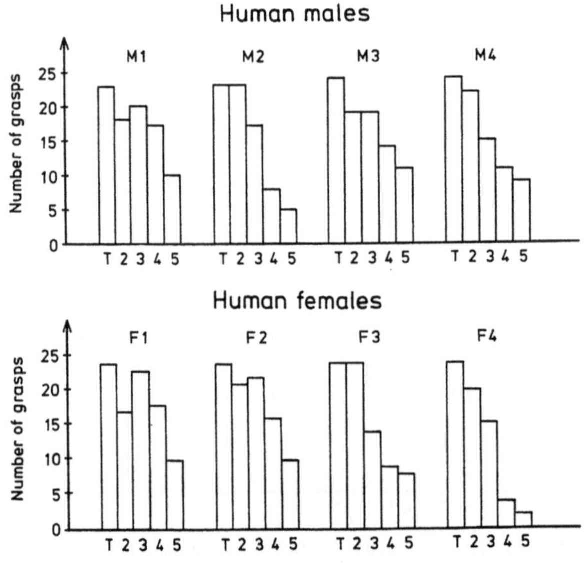

Figure 16c - Use of the individual terminal volar pads in grasping: human adults.

initially its volar pad had been used less often than that of the forefinger. Among the human infants there appeared to be relatively more use of the radial digits than in the gorillas, but in none of the human infants was the bias towards using the radial portion of the hand as evident as among the adult human subjects. Three of the adults showed a slight preference for using digit 3 in opposition to the thumb, rather than the forefinger, a pattern not observed among the infants. Despite this minor difference it is clear that among the adults the bias towards the use of the radial digits is brought about by a reduction in the number of digits involved in prehension. Digits 4 and 5 were used considerably less by the adults than the infants, even at sixty-two weeks of age. One explanation for this may lie in the size difference between the adult and infant hand. Obviously a small hand might be obliged to employ more digits for successful prehension. This may in part be an explanation for this behavioural difference, but it is not entirely adequate since the infants were at times observed to use all the digits even on the smaller objects. It seems that further refinements have yet to take place.

Relating hand shape to grip

In examining the link between the configuration of the hand just prior to contact, and the adjustments made during grasping behaviour, two aspects of prehension were considered. These focussed on the roles of the forefinger and the thumb.

A comparison was made of the differentiation of the forefinger before contact and in the grips used. The human infants tended to differentiate this digit at the end of a reach, before making contact, but then not to use a differentiated forefinger in grasping (*Figure 17*) perhaps indicating, as suggested previously, that differentiation by hyperextending this digit is an immature response. The gorillas behaved differently and showed a higher level of differentiation in the grips used than in the pre-grasping behaviour. When data from the six sessions were combined this was found to be a highly significant difference ($p = <0.001$). The human adults differentiated the forefinger similarly at both time points, much as the human infants did in the last session.

The relationship between thumb postures in which full abduction and opposition were observed before contact and the actual use of the terminal thumb volar pad in gripping, is illustrated in *Figure 18*. The pattern of thumb use was more uniform than that for the forefinger. Active use of the thumb volar pad in grasping clearly leads the appearance of abduction and opposition in pre-grasping in infants of both species. To start with, at twenty-two weeks, though the thumb was used quite frequently in grasping, relatively little anticipatory abduction and opposition was seen. But forty weeks later this had increased, such that the majority of prehensive episodes in which the thumb volar pad was used for grasping, were led by appropriate positioning of the thumb before the actual grasping took place.

A comparison of the grips used in prehension

In general, a considerable overlap was found in the kind of grips used by the human and gorilla infants. However, one difference lay in the greater extent to which the young gorillas used grips involving flexion at the proximal and distal joints of the digits (see *Plate 1d*). This may be a feature of prehension that is related to the compromise of having a hand that can be used for manipulative activity, as well as climbing and knuckle-walking.

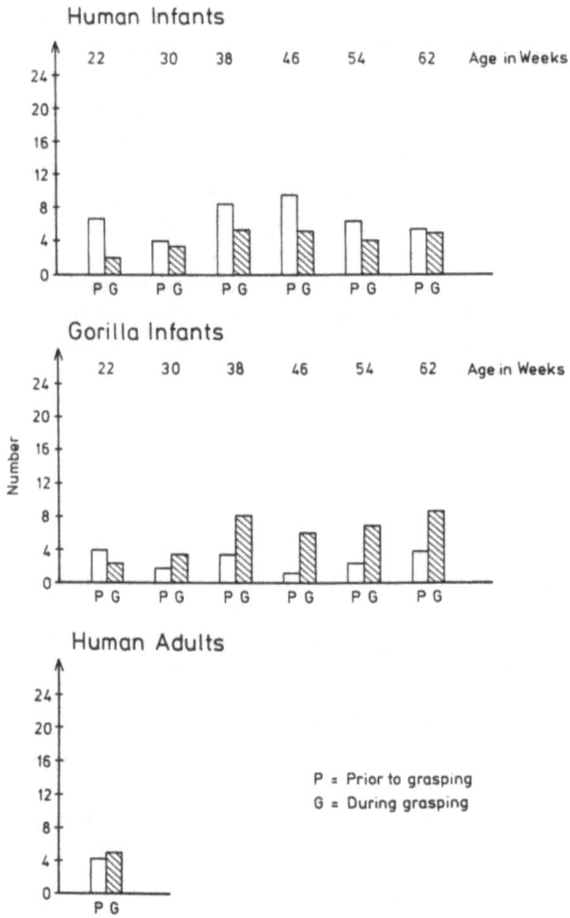

Figure 17. Differentiation of the forefinger before and during prehension.

In comparison with the adult subjects less mature grips were used by the infants and these tended to involve more use of the palmar surface of the hand and ulnar digits. With a pellet CASTNER (1932) found that a palmar grip predominated at thirty-two weeks, but HALVERSON (1931) using a larger one-inch cube found this grip common at twenty-four weeks, after which it was only used occasionally. In this experiment a palmar grip was seen relatively infrequently and then at the younger ages. A precise pincer grip occurred slightly later in the gorilla infants than the humans, being first recorded in the latter at thirty and thirty-eight weeks and in the former at forty-six and fifty-four weeks. It was also seen less often in young apes, for when grasping small or thin objects, they apparently preferred to use grips in which the flexed finger-tips rested on the table's surface and the thumb or forefinger moved towards the other for object capture. The gorilla infants could achieve a tip-to-tip precision grip, but a pulp-to-pulp contact of the kind described by NAPIER (1961) for humans was not observed, and due to the proportions of the thumb and forefinger may not have been possible. It seems as if in the ape infants' precision pincer

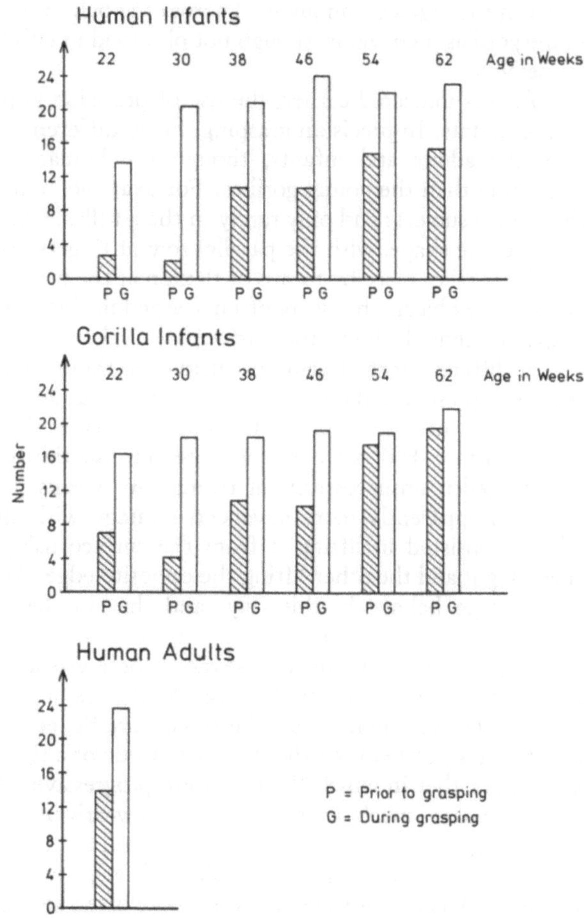

Figure 18. Abduction and opposition of the thumb before prehension and use of the terminal thumb volar pad during prehension.

grip a smaller surface area is covered where the digits meet and more force required to maintain this contact. (For further discussion of this point see CHRISTEL, this volume.) The gorillas and two of the human infants also used a grip that NAPIER (1960) observed in adult apes during prehension of small objects. This consists of a hook-like action of the forefinger in which an object is held almost entirely within the flexed forefinger, with the thumb taking no part at all.

Grips in which objects are held between the palmar surface and flexed digits without the active participation of the thumb, showed a clear decline in use by the infants, though this type of grip was still used on a few objects, even at sixty-two weeks of age. In contrast such grips were never actually seen in the adults' behaviour, though several employed some that resembled these, but with the thumb being active rather than passive. Grips in which just a minimal area of the palm adjacent to the fingers was used were seen more in the behaviour of the human infants, though recorded earlier for the gorillas.

Object capture of a «primitive» sort in which small or thin objects were caught between the digits or simply held across the palm was observed more frequently and later in the gorillas' behaviour, though not observed in either species during the last experimental session.

As was indicated earlier, the use of precision grips gradually increased over time for all the infants. In precision grasping, many different combinations of digits were used by both the adults and infants, though the humans used all the digits together more frequently than the young gorillas. For example, a grip that was observed quite often in the human subjects and only rarely in the gorillas, was one in which the hand is made into a wedge-like shape, with the parallel row of fingers opposed to the thumb and closure on the object takes place by means of flexion at the proximal joints in a kind of snapping shut action. The objects that were often grasped in this way were some of the rods and the tile shape, presented above the surface.

A different sort of grip, just using the thumb and fingertips was employed with the tile shape presented flat on the surface. This was levered off by the thumb and the fingers putting pressure on the object from opposite sides simultaneously, forcing it upwards and off the surface. Half the adults were seen to use this one-handed grip and though only one infant tried it unsuccessfully at twenty-two weeks, it was in more frequent use at later ages. An apparently more advanced strategy with the tile, mainly seen in the human adults, consisted in lifting it from the surface using the two hands jointly, one hand steadying it and the other lifting the opposite edge. Among the infants only one, a gorilla, was seen to behave in this way, and that in the last two sessions.

As was expected, the range of grips used by the infants changed over time, but even by the end of the experimental sessions there was a discrepancy between their behaviour and that of the adults in this regard. For example, no infants were seen to use the modified power grip in which the fingers are flexed and in contact with the palm and the object is grasped between the thumb and side of a tightly flexed forefinger (as is used with a key), nor that in which the digits are progressively flexed and an object is held within their compass, much in the way a screw-driver is held. The adults frequently used precision grips where the thumb was opposed to digit 2 and the side of digit 3, or else to 2 and 3 and the side of digit 4. while such grips were rarely, if ever used by the infants. Precision pincer grips were also seen more often among the responses of the human adults who used them on a wider range of objects than just the pellets.

More uniform grasping behaviour might have been observed with a smaller number of objects and repeated presentations, but as PIAGET (1952) and BRUNER (1974) have emphasised, with infants at this age, variation and experimentation are typical of their behaviour as they progress through the rapidly changing sensorimotor stage.

Use of the hands during prehension

Hand use during prehension can be very variable and adjustments may be made at many points in the sequence of behaviour that makes up a prehensive act. Each step in the string of events, from initiating a response, moving a hand towards an object, contacting and grasping it, may be achieved by one or both hands, and if only one, by the left or the right. Thus there is considerable potential for variation in performance. For example, the right hand might begin an approach, be overtaken by the left hand moving along a different trajectory which contacts the object first, but yet the right hand may be the first to grasp, though secure capture may require the use of the two hands jointly.

The development of consistency of hand use in prehension was examined with a view to assessing whether asymmetries develop at an early stage in the act prior to grasping, or if the approach, contact and grasping segments mirror one another in the hand or hands used. It was thought that the infants, as a function of immaturity, would show more variation and inconsistency in this regard than the human adults for whom hand preference was well established.

Hand use in initiating an approach, completing it, contacting and securely grasping a presented object was found to be quite closely linked for all the subjects and increasingly so with age. This was most clear among the adults. Two of the human infants behaved quite erratically in the earlier sessions though later their behaviour seemed to have become more coherent and organised in this respect. The other human infant and the two young gorillas, like the adults, were more consistent across the prehensive act, and though the distribution of hand use changed over time, the hand that began the approach was likely to be the one that completed the sequence.

The human adults used very little bimanual behaviour at all, ranging from 2% in the first approach to 6% of secure grasps. In contrast this was more frequently found among the responses of the infants, though one did approximate to the adult pattern, even from the earliest session. The pattern of use of two hands by the other infants, humans and gorillas, overlapped considerably.

All the adult subjects showed an exclusive hand preference that operated at all the stages of the prehensive sequence examined. Seven of the eight favoured the right hand, and one the left. It was expected that by sixty-two weeks the human infants would be showing a similar bias towards using the right hand preferentially. Combination of the sessional data indicates a significant right bias in the hand used to contact and the hand used to grasp the objects in both species of infant ($p = <0.001$). At sixty-two weeks this was only the case for the human infants ($p = <0.001$), though at the same age on another less structured manipulative task, the gorillas too showed a right hand preference ($p = <0.001$). Subsequently at two and three years of age the gorilla infants were found to have a continuing right hand bias in prehension of small items and in carrying out a range of manipulative tasks. Similar follow-up data are unfortunately not available on the human infants.

Summary and Discussion

Developmental changes were found at all stages in the prehension sequence, from approach to object capture. Over the experimental sessions both species of infant showed a progression in which prehension gradually became more smoothly co-ordinated and more dextrous. They moved from a stage in which the hand appeared to function much as a paw-like appendage that was hurled at the presented objects and which gripped the first part of the object contacted, often losing it in the process, to one in which the movements were better modulated. Thus, after entering their second year, the infants were more direct in their approaches, more appropriate in matching the configuration of their hands to object characteristics, and more often used the radial portion of the hand in grasping with little or no use of the palmar surface of the hand, except on the largest objects.

In general, the developing prehensive skills of the two species of infant were similar. The shorter fingers of the gorilla and more similar proportions of the hand compared with those of the longer palms and digits of the other two great apes no doubt contributed to this similarity (*Figure 19*). There were. however, some differences between the species.

Figure 19. Example of often used hand postures in a gorilla infant.

The young gorillas more frequently employed a hand in which the digits were flexed at the inter-phalangeal joints. At the same time they appeared less able to hyperextend the forefinger as the human adults and infants did at times, or to differentiate this digit and the adjacent finger jointly. Over all they used more power and fewer precision grips than the human infants, the precision pincer grip occurred slightly later and they often used grips involving the thumb moving to meet the side of the forefinger.

Each of the differences mentioned can be related to the gorillas' use of the hand in both locomotor and sensory manipulative activity. The habitual knuckle-walking posture which they develop and maintain, even while sitting, the use of the hands as efficient organs for climbing and suspending the body and as powerful tools for foraging, all appear to have contributed to the more robust appearance of gorilla prehension.

It is clear that radical changes take place in prehensive skill during great ape infancy, but at the same time it is evident that there are many other developments which take place subsequently as shown by the comparison with the manual behaviour of human adults. The latter, who were quicker to approach and grasp the objects, used fewer fingers and greater amounts of anticipatory hand closure, but did not abduct or oppose the thumb as much as the literature, with its emphasis on the primary roles of these two digits, would suggest. Precision movements in grasping reached only 50% for the infants in the last session, but for the adults the figure was over 80%. There were also less easily interpreted differences like the more indirect approaches made towards objects which may have been a way of allowing more accurate object capture.

One focus of interest concerned the manner and extent to which a hand may mimic the shape of the object that is to be grasped, before any tactile feed-back with the particular object is possible. It was hypothesised that only gradually would the infants of both species come to anticipate the shape of an object as shown by the configuration of their hands, but that the human adults would respond to the same objects with greater discrimination. This was largely found to be the case, for only 40% of of the infants' hand shapes were judged appropriate initially, rising to 70%. at sixty-two weeks. This still fell well below the adults' level of 94%.

The infants' skill in shaping the hand prior to contact and object capture certainly improved with age, but it should be emphasised that when first tested, at twenty-two weeks, both the young humans and gorillas showed some ability to match the hand to the object presented. This was particularly true with respect to orientation, for, more often than not. they were able to orient their hands differently towards vertical and horizontal objects.

Among the adults and infants large and small objects provoked different behaviour. With a large object two hands were more likely to be used, whereas with a small object one hand was employed, usually with greater hand closure and more differentiation of the radial digits. More even flexion of all the fingers and abduction of the thumb was observed with the various rods, whatever the orientation, but with the pellet the fingers were successively more flexed or extended and the thumb opposed.

Such was the consistency of the adults' performance and the markedly differential nature of the responses to the variably sized and shaped objects, that it would be tempting to predict a fair degree of success in guessing the shape or size of these, if photographs were taken at this point in the prehensive act. and the objects themselves blacked out. This, however, would be a different experiment.

At all stages of the prehensive act there was largely continuity and consistency in the sequence of behaviour, such that the hand that made the first approach was also more likely to contact and grasp the object securely. This was most marked amongst the adult subjects with a clear right preference. Only one infant, a human, showed such a clear-cut preference for the right hand, though overall the infants of both species showed a similar right bias. Data collected by FISCHER *et al.* (1982) on four adult female gorillas also indicate a right hand preference for grasping objects and food (see also BRENOT in this vol.), while SCHALLER (1963) reported a significant difference in favour of chest-beating sequences starting with the right hand.

Fortunately the area of prehension is one which was amenable to cross-species comparisons and the use of a human adult comparison group. The experimental design and procedure was sufficiently simple to be used with all three kinds of subject. Later work with the young gorillas on prehension was more difficult, owing to their strength and exuberance.

In essence this comparative study on the development of prehension confirmed that the acquisition of this skill is basically similar in human and gorilla infants. Some of the relatively few differences found may be related to structural anatomical differences in the hand and the behavioural adaptations and habits that go with these. Undoubtedly, the manipulative and locomotor functions that have evolved in the very different niches that the species occupy fit them for very different lives. However, the larger number of differences found between the behaviour of the infants of both species and that of the human adults indicates that yet further refinements take place in the development of prehension, even after the stage when infants and young children appear to be fairly proficient at capturing objects manually.

ACKNOWLEDGEMENTS — Many thanks are due to the parents of the human infants and those caring for the gorilla infants at Jersey Zoo. For discussion and continuing interest I would like to thank Alison Jolly, Bob Martin, Henry Plotkin and Chris West.

The research was funded by the Medical Research Council of the United Kingdom.

References

BISHOP A., 1962a. *Use of the hand in lower primates.* Harvard University, PhD Thesis.

BISHOP A., 1962b. *Control of the hand in lower primates.* In: Evolutionary and Genetic Biology of the Primates, Vol. II, Ed. J. Buettner-Janusch, Academic Press, London.

BRUNER J.S., 1974. *Nature and uses of immaturity.* In: The Growth of Competance, Eds. K. Connolly and J. Bruner, Academic Press, London.

CARTER S.F., 1973. *Comparison of baby gorillas with human infants at birth and during the post-natal period.* Jersey Wildl. Preserv. Trust Ann. Rep., 10: 29-33.

CASTNER B.M., 1932. *The development of fine prehension.* Genet. Psychol. Monogr., 12: 111-191.

FISCHER R.B., MEUNIER G.F. & WHITE P.J., 1982. *Evidence of laterality in the lowland gorilla.* Percept. Mot. Skills., 54: 1093-1094.

HALVERSON H.M., 1931. *An experimental study of prehension in infants by means of systematic cinema records.* Genet. Psychol. Monogr., 10: 110-286.

HALVERSON H.M., 1932. *A further study of grasping.* J. Genet. Psychol., 7: 34-63.

HOFER H., 1960. *Beobachtungen an Brüllaffen.* Natur und Volk 90: 241-251.

HUGHES J. & REDSHAW M., 1974. *Cognitive, manipulative and social skills in gorillas. Part I: The first year.* Jersey Wildl. Preserv. Trust Ann. Rep., 11: 53-60.

JENSEN G.D., 1961. *The development of prehension in a macaque.* J. Comp. Physiol. Psychol., 54: 11-12.

LORENZ R., 1970. *The functional interpretation of the thumb in the Hylobatidae.* Proc. 3rd Int. Congr. Primat., Zürich, Vol. 1: 130-136.

NAPIER J.R., 1960. *Studies of the hands of living primates.* Proc. Zool. Soc. Lond., 134: 647-657.

NAPIER J.R., 1961. *Prehensility and opposability in the hands of primates.* Symp. Zool. Soc. Lond., 5: 115-132.

NAPIER J.R. & NAPIER P.H., 1967. *Handbook of the Living Primates.* Academic Press, London.

NAPIER J.R., 1980. *Hands.* George Allen and Unwin, London.

REDSHAW M., 1975. *Cognitive, manipulative and social skills in gorillas. Part II: The second year.* Jersey Wildl. Preserv. Trust Ann. Rep., 12: 56-60.

REDSHAW M., 1978. *Cognitive development in human and gorilla infants.* J. Hum. Evol., 7: 133-141.

REDSHAW M., 1985. *The development of hand functions in human and gorilla infants.* Ph.D. Thesis, University of London.

REDSHAW M., 1989. *A comparison of neonatal behaviour and reflexes in the great apes.* J. Hum. Evol., 18: 191-200.

REDSHAW M. & LOCKE K., 1976. *The development of play and social behaviour in two lowland gorillas.* Jersey Wildl. Preserv. Trust Ann. Rep., 13: 71-85.

ROLLINSON J., 1975. *Interspecific comparisons of locomotor behaviour in eight species of African forest monkey.* Ph.D. Thesis, University of London.

SCHALLER G.B., 1963. *The Mountain Gorilla: Ecology and Behaviour.* University of Chicago Press, Chicago.

WOOD JONES F., 1944. *The Principles of Anatomy as Seen in the Hand.* 2nd ed., Bailliere, Tindall and Cox, London.

M. Christel

Grasping techniques and hand preferences in Hominoidea

Starting off from the generally known, but very differently evaluated, manipulatory skill of the hand of higher primates, the present investigation tests in detail the functional characteristics of precise grasping behavior and the hand preference of 22 apes (5 *Pan troglodytes*, 2 *Pan paniscus*, 4 *Gorilla gorilla*, 4 *Pongo pygmaeus*, 6 *Hylobates* (3 *lar* and 3 *moloch*) and 8 humans. The method was focused on the following: a) ad libitum picking up differently shaped and very small objects without task-specific set ups; b) many repeated actions; c) exact distinction of different areas of contact between the grasping fingers; e) a comparable experimental set up for both, apes and humans.
Small objects are picked up and held by a variety of grips. Apes have not only one, but many different precise grasping techniques. The «classical precision grip», the grip between thumb tip and index tip, was just one variant among many others, but it was frequently used in chimpanzees, gorillas and orangutans, and it more than met the criteria which can be postulated for precision grasping: quickness, secure hold and success of the first approach. The number of preferred grasping variants differs among species and individuals. The greatest variability of precision grasping was shown by orangutans, while gibbons used the fewest. Intraspecifically, chimpanzees showed the greatest individuality of preferred grips. While apes additionally to the fingertips used the lateral side of the index finger, humans exclusively preferred the distal area, in particular the fingerpads, in grasping small objects.
The left-right breakdown of hand usage for picking up objects varied individually from a pronounced preference of one hand to near ambidextrousness. More individuals of *P. troglodytes*, *P. paniscus* and *G. gorilla* showed more pronounced hand preference than *P. pygmaeus* and *H. lar* and *moloch*. The most frequently used grasping technique is often more or less confined to one, namely the «preferred» hand. The grasping modes next in frequency are often exerted by the other, or by both hands.

Key words: precision grip, contact areas, grasping variants, manipulation, hand posture, hand preference.

Introduction

«The acme of precision in man is the ability to pick up small objects between thumb and index finger and hold them delicately yet securely between the opposed pulp surfaces». (Napier, 1960, p. 652). According to Napier (1960), non-human primates are not capable of this precision, and in 1980 (p. 60) he reiterated this point using a figure showing a chimpanzee hand grasping with the thumb against the radial surface of the index finger, emphasizing: «... note the imperfection ...».

The precision grip of humans is seen as the measure of an optimal manipulatory skill of the hand. The definition of the precision grip is, however, very diverse in the literature; there exist semantic problems involving the precise grip types used by the non-human primates, and our view is too encumbered by the question of similarities with form and function of the human hand.

A fundamental definition of the grasping technique of humans is also provided by NAPIER (1956). He defined a «dual principle of the hand» in grasping and holding objects: the «precision grip» and the «power grip», a dichotomy which has become widely accepted until today.

Accordingly, the «precision grip» is the one in which the thumb tip can oppose one or more other finger tips. This contrasts with the «power grip», in which all of the digits, including the thumb, prehend simultaneously and press the object against the palmar surface. The criterion for the definition of the type of grip has until now been the opposition of the thumb to the other digits. NAPIER (1961) distinguished the non-opposable thumbs of the Tupaioidea, the pseudo-opposable thumb of the Prosimii and Ceboidea and the opposable thumb of the Cercopithecoidea. Tree shrews grip by jointly bending all of the digits toward the palma. They possess a so-called «whole-hand control» (BISHOP, 1964; see also LANDSMEER, this volume). Tarsius and the Cebinae have a «pseudo-opposable» thumb marked by greater abduction and adduction capability (NAPIER, 1960). Small objects are pressed against either the proximal phalanx of the digit or the thenar ball by means of flexion of thumb and index, or they are picked up by means of abduction and adduction movements of the index finger and middle finger (BISHOP, 1964). The thumbless *Ateles* grips small objects by rolling them into the three flexed phalanges of the index finger and at the same time executing a marked pronation movement of the lower arm, thus turning the radial side of the hand toward the substrate (BISHOP, 1964).

The Cercopithecoidea, with the exception of the thumbless Colobinae, possess an opposable thumb (SCHULTZ, 1930). This opposable thumb after Napier makes possible the «precision grip» (NAPIER, 1956, 1961).

The ability to execute an «optimal» precision grip as in NAPIER's definition is in humans largely the result of a combination of anatomical features, in particular a high relativ thumb length, which is significantly lower for pongids and hylobatids (orang-utan=43; chimpanzee=47; gorilla=54; hylobates=52; in contrast: human=67; SCHULTZ, 1956). Broad apical bone tufts of the distal phalanges and associated fleshy finger pads, densely vascularized and innervated, provide humans with an adhesive surface capable of holding objects tightly and securely (SHREWSBURY *et al.*, 1986; LEWIS, 1989; see also BUCK & BÄR, this volume). In the pongids, however, the apical bone tufts of the distal phalanges taper to a point, and the finger pads do not have the broad surface seen in humans (NAPIER, 1962; SUSMAN, 1988). This trait was shown by PREUSCHOFT (1969, 1973) to be related to the mechanical strength of the nail.

Compared with humans, the thumb of the pongids is weakly developed as illustrated by the total weight of the thenar muscles which is considerably lower (PREUSCHOFT, 1965). The M. flexor pollicis longus, which flexes the distal thumb joint, is always present in humans (and in the hylobatids), but rarely developed in the other hominoidea, and when present it is shorter. The *M.* opponens is also weaker and the deep head of the flexor pollicis brevis, which flexes the proximal joint, is frequently lacking in chimpanzees and gorillas or it is developed in a relatively slender form (JOUFFROY & LESSERTISSEUR, 1960; DAY & NAPIER, 1963; TUTTLE, 1964; PREUSCHOFT, 1965; MAIER, 1971; SWINDLER, 1973).

In contrast to the morphological differences between the hand of humans and the hand of apes, the descriptions available of the differing functional capabilities and grasping techniques used by non-human primates present a less uniform picture.

«Precision grasping» generally implies that small objects can be picked up and held up quickly and securely. This skill is doubtlessly highly refined in humans, although the higher classification emphasized in humans has led to the establishment of a hierarchical gradient in which the skills of the non-human primates drop off sharply. Although various authors have mentioned that a «precision grip» generally occurs in pongids, particularly in chimpanzees, and although SCHULTZ writes in 1961 that every chimpanzees can extract a thorn out of its skin with thumb and index finger more dexterously than most of us, none of the illustrations (NAPIER, 1960, Figure 3-9; TUTTLE, 1969, Figures pp. 351-352), with the exception of BOESCH and BOESCH (this volume), shows the hand of any non-human primate using a «precision grip» as characterised above.

NAPIER (1960), distinguished a number of different ways in which chimpanzees pick up small objects (p. 653): 1) by means of the pulp of the thumb and the side of the middle or terminal phalanx of the index finger; 2) by means of the side of the thumb and the side of the middle phalanx; 3) by means of the dorsum of the thumb and the dorsum of the terminal phalanx of the index. TUTTLE (1965, p. 120) agrees with NAPIER on most points and notes that «*Pan, Pongo* and *Gorilla* pick up small objects by opposing the tip of the thumb to the radial aspect of the finger, generally to the middle phalanx». This grip is capable of holding an object as small as a grain of rice, but it is not as strong as the precision grip of the human hand. Only in grooming behavior did he observe a chimpanzee to hold a fine particle of skin between thumb tip and index tip. Thus, the human precision grip is one in which objects are held securely and steadily between the finger pads of the thumb and other digits, while in the case of pongids «the nearest that the thumb can get to opposition in human sense is a tenuous and unstable grip between the extreme tip of the thumb and the extreme tip of the acutely flexed finger» (NAPIER, 1960, p. 654). Even within the hominoidea the capabilities involved in a precision grip has been judged differently. According to NAPIER & NAPIER (1967), the gorilla, due to its hand anatomy, has first place, but the orangutan does better when the goal is defined as making the best use of the thumb in fine manipulation. In the detailed study of REDSHAW (this volume) the precision grip in gorillas is emphasized and documented. The hylobatids are unable to place the thumb tip against the index tip. Their precision grip, achieved without using the thumb, comes about, according to LORENZ (1971, p. 133), «between the index finger and middle finger».

As opposed to humans, most non-human primates seem to exhibit no clear-cut handedness. However, this has not yet been proved on a population level. But strong preferences for one hand can nevertheless be observed frequently in grasping and other activities. (SCHALLER, 1963; REYNOLDS, 1975; WARREN, 1980; FISCHER et al., 1982; MARCHANT & STECKLIS, 1986; BRESARD et al., 1987; MACNEILAGE et al., 1987; OLSON, et al., 1990; BRENOT, REDSHAW, this volume).

Feeding, object exploration and social grooming require many non-human primates to touch and grip objects of various sizes, including small items like seeds and particles of skin. (SCHULTZ, 1961; SCHALLER, 1963; GOODALL, 1965; KUMMER, 1968; MAIER, this volume). Proceeding from the generally known, great manipulatory skill of the hand of higher primates, the present investigation tests the functional characteristics of these skills. Apes and humans were given refined grasping tasks, areas of the hand which came into contact by precision grasping were defined and the achievements accomplished are described.

Method

Subjects investigated

A total of 22 apes and 8 humans were investigated:
P. paniscus (1 juvenile female, 1 juvenile male, both born in captivity);
P. troglodytes (2 adult females, 3 adult males, 4 were captive-born, 1 was wild-born);
P. pygmaeus (2 adult females, 1 juvenile female, 1 juvenile subadult male, all captive-born);
G. gorilla (2 adult females, 1 juvenile female, 1 adult male, one male was wild-born, others were captive born);
H. l. lar (1 adult female, 1 adult male, 1 juvenile male, all captive-born);
H. l. moloch (1 adult female, 1 adult male, 1 juvenile male, two captive-born, one male wild-born);
Human (8 adults: 6 females and 2 males).

Procedure

The prehensive behavior of the hominoidea was observed, video-recorded and evaluated by means of slow-motion and still-frame analyses over a period of several months both inside a cage and in an open-air enclosure. A mixture consisting of grapes, peanuts, raisins, maize, pumkin seeds, oatflakes, rice and millet (object size ~10·7·4 mm) was strewn broadly across the ground. Animals were moving freely in their cages. Two males (one gorilla, one chimpanzee) were investigated while isolated, all the other individuals were kept and observed in groups. In focal animal sampling methods (ALTMANN, 1974) each individual was observed during 2 or 3 sessions, each lasting between 1 ½ and 3 hours. Grasping activities were noted and every 10-15 minutes sequences of 5 minutes duration were video-recorded and each frame analysed. A total of 3674 grips was investigated. The humans were given the same mixture spread out on the center of a table an arm's length away. A total of 688 grips was included. Indistinctly visible movements or grasps not in good focus on the screen were not included. The data were statistically evaluated with the aid of the Chi Square Test.

Results: Areas of the hand which come into contact in grasping

Pongids and hylobatids (Figure 1)

nineteen different contact areas were distinguished:
Id: at the distal tip of the thumb, volar.
IId: at the distal tip of the index finger, volar.
IIdr: at the radial aspect of the distal phalanx of the index finger.
IIdigr: at the radial aspect of the distal joint of the index finger.
IImr: at the radial aspect of the middle phalanx of the index finger.
IIpigr: at the radial aspect of the proximal joint of the index finger.
IIpr: at the radial aspect of the proximal phalanx of the index finger.
IIdu: at the ulnar aspect of the distal phalanx of the index finger.
IIdo: on the dorsal aspect of the distal phalanx of the index finger.
IIdigo: on the dorsal aspect of the distal joint of the index finger.

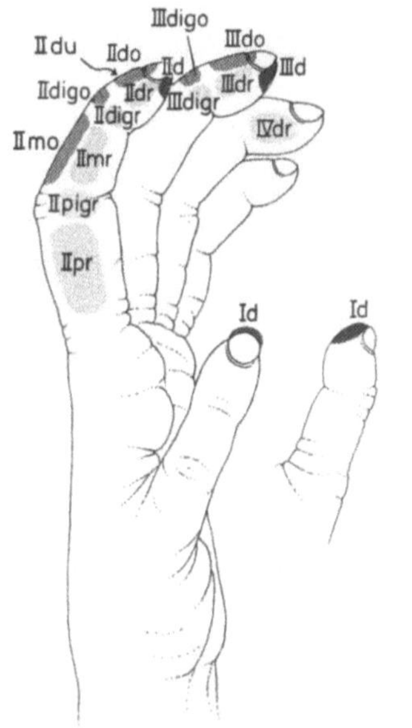

Figure 1 - Areas of contact at the fingers during precise grasping: pongids. The left picture shows the dorsal view of the thumb; whereas the insert on the right side shows its lateral view.

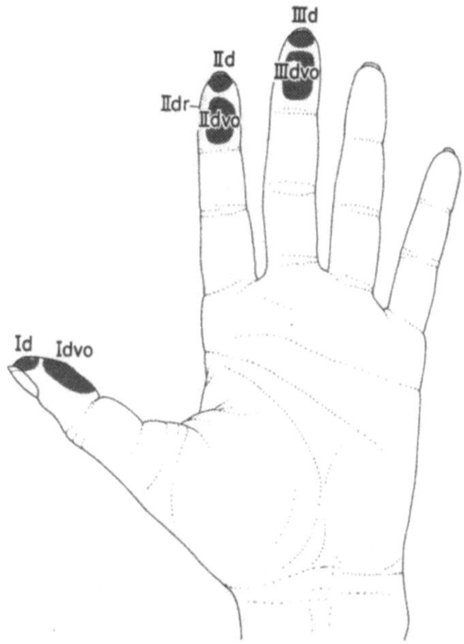

Figure 2 - Areas of contact at the fingers during precise grasping: humans.

IImo: on the dorsal aspect of the middle phalanx of the index finger.
IIdigu: at the ulnar aspect of the distal joint of the index finger.
IIId: at the distal tip of middle finger, volar.
IIdo: on the dorsal aspect of the distal phalanx of the middle finger.
IIIdr: at the radial aspect of the distal phalanx of the middle finger.
IIdigo: on the dorsal aspect of the distal joint of the middle finger.
IIIdigr: at the radial aspect of the distal joint of the middle finger.
IIIdu: at the ulnar aspect of the distal phalanx of the index finger.
IVdr: at the radial aspect of the distal phalanx of the ring finger.

Humans (Figure 2)

Eight contact areas used in grasping were distinguished:
Id: at the tip of the distal phalanx of the thumb.
Idvo: at the distal thumb pad.
IId: at the tip of the distal phalanx of the index finger, volar.
IIdvo: at the distal index finger pad.
IIdr: at the radial aspect of the distal phalanx of the index finger.
IIId: at the tip of the distal phalanx of the middle finger.
IIIdvo: at the distal pad of the middle finger.
IIIdr: at the radial aspect of the distal phalanx of the middle finger.

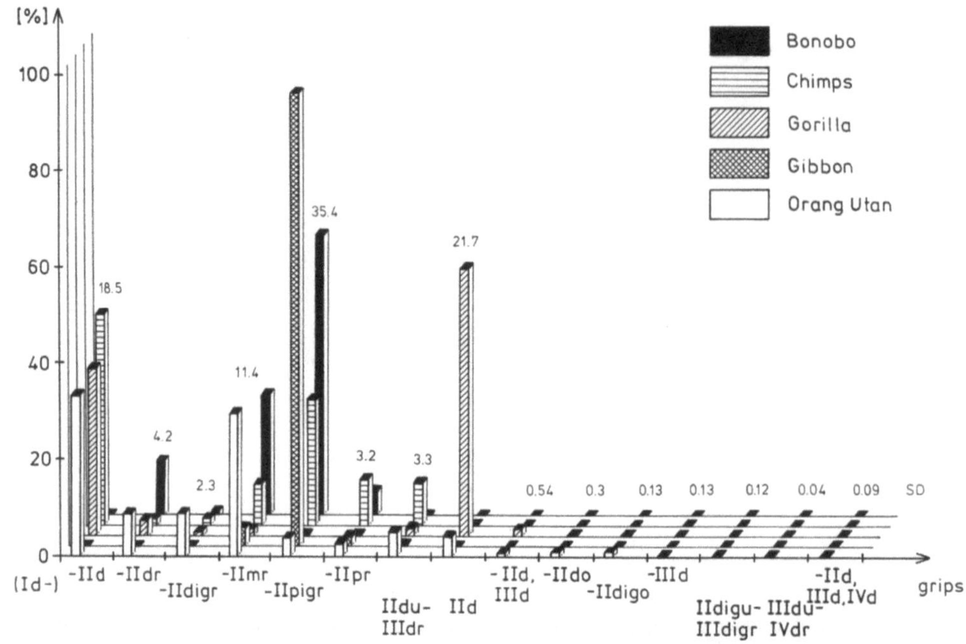

Figure 3 - Percentage of the frequency of all precision grips observed. chimpanzee (5 individuals), bonobo (2 individuals), gorilla (4 individuals), orangutan (5 individuals), gibbon lar (3 individuals), gibbon moloch (3 individuals); (SD = Standard deviation).

Results: Grasping variants of the Hominoidea

In picking up very small object of food a total of 15 different grasping variants were distinguished in the hominoidea. The data for apes and humans are discussed separately because in the humans only the distal areas of the index finger make contact and not the radial aspect as is typical for apes.

Pongids and hylobatids (Figure 3, 5)

The differences in frequency and preference for certain grasping variants are statistically highly significant both between and within species ($p = 0.0001$).

1. Grip Id-IId: thumb tip and index finger tip. This grip was the most frequent observed variant in three of the five species studied. Most frequently in *P. troglodytes,* followed by humans, gorillas and orangutans in descending order. It was not observed in *P. paniscus* and *H. l. lar* and *H. l. moloch.*

2. Grip Id-IIdr: thumb tip against the radial aspect of the distal phalanx of the index finger was, compared with other grips, used considerably less by all species. The highest levels were recorded in *P. paniscus* and *P. pygmaeus. Gorilla* and *P. troglodytes* employed this variant less frequently, and the hylobatids did not use it at all.

3. Grip Id-IIdigr: thumb tip against the radial aspect of the distal joint of the index finger was most common in *P. pygmaeus*, rare in *P. troglodytes*, *G. gorilla* and *P. paniscus*, and not recorded by *Hylobates* species.

4. Grip Id-IImr: thumb tip against the radial aspect of the phalanx of the middle finger, was a frequently employed grip in orangutans and bonobos, less so in *P. troglodytes* and only rarely was it used by *H. l. moloch* and *G. gorilla*.

5. Grip Id-IIpigr: thumb tip against the radial aspect of the proximal joint of the index finger was used by *P. paniscus* and hylobatids as their main grasping technique. *P. troglodytes* also used this grip frequently, though it was observed seldom in *Pongo* and *Gorilla*.

6. Grip Id-IIpr: thumb tip against the radial aspect of the proximal phalanx of the index finger was, among pongids, used most frequently by *P. troglodytes* and then by *P. panisicus* and *Pongo pygmaeus*.

7. The IIdu-IIIdr, «adduction grip»: the ulnar aspect of the distal phalanx of the index finger and the radial aspect of the distal phalanx of the middle finger grip against one another, was used predominantly by *P. troglodytes*, being infrequent or absent in the other species.

8. «Adhesion grip» IId, with the volar aspect of the index tip: This variant cannot, in a strict sense, be termed a «grasping» variant; it must rather be seen as a prehension technique using the volar index finger tip. It was almost exclusively observed in *Gorilla*, though occasionally in *P. pygmaeus* and in *P. troglodytes*.

The remaining seven grasping variants were observed only in gorillas and orangutans, with the exception of one variant (Id-IIId), which was also used repeatedly by chimpanzees. These grips were applied considerably less often than those listed above, that is to say, less frequently than 6% of the mean of all grips observed.

Humans (Figure 4,5)

In contrast to the non-human primates, whose contact areas in grasping were often located at the radial aspect of the index finger, the precision grasping techniques used in fine prehension by humans are limited to the distal area of the thumb, the index and middle finger. The human subjects used a total of 13 different grips:

1. Id-IId: grip with the thumb tip and the tip of the index finger.
2. Id-IIdvo: grip with the thumb tip and the distal pad of the index finger.
3. Idvo-IId: grip with the distal pad of the thumb and the tip of the index finger.
4. Idvo-IIdvo: grip with the distal pads of thumb and index finger.
5. Idvo-IId,IIId: grip with the distal pad of the thumb and the tip of index and middle finger.
6. Id-IIId: grip with the thumb tip and the tip of the middle finger.
7. Id-IIIdvo: grip with the thumb tip and the distal pad of the middle finger.
8. Idvo-IIIdvo: grip with the distal pads of thumb and middle finger.
9. Id-IIdvo,IIIdvo: grip with the thumb tip and the distal pads of index and middle finger.
10. Id-IId,IIId: grip with the tips of thumb, index and middle finger.
11. Idvo-IIdvo,IIIdvo: the grip with the distal pads of thumb, index and middle finger.
12. Id-IIdr: grip with the thumb tip against the radial aspect of the tip of the index finger.

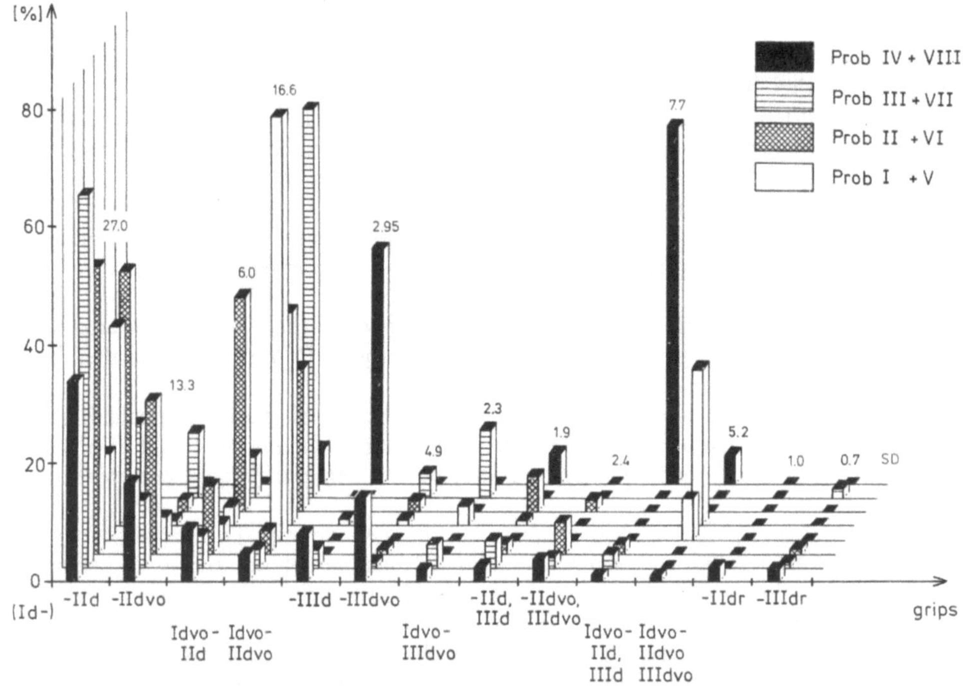

Figure 4 - Percentage of the frequency of all precision grips observed: humans (8 individuals) (SD= Standard deviation).

13. Id-IIIdr: grip with the thumb tip against the radial aspect of the tip of the middle finger.

The grip tip of thumb - tip of index finger, Id-IId, was used less frequently in humans than in chimpanzees and approximately as often as in gorillas and orangutans. For precision grasping, humans above all used the volar aspect of the distal phalanx of thumb and index finger, roughly in the middle of the finger pad (dvo). This grasping technique was never observed in the non-human primates. Like the non-human primates, the humans used grasping variants Id- IId, IIId; Id-IIId and Id-IIIdr; though generally more frequently than the apes. In contrast, grasping variant Id-IIdr was recorded almost exclusively in the non-human primates, for in the humans, it only occurred once with one test person. Of all the grips used, the humans most frequently employed grip Id-IId. The next frequent variant of humans was the grip between the distal pads of thumb and index finger (Idvo-IIdvo). The other nine variants were distributed uniformly between 4.7% and 0.4%.

The number of the grips used differs among the species: The greatest number of different grasping variants was used by *P. pygmaeus* (15 variants), this was followed by *Homo s. sapiens* (13 variants) and then by *G. gorilla* (12 variants), *P. troglodytes* (9 variants), *P. paniscus* (4 variants), *H. l. moloch* (4 variants) and lastly *H. l. lar* (2 variants). Intraspecifically, only few grasping variants occured more often than 10% of all grips, and therefore we regard them as «preferred» grips. Thus, the chimpanzees used only 2

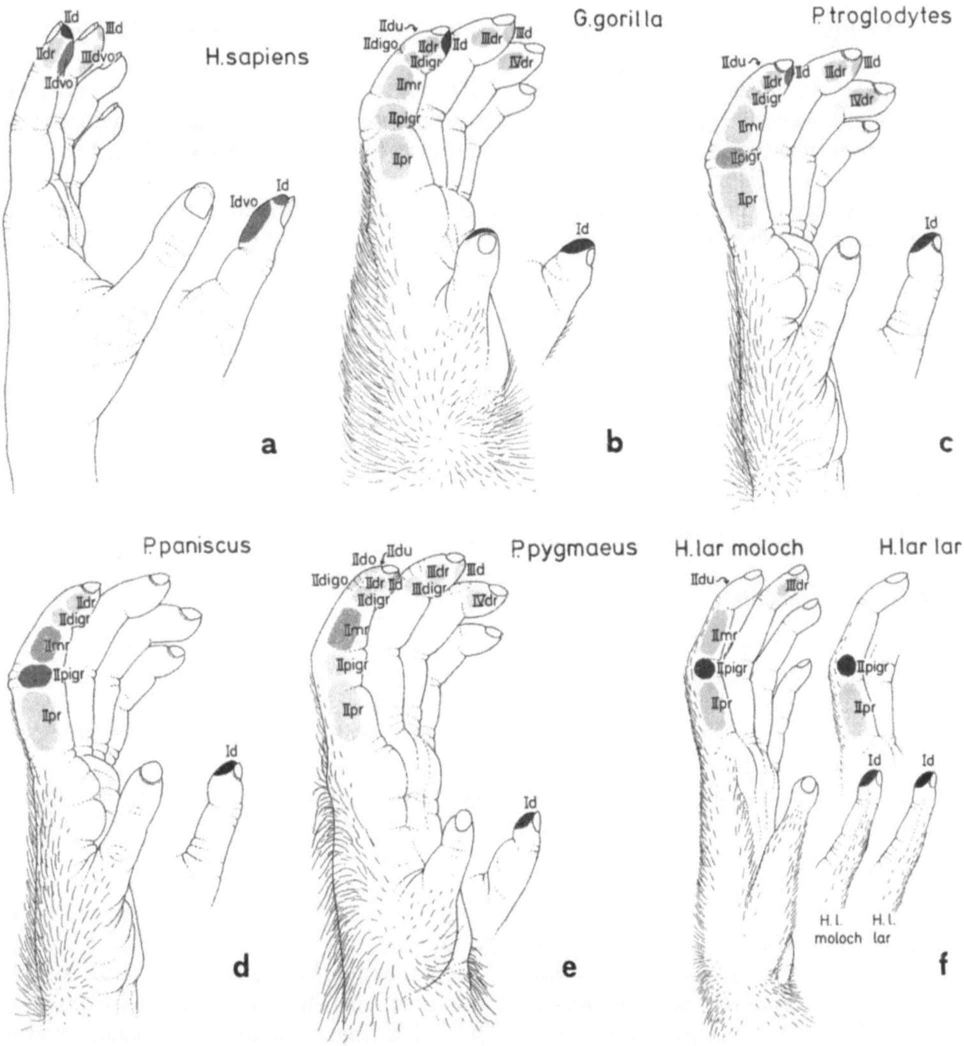

Figure 5 - Areas at the hand which came into contact in precise grasping. a= human, b=gorilla, c=chimpanzee, d=bonobo, e=orangutan, f=gibbon lar + moloch (darker areas are more often used than lighter areas).

preferred grasping techniques, bonobos only 3, gorillas 2, orangutan 2, gibbons (*lar* and *moloch*) 1 and humans 3.

Intraspecific differences were found in the preference for certain grasping variants. Although in an interspecific comparison the high frequency of the grip thumb tip against tip of index finger in chimpanzees was obvious (see *Figure 3*), the preference for this grip among individuals differs greatly; three of the five individuals did not use this grip at all. One male, Jonny, on the other hand, was the only one to use the «adduction grip» (IIdu-IIdr) as his main prehension technique.

The two bonobos used four different grasping variants with just slight individual variations.

All of the gorillas employed their distal finger areas significantly and relatively uniformly. However, Mpenzi, a two-year-old, rarely used the grip thumb tip against tip of index finger, she tended more often than the others to use the radial aspect of the tip of the index finger (IIdr).

The grasping profile for the five orangutans showed for all a relatively uniform pattern from pronounced preference for thumb tip against tip of index finger and along the radial aspect of the index finger.

For all six gibbons, the grasping area is concentrated at the proximal joint of the index finger but the two juveniles did not favor these contact areas so much as the adults. The variability of grasping was greater for *H. l. moloch* than for *H. l. lar.*

Individual preferences for particular grasping variants were also observed in humans. Test persons V and VIII, for instance, preferred grips rarely used by the other humans.

Hand posture and behavior in grasping *(Figures 6, a-h; 7, a-h)*

Pongids and hylobatids

A flexion of the distal joint of the thumb was observed only in *P. paniscus, Hylobates* and of course in humans. The distal thumb joint remained unflexed in *P. troglodytes, Gorilla* and *Pongo* and the finger posture looked quite rigid. In most cases no changes of the positions of the finger joints took place between positioning the hand in reaching out for food and the grip itself.

In the grip thumb tip against tip of index finger, the carpal joint of the hand is slightly dorsiflexed, the forearm being parallel to the ground surface (*Figure 6, g-h*). In picking up objects, the hand is either placed on the ground with the dorsal aspect of the terminal or middle phalanges of digits III-V, or remains unsupported. Furthermore, finger posture differs in the species. The basal phalanges of digits III-V are often abducted in the chimpanzees (*Figure 6, a+b*), in the orangutans they are mostly dorsiflexed (*Figure 6, c+d*), while the basal phalanges of gorillas are adducted, kept straight or show a slight volar flexion (*Figure 6*; see also REDSHAW, this volume). The other interdigital joints of fingers III-V show marked volar flexion. The index finger, on the other hand, exhibits only moderate volar flexion in the basal joint or remains straight. In chimpanzees and orangutans, it rather forms an arch, in gorillas it is mostly flexed and tends toward a right angle. The thumb moves in the metacarpo-phalangeal joint, giving rise to an opposition (DAY & NAPIER, 1963). Index finger and thumb tip for the most part move the object synchronously against each other; but it was often observed in chimpanzees that the tip of the index finger alone moved the object against the thumb tip.

In contrast to the apes, all humans using the grip thumb tip against tip of index finger, flex and extend the distal interdigital joints in order to exert pressure between finger pads or between one finger tip and the finger pad of the other digit (*Figure 6, f, g, h*), (NAPIER, 1952). The act of picking up objects appears more agile. In contrast to the grip of the pongids, humans perform flexing and extending movements of the distal joints between grasping and holding (LONG *et al.*, 1970).

When grasping against the radial aspect of the index finger, the hand in the gorillas for the most part remains in the hook hand or knuckle-walking posture. The dorsal

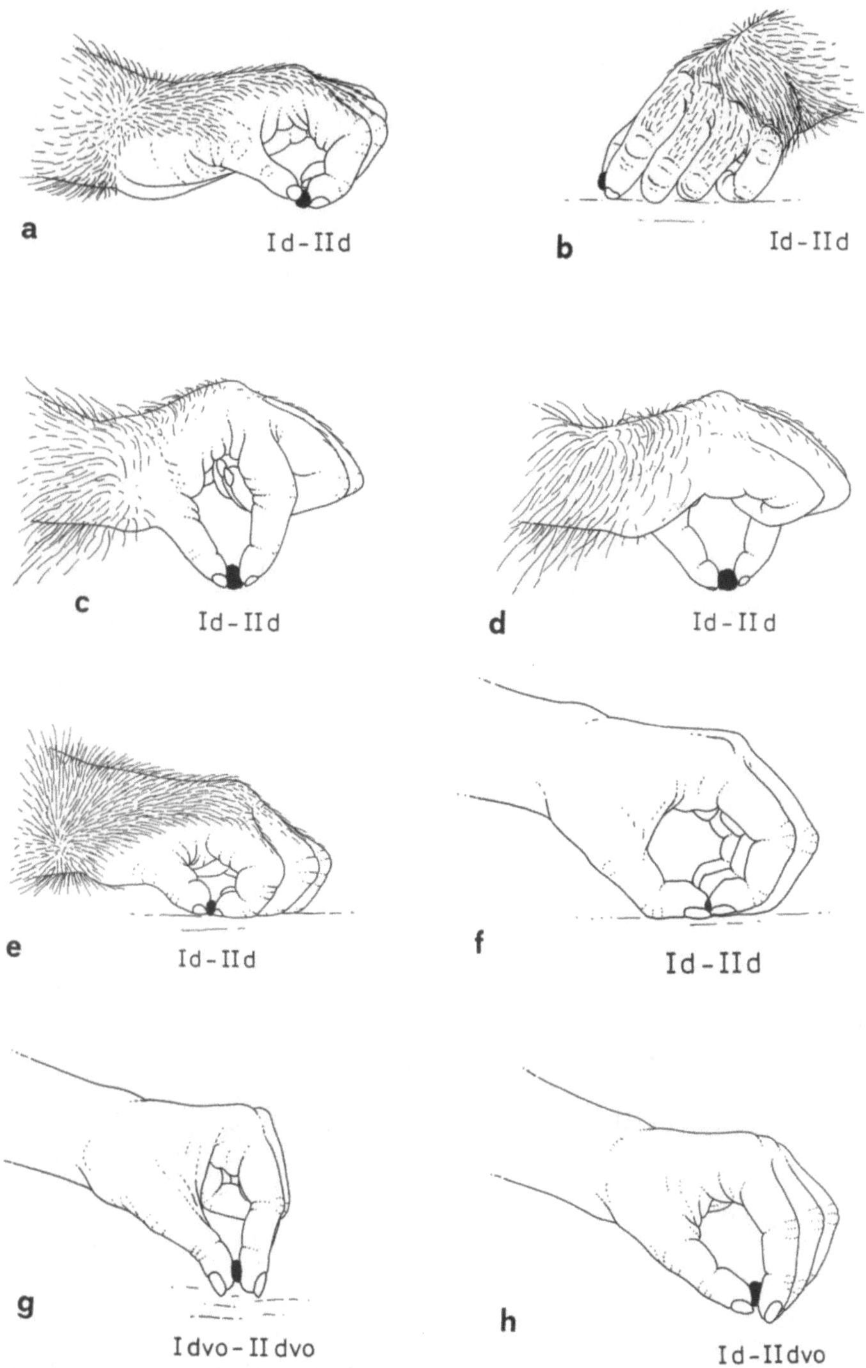

Figure 6 - Hand postures in grasping. Grips with the tip of index and the tip of thumb. a,b = chimpanzee, c,d = orangutan, e = gorilla, f,g,h = human.

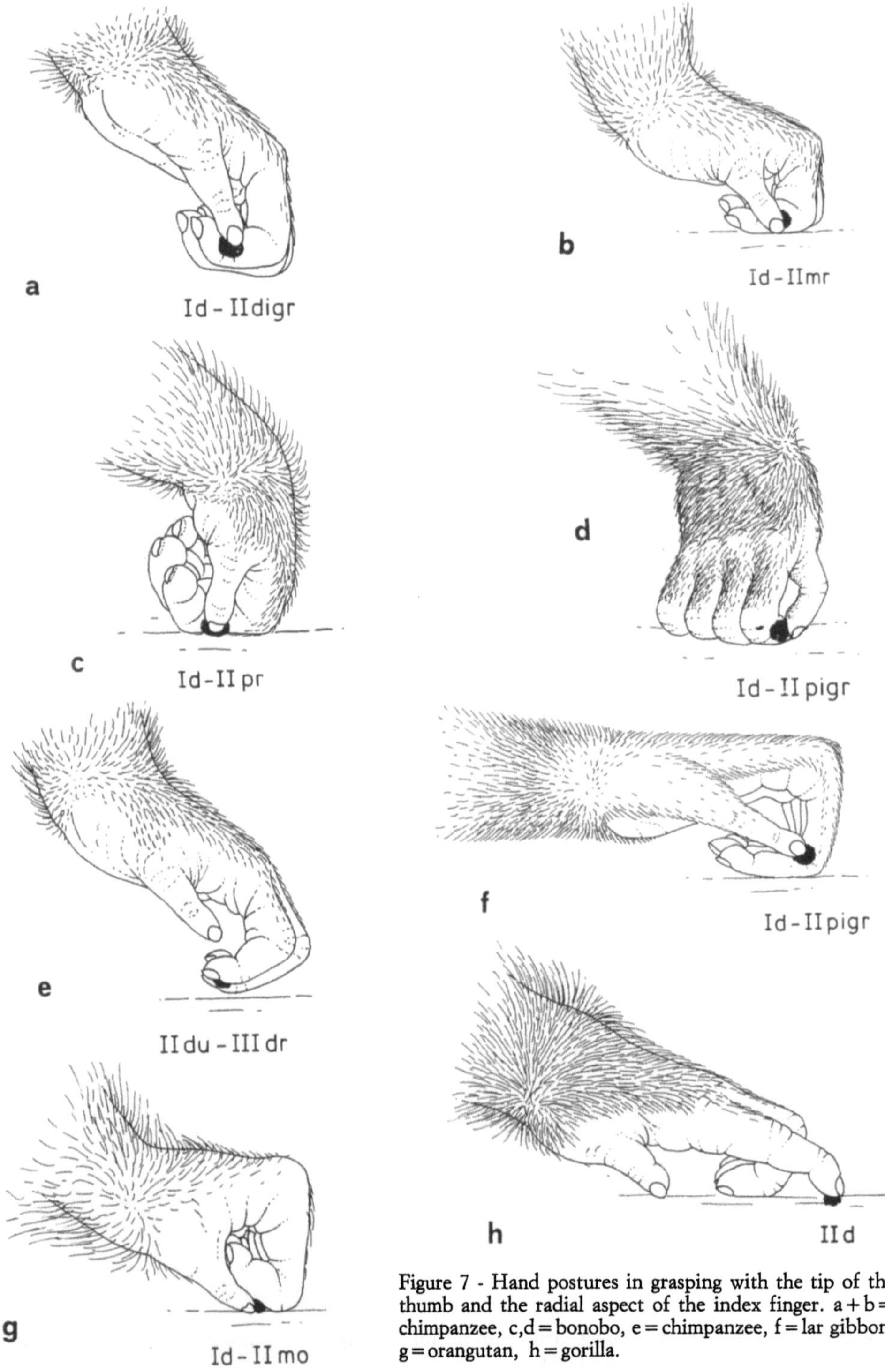

Figure 7 - Hand postures in grasping with the tip of the thumb and the radial aspect of the index finger. a + b = chimpanzee, c,d = bonobo, e = chimpanzee, f = lar gibbon, g = orangutan, h = gorilla.

surfaces of the flexed digits are placed besides the object on the ground and the thumb tip pushes the object against the index finger. The object is picked up in ulnar abduction by using pronating movement of the arm and it is taken to the mouth by supination movements. If the thumb grips more proximally against the radial aspect of the edge of the index finger, the ulnar abduction of the carpal joint of the hand and the pronation of the arm are more marked (*Figure 7 c*).

In grasping small objects, the two bonobos kept their arm more or less extended and performed pronating movements. In contrast to all other ape species, their distal thumb phalanx (and that of the hylobatids) remained flexible.

In the «adhesion grip», the carpal joint remains straight or exhibits a slight dorsiflexion. The volar surface of the tip of the index finger is pressed onto the object. The index finger is extended, digits III-V exhibit volar flexion and the dorsal surface of the distal or middle phalanges rest on the ground (*Figure 7 h*).

Picking up objects with the mouth

In 4.5% (average) of all actions the individuals of the different non-human species used their mouth to pick up food items. Individual variation ranged from 10.9% in orangutan to none in gibbons.

Hand preference

In contrast to about 90-93% right-handedness in humans (ANNETT, 1970), the non-human primates exhibited marked variation in preference. While some show an accentuated hand preference, others have a tendency towards ambidextrousness. In the species investigated, hand preference is represented by an index of handedness (IH), the ratio of left-hand preference (l) over the total number (T) of grips ($IH = l/T$), whereby $IH = 1$ indicates exclusive use of the left hand, and $IH = 0$ absolute right-hand preference (*Figure 8*) (see also BRENOT, this volume).

In an interspecific comparison, among the nineteen individuals four showed marked hand preference, two favouring the right hand (IH between 0 - 0.15) and two the left hand (IH between 0.85-1; $p = 0.001$). Ten animals were ambidextrous (IH between 0.35 - 0.65; $p = 0.001$) and the other five preferred one hand about 2/3 more often than the other hand (three the right and two the left; $p = 0.001$; *Figure 8*).

In chimpanzees, two individuals strongly preferred one hand: one of them preferred the left ($IH = 0.85$) the other the right hand ($IH = 0.15$). One animal used the left hand about 2/3 more often than the right hand, and the fourth was ambidextrous (see also BRESARD *et al.*, 1987).

In bonobos one prefered the right hand ($IH = 0.30$) and the other tended to ambidextrousness ($IH = 0.38$).

Two gorillas out of four showed strong right-hand preference ($IH = 0.07$), one used the right hand in 2/3 of the cases; one other was ambidextrous ($IH = 0.57$) (see also SCHALLER, 1963, and REDSHAW, this volume).

The results presented here for orangutans concur with BRENOT (this volume), with a mean value of $IH = 0.51$. They showed more even usage of hands than the other species. Two out of four gibbons were ambidextrous; one used the left hand more often than the right and one female H. lar moloch showed strong left hand preference ($IH = 0.84$) (see also CHIVERS (1974) for *H. syndactylus*).

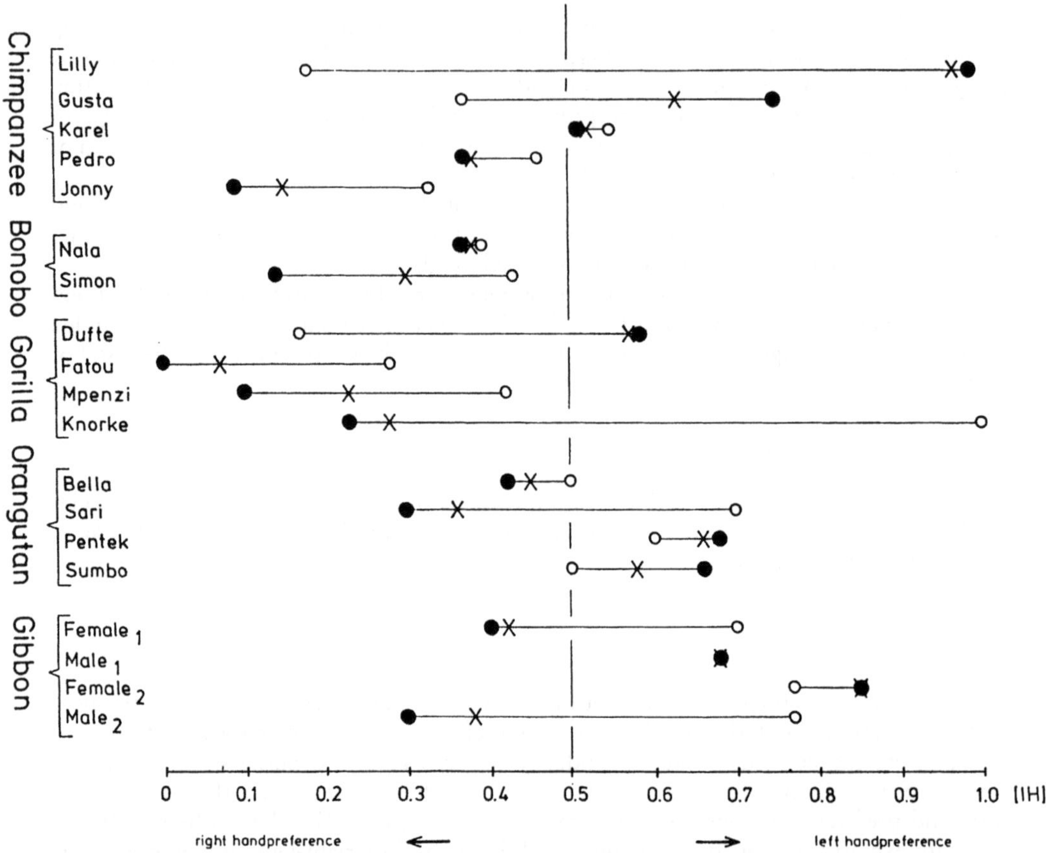

Figure 8 - Index of hand preference (IH) according to precise grasping techniques. Chimpanzee, bonobo, gorilla, lar gibbon, moloch gibbon, orangutan [1 = left hand preference, 0 = right handedness, between 0.25-0.75 = defined as being both handed. (* mean of all grips; • mostly used grips (mean out of 80% of all grips); o rarely used grips (mean out of 20% of all grips)]. Gibbon 1 = *lar*; gibbon 2 = *moloch*.

Preferred grasping techniques and hand preference

A striking characteristic of one female chimpanzee, «Lilly», was the overwhelming dominance of one hand and one preferred grasping variant. She showed the strongest hand preference out of all individuals. Moreover, in 99% of the cases she applied the same grip (Id-IId) with the preferred hand. In the rare cases in which she changed grasping technique, she also used the alternate hand. So the question is raised, whether or not hand preference is more accentuated in preferred grips than in other grips. Hand preference in the commonest grasping technique (mean of 80% out of all grips used) with 4.7% is more accentuated than in the mean of all actions. In seven cases out of nineteen (three chimpanzeees, two gorillas, one bonobo and one female of *H. l. moloch*), this preponderance amounts to 53%. However, hand preference in the other grasping techniques differs from that in the most preferred grips by about 25%. That means some individuals constantly changed their hand in less preferred grips.

Discussion

Precision grip

The presentation of very small, but attractive pieces of food, which was first tested by SPRANKEL (1964; published 1984), unveiled dexterity and therefore the use of precise grasping techniques. In all hominoidea, with the exception of bonobos and the gibbons, the grip thumb tip and index tip was observed. The way in which this grip was used by the pongids did not differ from the same grip in humans, as far as the technique and standards of precision grasping (speed, success of the initial grasp and tight hold) are concerned. That the apes so frequently and elaborately grasped with the thumb-tip against the index-tip in picking up small objects was not to be expected from the literature. Differences between apes and humans did, however, become apparent in the process of grasping.

Humans execute manipulative movements, flexion and extension, between picking up very small pieces and coming to hold them securely; these movements are employed to position the object securely and tightly between the finger pads. In apes, on the other hand, no «intermediate movement» appeared to take place between first picking up the object and the fully secure grip. The grasping movements of the thumb and the index finger thus appear mechanical. LANDSMEER (1962) writes that the concept of «precision grip» (NAPIER, 1960) tends more to designate a static final state. As he notes, it should be termed «precision handling» due to the manipulative movements involved (see also LONG *et al.*, 1970). In accordance with our observations, this designation might be applicable for humans, but for pongids NAPIER'S definition of the «precision grip» characterises the process of grasping better.

The frequently cited reason for an alleged insufficient «precision grip» of the pongids is seen by and large in a too low thumb-forefinger index. However, gorillas and gibbons are the species having the highest thumb-forefinger index amongst the hominoids, while the orangutans have the lowest (SCHULTZ, 1930). Gorillas and orangutans applied a high rate of precise grasping with tip of the thumb and tip of the index finger, while gibbons did not use this technique at all. So from this study one could conclude that a low thumb-forefinger index does not necessarily entail a limitation on this type of grip, nor does a high index necessarily favor it.

For the grip in which the distal pad of the thumb and the index finger are used, Idvo-IIdvo («pad-to-pad», as per NAPIER, 1960), a greater mobility of the distal phalanges, in particular a dorsiflexion (see *Figure 6, f.g.h.*), is required which enables humans to press the entire surfaces of these phalanges against one another. Pongids are not able to execute this grip, since the passive hyperextension of the distal phalanges is restricted by relatively short deep flexor tendons (PREUSCHOFT, 1965; TUTTLE, 1967), which seem to prevent this type of grip.

The lack or the presence of M. flexor pollicis longus, which flexes the distal thumb phalanx, has also been discussed as a 'diagnostic feature' for the precision grip (SUSMAN, 1988). The development of this muscle differs individually in the ape species; its tendon can extend up to the middle of the distal phalanx or insert more proximally (PREUSCHOFT, 1965; MAIER, 1971). Most of the hylobatids possess this muscle (JOUFFROY & LESSERTISSEUR, 1960), and they can also flex the distal thumb phalanx, but do not use this ability to execute the precision grip with thumb tip and index tip. As yet unsystematic observations show however, that their thumb is very mobile in opening pumpkin seeds. The two bonobos were also able to flex the distal thumb phalanx, but they do not employ the grip with the thumb tip against index tip either. It must be assumed that they are not less able to execute this grip than chimpanzees, with whom they share a high degree of anatomical similarity (STERN & SUSMANN, 1983) and that there are other reasons why this type of grip was not used. It remains doubtful whether one muscle on its own can be responsible for a complex function like flexing actively the distal thumb phalanx (TUTTLE, 1967; MAIER, 1971). For this reason and on the grounds of the present findings, it is questionable whether the M. flexor pollicis longus can be used as a reliable indicator for the existence of an applied precision grip in fossil hominids, as is discussed by SUSMAN (1988).

All the individuals of the various species tested exhibited, more or less markedly, individual preferences for different grips. The body posture in taking up objects was clearly distinguished in the different grasping variants. This was particularly striking in the chimpanzees. The thumb tip against index tip grip required a high degree of visual control, the head was positioned very close to the ground surface in a posture which appeared myopic (see also SCHALLER, 1976). Picking up objects with the thumb tip and the radial aspect of the index finger was usually done without prolonged visual control, the object was located, and as soon as the hand was positioned beside it, the head was again directed toward the environment. Characteristic body posture for these two grips were pronounced in two of the chimpanzees, «Lilly» and «Pedro». Lilly, who executed grips almost exclusively with thumb tip and index tip, hardly directed her attention to her environment, she seemed to be absorbed in her task, while Pedro, possibly due to his dominant behavior, hardly took his eyes from the group and selected exclusively the grip at the radial aspect of the index finger. This grip, in addition, as executed with a markedly pronated and streched arm. So there is, of course, a relationship between ongoing social behavior, and grasping behaviour.

Hand preference

Extensive discussions of the evolutionary significance of handedness and cerebral asymmetry in non-human primates are provided by WARREN (1980), MACNEILAGE *et al.* (1987) and BRENOT (this volume).

These previous studies are largely confirmed by the present paper. Apes, some cercopithecoid monkeys and species of prosimians show a pronounced tendency to prefer one hand. This preference follows a bimodal U-shaped distribution and exhibits task-specific and time-dependent fluctuations. Either the right hand or the left hand may be preferred, with a variable percentage of animals without any preference (FINCH, 1941; LEHMANN, 1971; FISHER et al., 1982; BRESARD et al., 1987). However, they are also able to master complex grasping tasks with both hands. This may well be because, due to their mode of life they, as TODT *et al.* (1981) stated for prosimians, «cannot afford an extreme specialization of only one hand».

It may be that a first step toward specialization is the association of one particular grasping mode with one or the other of the hands (see MACNEILAGE et al., 1987, who postulates that manipulative skills favor the preference of the right hand); indeed some individuals executed the preferred grip primarily with only one hand. There was also a slight dominance of the right hand in the average of all individuals. If the hand was changed after an interruption of the movement sequence, the grasping technique was often also changed. On the other hand, this phenomenon rather might indicate stereotyped repetitive motions (as in the case of Lilly, see above), than advanced specialisation (LEHMANN, 1970). But this will be discussed in more detail elsewhere.

The preliminary evaluation of additional data, on the other hand, seems to indicate a preference of different grips under different substrate conditions. The number of graspings with thumb and forefinger tip grew in many individuals on gras and earthy ground, while on flat surfaces predominated the grasping technique in which the dorsum of the fingers rested on the ground. In this case too, more details will be given elsewhere.

ACKNOWLEDGEMENTS — For support and encouragement, for discussion and criticism of this work, I would like to thank H. Preuschoft. I also would like to thank C. Niemitz for encouraging me to investigate this topic. B. Demes and M. Günther provided critical readings of the manuscript. H. Schulze has drawn the hands, M. Wischniwetzki has drawn all the other figures. W. Dewitz, Zentraleinrichtung für Audiovisuelle Medien (ZEAM), Berlin (West), and B. Demes generously provided video equipment. P. Rahn, W Haase and K. Walter of the Zoologischer Garten, Berlin (West) provided access to the apes, and P. Knowton translated the manuscript. All I sincerely like to thank. My special thanks to L. Schachtschneider and R. Jäkl.

References

ALTMANN J., 1974. *Observational study of behavior. Sampling methods.* Behavior 49: 227-267.

ANNETT M., 1970. *A classification of hand preference by association analyses.* British Journal of Psychology. Vol. 61, 303-321.

BISHOP A., 1964. *Use of the hand in lower primates.* In: J. Buettner-Janusch (ed.) Evolutionary and Genetic Biology of Primates, Vol. 2, 133-225.

BUCK CH. & BÄR H., this volume.

BRENOT P.H., this volume.

BRESARD B. & BRESSON F., 1987. *Reaching or manipulation: Left or right?* Behavioral and Brain Sciences, **10**: pp. 247-303.

CHIVERS D.J., 1974. *The Siamang in Malaya. A field study of Primate in Tropical Rain Forest.* Contributions to Primatology. Vol. 4, S. Karger, Basel.

DAY M.H. & NAPIER J.R., 1963. *Functional significance of the head of the flexor pollicis brevis in primates.* Folia Primatologia **1**: 122-134.

FISCHER R., MEUNIER G.F. & WIRTHE D.J., 1982. *Evidence of laterality in the lowland gorilla.* Perceptual and Motor Skills, **54**: 1093-1094.

FINCH G., 1941. *Chimpanzee handedness.* Science, **94**. pp. 117-118.

GOODALL J., 1965. *Chimpanzee of the Gombe Stream Reserve.* In: DeVore (ed.) Primate Behavior, New York.

JOLLY A., 1975. *Die Entwicklung des Primatenverhaltens.* Fischer. Stuttgart.

JOUFFROY F.K. & LESSERTISSEUR J., 1960. *Les specialisation anatomique de la main chez les singes a progression suspendue.* Mammalia; **24**: 93-151.

KUMMER H., 1968. *Social organization of hamadryas baboons. A field study.* Bibliotheca Primatologia, No.6. Univ. Chicago, Press.

LANDSMEER J.M.F., 1962. *Power grip and precision handling.* Annals of Rheumatic Diseases, **21**, 141: 164-170.

LANDSMEER J.M.F., this volume.

LEHMANN R.A.W., 1970. *Hand preference and cerebral predominance in 24 rhesus monkeys.* Journ. of the Neurological Science, **10**: 185-192.

LONG CH., CONRAD P. W., HALL C. A. & FURIER M. S., 1970. *Instrinsic-extrinsic muscle control of the hand in power and precision handling.* The Journal of Bone and Joint Surgery. A., Vol., 52-A, Nr. 5.

LORENZ R., 1971. *The functional interpretation of the thumb in the Hylobatidae.* In: ed. Proceedings of the 3rd Int. Congress of Primatology, Zürich, 1970, Karger, Basel, Vol. **1**, pp. 130-136.

LEWIS O.J., 1989. *Functional morphology of the evolving hand and foot.* Clarendon Press. Oxford.

MACNEILAGE P.F., STUDDERT-KENNEDY M.G. & LINDBLOM B., 1987. *Primate handedness reconsidered.* Behavioral and Brain Sciences, **10**: 247-303.

MAIER W., 1971. *Vergleichende und funktionell-anatomische Untersuchungen an der Vorderextremität von Theropithecus gelada.* Abh. Senckenbergischen Naturforschenden Gesellschaft. Kramer, Frankfurt, Nr. 527.

MARCHANT L.F. & STEKLIS H.D., 1986. *Hand preference in a captive island group of chimpanzees* (Pan trogl.). Am. J. Prim. **10**: 301-313.

NAPIER J.R., 1952. *The attachment and function of abductor pollicis brevis.* Journal of Anatomy, London, **86**: 335-341.

NAPIER J.R., 1960. *Studies of the hands of living primates.* Proceedings Zoological Society London, **134**(4): 647-657.

NAPIER J.R., 1961. *Prehensility and opposability in the hands of primates.* In: Harris, (ed.) Vertebrate Locomotion, Symposium of the Zoological Society London, **5**: 115-132.

NAPIER J.R., 1962. *The evolution of the hand.* Scientific American, **12**: 49-115.

NAPIER J.R., 1980. *Hands.* G. Allen Unwin. London, Boston, Sydney.

NAPIER J.R. & NAPIER P.H., 1967. *Handbook of Living Primates.* New York, Academic Press.

OLSON D.A., ELLIS E.J. & NADLER D.R., 1990. Hand preference in captive gorillas, orang-utans and gibbons. Am. J. Prim. **20**: 83-94.

PREUSCHOFT H., 1965. *Muskeln und Gelenke der Vorderextremität des Gorillas.* Morphologisches Jahrbuch, **107**: 99-182.

PREUSCHOFT H., 1969. *Statische Untersuchungen an den Füßen der Primaten. Teil I: Statik der Zehen und des Mittelfußes.* Zeitschrift für Anatomie und Entwicklungs- Geschichte **129**.

PREUSCHOFT H., 1973. *Functional anatomy of the upper extremity.* In: G. Bourne, (ed.). The Chimpanzee, Vol. 6, Karger, Basel, 34-120.

RAUWERDINK P., this volume.

RAWLINS R.G., this volume.

REDSHAW M., this volume.

REYNOLDS P.C., 1973. *Handedness and the evolution of the primate forelimb.* Neuropsychologia, Vol. **13**, Pergamon Press.

ROTHE H., 1973. *Handedness in the common marmoset.* American Journal of Physical Anthropology, **38**.

SCHALLER G.B. 1963. *The mountain gorilla: Ecology and behavior.* University of Chicago Press.

SCHULTZ A.H. 1930. *The skeleton of the trunk and limbs of higher primates.* Human Biology, **2**: 303-438.

SCHULTZ A.H. 1961. *Vertebral column and thorax.* In: Hofer, Schultz, Starck, (eds.) Primatologia, Karger Basel, Vol. 4,: 1-68.

SHREWSBURY M.M. & SONEK A., 1986. *Precision holding in humans, non-human primates, and Plio-Pleistocene hominids.* Human Evolution, Vol. 1, No. 3, pp. 233-242.

SPRANKEL H., 1984. *Lagothrix lagotricha (Cebidae) — Ergreifen kleiner Objekte;* — Film E 1931 (v)/ Publ. Inst. Wiss. Film (IWF), Göttingen.

SPRANKEL H., 1984. *Papio hamadryas (Cercopithecidae) — Ergreifen kleiner Objekte;* — Film E 1932 (v)/ Publ. Inst. Wiss. Film (IWF), Göttingen.

SPRANKEL H. & LORENZ R. 1984. *Ateles geoffroyi (Cebidae) — Ergreifen kleiner Objekte;* — Film E 1797/ Publ. Inst. Wiss. Film (IWF), Göttingen.

STERN J.T. & SUSMAN R.L., 1983. *The locomotore anatomy of Australopithecus afarensis.* American Journal of Physical Anthropology, **60**: 279-317.

SUSMAN R. 1988. *The hand of Paranthropus robustus from Member 1, Swartkrans: Fossil evidence for tool behavior.* Science, **240**, pp. 781-784.

SWINDLER D.R. 1973. *Atlas of primates gross anatomy -baboon, chimpanzee, and man.* Washington Press.

TODT D., KRABERGER A. & HEERMANN P., 1981. *The lateral balancing of handedness tested in slowly and rapidly moving lorisidae.* In: Chiarelli et *al* (eds.), Primate Behavior and Sociobiology, Springer-Verlag, Heidelberg, pp. 36-42.

TUTTLE R. 1967. *Knuckle-walking and the evolution of hands.* American Journal of Physical Anthropology, **26**: 171-296.

WARREN J.M., 1980. *Handedness and laterality in human and other animals.* Physiological Psychology, **8** (3), p. 351.

M.M. Günther
C. Boesch

Energetic cost of nut-cracking behaviour in wild chimpanzees

One of the major hypotheses to explain the use of tools is the access to food sources that would otherwise not be exploitable for the animal species. The resulting assumption is that such additional food sources supply this species with a net benefit. Many reports from various sites describe a variety of tool-using behaviour in wild chimpanzees. The cracking of nuts is one of them. Here, a first attempt is made to measure the energy budget of nut-cracking behaviour due to the use of natural hammers employed by the wild chimpanzees of the Taï National Park in Cote d'Ivoire.
The calculations for the pounding of Coula nuts are mainly based on film analyses of the adolescent male 'Darwin'. Statistical data on the amount of cracked nuts, the transportation distances used for collecting nuts and/or hammers, and the duration of each activity are all based on the observations of adult chimpanzees during the Coula season. Additional data on the energetic cost of each activity have been taken from the literature and added to our calculations. As a result four major fractions related to the nutcracking behaviour contribute almost equally to the overall energy expenditure: I. nut-cracking, II. eating while sitting, III. collecting and transporting nuts, and IV. collecting and carrying nuts and hammers. By looking at the relationship of gross energy output to gross energy digestibility we calculated a ratio of 1:9. Despite of the fact that due to specific circumstantial reasons the real energy expenditure might exceed this calculated figure, our result of an energy gain which is nine times the energy output can only be affected to a negligible degree. Thus, the hypothesis on the assurance of a net benefit due to tool-use behaviour can be confirmed at least for the chimpanzees in the Tai forest. Further similar investigations should help to shed light on the evaluation of tool-use behaviour in general, and the development of traditions and its differences in various populations of chimpanzees in particular.

Key words: Biomechanics, work metabolism, impediment factor, profitability, tool-use, energy budget.

Introduction

Tool-use has been observed in many different animal species (for a review see Beck, 1980). The largest variety of tool-use is found within the chimpanzees, which have been observed in a feeding context (nut-cracking: Boesch & Boesch, 1984a, 1984b; Sugiyama & Koman, 1979; Sugiyama, 1981; termite-fishing and ant-dipping: Goodall, 1970, 1986; McGrew, 1977, 1979; McGrew *et al.*, 1979; Teleki, 1974, 1975; Uehara, 1982; wood-boring and ant-fishing: Nishida, 1973; Nishida & Hiraiwa, 1982), in an aggressive context (Albrecht & Dunnet, 1971; Goodall, 1970, 1986; Kortlandt, 1972; Sugiyama & Koman, 1979), as well as in body care (Goodall, 1970; McGrew & Tutin, 1972) or in reaching and inspecting certain objects (Goodall, 1970, 1986; Sugiyama & Koman, 1979). Some of these tool-uses can be important activities and yield a large amount of the chimpanzee's diet (according to BOESCH & BOESCH, 1984 b, the nut-cracking behaviour can be performed for 1 hour and 30 min. per day and yield 10953 kJ per day). Such a variety and importance of tool-use in a non-human species has challenged the idea of 'Homo faber', being the only animal using and making tools (LEAKEY & LEWIN, 1977; JOHANSON & EDEY, 1981).

Different hypotheses have been forwarded to explain the use of tools: The major proposition being that tool-use allows the access to food sources that would otherwise not be exploitable for this animal species (ALCOCK, 1972; BECK, 1980; PARKER & GIBSON, 1977). Many observations on chimpanzees clearly illustrate this and it remains to prove that through tool-use such access to an unexploited food source assures that species with a net benefit. If this would not be the case, it is difficult to explain how natural selection could favour such a behaviour.

In this paper we shall present a first attempt to measure the cost and the benefit of nut-cracking behaviour due to the use of natural hammers employed by the wild chimpanzees of the Taï National Park in Cote d'Ivoire.

Methods

The chimpanzees of the Taï National Park are studied since September 1979 and have been habituated to human observers without the use of any artificial provisioning (see BOESCH & BOESCH, 1981, 1983, 1984a, 1984b, for more details on the habitat, the habituation process and the observational methods). Taï chimpanzees live in a tropical rain forest and use clubs and stones to crack five different species of nuts.

For the purpose of the present calculation, we could collect enough data only for one of the nut species: *Coula edulis*. It is the easiest to crack of the five species and is pounded with wooden clubs in 92% of the observations (BOESCH & BOESCH, 1983). Coula nuts are very abundant and eaten daily during 4 months of the year.

As it was impossible to collect all the necessary data with one single individual, we proceeded as follows:

— Detailed pounding measures were made with a 16 mm-colour-film on 'Darwin', at that time a late adolescent male *(Figure 1)* and on 'Le Chinois', an adult male. The visibility in the Taï primary tropical rain forest is about 20 m, a distance which is by far too short a wild chimpanzee would tolerate an observer. Only after 2½ years of habituation, in January 1982, the observation of nut-cracking chimpanzees was undisturbed (BOESCH & BOESCH, 1984b). From 1983 - 85 some 16 mm-colour-films at 24 f/s have recorded the wild chimpanzees while pounding the nuts at various 'ateliers'. The scenes were taken with a 20× zoom lens at a distance of about 12 m. Altogether about 70 sequences of pounding a hammer on a nut have been taken into consideration for further analysis, depending on the overall light conditions. The most important factors for the final selection of sequences was first, the minimization of probable errors and second, the continuity of the sequence, that is, all hits on the nut until its single kernel was taken out of its shell.

The requirements essential for a reliable kinematic analysis was reached only in one individual who pounded a hammer 15 times on three different nuts. 'Darwin' used a 2.3 kg wooden hammer in order to crack Coula nuts on a ground anvil. Altogether three nuts have been cracked by 'Darwin' (4 hits on a nut in sequences no. 1-4, 6 hits on a nut in sequences no. 5-11, 4 hits on a nut in sequences no. 12-15). The cracking of one Coula nut

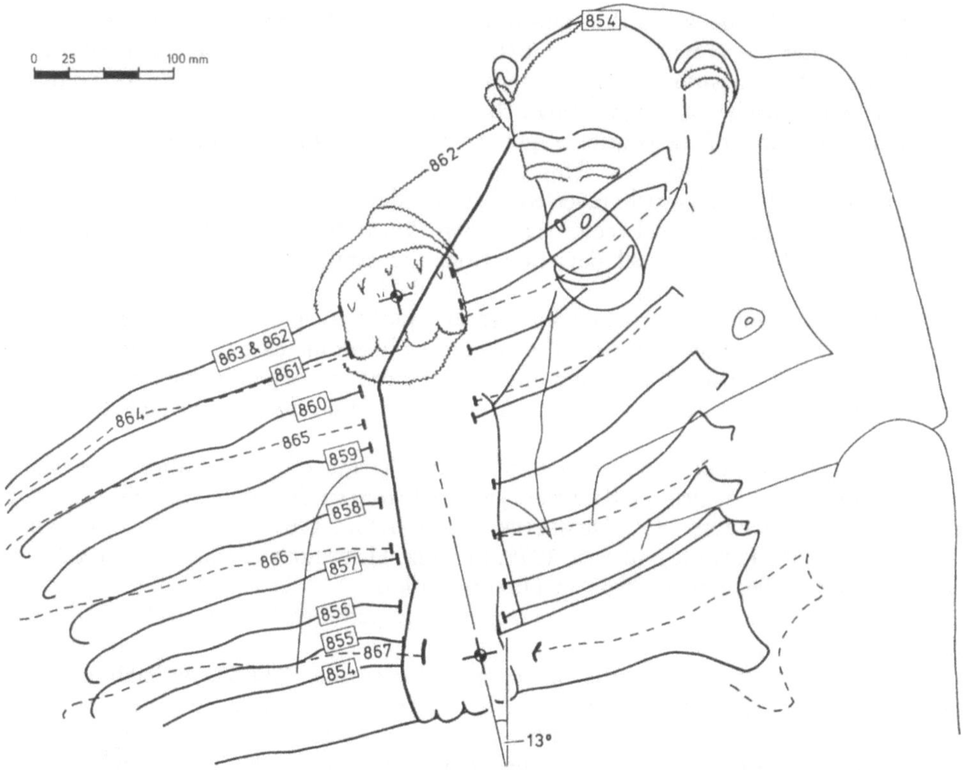

Figure 1 - Outline-drawings of 'Darwin' cracking a Coula nut, to determine the hammer's velocity at impact on the Coula nut (sequence 15).

Frame-no.	854		Position of 'Darwin' at the beginning of hammer lifting
Frame-no.	862	& 863	Highest position (height = 0.28 m) of raised hammer
Frame-no.	864	866	Downwards pushing of hammer
Frame-no.	867		Shortly after moment of impact [for estimation of velocity *at* impact (= – – –X in *Figure* 2) see text]

done by 'Le Chinois', an adult male, was also analysed (2 hits in sequences no. 16-17) as a control. The heavy wooden hammer (5.0 kg) in this case was raised to a lesser height. Due to this fact, the error was larger than in the analyses of 'Darwin' and subsequently not used for the calculation of the energy balance. The calibration of lenghts is based on meter-scales which were positioned at the sites where the chimpanzees had been sitting. A 'Photo Optical Data Analyzer' (Modell 224A/MK5, L-W International, Athena Co.) and a system of three mirrors for deflection gave an optimal 46× enlargement of the motion picture, which depicted the chimpanzee as large as 1 : 2.5 of his actual size. All 17 sequences, each consisting of at least 15 frames, have been analyzed in detail by tracing the outline of the chimpanzee at the moment before lifting the hammer upwards and again when the hammer was at its highest point *(Figure 1)*. For each frame the upper outline of the hammer and the position of the hand was drawn in order to determine the point of reversal and its change of height above the nut.

All displacements of the centres of gravity of hammer and arm segments have been treated as translatory movements. This was based upon the analysis of a nut-cracking sequence in lateral view, which revealed that the deviation between the direction of movement and the gravitational vector is less than 15°. In the frontal plane this angle also never exceeded 13°. This small deviation from the straight line could be neglected, because the true distance follows the cosine of the deviation which is no less than 0.9659 at 15°. Since the hammer was pushed downwards rapidly, the moment of and the velocity at impact had to be extrapolated from the preceeding linear increase in velocity (cf. *Figure 2*). The outline of the back of the hand one frame before and one after impact (extension versus degree of dorsiflexion) gave a valuable additional indication and certainly increased the precision.

— Detailed anatomical and physiological data relevant to the calculation were taken from the literature, in which the data were gathered from studies on captive chimpanzees (e. g. BRUHN & BENEDICT, 1936; DEMPSTER & GAUGHRAN, 1967; TAYLOR & ROWNTREE, 1973; PREUSCHOFT, FRITZ & NIEMITZ, 1979; SPITZER, HETTINGER & KAMINSKY, 1982; for more details see the various steps in our calculation).

— Average data on the Coula nut-cracking were gathered for a typical day. Except when mentioned otherwise, all data used for the present study are unpublished and collected with identified individuals during the Coula season 1984-85. As habituation of the animals had progressed, these individuals were not disturbed by the presence of C. and H. Boesch when nut-cracking and both investigators could follow their movements between different nut-cracking sites: Therefore both factors, the duration of nut-cracking at one site and the transportation distances of hammers and/or nuts, are more representative than earlier observations with non-habituated individuals. For the present calculation we selected those individuals that cracked nuts without being disturbed in their activity neither by joining the chimpanzees nor by the group itself leaving a nut-tree region. However, we suspect that the final result still underestimates the real chimpanzees performance.

Results

The body mass of the adolescent male 'Darwin' was estimated to be 35.0 kg (criteria for estimates in GOODALL, 1968, 1975). Additionally, the mass of each arm segment had to be known in order to determine the kinetic energy at the moment of impact. Because of the large quantity of hits per day (on average: 6.71 hits/nut and 274 nuts/day = 1838 hits/day) the insertion of reasonable figures into the calculations was important. From two different studies we know of just two specimens of *Pan troglodytes* which had been analyzed to determine the mass distribution of various body segments. The results, however, differ considerably. ZIHLMAN (1984) dissected one adult female in which the mass of one arm was just 8% of the total body mass. As for PREUSCHOFT, FRITZ & NIEMITZ (1979) and PREUSCHOFT (personal communication) the arm of an adult male constituted 11% of the total body mass (upper arm 5.1%, forearm 3.9%, hand 2.0%). Applied to the size of 'Darwin' this divergence leads to a difference of at least 1 kg. We

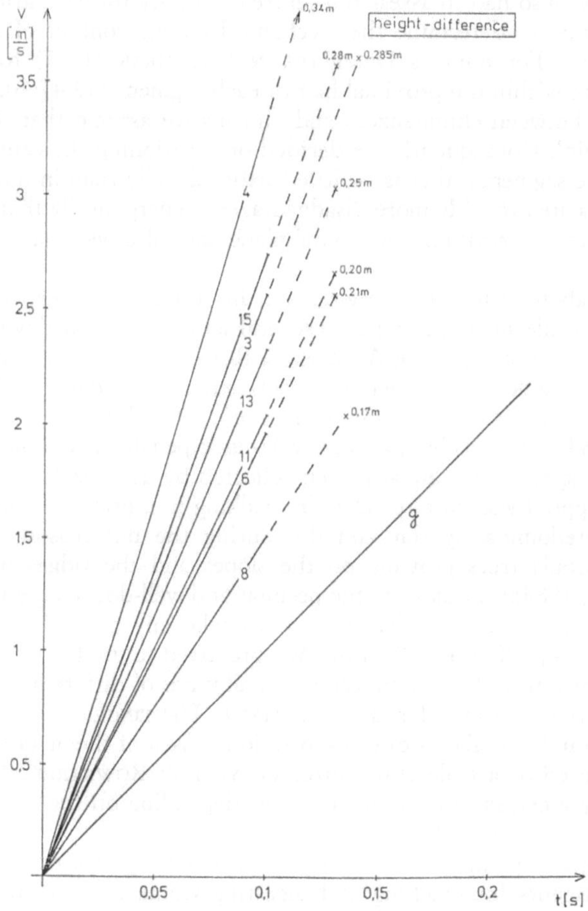

Figure 2 - Seven sequence-curves as examples for the estimation of the hammer's velocity at impact on Coula nuts.

g = velocity for gravity

——┤ = calculated velocity based on data from outline-drawings (cf. *Figure 1*)

- - - -X = extrapolated moment of impact with height to which the hammer was raised

v g = additional acceleration a ($^m/_{s^2}$)

decided to insert the figures of Preuschoft, Fritz & Niemitz (1979), because 'Darwin' is also a male, and at least part of the difference might be attributed to a greater muscular mass as compared with the arm of a female chimpanzee. Nonetheless, both studies agree on the distribution of mass within the arm by obtaining a ratio of arm : upper arm : fore-arm : hand to about 6 : 3 : 2 : 1.

Furthermore, we also had to assess the centre of gravity for each arm segment. This was necessary in order to determine the mechanical energy content of the arm system during its movement. For humans DEMPSTER & GAUGHRAN (1967) found that these centres of gravity are within the proximal half of each segment (42-47%). Because of the muscular similarity between chimpanzees and humans we assume that these results are applicable to 'Darwin'. Consequently we decided on positioning the centre of gravity in the middle of all the segments; that is, a little further distally than in humans, accepting that our calculations are probably more disadvantageous energetically than in reality. The position of the centres of gravity in the frontal plane have also been based on the analysis of the lateral plane.

The basal metabolic rate for 'Darwin', and his total energy expenditure for nut-cracking and eating while sitting, depend on several sources concerned with human sports and work physiology (NÖCKER, 1976; ASTRAND & RODAHL, 1977; SPITZER, HETTINGER & KAMINSKY, 1982; ROHMERT & RUTENFRANZ, 1983) and on some data for chimpanzees and other primates (BRUHN & BENEDICT, 1936; DALE *et al.*, 1970; MÜLLER, 1985). Data on the gross energy expenditure for chimpanzees walking bipedally and quadrupedally on a treadmill at various speeds have already been collected by TAYLOR & ROWNTREE (1973). Again, we had to apply these data to 'Darwin' walking in a primary tropical rain forest, which in fact is predominantly flat and dry during the nut season, but consists of undergrowth and Coula trees growing on the slopes and the ridges between streams (BOESCH & BOESCH, 1984a). Because of the peculiar and well-documented situation in the case of the Tai chimpanzees, we decided not to take any general formula published elsewhere (e.g. CALDER, 1984: pp. 197-198). We preferred to use the extensive investigations on humans which include data on walking in all kinds of different environments and situations (e.g. carrying various loads; cf. SPITZER, HETTINGER & KAMINSKY, 1982). Finally, our results on the total cost of transportation consist of the metabolic power input at the calculated speed of a male chimpanzee (TAYLOR & ROWNTREE, 1973; HEGLUND, 1985) multiplied by a certain 'impediment factor' depending on the specific task of the animal respectively (cf. *Figure 4: steps «5»-«7»*).

A male chimpanzee spends 137 minutes per day on average nut-cracking. The total or gross energy investments for cracking and digesting Coula nuts at an 'atelier' can be divided into four major fractions (cf. *Figure 4*: *step «8»*):

I. The total amount of energy necessary for actually cracking the nuts ($GE_{\Sigma Ncr}$).

II. The total amount of energy required for eating while sitting (GE_{es}). Both, $GE_{\Sigma Ncr}$ and GE_{es}, were done over a period of about 92 min. (i.e. 67% of the total time).

III. The gross energy requirements for carrying the nuts to an 'atelier' (GE_{Nct}) during 18 min. (i.e. 13% of the total time).

IV. The gross energy requirements for carrying both, hammers and nuts, together to an 'atelier' (GE_{NcHt}) during 27 min. (i.e. 20% of the total time).

Two different approaches for the average daily total energy output on nut-cracking have been pursued: The data of fraction I «nut-cracking» and parts of fraction II «sitting while eating» were taken from the analyses of the one individual 'Darwin'. As for fraction III and IV «cost of transport», all data are based upon the statistical analyses of a large number of field observations. Nonetheless, even for these fractions the necessary specific data on the individual's level (e.g. absolute basal metabolic rate, oxygen uptake) were applied to 'Darwin'. After all, we consider the results of the nut-cracking energy budget on 'Darwin' to represent the lowest energetic surplus margin for all the other adult chimpanzees in the Taï forest.

Figure 4 shows at various steps the calculation of different energy components which need to be considered for 'Darwin's' total energy output of nut-cracking per day.

Fraction I. «Nut-cracking»

The linear distribution of speed in seven nut-poundings *(Figure 2)* reveals the general tendency that the greater the height to which the hammer is raised, the higher is the acceleration added to gravity. For further analysis we decided to choose one pounding on the Coula nut which is intermediate in the power output (No. 15 in *Figure 2*).

Step «1»

Before calculating the total amount of 'invested' muscular work ($W_{musc.}$), it is necessary to determine the accelerating forces for each segmental mass of the analysed system (H = hammer, RH = hand, RF = forearm, RU = upper arm). These forces (F_s) are the result of the multiplication of each segment's mass (m_s) times its absolute acceleration (a_s), subtracting the effect of gravity:

$$F_s = m_s \cdot (a_s - g)$$

The shoulder is the centre for the rotation of the 'rigid' arm. Therefore, the absolute accelerations of the arm's proximal segments are proportional to the relative distances from their centre of gravity (COG) to the shoulder. As a result of the described analysis we know, that this absolute acceleration of the hammer (a_H) is 27.3 m/s². The absolute accelerations of the other segments need to be calculated as fractions of this value. The factor 'f' for this is the following ratio:

$$f = \frac{\text{distance COG of segment to shoulder}}{\text{distance COG of hammer to shoulder}}$$

Subtracting the earth accelaration 'g' from these values leads to the segment's acceleration produced by muscular work. Finally, the multiplication by the segment's masses completes the calculations of the forces necessary to produce these accelerations:

$$F_s = m_s \cdot (f \cdot a_H - g) = m_s \cdot (a_s - g)$$

$$F_H = 2.3\,\text{kg} \cdot (1 \cdot 27.3\,\text{m/s}^2 - 9.8\,\text{m/s}^2) = 40.25\,\text{N}$$

$$F_{RH} = 0.7\,\text{kg} \cdot (0.96 \cdot 27.3\,\text{m/s}^2 - 9.8\,\text{m/s}^2) = 11.5\,\text{N}$$

$$F_{RF} = 1.4\,\text{kg} \cdot (0.46 \cdot 27.3\,\text{m/s}^2 - 9.8\,\text{m/s}^2) = 3.8\,\text{N}$$

$$F_{RU} = 1.8\,\text{kg} \cdot (0.25 \cdot 27.3\,\text{m/s}^2 - 9.8\,\text{m/s}^2) = -5.4\,\text{N}$$

Each force multiplied by the corresponding linear displacement of its centre of gravity yields its mechanical work:

$$W_{H} = 40.25\ N \cdot 0.28\ m = 11.3\ J$$
$$W_{RH} = 11.5\ N \cdot 0.268\ m = 3.1\ J$$
$$W_{RF} = 3.8\ N \cdot 0.13\ m = 0.5\ J$$
$$W_{RU} = -5.4\ N \cdot 0.07\ m = -0.4\ J$$
$$W_{musc.} = 14.5\ J$$

In order to find out the overall kinetic energy investment in nut-pounding (Np), we have to add the work invested in potential energy when lifting both, hammer and arm. This energy is calculated as the product of segmental mass, earth acceleration, and height difference between the segment's centre of gravity before and after the lifting.

$$E_{pot(H)} = 2.3\ kg \cdot 9.8\ m/s^2 \cdot 0.28\ m = 6.3\ J$$
$$E_{pot(RH)} = 0.7\ kg \cdot 9.8\ m/s^2 \cdot 0.268\ m = 1.8\ J$$
$$E_{pot(RF)} = 1.4\ kg \cdot 9.8\ m/s^2 \cdot 0.13\ m = 1.8\ J$$
$$E_{pot(RU)} = 1.8\ kg \cdot 9.8\ m/s^2 \cdot 0.07\ m = 1.2\ J$$
$$E_{pot(Np)} = 11.1\ J$$

The work invested in one hit for cracking a nut, that means nut-pounding, is equal to the amount of potential energy and the kinetic energy added by muscular work (cf. *Figure 3*):

$$E_{kin(Np)} = E_{pot(Np)} + W_{musc.} = 11.1\ J + 14.5\ J = 25.6\ J$$

An alternative way of calculating the amount of kinetic energy also rests on the interpretation to consider the whole arm as a rigid system: The velocity of the hammer at impact is 3.55 m/s. Thereby the velocities of the other segments may be calculated as fractions 'f' (see above) of this value:

$$V_{H} = 3.55\ m/s$$
$$V_{RH} = 3.55\ m/s \cdot 0.96 = 3.41\ m/s$$
$$V_{RF} = 3.55\ m/s \cdot 0.46 = 1.63\ m/s$$
$$V_{RU} = 3.55\ m/s \cdot 0.25 = 0.89\ m/s.$$

By using these calculated values, the kinetic energy ($\frac{1}{2}\ mv^2$) of the system 'right arm + hammer' just before impact is equal to:

$$E_{kin(Np)} = 1/2 \cdot [2.3\ kg \cdot (3.55\ m/s)^2 + 0.7\ kg \cdot (3.41\ m/s)^2 + 1.4\ kg \cdot (1.63\ m/s)^2 + 1.8\ kg \cdot (0.89\ m/s)^2] = 21.2\ J$$

The difference between the values resulting from these two alternative calculations may have two main reasons:

1) We do not exactly know, what is happening between the situations described by the four or five frames of the analysed film sequences (cf. *Figure 1*). Therefore, the duration of the hits may be over- or underestimated. If the duration of one hit is not 'four frames' but 'five frames', then the invested energy changes to 21 J, that is about the same value as calculated for the kinetic energy.

2) The analysed system is not ideally rigid.

To be sure not to underestimate the energy necessary for the cracking of a nut, we shall further use the greater value (i.e. $E_{kin(Np)} = 25.6$ J) out of those computed above.

Step «2»

We have the crucial problem of determining the mechanical efficiency (ME), which is the quotient of the external (physical) work performed and the extra energy production expressed as a percentage (ASTRAND & RODAHL, 1977). Because of its wide range it is difficult to give a good estimate on mechanical efficiency: during bicycling the ME is about 27%, which is close to the maximal ME of 35% in an isolated muscle, whereas in the human industrial worker the ME is 5% or even less (ULMER, 1985; STEGEMANN, 1984). For a man kneeling, the metabolic cost of working with a 2 kg-handhammer is 125 J/hit at 34 hits/min. (cf. SPITZER, HETTINGER & KAMINSKY, 1982: p. 83). If directly applied to 'Darwin' this would yield a ME of 20% (25.6 J/'D'-hit versus 125 J/man's-hit). But this interspecific comparison does not take into consideration that the *absolute* arm mass of a man (100% m_B = 75 kg : arm mass 4.8% m_B = 3.6 kg; DEMPSTER & GAUGHRAN, 1967) is still about 7% lighter than that of 'Darwin' (100% m_D = 35 kg : arm mass 11% m_D = 3.9 kg). This difference is even greater in the more distal arm segments which decreases the amount of kinetic energy for a man hammering. The peak frequency of pounding Coula nuts in adolescent male chimpanzees like 'Darwin' is less than 14 hits/min. (BOESCH & BOESCH, 1984b: 6.71 hits/nut, 2 nuts/min.). This is only about $^2/_5$ if compared to the human subject. Kneeling as opposed to sitting almost doubles the amount of energy expenditure in humans (MAINZER, 1983: fig. 5.20).

In general, even though the pounding of nuts is an energy consuming discontinuous movement, the chimpanzees do it in a fairly relaxed sitting position, while often eating the previously cracked nuts. The arm with the hammer is never completely extended or raised above shoulder-height. In conclusion, by considering all the pros and cons of this interspecific comparison (frequency of hits, posture, mass of arm and hammer) an assumed mechanical efficiency of 16% is a low estimate. It means that the total energy output for pounding or hitting a nut is about six times higher than the invested kinetic energy ($E_{kin(Np)}$: E_{Np} = 25.6 J : 160 J). This ratio (resp. ME = 16%) is acceptable (ULMER, personal communication) and subsequently used, despite a ME of 20% in the interspecific comparison of man and 'Darwin'. The lower the mechanical efficiency the higher the overall energy expenditure. By using a value of ME = 16% we are sure not to underestimate such kind of energy investment. Thus, the total energy required for each nut-pounding E_{Np} is 160 J or 0.160 kJ (*Figure 3*).

Step «3»

The highest percentage of pounding frequency for adolescent male chimpanzees to open a Coula nut is 6.71 hits/nut. This figure is much above the average of our frame-by-frame analyses of 'Darwin' (5 hits/nut). It means that in these cases 'Darwin' was more efficient than expected from the overall statistical data. Using the statistical figure of 6.71 hits/nut therefore results in a safe over-estimation on energy output. Thus, the energy of pounding a nut 6.71 times is equivalent to the energy for cracking one nut:

$$E_{Ncr} = 6.71 \times 0.160 \text{ kJ} = 1.074 \text{ kJ}.$$

Step «4» (cf. Figure 3)

On average a male chimpanzee cracks 274 Coula nuts a day, or in other words, a chimpanzee needs 274 × 1.074 kJ = $E_{\Sigma Ncr}$ = 294.2 kJ in order to crack all the nuts he wants to eat during a single day.

Basal * Metabolic Rate (Pan troglodytes)	adult ♀♀ [m = 38.5 kg]	4657 kJ/day (121 kJ/kg·day)
	adult ♂♂ [m = 42.2 kg]	4813 kJ/day (114 kJ/kg·day)
	'Darwin' ♂ (estimate from*) [m = 35 kg]	4511 kJ/day (129 kJ/kg·day)
Thermal Neutral Zone *		20–29 °C

Fraction-No.	Step.-No. of Calculation		
I.	<1>	$W_{musc.}$	14.5 J
		$E_{pot\,(N_p)}$	11.1 J
		$\left[\frac{m_H}{2} v_H^2 + E_{kin\,(A)}\right.$	$\left.21.2\ J\right]$
	<2>	$E_{kin\,(Np)}$	25.6 J
		ME (mechanical efficiency)	16 %
	<3>	E_{Np} (× 6.71)	0.160 kJ
	<4>	E_{Ncr} (× 274)	1.074 kJ
		$E_{\Sigma Ncr}$	294.2 kJ
	<8>	$GE_{\Sigma Ncr}$	390.2 kJ
II.	<3>	$P_{es} \cdot t$	3.44 kJ/min.
		E_{es}	212 kJ
	<8>	GE_{es}	404 kJ

* Bruhn & Benedict (1936)

Figure 3 - General physiological data on chimpanzees and 'Darwin's' average energy output for nut-cracking (fraction I) and eating while sitting (fraction II) in a day.

Step «8»

For the calculation of the gross energy requirements (ΣGE_{output}; see *Figure 4*: *step «9»*) it was important to add 'Darwin's basal metabolic rate (BMR) during nut-cracking and eating while sitting. Fortunately we are able to use data which «are the best approximation of the basal metabolic rate in this species» (Dale *et al.*, 1970: p. 101). The results of Bruhn & Benedict (1936) give the BMR of two male chimpanzees with a body mass of

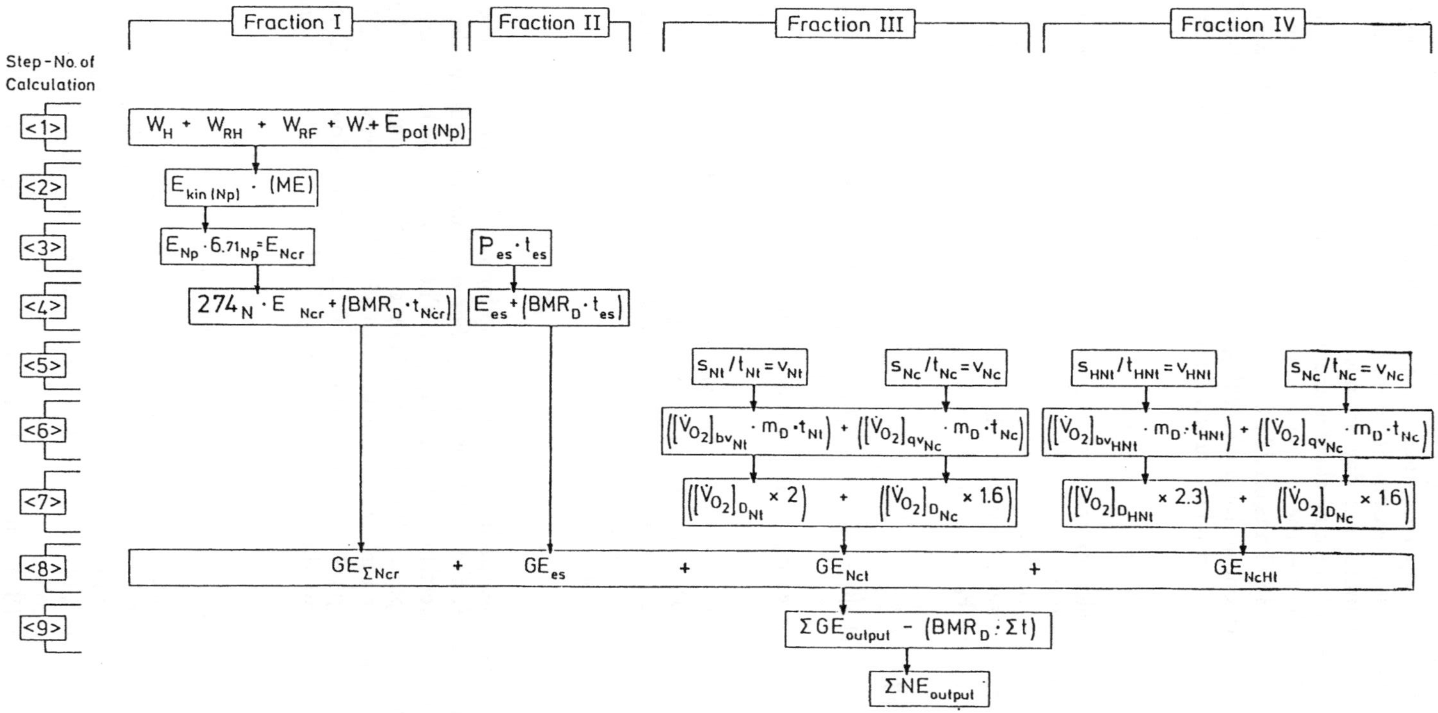

Figure 4 - Schematic representation for calculating the average total energy output due to nut-cracking behaviour of 'Darwin' in a day (for results of fraction I and II see *Figure 3*, for fraction III and IV see *Figure 5*; see lists for symbols and abbreviations).

33.2 kg and 37.1 kg = 129 kJ/kg/day on average, i.e. about 4500 kJ/day for 'Darwin'. The problems involved in applying these experimental values to certain ecological conditions will be discussed later, because it does not change our results significantly for the short period of 137 minutes per day at most.

The net time for nut-cracking alone is about 30:30 minutes. It is based on 'Darwin's sequence-no. 1-15. In all cases the hits on a Coula nut are done within one second. It sums up to 30:38 minutes for 1838 hits on 274 Coula nuts and amounts to a $BMR_D \cdot t_{Ncr}$ of 96 kJ. Added to the energy output for nut-cracking $E_{\Sigma Ncr}$ = 294.2 kJ, we get the daily gross energy output of 'Darwin' spend on nut-cracking during this period $GE_{\Sigma Ncr}$ = 390.2 kJ.

Fraction II. «Eating while sitting», step «3»

Now we have to calculate the work metabolism for eating while sitting (P_{es}) during 61 : 22 minutes, because the whole period for fraction I and fraction II is 92 minutes. In humans the P_{es} amounts to an additional 106% on top of the basal metabolism (data taken from SPITZER, HETTINGER & KAMINSKY, 1982). We inserted 110% in case of 'Darwin' to result in his complete $P_{es} \cdot t_{es}$ or E_{es} of 212 kJ.

Steps «4» and «8»

If the work metabolism for eating while sitting (P_{es}) is added to the corresponding $BMR_D \cdot t_{es}$ we come to 212 kJ + 192 kJ = 404 kJ = GE_{es}, i.e. 'Darwin's' total amount of energy required for eating while sitting during 61 : 22 minutes.

General statements concerning the calculations for *fraction III and IV «cost of transport»*: TAYLOR & ROWNTREE (1973) have analyzed the cost of locomotion in bipedal and quadrupedal walking chimpanzees on a treadmill, measuring the steady-state oxygen consumption ($mlO_2/g \cdot h$) at various running velocities (km/h). Their results can be expressed as a linear equation of the form

$$\mathring{V}_{O_2}/m_B = \text{intercept} + \text{slope} \cdot \text{velocity}.$$

where V_{O_2}/m_B is the mass-specific oxygen consumption rate and intercept the extrapolated zero-velocity cost (cf. also HEGLUND, 1985). For quadrupedal running the slope was 0.25 ml $O_2/g \cdot km$ and intercept at 0.79 ml $O_2/g \cdot hr$. The slope of bipedal running was 0.23 ml $O_2/g \cdot km$ and intercept at 0.90 ml $O_2/g \cdot hr$ (TAYLOR & ROWNTREE, 1973: fig. 1). The 'equivalent of energy' for 1 ml oxygen uptake ($\mathring{V}_{O_2}$) with the least error is 20.2 J assuming an average metabolic respiratory quotient (RQ) of 0.82 (for discussion of RQ see ULMER, 1985). To avoid confusion, the data for the calculation of fraction III and IV are given in kilojoules only.

Fraction III. «Collecting and transporting nuts», step «5» (cf. Figure 5)

Out of 274 Coula nuts brought daily to an 'atelier', 176 nuts have been carried in about 29 runs (i.e. 6 nuts/run) over an average distance of 7 m in altogether 18 minutes: 29.3 (176 nuts/6 nuts) $\cdot$ 7 m = 205 m for s_{Nt}. However, the chimpanzee has to cover this distance about two times, because he has to walk (usually quadrupedally) from his 'atelier' to the place where he collects his nuts: 0.205 km (s_{Nt}) + 0.205 km (s_{Nc}) = 0.410 km. Walking 0.410 km in 18 minutes yields an average speed (v_{Nt}; v_{Nc}) of 1.37 km/hour.

Step.-No. of Calculation		Fraction III			Fraction IV		
<5>		$\Sigma_{\text{Coula nuts (N)}}$	176 pc.		$\Sigma_{\text{Coula nuts (N)}}$ $\Sigma_{\text{Hammers (H)}}$	98 pc. 18 pc.	
		$1s_{Nt\ (6\ nuts)}$	7m		$1s_{HNt}$	26.97m	
		$\Sigma s_{Nt} + \Sigma s_{Nc}$	2×205m = 410m		$\Sigma s_{HNt} + \Sigma s_{Nc}$	485.46m	
		t_{Nc}	14:46 min.		t_{Nc}	22:08 min.	
		t_{Nt}	3:14 min.		t_{HNt}	4:52 min.	
		$v_{Nt} \triangleq v_{Nc}$	1.37 km/h		$v_{HNt} \triangleq v_{Nc}$	1.08 km/h	
collecting	<6>	* $[\dot{V}_{O_2}]_q + [\dot{V}_{O_2}]_{v_{Nc}}$	22.9 kJ/(kg·h)		* $[\dot{V}_{O_2}]_q + [\dot{V}_{O_2}]_{v_{Nc}}$	21.4 kJ/(kg·h)	
		$\times m_D$	13.3 kJ/min.		$\times m_D$	12.5 kJ/min.	
		$\times t_{Nc}$	197 kJ		$\times t_{Nc}$	276.4 kJ	
	<7>	⊥ impediment factor $[1.6\times]q_{Nc}$	197 kJ × 1.6	= 315 kJ	⊥ impediment factor $[1.6\times]q_{Nc}$	276.4 kJ × 1.6	= 442 kJ
transporting	<6>	° $[\dot{V}_{O_2}]_b + [\dot{V}_{O_2}]_{v_{Nt}}$	24.5 kJ/(kg·h)		° $[\dot{V}_{O_2}]_b + [\dot{V}_{O_2}]_{v_{HNt}}$	23.2 kJ/(kg·h)	
		$\times m_D$	14.3 kJ/min.		$\times m_D$	13.5 kJ/min.	
		$\times t_{Nt}$	46.5 kJ		$\times t_{HNt}$	65.9 kJ	
	<7>	⊥ impediment factor $[2\times]b_{Nt}$	46.5 kJ × 2	= 93 kJ	⊥ impediment factor $[2.3\times]b_{HNt}$	65.9 kJ × 2.3	= 152.kJ
Energy output		GE_{Nct}		408 kJ	GE_{NcHt}		594 kJ

Figure 5 - Estimated daily average gross energy output for 'Darwin' walking in the Taï rain forest and transporting nuts and hammers to 'ateliers'.

° $[\dot{V}_{O_2}]_b$ = 0.90 ml O_2/(g·h)
$[\dot{V}_{O_2}]v_b$ = 0.23 ml O_2/(g·km)

* $[\dot{V}_{O_2}]_q$ = 0.79 ml O_2/(g·h)
$[\dot{V}_{O_2}]v_q$ = 0.25 ml O_2/(g·km)
data from Taylor & Rowntree (1973)
(1 ml O_2 = 20.2 J / RQ = 0.82; cf. Ulmer, 1985)

Calculated from Spitzer, Hettinger & Kaminsky (1982)

Step «6»

The metabolic rate for 'Darwin' (m_D = 35 kg) walking 74 m bipedally or tripedally at 1.37 km/hour for about 3:14 minutes (i.e. 18% of 18 min.) would amount to 858 kJ/hour. Walking quadrupedally at 1.37 km/hour for 336 m and 14:46 minutes (i.e. 82% of 18 min.) results in a metabolic rate for 'Darwin' of 800 kJ/hour. Therefore he spends 197 kJ for $[\dot{V}_{O_2}]\cdot_{D}\cdot_{Nc}$ and 46.5 kJ for $[\dot{V}_{O_2}]\cdot_{D}\cdot_{Nt}$ respectively.

Step «7»

Up to this point, both results refer to 'Darwin' walking along a treadmill for the time and speed in question. This does not give us more than an indication, however, of the actual metabolic rate of 'Darwin' walking in a forested environment. SPITZER, HETTINGER & KAMINSKY (1982) listed the work metabolism of many different conditions for human walking. Selecting the comfortable speed of 4 km/hour for a man (m_B = 75 kg) we calculated his overall metabolic rate for walking on a treadmill and terrains of various difficulties. The resulting specific 'impediment factors' are used as multipliers of the treadmill value, i.e. an increase of 1.6 × for 'Darwin' walking on a difficult heterogenous soil. This factor has been rounded up to 2× because of the difficulties involved in carrying the nuts. If the amount of bipedal or tripedal walking for the transport of nuts ($[\mathring{V}_{O_2}]_{\text{'D'Nt}}$ × 2) is added to the amount of quadrupedal walking for collecting the nuts ($[\mathring{V}_{O_2}]_{\text{'D'Nc}}$ × 1.6), we get the total amount of energy necessary to bring the Coula nuts to an 'atelier' which is GE_{Nct} = 408 kJ.

Fraction IV. «Collecting and carrying hammers and nuts», step «5» (cf. Figure 5)

As a daily average, another 98 Coula nuts and 18 hammers are carried together on 18 different occasions over a distance of 26.97 m in altogether 27 minutes: 18 hammers and 5.4 nuts each × 26.97 m = 485.46 m (s_{HNt} *and* s_{Nc}). This 0.485 km is the total distance, because usually the chimpanzees walk only one way as they move towards another anvil while carrying the hammer along to collect more nuts. Thus, for walking 0.485 km during 27 minutes the average speed is 1.08 km/hour.

Step «6»

'Darwin's' absolute bipedal energy consumption $[\mathring{V}_{O_2}]_{\text{'D'NHt}}$ for a period of 4:52 minutes (i.e. 18% of 27 min.) at a speed of 1.08 km/hour (v_{HNt}) makes 65.9 kJ. The same factors applied to 'Darwin's' absolute quadrupedal energy uptake $[\mathring{V}_{O_2}]_{\text{'D'Nc}}$ results in 276.4 kJ for 22:08 minutes (i.e. 82% of 27 min.).

Step «7»

As in fraction III we inserted an 'impediment factor' of 2.3 × for 'Darwin' carrying various loads and walking bipedally, while walking quadrupedally once again surely does not exceed an increase by 1.6 × greater than its corresponding treadmill energy value. Consequently the total amount of energy which is necessary to collect and carry hammers and nuts to an 'atelier' is GE_{NcHt} = 594 kJ.

All fractions, step «8» (cf. Figure 6)

The daily gross energy for 'Darwin' engaged in nut-cracking is the sum of the 4 fractions contributing to the total energy expenditure *(Figure 4)*. The daily energy intake on the other side of the energy balance is based solely on the average amount of digested Coula nuts (nutritional values for Coula nuts; see BOESCH & BOESCH, 1983). Because one nut in ten on average is already decayed when opened, we had to subtract its nutrient content from the overall amount of 274 collected nuts. These nuts are included in the total energy output, however, since they were collected and cracked before the chimpanzees

Fraction	Name of fraction	Duration of fraction	Gross-energy of fraction	Basal metabolic rate of fraction	Net-energy of fraction	Power-output of fraction
		(t)	(GE/t)	(BMR·t) *	(GE−BMR·t)•	(NE/t)
I.	nut-cracking (Ncr)	30:30 min.	390 kJ	96 kJ	294 kJ	161 W
II.	eating while sitting (es)	61:30 min.	404 kJ	192 kJ	212 kJ	58 W
III.	collecting and transporting nuts (Nct)	18:00 min.	408 kJ	56 kJ	352 kJ	326 W
IV.	collecting and carrying nuts and hammers (NcHt)	27:00 min.	594 kJ	85 kJ	509 kJ	314 W
I-IV.	nut-cracking behaviour	137:00 min.	1796 kJ	429 kJ	1367 kJ	166 W

Gross energy output	versus	Gross energy digestibility (= daily energy intake of 274 Coula nuts)
1796 kJ	versus	15752 kJ
Energy loss = 1	versus	9 = Energy gain

* for Darwin (m = 35 kg ; 4511 kJ/day) • work metabolism

Figure 6 - Summary of fractional and complete durations, various energy components and power outputs estimated for 'Darwin's average daily nut-cracking behaviour.

had been aware of their edibility. Thus, the energy intake for 1200 g net weight of Coula nuts is 17900 kJ instead of 19889 kJ. This taken into account, the energy intake is 10× higher than the energy expenditure, because the total energy output (ΣGE_{output}) amounts to just 1796 kJ.

For the energy budget in the strict sense we still need to go one step further in our calculation and this, in fact, on both sides of our equation.

Step «9»

Energy investment (cf. *Figure 6*): So far we have listed the complete metabolic requirements related to nut-cracking, including the BMR during the time concerned. This was necessary because the results based on oxygen uptake never separate the BMR from the specific energy expenditure of a certain task.

But for an energy balance on the input/output of a certain food item, only the amount of invested energy is important which is related to the accessibility of its energy components. We call this the net energy expenditure (ΣNE_{output}), otherwise labelled as work metabolism (ASTRAND & RODAHL, 1977; SPITZER, HETTINGER & KAMINSKY, 1982) or profitability (CURIO, 1983). Only if we exclude the BMR in our calculation, we get a comparable result which is relatively independent of body weight and thus applicable to other chimpanzees as well. Subsequently we have to subtract 429 kJ from the overall gross energy output (ΣGE_{output}) of 1796 kJ, because it is the calculated BMR of 'Darwin' for a period of 137 minutes, i.e. the time spent on all fractions together. Doing this, the overall net energy expenditure (ΣNE_{output}) is 1367 kJ.

Energy intake

Despite the amount of ingested nuts it is more important to consider the relevant gross energy digestibility, since it is the actual source which supplies an organism with energy. In humans about 12% of the assimilated food is lost, due to incomplete resorption and the specific-dynamic action (SDA) of the nutrients (ASTRAND & RODAHL 1977, ULMER 1985). In animals, such as porcupines, this energy digestibility can be as low as 80% (ALKON *et al.*, 1986). We decided to use the data for humans, mainly because of their close physiological resemblance to chimpanzees (e.g. almost the same BMR in both species). If the total energy intake of 17900 kJ is reduced by 12% for energy loss, then the total metabolizable and digestible energy is 15752 kJ.

The actual energy gain due to nut-cracking behaviour (gross energy digestibility = 15752 kJ) is 11× greater than the required energy output (net energy expenditure = 1367 kJ). The profitability defined as the Coula nut's energy yield per unit handling time (CURIO, 1983) is 131 kJ/min. In conclusion, since 'Darwin' cannot exist without his individual BMR we subsequently include the BMR for the time 'Darwin' is busy with nut-cracking behaviour. Then the ratio of gross energy output of 1796 kJ (cf. *Figure 6*) to gross energy digestibiliy is about *1 : 9*.

Discussion

As outlined before, the results clearly indicate an enormous energetic advantage due to tool use in the Coula nut-cracking of Taï chimpanzees. To fully appreciate the scope of this behaviour, we went beyond the mere energy budget of expending and passing kilojoules during little more than two hours a day. In other words, if we distribute the energy surplus of about 14400 kJ (15752 kJ ($\Sigma GE_{digestibility}$) – 1367 kJ ($\Sigma NE_{output}$)) throughout the day, we will know whether the Tai chimpanzees could subsist on Coula nuts at least in terms of energy supply and during the nut-season, of course. The basal metabolic rate (BMR; see *step «4»*) for an adult male chimpanzee ($m_{ad.♂}$ = 48.1 kg) is nearly 5150 kJ/day (cf. BRUHN & BENEDICT, 1936). If we add the net energy output (ΣNE_{output} = 1367 kJ) to this BMR, then the chimpanzee has still almost 8000 kJ to spend during the day.

The calculation presented here shows an important net benefit for the nut-cracking chimpanzees. This result supports the hypothesis that tool-use can be considered as an optimal foraging strategy, as it allows a benefit by exploiting a food source which is otherwise not accessible to that species.

However, we have to further discuss this result, because it implies different assumptions: First of all, 'Darwin', the adolescent male chimpanzee in the analysed film, used an optimal hammer with which he cracked very dry nuts (due to an unusually prolonged dry season). In the rain forest, this situation is not the rule but the exception: wooden clubs have all possible irregular forms and the pounding may be much more difficult. Then, during the first two months of the Coula season, the nuts are green and consequently 22% harder to open than when dried (see BOESCH & BOESCH, 1984a, 1984b). These facts are reflected by the film in which 'Darwin' cracks nuts with an efficiency which is above the average (5 hits per nut for 'Darwin', instead of 6.71 for all adolescent males). Considering these two factors, we think that the energy necessary for cracking nuts (E_{Ncr}) could increase significantly in many cases. Secondly, 'Darwin' suffers from two handicapped feet and has therefore a delayed physical development. Thus, 'Darwin' may not be the best representative of wild chimpanzees. Thirdly, 'Darwin' was a late adolescent at the time of filming and the data we collected for the fraction III and fraction IV include only adults of both sexes. Thus, we should consider this calculation as a mixture of data, both from adolescent and adult chimpanzees cracking nuts in optimal situations. We should correct the result towards a decrease of the net benefit to adjust it to more real situations. However, our result is so clearly positive that such a correction should barely affect the outcome.

One more general assumption is based upon the important and widely accepted value of the BMR in chimpanzees found by BRUHN & BENEDICT (1936). Usually the unknown BMR of a mammal is estimated by using the KLEIBER-relationship (1961): BMR (kcal/day) = 70 $m<kg>^{0.75}$. HAYSSEN & LACY (1985) and MCNAB (1988) heavily criticize this single equation and MÜLLER (1985) refers to possible ecological adaptations of the BMR even among higher primates. The results for KLEIBER's equation applied to 'Darwin' is indeed 4% less than the experimental BMR found by BRUHN & BENEDICT. This deviation becomes even larger since the measurements of BRUHN & BENEDICT were made under conditions conducive to sleep, a state acknowledged to be something less than basal. The data of DALE *et al.* (1970) based upon awake chimpanzees, however, resulted in a completely heterogeneous BMR due to some uncontrolled muscular activity. Another difficulty to apply the results of BRUHN & BENEDICT to wild chimpanzees is that the measurements were made at a chamber temperature which is within a chimpanzee's thermal neutral zone of 20°C to 29°C. However, the ambient temperature in the Tai forest during the dry Coula nut-season (November to March) is indeed 24°-28°C during the day and about 20°-22°C at night. Because of the dry season an increase of the BMR due to high humidity can most likely be neglected as well. It should be noted that the overall BMR of the adult female chimpanzee is 3% less if compared to that of the males with their heavier body build (cf. *Figure 3*; BRUHN & BENEDICT, 1936). This is another point which creates an even more advantageous energy budget for the already very efficient nut-cracking females (BOESCH & BOESCH, 1984b). Overall we may assume that the individual BMR of chimpanzees is probably higher in their natural environment than under experimental conditions. Even though this increase is still unknown, it would definitely have no considerable effect on the energy budget in the analyzed nut-cracking behaviour. For example, a huge 10% increase of the BMR would add less than 1% to our calculated gross energy output.

Now, in order to further evaluate the total daily energy balance in chimpanzees we lack the information of a) a complete record of the chimpanzee's locomotor behaviour during any season, and b) their metabolic rates. The laborious procedure would then be to collect all these data on each chimpanzee seperately, because of the enormous individual versatility generally observed in all primates (FLEAGLE, 1984). Supposing we have these data we would still need to find out the metabolic rates for the different activities. For example, vertical tree climbing requires a high energy investment, since the body mass has to be lifted against the gravitational pull (TAYLOR, 1973). But in order to determine the exact metabolic rate of climbing many cases have to be analyzed due to different conditions (e.g. thickness of tree-trunks, availability of branches for body support). Even if there were some general models for a variety of locomotor activities, we could hardly adopt them to such a specific case as the estimation of the daily energy output in chimpanzees. The problems involved in such a general-to-specific transfer can be seen in the model of WUNDER (1975) on the metabolic rate of active mammals moving horizontally. Like with other general formulas such as developed in TAYLOR & HEGLUND (1982) and HEGLUND (1985), the energy consumption in the terrestrial locomotion of chimpanzees is considerably higher than expected and not significantly different for the quadrupedal and bipedal locomotor mode. This phenomenon might be explained by the chimpanzee's particular morphology causing a forelimb steering and hindlimb driving locomotion for quadrupedal walking (KIMURA, OKADA & ISHIDA, 1979; KIMURA, 1985) and disadvantageous energy consuming lateral movements in bipedal walking (JENKINS, 1972), which, together with mechanical considerations, result in a potential ability of 28% for bipedal walking if compared to humans (YAMAZAKI, 1985). It is possible that this low value may have some impact on the further development of nut-cracking behaviour and/or hindlimb morphology.

In conclusion, to understand the evolution of tool-use, more field data are needed to make such cost/benefit evaluations in different sub-optimal situations (BOESCH & BOESCH, in prep.).

Not all chimpanzees living in habitats with edible nuts and tools to crack them, do adopt this tool-use in order to be able to eat them: In the primary forest of Gabon, at least two species of nuts cracked by Taï chimpanzees occur (including Coula), but this population of chimpanzees does not crack them although wooden clubs seem to be accessible in abundance (TUTIN & FERNANDEZ, 1983). This situation is exactly the same in the Cameroun forests (SUGIYAMA, 1985). In Gombe, oil palm nuts are abundant as well as quartzite stones, but the chimpanzees never cracked them (GOODALL, 1986), whereas Bossou chimpanzees in Guinea do so (SUGIYAMA & KOMAN, 1979). An energetical approach, aiming at an understanding of the costs of nut-cracking at these sites could bring some explanations why these differences exist and what factors have to be attributed to traditional differences.

List of Abbreviations

Σ = total of-
N = net-
G = gross-

b = bipedal walking
q = quadrupedal walking

es = eating while sitting

Np = Nut pounding (= hit)
Ncr = Nut cracking
Nc = Nut collecting
Nt = Nut transport

H = Hammer
Ht = Hammer transport

HNt = Hammer and Nut transport

RH = Right Hand
RF = Right Forearm
RU = Right Upper Arm
A = Arm
B = Body
D = 'Darwin' ♂

List of Symbols and Units

m [kg] = mass
s [m] = distance
t [s] = time
v [ms^{-1}] = velocity
a [ms^{-2}] = acceleration
g [ms^{-2}] = gravity
F [N] = force
W [Nm] = work
E [J] = energy
P [$J \cdot s^{-1}$] = power
$\dot{V}o_2$ [$ml \cdot min^{-1}$] = oxygen uptake
BMR [$J \cdot day^{-1}$] = basal metabolic rate
(ME) [%] = mechanical efficiency

ACKNOWLEDGEMENTS — We thank the «Ministère de la Recherche Scientifique» and the «Ministère des Eaux et Forêts» of Cote d'Ivoire for permitting this study and the Swiss National Foundation, the Schultz Foundation and the Messerli Foundation for financing it. This project is integrated in the UNESCO project Tai-MAB under the supervision of Dr. Henri Dosso. We are most grateful to A. Aeschlimann, F. Bourlière, H. Kummer and H. Preuschoft, for their constant encouragement, P. and B. Lehmann, directors of the CSRS in Abidjan for their efficient logistic help and T. Tiépkan of the station IET in Tai for his friendly support in our daily camp life. We are indebted to H.V. Ulmer for his careful notes of our physiological calculations. We also wish to thank M. Wischniwetzki for the drawing of all figures.

References

ALBRECHT H. & DUNNET S.C., 1971. *Chimpanzees in Western Africa.* München: Piper.

ALCOCK J., 1972. *The evolution of the use of tools by feeding animals.* Evolution 26: 464-473.

ALKON P.U., DEGEN A. A., COHEN A. & POLLAK H., 1986. *Seasonal energy requirements and water intakes of indian crested porcupines (Hystrix indica) in captivity.* J. Mamm. 67: 333-342.

ASTRAND P.-O. & RODAHL K., 1977. *Textbook of Work Physiology.* New York: McGraw-Hill.

BECK B.B., 1980. *Animal Tool Behavior.* New York: Garland STPM.

BOESCH C., 1978. *Nouvelles observations sur les chimpanzes de la foret de Tai (Cote d'Ivoire).* Terre et Vie 32: 195-201.

BOESCH C. & BOESCH H., 1981. *Sex differences in the use of natural hammers by wild chimpanzees: a preliminary report.* J. Hum. Evol. 10: 585-593.

BOESCH C. & BOESCH H., 1983. *Optimization of nut-cracking with natural hammers by wild chimpanzees.* Behaviour 3/4: 265-286.

BOESCH C. & BOESCH H., 1984a. *Mental map in wild chimpanzees: an analysis of hammer transports for nut cracking.* Primates 25: 160-170.

BOESCH C. & BOESCH H., 1984b. *Possible causes of sex differences in the use of natural hammers by wild chimpanzees.* J. Hum. Evol. 13: 415-440.

BOESCH C. & BOESCH H. (this volume).

BOURNE G.H., 1971. *Nutrition and diet of chimpanzees.* In: G.H. Bourne, ed. The Chimpanzee, vol. 4, pp. 373-400. Basel: Karger.

BRUHN J.M. & BENEDICT F.G., 1936. *The respiratory metabolism of the chimpanzee.* Proc. Amer. Acad. Arts Sci. 71: 259-326.

CALDER W.A., 1984. *Size, Function, and Life History.* Cambridge, Mass.: Harvard University Press.

CANT J.G.H. & TEMERIN L.A., 1984. *A conceptual approach to foraging adaptations in primates.* In: P.S. Rodman & J.G.H. Cant, eds. Adaptations for Foraging in Nonhuman Primates, pp. 304-342. New York: Columbia Univ. Press.

CURIO E., 1983. Time-energy budgets and optimization. Experientia 39: 25-34.

DALE H.E., SHANKLIN M.D., JOHNSON H.D. & BROWN W.H., 1970. *Energy metabolism of the chimpanzee.* In: G.H. Bourne, ed. The Chimpanzee, vol.2, pp. 100-122. Basel: Karger.

DEMPSTER W.T. & GAUGRHAN G.R.L., 1967. *Properties of body segments based on size and weight.* Am. J. Anat. 120: 33-54.

FLEAGLE J.G., 1984. *Primate locomotion and diet.* In: D.J. Chivers, B.A. Wood & A. Bilsborough, eds. Food Acquisition and Processing in Primates, pp. 105-117. New York: Plenum.

GOODALL J., 1968. *The behavior of free-living chimpanzees in the Gombe Stream Reserve.* Animal Behavior Monographs 1: 161-311.

GOODALL J., 1970. *Tool-using in primates and other vertebrates.* In: D.S. Lehmann, R.A. Hinde & E. Shaw, eds. Advances in the Study of Behavior, vol. 3, pp. 195-249. New York: Academic Press.

GOODALL J., 1975. *The behaviour of the chimpanzee.* In: G. Kurth & I. Eibl-Eibesfeldt, eds. Hominisation and Behavior, pp. 74-136. Stuttgart: Fischer.

GOODALL J., 1986. *The Chimpanzees of Gombe: patterns of behavior.* Cambridge: The Belknap Press.

HAYSSEN V. & LACY R.C., 1985. *Basal metabolic rates in mammals: taxonomic differences in the allometry of BMR and body mass.* Comp. Biochem. Physiol. 81A: 741-754.

HEGLUND N.C., 1985. *Comparative energetics and mechanics of locomotion: how do primates fit in?* In: W.L. Jungers, ed. Size and Scaling in Primate Biology, pp. 319-335. New York: Plenum.

JENKINS F.A., 1972. *Chimpanzee bipedalism: cineradiographic analysis and implications for the evolution of gait.* Science 178: 877-879.

JOHANSON D. & EDEY M., 1981. *Lucy: the beginnings of humankind.* New York: Simon & Schuster.

KIMURA T., 1985. *Bipedal and quadrupedal walking of primates: comparative dynamics.* In: S. Kondo, ed. Primate Morphophysiology, Locomotor Analyses and Human Bipedalism, pp. 81-104. Tokyo: Univ. of Tokyo Press.

KIMURA T., OKADA M. & ISHIDA H., 1979. *Kinesiological characteristics of primate walking: its significance in human walking.* In: M.E. Morbeck, H. Preuschoft & N. Gomberg, eds. Environment, Behavior, and Morphology: Dynamic Interactions in Primates, pp. 297-311. New York: Fischer.

KLEIBER M., 1961. *The fire of life.* New York: Wiley.

KORTLANDT A., 1972. *New perspectives on ape and human evolution.* Amsterdam: Stichting voor Psychobiologie.

LEAKEY R. & LEWIN R., 1977. *Origins.* London: McDonald & Jones.

MAINZER J., 1983. *Zulässiger Energieumsatz.* In: W. Rohmert & J. Rutenfranz, eds. Praktische Arbeitsphysiologie, pp. 165-172. Stuttgart: Thieme.

MCGREW W.C., 1977. *Socialization and object manipulation of wild chimpanzees.* In: S. Chevalier-Solnikoff & E. Poirier, eds. Primate Bio-social Development: Biological, Social and Ecological Determinants, pp. 159-187. New York: Garland Publishing.

MCGREW W., 1979. *Evolutionary implications of sex differences in chimpanzee predation and tool-use.* In: D. Hamburg & E. McCown, eds. The Great Apes — Perspectives on Human Evolution, vol.5, pp. 441-463. Menlo Park,Ca.: Benjamin/Cummings.

MCGREW W.C., BALDWIN P.J. & TUTIN C.E.G., 1979. *Chimpanzees, tools and termites: cross-cultural comparisons of Senegal, Tanzania and Rio Muni.* Man 14: 185-214.

MCGREW W.C. & TUTIN C.E.G., 1972. *Chimpanzees dentistry.* JADA 85: 1198-1244.

MCNAB B.K., 1988. *Complications inherent in scaling the basal rate of metabolism in mammals.* Q. Rev. Biol. 63: 25-54.

MÜLLER E.F., 1985. *Basal metabolic rates in primates — the possible role of phylogenetic and ecological factors.* Comp. Biochem. Physiol. 81A: 707-711.

NISHIDA T., 1973. *The ant-gathering behaviour by use of tools among wild chimpanzees of the Mahale Mountains.* J. Hum. Evol. 2: 357-370.

NISHIDA T. & HIRAIWA M., 1982. *Natural history of a tool-using behaviour by wild chimpanzees in feeding upon wood-boring ants.* J. Hum. Evol. 11: 73-99.

NÖCKER J., 1976. *Physiologie der Leibesübungen.* Stuttgart: Enke.

PARKER S.T. & GIBSON K.R., 1977. *Object manipulation, tool-use and sensorimotor intelligence as feeding adaptations in cebus monkeys and great apes.* J. Hum. Evol. 6: 623-641.

PREUSCHOFT H., FRITZ M. & NIEMITZ C., 1979. *Biomechanics of the trunk in primates and problems of leaping in Tarsius.* In: M.E. Morbeck, H. Preuschoft & N. Gomberg, eds. Environment, Behavior, and Morphology: Dynamic Interactions in Primates, pp. 327-345. New York: Fischer.

RAHM U., 1971. *L'emploi d'outils par les chimpanzes de l'ouest de la Cote d'Ivoire.* Terre et Vie 25: 506-509.

ROHMERT W. & RUTENFRANZ J. eds., 1983. *Praktische Arbeitsphysiologie.* Stuttgart: Thieme.

SPITZER H., HETTINGER T. & KAMINSKY G., 1982. *Tafeln für den Energieumsatz bei körperlicher Arbeit.* Berlin: Beuth.

STEGEMANN J., 1984. *Leistungsphysiologie.* Stuttgart: Thieme.

STRUHSAKER T. & HUNKELER P., 1971. *Evidence of tool-using by chimpanzees in the Ivory Coast.* Folia Primatol. 15: 212-219.

SUGIYAMA Y., 1981. *Observations on the populations dynamics and behavior of wild chimpanzees at Bossou, Guinea, in 1979-1980.* Primates 22: 435-444.

SUGIYAMA Y. & KOMAN J., 1979. *Tool-using and -making behavior in wild chimpanzees at Bossou, Guinea.* Primates 20: 513-524.

TAYLOR C.R., 1973. *Energetic cost of animal locomotion.* In: L. Bolis, K. Schmidt-Nielsen & S.H.P. Maddrell, eds. Comparative Physiology, pp. 23-42. Amsterdam: North-Holland.

TAYLOR C.R. & HEGLUND N.C., 1982. *Energetics and mechanics of terrestrial locomotion.* Ann. Rev. Physiol. 44: 97-107.

TAYLOR C.R. & ROWNTREE V.J., 1973. *Running on two or on four legs: which consumes more energy?* Science 179: 186-187.

TELEKI G., 1974. *Chimpanzee subsistence technology: materials and skills.* J. Hum. Evol. 3: 575-594.

TELEKI G., 1975. *Primate subsistence patterns: collector-predators and gatherer-hunters.* J. Hum. Evol. 4: 125-184.

TUTIN C.E.G. & FERNANDEZ M., 1983. *Recensement des gorilles et des chimpanzes du Gabon.* C.I.R.M.F.

UEHARA S., 1982. *Seasonal changes in the techniques employed by wild chimpanzees in the Mahale Mountains, Tanzania, to feed on termites.* Folia Primatol. 37: 44-76.

ULMER H.V., 1985. *Energiehaushalt; Ernährung; Arbeitsphysiologie - Umweltphysiologie.* In: R.F. Schmidt & G. Thews, eds. Physiologie des Menschen, pp. 574-582; pp. 602-641. Berlin: Springer.

WUNDER B.A., 1975. *A model for estimating metabolic rate of active or resting mammals.* J. Theor. Biol. 49: 345-354.

YAMAZAKI N., 1985. *Primate bipedal walking: computer simulation.* In: S. Kondo, ed. Primate Morphophysiology, Locomotor Analyses and Human Bipedalism, pp. 105-130. Tokyo: Univ. of Tokyo Press.

ZIHLMAN A.L., 1984. *Body build and tissue composition in Pan paniscus and Pan troglodytes, with comparisons to other hominoids.* In: R.L. Susman, ed. The Pygmy Chimpanzee: Evolutionary Biology and Behavior, pp. 179-200. New York: Plenum.

II

HAND FUNCTION

F.K. Jouffroy
M. Godinot
Y. Nakano

Biometrical characteristics of primate hands *

The length proportions of the primate hands and their elements are analyzed in 43 extant genera and 6 fossil genera. The length of the hand in relation to the forelimb length does not characterize taxonomic groups, but rather locomotor modes, such as vertical-clinging-and-leaping, claw-climbing, and terrestrial quadrupedalism, as opposed to arboreal quadrupedalism. The relative lengths of the carpus, metacarpus, and digits appear mainly related to the phylogenetic history of the primates. Paraxony, instead of mesaxony, is the most frequent pattern of the simiiform hand, whereas hypermesaxony characterizes humans, hylobatids, and tarsiers. The proportions of the primitive euprimate hand are discussed in the light of the hand proportions of extant primates. Proportions drawn from hand remains of *Megaladapis edwardsi*, *Proconsul africanus*, *Mesopithecus pentelici*, *Notharctus*, *Adapis parisiensis*, and *Plesiadapis insignis* are compared to those of extant primates, and discussed from both phylogenetical and functional points of view.

Key words: forelimb length, fossil primates, hand length, hand proportions, ectaxony, mesaxony, paraxony.

Introduction

That primates have hands, rather than paws or hooves, has enabled them to develop a wide diversity of modes of locomotion, as well as handling activities in relation to their unique brain development. With few exceptions, such as *Ateles*, *Colobus*, and *Perodicticus*, the primate hand has retained the basic pattern of the five-rayed manus of the early land vertebrates. By comparison with the pentadactyl hand of non-primate mammals, the outstanding feature of the primate hand is the capability of the first digit (the thumb) to abduct and adduct in relation to the other four digits, which allows grasping.

As early as the 4th century B.C., the similarity of primate hands and feet with the human hand had been noted by Aristotle (History of Animals, II,8). At the dawn of modern zoology, BUFFON (1770) emphasized the distinction between humans, who are «bimanes et bipèdes», and monkeys, who are «quadrumanes». BLUMENBACH (1790) established these terms as taxa when defining the order «Bimanus» for the single genus *Homo*, as opposed to «Quadrumanus» for all other primates.

Within the framework of the thumbed/pentadactyl pattern, primates display a wide diversity of hand morphology and function (MIDLO, 1934; WOOD-JONES, 1941; BISHOP, 1964; TUTTLE, 1970; SCHULTZ, 1972; ETTER, 1974). Due to its significance for hand dexterity and human evolution, the best known and most extensively studied primate

* To the memory of our colleague Jacques Lessertisseur.

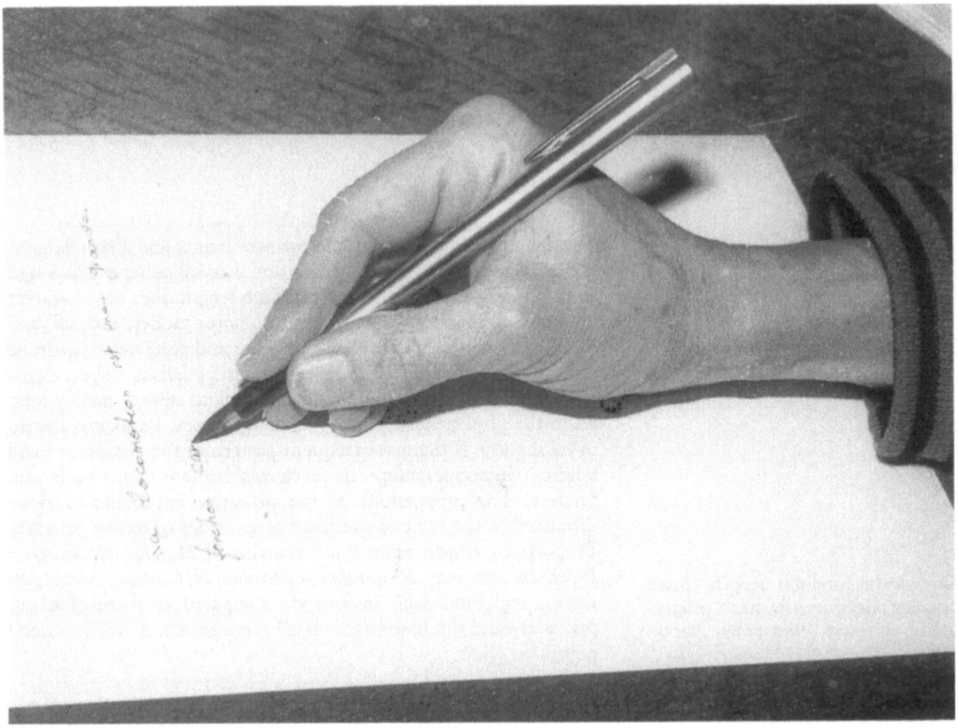

Figure 1 - Human hand writing, its most elaborate prehensile activity.

hand feature is thumb opposability. Besides abducting and adducting, the first metacarpal of humans can rotate inwards towards the palm, so that the ventral side of the thumb comes to face the palm and the ventral side of the other digits. Such a movement, combined with varied degrees of flexion-extension and abduction-adduction, allows the unique circumduction movement of the human first metacarpal. With the flexion of the interphalangeal joints, these movements result in very accurate and fine handling. Most anatomical studies of the human hand have focused on the saddle-shaped trapezo-primometacarpal joint which allows the combined medial rotation/flexion/adduction involved in thumb opposition (HAINES, 1944; LANDSMEER, 1955, 1976; NAPIER, 1955, 1956, 1960, 1961, 1966; STACK, 1962; KAPLAN, 1965; KAPANDJI, 1966; PIERON, 1973; KUCZYNNSKI, 1974, MCCONAILL & BASMAJIAN, 1977).

The observation of grasping activities by human hands shows that prehensile capability also depends on the relatively great length of the thumb, the tip of which can reach the tips of other fingers, even when the latter are almost completely extended at their interphalangeal joints, as in writing (*Figure 1*). As shown by SCHULTZ's (1972) drawings of primate hands, the human thumb is much longer than that of the other Simiiformes HOFFSTETTER 1982 (=Anthropoidea). Thus, biometric data about the relative length of the thumb in the entire Primates order are essential for a better understanding of primate hand (and especially human hand) evolution.

SCHULTZ's (1972) comparative illustrations also demonstrated the most striking specializations of primate hand morphology. For example (*Figure 2*): (1) the thumb is extremely short in the colobines *Presbytis* and *Nasalis*, and it is absent in *Colobus* and *Ateles*; (2) the second finger is very short in the lorisines *Nycticebus* and *Perodicticus*; and (3) the middle finger is the longest in *Tarsius*, *Hylobates* and *Homo*, whereas the fourth finger is by far the longest in Strepsirhini (e.g. *Lemur*).

Long before SCHULTZ's (1972) illustration of the diversity of primate hand architecture, numerous biometrical data concerning the hand skeleton had been published, either as a part of comparative osteometrical studies of the primate skeleton (MIVART, 1867; MOLLISON, 1911) or as limited comparative data required as a basis for studying some fossil or extant primate genera or higher taxa (LUCAE, 1865; SCHULTZ, 1930; STRAUS, 1942; NAPIER & DAVIS, 1959; WALKER, 1967). Unfortunately, these data are too disparate and cannot be pooled together for a synthetic approach.

MIVART (1867) was the first to provide an impressive set of 18 ratios quantifying the extrinsic and intrinsic proportions of the manus in 29 primate genera, including strepsirhines, tarsiers and simiiforms. Unfortunately, these ratios were calculated from the measurements of only one specimen for each genus, which could only provide rough results in terms of generic characteristics, as noted by Mivart himself «as there is considerable individual variation, the proportions here given are offered merely as approximative to the true standard of each genus. An average, drawn from the comparison of a considerable number of specimens of each genus, would have been more satisfactory; but, first, materials for such an estimate are not yet accessible, and second, even were they so, the expenditure of time would have been out of proportion to the result» (p. 371).

In the course of the last three decades, several papers offered abundant new osteometrical data on the primate hand (NAPIER & DAVIS, 1959; NAPIER & NAPIER, 1967; ETTER, 1974, 1978; JOUFFROY & LESSERTISSEUR, 1977, 1978, 1979. ETTER's (1974) detailed study focused mainly on the length of the hand and its different parts as related to the length of the trunk, from an interspecific allometric viewpoint, inspired by BIEGERT & MAURER's (1972) method. ETTER (1974) disregarded the relative length variation of the carpus as being too small and not characteristic of hand shape. As he defined it, the total length of the hand was measured by the length of the longest digital ray (metacarpal + the three phalanges). JOUFFROY & LESSERTISSEUR (1978, 1979) carried out an extensive study of length proportions of strepsirhine and tarsier hands (19 genera), but gave only limited comparative data for six simiiform genera representing the different families.

Being invited to participate in the Reisensburg symposium on the evolution of the hand, we faced the situation that a comparative osteometrical study dealing with the entire order of Primates was still lacking. This prompted us to process and present in ratio form the 5,300 measurements that had been gathered previously in European and American collections over a period of more than ten years from 43 primate genera representing all the extant families. Calculators and computers allowed us to fulfill what MIVART (1867) had expressly regretted not being able to do.

The aim of our osteometrical approach is to provide a basis for functional analysis of primate hand morphology as well as for a well-founded interpretation of fossil hand bones. This study is composed of three parts. The first deals with the length of the hand as a whole, i.e. as an element of the body. The hand is one among the series of elements that make up the locomotor system. Its length is an essential characteristic of the locomotor and postural mechanism. Hand length in relation to trunk length was studied by MIVART (1867), MOLLISON (1911) and ETTER (1974). From a biomechanical point of view, it is difficult to interpret the variation of any kind of hand/trunk ratio, because such ratios

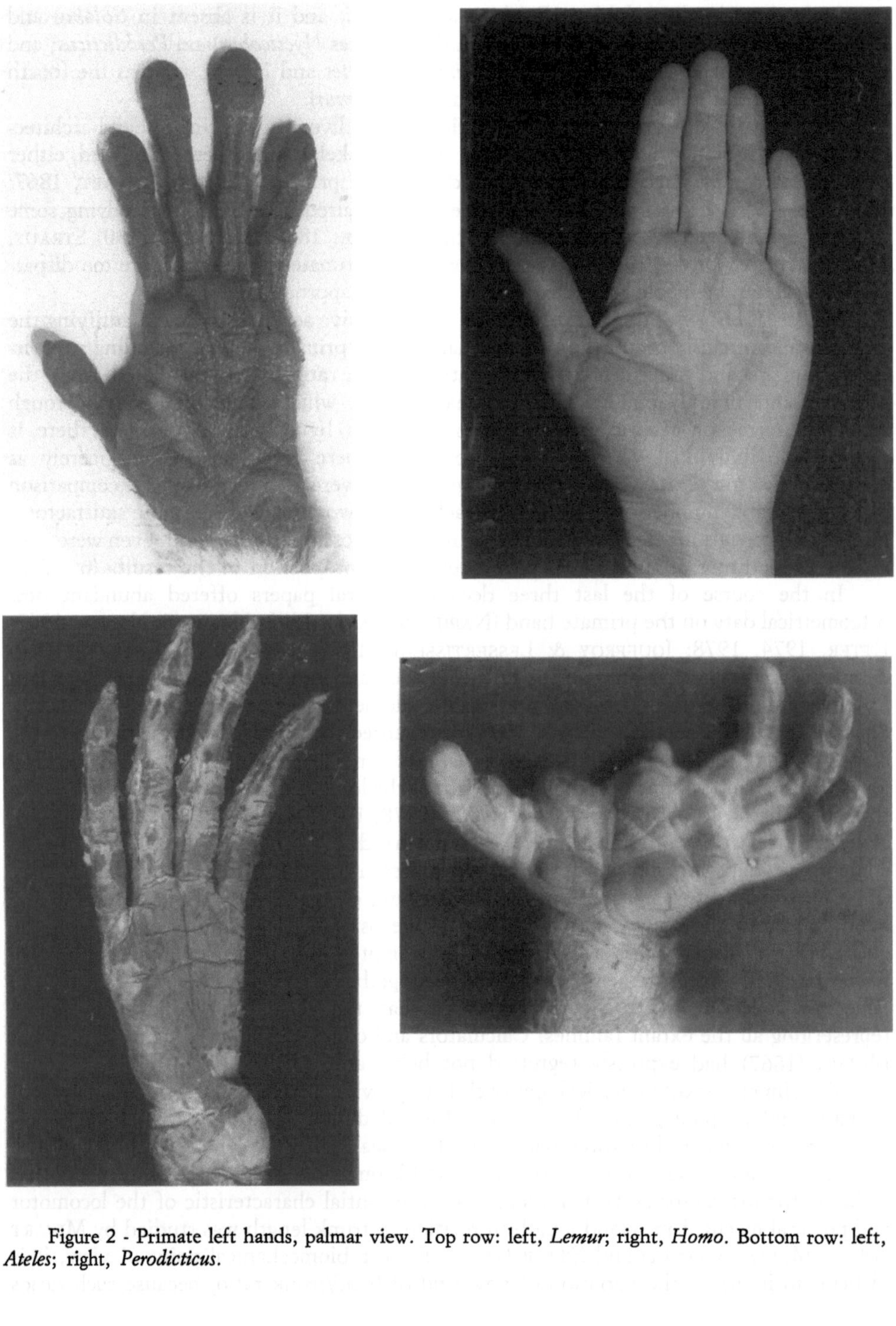

Figure 2 - Primate left hands, palmar view. Top row: left, *Lemur*; right, *Homo*. Bottom row: left, *Ateles*; right, *Perodicticus*.

depend on two strongly correlated factors: (a) the length of the hand in relation to forelimb length; and (b) the length of the forelimb in relation to trunk length. Both factors show a wide range of variation among primates. The forelimb length in relation to trunk length, for example, varies from 80% in *Lemur* to 215% in *Hylobates* (JOUFFROY & LESSERTISSEUR, 1978). On the other hand, primates having forelimbs of almost the same relative length in relation to the trunk (e.g. 104% in *Loris* and 114% in *Daubentonia*) can display hand lengths varying by a factor of two in relation to their forelimb length (20% in *Loris*; 43% in *Daubentonia*). In order to focus on the importance of the hand as a part of the forelimb architecture and locomotor system, we first studied the length of the hand (Hd) in relation to the total forelimb (FL) by means of a ratio (Hd/FL).

The second part of this study deals with the comparative morphology of the hand, as quantified by its intrinsic length proportions. Firstly, the maximal lengths of the three main sets of bones that proximodistally make up the hand (carpus, metacarpus and phalanges) are related to the total hand length. Secondly, the total phalangeal lengths of the fingers 1 to 4 are compared to one another by studying successively the lengths of the digital rays (metacarpal plus set of the two or three phalanges together) and the lengths of the metacarpals. The fifth finger is not included in this study because our measurements were initially gathered to analyze the axony pattern, which concerns mainly fingers 2 to 4. Due to its relatively small size, the fifth ray is often incomplete in some skeletons. Disregarding its length, however, is incidental with regard to the questions discussed here.

The last part is an attempt to reconstruct some fossil hands and to analyze their functional characteristics in light of the comparative biometry of extant primates.

Material

Extant primates

Our sample of extant genera is made up of 441 forelimb skeletons, representing 43 genera (see list of symbols) and belonging to the following collections: Muséum national d'Histoire naturelle (Paris); Département de Psycho-physiologie, Université Paris X, Nanterre; British Museum (Natural History), London; Anthropologisches Institut, Zürich; Musée Royal de l'Afrique centrale, Tervüren; Institut royal des Sciences naturelles de Belgique, Brussels; Rijksmuseum van Natuurlijke Historie, Leiden; American Museum of Natural History, New York; Field Museum of Natural History, Chicago; Smithsonian Institution (National Museum of Natural History), Washington, D.C.; Department of Anatomical Sciences, State University of New York, Stony Brook; and some private collections. Zoo specimens are occasionally included in this sample.

The data consisted of about 5,300 length measurements. They were taken from a set of about 12,000 limb length measurements that have been compiled during the last fifteen years by one of the authors (F.K. JOUFFROY), in collaboration with the late Professor J. LESSERTISSEUR.

Additional and complementary data were obtained by studying a series of hand radiographs made at the Laboratoire d'Anatomie comparée (Muséum national d'Histoire naturelle, Paris), the Department of Human Sciences (University of Osaka) and the Primate Research Institute (Kyoto University) ; radiographs of tarsiers were kindly provided by C. NIEMITZ (Freie Universität, Berlin).

This list of extant primate specimens calls for some additional remarks: due to the extensive nature of our sample, we could not take into consideration intrageneric variations. A comparative study of interspecific variability would be feasible only within smaller taxonomic groups, such as galagines or tarsiers, as has been done for the foot (JOUFFROY & GÜNTHER, 1985). Such studies require numerous specimens of each species, because the more tenuous the differences to be detected, the larger the sample needed. Furthermore, they also require that the systematics of the group be clearly established, and that the species of each skeleton should be unambiguously defined, which does not always occur. In many collections, primate skeletons have been labelled at different times according to different systematics ; it is almost impossible to ascertain the species of some specimens. The same problem exists for the genera of the subfamily Colobinae : due to systematic uncertainties, we considered five genera (*Semnopithecus*, *Colobus*, *Presbytis*, *Rhinopithecus*, *Trachypithecus*) in order to classify all the colobine skeletons available. Such a classification comes only from practical considerations and leaves the systematic problem of the colobines open. In addition, due to significant differences between the siamang and other gibbons in many body proportions, we have chosen to analyse data separately for these two groups (*Symphalangus* = *Hylobates syndactylus*).

Fossil material

Unbroken limb bones are known for only a few fossil primate genera. It is quite exceptional to find several intact limb bones belonging to the same individual. Consequently, lengths that were used to plot tentatively a few fossil primates in the *Figures 3* to *6, 11* and *12* are from GODINOT & BEARD'S (this volume) analysis, drawn from data in the literature synthesized with their own measurements.

Methods

Humerus length was measured between the apex of the head and the most salient part of the trochlea. Radius length was measured between the head and the carpal articular surface. Lengths of the elements of the hand, i.e.carpus, metacarpals, and digits (the two or three phalanges all together) were measured on the dorsal side of the bones, through the axis of the longest finger.

In this paper, «digit» refers to the phalangeal part of each finger, whereas «digital ray» refers to a single set of one metacarpal and the corresponding phalanges.

The calculated ratios (in percentage) are listed in *Tables 1* to *10.*

Results

Length of the hand in relation to forelimb length.

TAXONOMIC ANALYSIS

In most primates, including all the Cercopithecoidea and Ceboidea of our sample, the length of the hand amounts to one fourth to one third (24% to 33%) of the forelimb length (*Figure 3)*. Some prosimian genera, however, are not included within this range: (1) the slender lorisines *Arctocebus* and *Loris* are endowed with very short hands (about 20%

TABLE 1 - *Lengths of the humerus (Hu), radius (Ra), and hand (Hd) in relation to the entire forelimb length (FL) in simiiforms. Mean and standard deviation.*

Genera		N.	Hu/FL		Ra/FL		Hd/FL	
			x̄	s.d.	x̄	s.d.	x̄	s.d.
Callithrix	(CA)	6	36.9	1.0	32.7	0.8	30.4	1.7
Leontocebus	(LC)	7	34.3	1.3	33.5	1.1	32.2	1.7
Cebus	(CE)	11	37.3	0.5	35.2	0.5	27.5	0.7
Saimiri	(SM)	8	38.4	1.1	34.4	1.0	27.2	1.0
Aotus	(AO)	8	37.2	0.8	33.6	0.4	29.2	0.8
Callicebus	(CC)	3	38.2	0.8	31.3	0.3	30.5	0.6
Chiropotes	(CR)	2	39.4	0.1	32.8	0.3	27.8	0.5
Alouatta	(AL)	8	38.2	0.7	34.3	0.7	27.5	1.1
Ateles	(AT)	9	36.2	0.6	36.2	1.5	27.2	1.1
Brachyteles	(BT)	1	37.1	—	35.6	—	27.3	—
Lagothrix	(LA)	6	39.2	0.6	34.4	0.7	26.4	0.5
Cercopithecus	(CP)	11	38.0	1.8	35.5	2.2	26.5	1.5
Miopithecus	(MP)	7	36.4	0.5	36.2	0.5	27.4	0.5
Papio	(PP)	10	36.7	0.8	39.0	1.0	24.3	1.6
Macacus	(MA)	12	36.2	0.4	36.0	0.7	27.8	0.8
Semnopithecus	(SP)	5	34.2	0.5	37.0	1.0	28.8	1.0
Colobus	(CO)	10	35.9	0.4	35.0	1.2	29.1	1.1
Presbytis	(PB)	10	34.8	1.1	36.4	1.1	28.8	1.0
Rhinopithecus	(RH)	2	36.6	0.5	36.7	0.1	26.7	0.5
Trachypithecus	(TR)	2	36.3	1.0	34.5	0.7	29.2	0.2
Hylobates	(HY)	8	35.1	1.0	39.8	0.7	25.1	1.0
Symphalangus *	(SY)	3	34.4	0.6	39.4	1.0	26.2	1.4
Pongo	(PO)	9	35.7	0.9	36.4	1.2	27.9	1.2
Pan	(PA)	14	36.7	0.8	34.0	0.5	29.3	0.9
Gorilla	(GO)	6	41.3	1.0	33.5	0.5	25.2	1.6
Homo	(HO)	3	44.0	0.2	31.7	0.8	24.3	0.5

* *Symphalangus* = *Hylobates syndactylus*

of the forelimb); and (2) indriids, *Euoticus* among galagines, *Tarsius*, and, above all, *Daubentonia* (41%) have very long hands.

Because forelimbs consist of three segments (upper arm, forearm, and hand), any relative lengthening or shortening of the hand, in relation to the entire forelimb, is associated with an inverse variation in the combined lengths of the other two segments. This is illustrated by means of a triangular set of axes in *Figures 3* and *4*. In the slender lorisines *Loris* and *Arctocebus*, for example, the extreme shortness of the hands is compensated mainly by very long forearms and long upper arms. Both genera have the shortest hands (about 21%) and the longest radius (about 41%) among primates. Although the humeri of these two genera are the longest among prosimians, they are equalled by those of several cebids (38/39%) and even exceeded by those of *Gorilla* (41%) and, above all, *Homo* (44%).

In other prosimians, the great length of the hand is mainly compensated by the shortness of the humerus (about 30% in indriids and *Daubentonia*; 27% in *Tarsius*). In most long-handed strepsirhines, the forearm length falls within the average range for

TABLE 2 - *Lengths of the humerus (Hu), radius (Ra), and hand (Hd) in relation to the entire forelimb length (FL) in strepsirhines and Tarsius. Mean and standard deviation.*

Genera		N.	Hu/FL		Ra/FL		Hd/FL	
			x̄	s.d.	x̄	s.d.	x̄	s.d.
Varecia	(VA)	9	35.6	0.9	34.3	0.9	30.1	1.6
Lemur	(LE)	38	34.9	0.7	37.3	1.0	27.8	1.1
Lepilemur	(LL)	11	31.8	0.6	35.6	0.7	32.6	0.9
Hapalemur	(HL)	9	33.3	0.7	36.3	1.3	30.4	1.7
Microcebus	(MI)	16	33.9	1.1	37.3	1.0	28.8	1.0
Cheirogaleus	(CH)	8	34.9	0.7	34.5	1.0	30.6	1.4
Phaner	(PH)	2	33.2	1.2	34.8	0.5	32.0	1.7
Propithecus	(PR)	6	31.8	1.0	35.0	1.0	33.2	1.2
Indri	(IN)	7	29.4	1.4	35.8	1.3	34.9	0.9
Avahi	(AV)	6	30.0	1.0	36.0	0.9	34.0	1.5
Daubentonia	(DB)	9	29.2	0.5	29.6	0.6	41.2	1.0
Loris	(LO)	10	38.2	1.2	41.3	1.7	20.5	1.4
Nycticebus	(NY)	22	37.4	0.8	37.1	1.4	25.5	1.8
Perodicticus	(PE)	7	35.2	0.8	36.6	1.0	28.2	1.5
Arctocebus	(AR)	14	38.8	0.9	40.0	1.3	21.2	1.6
Galago	(GA)	55	33.7	1.4	35.7	1.7	30.6	2.5
Euoticus	(EU)	20	32.5	0.5	34.2	0.6	33.3	0.8
Tarsius	(TA)	11	27.3	0.8	35.0	0.7	37.7	1.4

primates. *Daubentonia* is the only exception, insofar as the extreme length of its hand in relation to the forelimb goes together not only with a short humerus but also with an exceptionally short radius (less than 30%, i.e., the shortest value among primates).

It appears that the moderate values of relative hand lengths (from 24% to 33%; *Figure 4*) characterize not only all cercopithecids and cebids, but also all lemurines and cheirogaleines, as well as *Galago* and the two thick-set lorisines, *Nycticebus* and *Perodicticus*. A set of closely-grouped genera make up a central cluster, indicated by a dashed polygonal line in *Figures 3* and *4*. Primates in this cluster have very similar forelimb proportions: Hd/Fl = 26-30%; Ra/FL = 33-38%; Hu/FL = 34-40%.

A few catarrhine and ceboid genera do not belong to this cluster:(1) the hylobatids and the baboons with shorter hands and longer radius; (2) the gorilla and *Homo (Figure 3)* with shorter hands and longer humerus; and (3) the callitrichids and *Callicebus*, with longer hands and shorter radius. All these genera are adapted to peculiar modes of locomotion: arm-swingers (hylobatids), ground-walkers (baboon, gorilla) and claw-climbers (callitrichids), which suggests that the intrinsic proportions of the forelimbs are related to locomotor modes (see below).

Three strepsirhine genera are also included in the central cluster: *Microcebus, Lemur*, and *Perodicticus.* The former genus is among the least specialized strepsirhines. The distance between *Lemur* and *Varecia* in *Figures 3* and *4* was an unexpected result of this study, because the two genera are systematically so close that *Varecia* has for a long time been considered as a subgenus of *Lemur*. The small standard deviations calculated from 38 *Lemur* and 9 *Varecia* (*Table 2*) confirm the dissimilarity of both genera.The inclusion of the nonprimate *Tupaia* in the cluster is also to be noted.

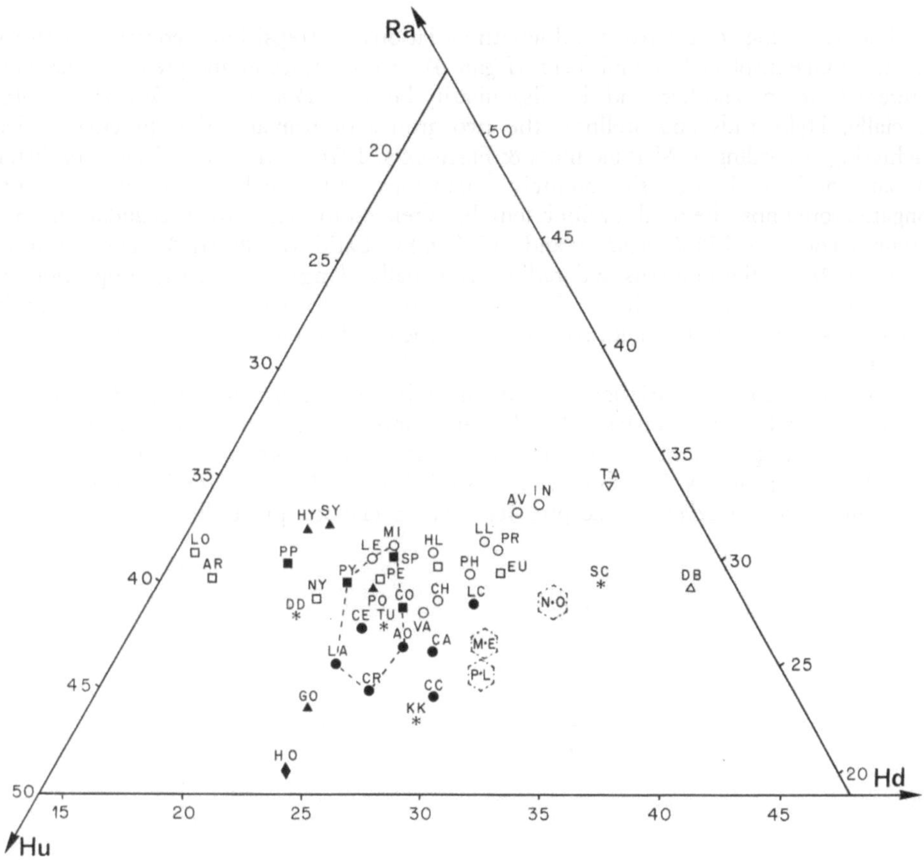

Figure 3 - Lengths of the humerus (Hu), radius (Ra) and hand (Hd) in relation to the total forelimb length (in percentage). The 3-dimensional coordinates along the sides of the triangle sum to 100%. List of genus symbols:

AD *Adapis* +
AL *Alouatta*
AO *Aotus*
AR *Arctocebus*
AT *Ateles*
AV *Avahi*

BT *Brachyteles*

CA *Callithrix*
CC *Callicebus*
CE *Cebus*
CH *Cheirogaleus*
CO *Colobus*
CP *Cercopithecus*
CR *Chiropotes*

DB *Daubentonia*
DD *Didelphis*

EU *Euoticus*

GA *Galago*
GO *Gorilla*

HL *Hapalemur*
HO *Homo*
HY *Hylobates*

IN *Indri*

KK *Potos (Kinkajou)*

LA *Lagothrix*
LC *Leontocebus*
LE *Lemur*
LL *Lepilemur*
LO *Loris*

MA *Macacus*
ME *Megaladapis* +
MI *Microcebus*
MP *Miopithecus*
MS *Mesopithecus* +

NO *Notharctus* +
NY *Nycticebus*

PA *Pan*
PB *Presbytis*
PC *Proconsul* +
PE *Perodicticus*
PH *Phaner*
PL *Plesiadapis* +

PO *Pongo*
PP *Papio*
PR *Propithecus*
PY *Pygathrix*

RH *Rhinopithecus*

SC *Sciurus*
SM *Saimiri*
SP *Semnopithecus*
SY *Hylobates syndactylus*

TA *Tarsius*
TR *Trachypithecus*
TU *Tupaia*

VA *Varecia*

The wide range of relative hand length variations in strepsirhines contrasts with the narrower range displayed by simiiforms (*Figure 4*). However, lesser and great apes are very different from one another, and the dissimilarity between *Pan* and *Gorilla* is to be noted especially. Hylobatids and atelines, the two groups of primates that practice typical brachiating (according to MITTERMEIER & FLEAGLE'S, 1976, definition), show very different patterns in their intrinsic forelimb proportions, although both are endowed with elongated forelimbs; the total forelimb length, as related to that of the precaudal vertebral column, amounts to 175% in *Ateles* and 210% in hylobatids (JOUFFROY & LESSERTISSEUR, 1978). In *Ateles*, the humerus and radius are equally elongated, and the proportions of each segment in relation to the forelimb length are close to that of *Cebus*. In hylobatids, the radius is more elongated and, consequently, the hand is relatively shorter than in most simiiforms.

The wide range of variations within the subfamily Lorisinae (*Figure 3*) is also an unexpected result of this analysis. Its slender members, the African *Arctocebus* and the Asian *Loris* are quite similar and show a unique intrinsic proportion pattern of their forelimb among primates: i.e., shortest hands (20/21% of the entire forelimb) and longest radius (40/41%). *Perodicticus*, despite its very specialized pincer-like hand, does not

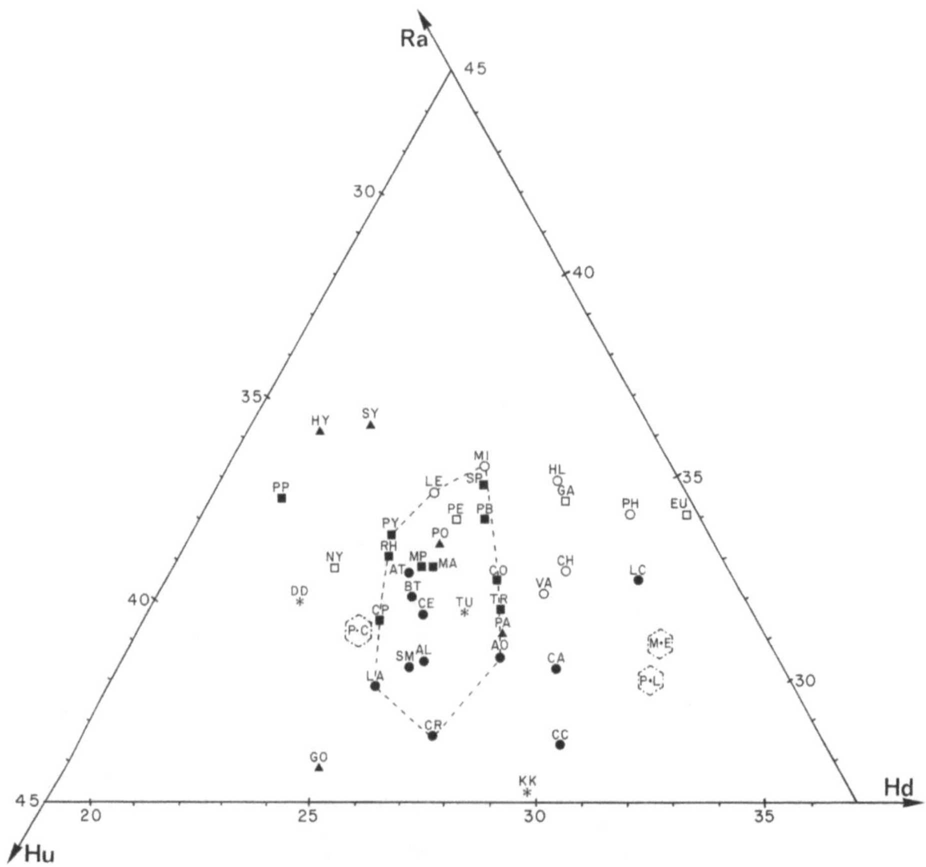

Figure 4 - Enlargement of the central area of *Figure 3*. Same symbols as *Figure 3*.

distinguish itself from the basic pattern, corresponding to the above described central cluster. *Nycticebus* shows a marked shortness of the hand. Here again, the differences in the intrinsic proportions of the forelimb are not related to changes in the length of the forelimb as related to body length. Among the two short-handed lorisines, *Arctocebus* has the relatively shortest forelimb and *Loris* the relatively longest one within the lorisine subfamily (respectively 81% and 104% of the precaudal vertebral column, according to JOUFFROY & LESSERTISSEUR, 1979).

In short, this analysis shows that the intrinsic proportions of the forelimb and especially the relative length of the hand do not characterize taxonomic groups of primates, with the exception of hylobatids and indriids, the members of which show very similar proportions.

FUNCTIONAL ANALYSIS

The location of the different genera in *Figure 3* shows that the relative length of their hands is to some extent related to their locomotor modes, more so than to their taxonomic position. Away from the central cluster, we find: (1) a trend toward longer hands for prosimian leapers («vertical-clingers-and-leapers», NAPIER & WALKER, 1967), as well as claw-climbing callitrichids, and (2) a trend toward shorter hands, including slow crawlers (lorisines), brachiators, and ground dwellers.

We observe different degrees of correlation between locomotor modes and hand relative length: (1) Among the least correlated are the brachiators (by MITTERMEIER & FLEAGLE'S, 1976, definition). Although they all use their hands similarly during suspended locomotion, their relative hand lengths differ. The atelines *Ateles* and *Brachyteles* are located in the central cluster, but the hylobatids distinguish themselves by their shorter hands (*Figure 4*). These two groups exhibit very different patterns in their intrinsic forelimb proportions (although forelimbs are elongated in relation to the trunk length in both cases). Yet, in *Ateles* the three segments of the forelimb are equally elongated, and the proportions of each segment in relation to the forelimb length are very similar to that of *Cebus*, for example. In hylobatids, the radius is much more elongated and consequently the hand is relatively shorter than in most simiiforms. (2) As regards the climbers (slow crawlers and arm-swingers), many are spread between the primates located in the central cluster (*Perodicticus*, *Pan*, *Pongo*) and the short-handed lorisines *Loris* and *Arctocebus*. Some genera, characterized by their relatively short hands, related to brachiating and climbing specialization, are set apart from the central cluster. However, this does not apply to all genera: climbers and brachiators are endowed either with relatively short hands or with hands of the same length as those of generalized arboreal quadrupeds (see below).

Most arboreal quadrupeds are located in the middle of the graph (i.e. the central cluster plus *Varecia* and *Cheirogaleus*) together with the above mentioned slow crawlers, and brachiators. Therefore no locomotor mode can be inferred from the location of a genus within or close to the central cluster. However, it should be noted that all the so-called arboreal quadrupeds, whether lemuriforms, ceboids, or cercopithecoids, possess generalized proportions, i.e. a moderately long hand, amounting to 26 to 31% of the forelimb (*Figure 4*). The location of *Callicebus* apart from this group is an exception and calls for additional studies of its locomotor behaviour, insofar as this genus displays «well developed jumping and leaping ability», according to NAPIER & NAPIER (1967).

Terrestrial walkers are too poorly represented in our sample to allow any general assessment. It should be noted, however, that both *Gorilla* and *Papio* possess very short hands relative to total forelimb length. Further studies of other genera such as *Erythrocebus* are required to know whether all quadrupedal ground walkers have relatively short hands, as in *Papio*.

The relationship between mode of locomotion and hand morphology is very much in evidence when considering the «vertical clingers-and-leapers» (NAPIER & WALKER, 1967), whether considering lemurids (*Hapalemur* and *Lepilemur*), indriids, galagines or tarsiers. These genera, which fall approximately on a line in *Figure 3*, are characterized by their relatively long hands and short humerus, possibly related to biomechanical constraints for landing. Between *Galago* and *Euoticus*, we find *Phaner*, a species differing from other cheirogaleids by its ability to leap frequently (PETTER, 1962; PETTER *et al.*, 1971), and even occasionally between vertical supports (TATTERSALL, 1982).

As could be expected, *Daubentonia* stands far apart from all other primates, due to its extremely long hand (*Figure 3*). This might be related in part to the presence of claws, among other factors, as we can observe that the clawed callitrichids differ from the nailed cebids by being endowed with definitely longer hands. In considering nonprimate arboreal mammals indicated by a star in *Figure 3*, it is should be noted that the claw-climber *Sciurus* (Rodentia) is located between the callitrichids and *Daubentonia* (as are also several fossil primates, discussed below).

Thus, we see that the length of the hand in relation to the total forelimb length is characteristic of such locomotor modes as vertical-clinging-and-leaping or claw-climbing. Terrestrial quadrupedalism is performed by a relatively shorter hand than arboreal quadrupedalism. The severe constraints of brachiating are compatible with different hand proportions, reflecting differences between hylobatid and ateline modes of brachiating, including the use of the tail in the latter (STERN *et al.*, 1980; PREUSCHOFT & DEMES, 1984a).

EVOLUTIONARY ANALYSIS

These results raise the question concerning the hand proportions in relation to forelimb that existed in the ancestral euprimate morphotype (*sensu* SZALAY & DAGOSTO, 1980). If we consider the cluster which includes the greatest number of families in *Figures 3 and 4*, a parsimony hypothesis would suggest that this cluster also comprises the euprimate morphotype. If so, then a minimum number of locomotor adaptational changes would explain the complexity of primate locomotor evolution. According to our results, the generalized arboreal quadruped locomotory pattern would appear to be the best candidate for the euprimate functional morphotype (see an analogous deductive reasoning in STERN & OXNARD, 1973). However such reasoning is too simplistic, due to the diversity, the long history, and the complexity of the evolution of primate locomotor modes.

Other considerations may be taken into account; these are related to the fact that many supposed primitive extant primates have long hands (strepsirhines, tarsiers and callitrichids). If claws are primitive for euprimates (LE GROS CLARK, 1936; CARTMILL, 1974; HERSHKOVITZ, 1977), then claw-climbers should include the euprimate morphotype. The callitrichid claws, however, have been alternatively considered as reverted nails instead of primitive structures (ROSENBERGER, 1977; FORD, 1987). Also, if it were true that vertical clinging-and-leaping was a primitive mode of locomotion (NAPIER & WALKER,

1967), the euprimate morphotype would correspond to the hand proportions of the vertical-clingers-and-leapers. Both the vertical-clingers-and-leapers and claw-climbers have relatively long hands, so that whatever group comprised the ancestral euprimates, it would be likely that they possessed relatively long hands. This hypothesis of a long hand characterizing the euprimate morphotype seems to us the most plausible; it is reinforced by the location of the Eocene adapid *Notharctus* in *Figure 3*. As a consequence of this hypothesis, a shortening of the hand in relation to forelimb length would have been a common trend of euprimate hand evolution.

Intrinsic proportions of the hand

Hand shape has been quantified by considering: (1) the length of the three main parts of the hand, i.e.carpus (c), metacarpus (m) and digits (d), in relation to the total length of the hand (here, «digits» refers to the phalangeal part of the longest finger); (2) the length of the digital rays 1 to 4 in relation to one another; and (3) the length of metacarpals 1 to 4 in relation to each other.

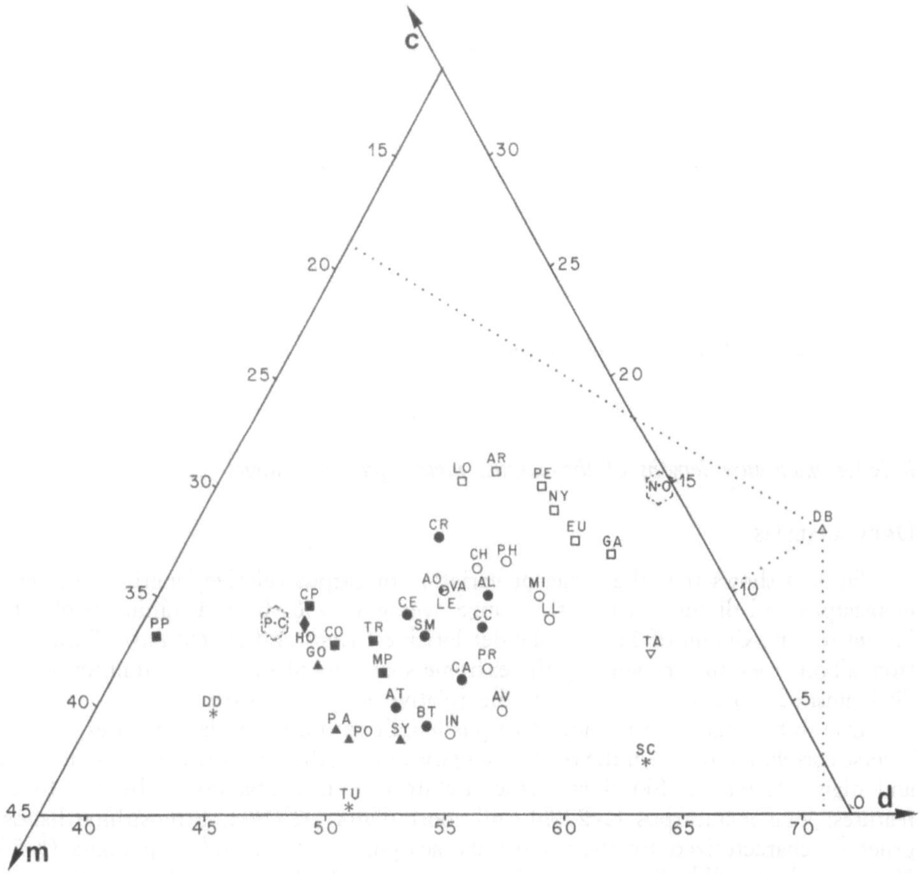

Figure 5 - Relative lengths of the carpus (c), longest metacarpus (m) and longest digit (d) in relation to the total hand length (in percentage). Same symbols as *Figure 3*.

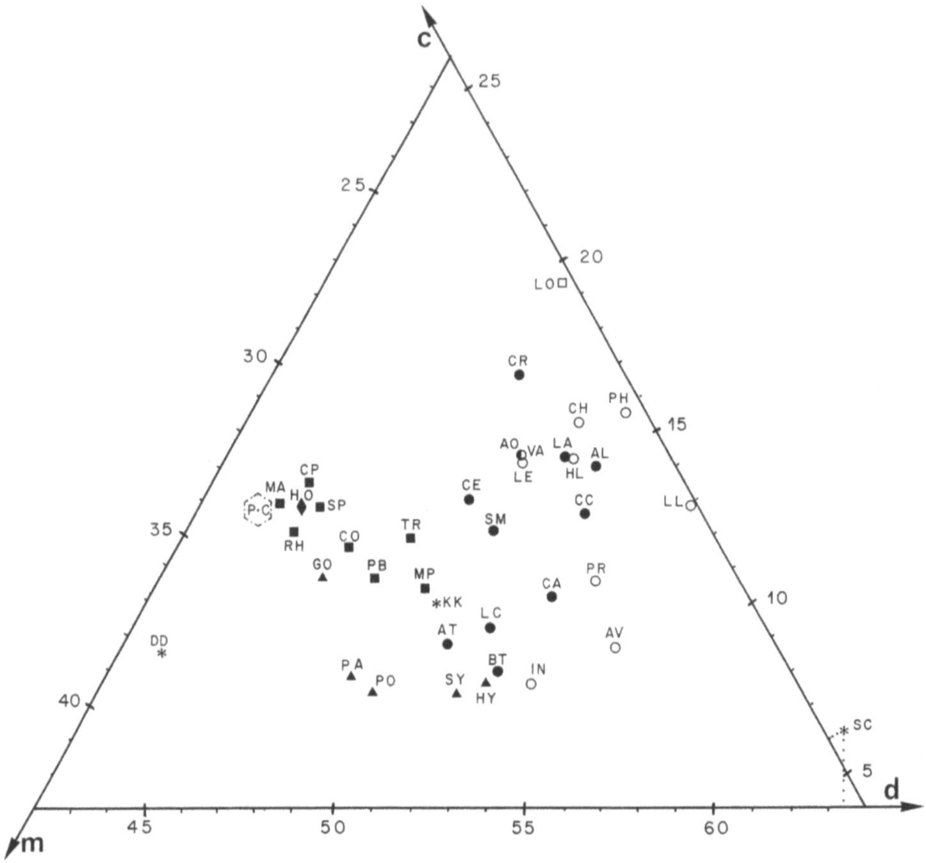

Figure 6 - Enlargement of the central area of *Figure 5*. Same symbols as *Figure 3*.

Relative maximum lengths of the carpus, metacarpus, and digits

DATA ANALYSIS

Figure 5 shows that the range of variation of carpus relative length is smaller than metacarpus and digital lengths: this range averages 15%, from a minimum of 10% in *Tarsius* to a maximum of 19% in slender lorisines(*Loris* and *Arctocebus). Tarsius* differs from all other extant primates by the extreme shortness of its carpus. Simiiforms are not distinguished from strepsirhines by the relative length of their carpus.

Due to the quasi-consistency of carpus length within the order Primates, there is an inverse correlation between the relative lengths of the other two segments, i.e. metacarpus and digits (*Figure 5*). Simiiform hand architecture is characterized by the following features: long metacarpus (>27%) and short digits (<57%). Strepsirhine hands are generally characterized by their short metacarpus (<29%) and long digits (>56%). However, the indriids differ from other strepsirhines by having hand proportions similar to those of simiiforms (*Figure 6*).

The transition zone between simiiform (and indriids) and strepsirhine areas in *Figure 5* (carpus about 14/16%; metacarpus 27/29%; digits 55/57%) includes cebids (*Alouatta, Aotus, Chiropotes, Callicebus*), lemurids (*Lemur, Varecia*) and a cheirogaleid (*Cheirogaleus*). It should be noted that these are proportions occurring in relatively unspecialized quadrupeds; they very likely correspond to the basic pattern (see below the discussion about the euprimate morphotype concerning intrinsic hand proportions).

In *Figure 5*, three genera stand apart. *Papio* has the longest carpus (20%) and metacarpus (36.5%) and the shortest third digit among primates. These proportions are very different from those of other cercopithecids. On the opposite side of *Figure 5*, *Tarsius* is characterized by the shortest carpus among primates (about 9%) and by its very long digits (about 63%). *Daubentonia*'s hand is unique among primates: very short carpus (almost as short as that of *Tarsius*), by far the shortest metacarpus (19%), and extremely long digits (about 71%).

Lorisids as a whole are characterized by their short metacarpus (about 24%). Within this family, the two subfamilies are distinct (*Figure 5*): lorisines have a longer carpus (about 19%) and shorter digits (about 56/58%), and galagines have a shorter carpus (14/15%)

TABLE 3 - *Lengths of the carpus (c), metacarpus of the longest finger (m), and digit of the longest finger (d) in relation to the hand length (Hd) in simiiforms. Mean and standard deviation.*

Genera		N.	c/Hd		m/Hd		d/Hd	
			x̄	s.d.	x̄	s.d.	x̄	s.d.
Callithrix	(CA)	6	12.7	1.0	31.6	0.7	55.7	0.7
Leontocebus	(LC)	7	12.9	1.7	33.0	2.6	54.1	2.4
Cebus	(CE)	11	16.0	1.6	30.5	1.2	53.5	1.7
Saimiri	(SM)	8	14.9	1.3	30.9	1.0	54.2	1.5
Aotus	(AO)	8	16.2	0.8	28.9	1.2	54.9	1.5
Callicebus	(CC)	3	14.1	0.1	29.3	0.2	56.6	0.1
Chiropotes	(CR)	2	18.1	0.5	27.1	0.4	54.8	0.9
Alouatta	(AL)	8	15.0	1.2	28.2	1.1	56.8	1.1
Ateles	(AT)	9	13.1	1.0	33.9	1.5	53.0	1.4
Brachyteles	(BT)	1	11.8	—	33.9	—	54.3	—
Lagothrix	(LA)	6	15.6	0.9	28.3	1.2	56.1	1.0
Cercopithecus	(CP)	11	18.5	1.4	32.2	0.9	49.3	1.5
Miopithecus	(MP)	7	14.6	1.4	33.0	1.4	52.4	1.4
Papio	(PP)	10	20.6	1.2	36.4	1.4	43.0	1.7
Macacus	(MA)	12	18.3	1.8	33.1	1.7	48.5	1.7
Semnopithecus	(SP)	5	17.8	0.3	32.6	2.0	49.6	1.8
Colobus	(CO)	10	16.5	1.5	33.1	2.2	50.4	2.4
Presbytis	(PB)	10	15.4	1.6	33.5	1.5	51.1	2.5
Rhinopithecus	(RH)	2	17.6	0.4	33.5	2.7	48.9	2.4
Trachypithecus	(TR)	2	15.9	1.5	32.1	0.9	52.0	0.6
Hylobates	(HY)	8	11.7	0.7	34.3	1.3	54.0	1.3
*Symphalangus**	(SY)	3	11.9	1.3	34.9	1.8	53.2	1.5
Pongo	(PO)	9	13.0	1.0	36.0	1.3	51.0	1.7
Pan	(PA)	14	13.6	0.7	35.9	0.7	50.5	1.0
Gorilla	(GO)	6	16.2	1.6	34.1	3.4	49.7	3.2
Homo	(HO)	3	18.0	1.3	32.9	2.9	49.1	2.8

* *Symphalangus* = *Hylobates syndactylus*

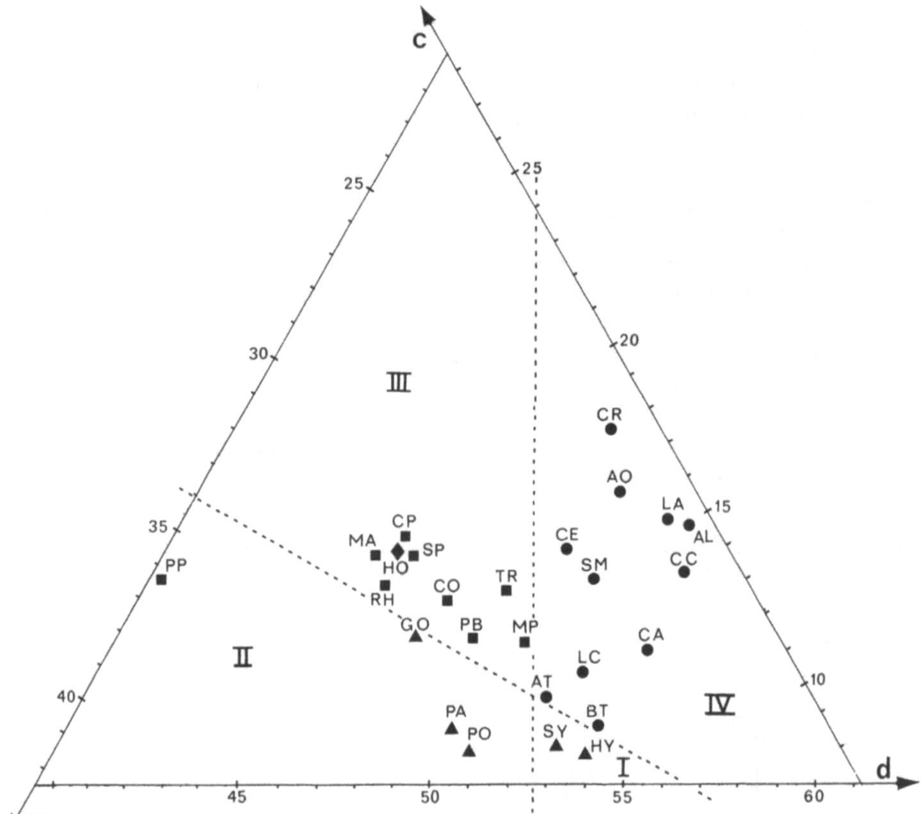

Figure 7 - Intrinsic proportions of simiiform hands (in percentage of the total hand length). Same symbols as *Figure 3*.

and longer digits (60/62%). Among lorisines, the two slender *Loris* and *Arctocebus* have shorter digits (56/57%) and longer carpus (19%) than the heavy-set *Perodicticus* and *Nycticebus* (longest digit measuring more than 59% versus carpus less than 18%).

Figure 7 shows that the intrinsic proportions of their hands characterize the different groups of simiiforms quite well.

(1) Apes differ from monkeys (*Papio* excepted) by having a longer metacarpus (m>34%; areas I and II in *Figure 7*). (2) Among monkeys, cercopithecoids are distinct from ceboids by having a shorter digit (m<52%; area III in *Figure 7*). (3) Among cercopithecoids, the unique architecture of *Papio*'s hand (extremely short digit, very long metacarpus and carpus) stands out. The marked dissimilarity between the dwarf guenon *Miopithecus* (longest digit and shortest carpus among cercopithecids) and the Allen's guenon *Cercopithecus* was also an unexpected result of this study. The dispersion of both genera (see standard deviation, *Table 3*) do not overlap. The dwarf guenon's hand is significantly different from that of other guenons. It is interesting because, among cercopithecids, *Miopithecus* is located at the opposite of *Papio*, the most terrestrial genus (along «d» axis, *Figure 7*). *Miopithecus* is believed to have «only recently become predominantly arboreal» (Rollinson & Martin, 1981). Its hand proportions may well reflect rapid

changes in hand function and morphology associated with this environmental shift, before more complex anatomical readjustments could have taken place. Digital elongation may be one of the first and simplest changes allowing a shift from a more terrestrial milieu to a more arboreal one.

(4) Among hominoids, great apes (area II in *Figure 7*) differ from lesser apes (area I in *Figure 7*) by having a shorter digit (<51%) and longer carpus (>13%).

As regards the functional specializations of the hand, it should be noted that all the true brachiators (atelines and hylobatids) display very similar proportions: carpus 12/13%; metacarpus 34/35%; digit 53/54%.

Unlike hand length in relation to the forelimb, the intrinsic proportions of the three main elements of the hand are more a reflection of primate phylogeny than of locomotor modes, because many families appear quite well clustered (only the Cebidae and Lemuridae overlap broadly). In general, hand architecture appears related to the phylogenetic (and adaptational) history of each group. This is not to say that locomotion is not involved, but only that the relationship is less apparent. Brachiating and terrestrial quadrupedalism are major exceptions: *Ateles* and *Brachyteles* are located far from other cebids and close to the hylobatids in *Figure 6*, and *Papio* is located completely apart from other cercopithecids. Obviously, locomotor behaviors that are particular to a single taxonomic group, such as the slow-crawling lorisines or claw-climbing callitrichids, are well clustered. In contrast, genera that practice similar modes of locomotion, although they belong to different families, are scattered: the vertical clingers-and-leapers are spread on the right of *Figure 5* with *Hapalemur* within a group of generalized arboreal quadrupeds. Many of the latter are clustered in the middle zone, but some others are located farther on the left.

Speculations concerning intrinsic proportions of the primitive euprimate hand

The diversity of the intrinsic proportions of primate hands raises the question of the ancestral euprimate morphotype. Analyzing the data from *Figure 5*, it is tempting to assume that the transition zone between simiiforms and lemuriforms includes the proportions of the ancestral euprimate morphotype. However, this presumes that this ancestor was a generalized arboreal quadruped. In other respects, the very short carpus of *Tarsius* is often considered as primitive among extant primates, because it evokes nonprimate hands (note *Tupaia*, with its similarly short carpus, and *Sciurus*, with its shorter one, in *Figure 5*). The transition zone defined above does not include genera with such a short carpus. If we try to detect a trend in *Figure 5* from more «primitive» to more «evolved» hands among extant primates, this would involve a transition from hand morphology with long digits and short metacarpals toward shorter digits and longer metacarpals. To generalize such a broad trend and assume it is the true path of evolution is tentative and oversimplified. But it is interesting to note that such a trend would point to the galagine hand as possibly the most primitive one among extant primates with regard to intrinsic hand proportions. The galagine hand is also considered as primitive among extant primates, with regards to prehension. This hypothetical trend suggests that the primitive euprimate hand proportions should be in the right part of the figure, somewhere in an area including the galagines and *Tarsius*. This hypothesis needs to be tested with the available fossils (see below). The positions of *Tarsius* and *Daubentonia* are very peculiar. It appears that *Daubentonia* is separated from other strepsirhines by essentially longer digits. Similarly,

Tarsius is separated from other haplorhines by its much longer digits. This recalls the position of *Miopithecus* compared to the other cercopithecines (although the phylogenetic relationships are much less close). It is thus tempting to propose the hypothesis that both *Tarsius* and *Daubentonia* secondarily acquired longer hands; this shift would have been accomplished most easily by increasing digital length. It should be possible to test these tentative hypotheses through the study of other parts, pending the recovery of relevant fossils.

Lengths of digital rays 1 to 4 in relation to each other

DATA ANALYSIS

The shape of the hand depends on the lengths and directions of the digital rays (metacarpus + phalanges) in relation to one another. Previous studies (JOUFFROY & LESSERTISSEUR, 1959, 1960, 1977, 1978, 1979; JOUFFROY, 1962, 1971, 1975) revealed that primate hand shapes can be classified into the following three «axonic» groups on the basis of the relative length of digital rays 2 to 4:

(1) the «mesaxonic» type corresponds to the basic mammalian pattern. Disregarding the thumb abductibility and focusing on the lengths of the digital rays in relation to each other, the hand is roughly symmetrical, with its axis of symmetry along the third ray, which is the longest. The fourth and the second rays are almost equal in length, and slightly shorter than the third. In the absence of any voluntary ulnar or radial deviation,

TABLE 4 - *Lengths of the carpus (c), metacarpus of the longest finger (m), and digit of the longest finger (d) in relation to the hand length (Hd) in strepsirhines and Tarsius. Mean and standard deviation.*

Genera		N.	c/Hd		m/Hd		d/Hd	
			x̄	s.d.	x̄	s.d.	x̄	s.d.
Varecia	(VA)	9	16.2	1.0	28.9	1.6	54.9	1.6
Lemur	(LE)	39	16.1	1.4	29.0	1.6	54.9	1.9
Lepilemur	(LL)	11	12.9	1.8	27.7	0.9	59.4	1.9
Hapalemur	(HL)	9	15.5	1.7	28.2	1.0	56.3	1.4
Microcebus	(MI)	16	14.0	2.3	27.0	1.9	59.0	2.6
Cheirogaleus	(CH)	7	16.2	1.9	27.2	1.3	56.4	2.8
Phaner	(PH)	2	15.8	1.0	26.6	0.6	57.6	1.7
Propithecus	(PR)	6	12.5	1.3	30.7	1.3	56.8	0.4
Indri	(IN)	7	11.1	0.6	33.7	0.7	55.2	0.8
Avahi	(AV)	6	10.8	2.0	31.8	1.1	57.4	2.1
Daubentonia	(DB)	9	10.2	0.9	18.9	0.8	70.9	1.1
Loris	(LO)	10	19.5	1.5	24.6	1.6	55.9	2.0
Nycticebus	(NY)	21	16.6	2.5	23.8	1.6	59.6	1.8
Perodicticus	(PE)	7	17.7	2.2	23.3	1.4	59.1	1.6
Arctocebus	(AR)	14	19.1	2.2	23.7	1.3	57.2	2.5
Galago	(GA)	54	13.9	2.0	24.2	1.6	61.9	2.3
Euoticus	(EU)	20	15.1	1.2	24.5	1.4	60.4	1.2
Tarsius	(TA)	10	9.4	2.4	27.5	1.4	63.1	2.7

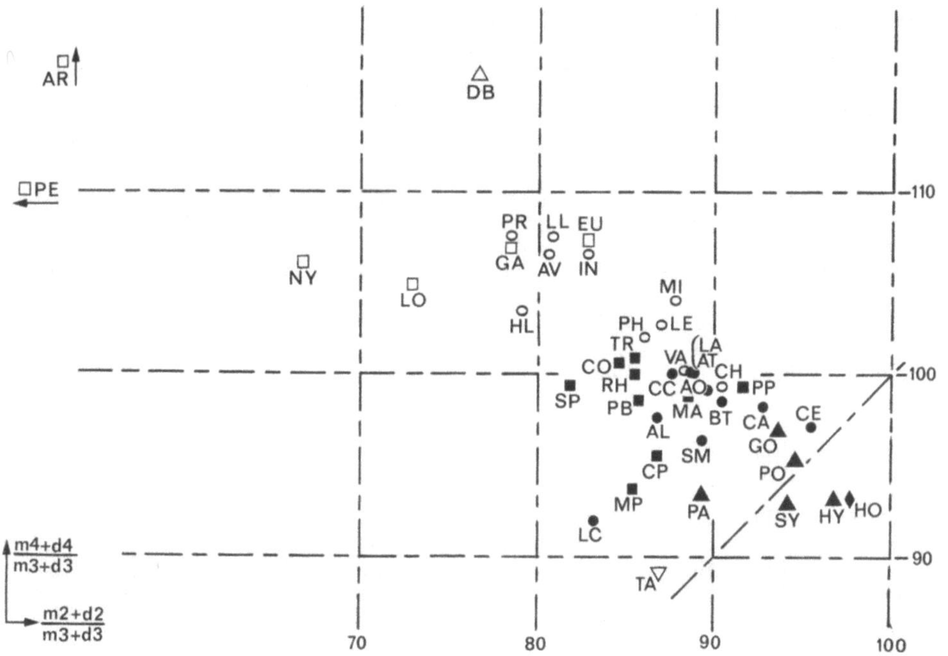

Figure 8 - Lengths of the digit rays 2 and 4 in relation to the digital ray 3 (in percentage). Same symbols as *Figure 3*.

the hand assumes a «reference» position (ZIEMER, 1978), with its symmetry axis parallel to the forearm axis. The hands of *Tarsius*, the hylobatids and *Homo* are the best examples of the mesaxonic type.

(2) Third and fourth digital rays of the same length characterize the «paraxonic» type, with the axis of symmetry going through the interdigital space between rays 3 and 4. This type is to be found in short-thumbed hands, such as those of *Colobus* and *Ateles*.

(3) The «ectaxonic» type is characterized by the prominence of the fourth ray, the longest one. Among the most striking examples of ectaxonic hands are the pincer-like hands of lorisines. The hypertrophied thumb makes up by itself one jaw of the pincer, whereas the four ulnar rays make up the other jaw. Three axes can be defined in such hands : the mechanical axis of the pincer, in the same plane as the forearm axis in the »reference» position of the hand, bisects the angle formed by the two jaws; and the two morphological axes of each jaw, which go respectively along the thumb and the fourth ray.

In order to classify the primate hands within these types, we have related the lengths of digital rays 1,2, and 4 to that of the middle ray 3. *Figure 8* shows that the fourth ray is longer than the third one (ectaxony) in most strepsirhines, whereas the third ray is the longest (mesaxony) in most simiiforms and *Tarsius*. The rays 3 and 4 are the same length (paraxony) in such strepsirhines as *Varecia* and *Cheirogaleus*, as well as in some simiiforms, such as the colobines, *Ateles*, *Cercocebus*, and *Lagothrix*. In 14 nonhominoid simiiform genera out of the 19 studied here, the length of the fourth digit ray in relation to the third averages 98% to 101%. When taking into account the individual variability of this ratio, as shown by the standard deviation (*Table 5*), it appears that paraxony, instead of mesaxony, is the most frequent pattern of the simiiform hand. A marked mesaxony

TABLE 5 - *Lengths of the digital rays 1 (m1 + d1), 2 (m2 + d2), and 4 (m4 + d4) in relation to the digital ray 3 (m3 + d3) in simiiforms. Mean and standard deviation.*

Genera		N.	(m1 + d1)/(m3 + d3) x̄	s.d.	(m2 + d2)/(m3 + d3) x̄	s.d.	(m4 + d4)/(m3 + d3) x̄	s.d.
Callithrix	(CA)	6	61.9	2.4	92.9	4.5	98.2	4.6
Leontocebus	(LC)	7	54.8	5.6	83.4	5.0	92.0	3.8
Cebus	(CE)	11	62.5	4.2	95.3	4.1	97.4	2.1
Saimiri	(SM)	8	58.7	5.9	89.3	3.4	96.4	2.8
Aotus	(AO)	8	55.6	1.4	89.6	3.9	99.1	3.0
Callicebus	(CC)	3	53.2	2.9	87.8	1.4	100.1	0.6
Alouatta	(AL)	8	53.8	1.6	86.4	1.8	97.7	5.7
Ateles	(AT)	9	19.7	2.4	88.9	1.7	100.0	1.9
Brachyteles	(BT)	1	25.4	—	90.2	—	98.5	—
Lagothrix	(LA)	6	55.5	3.2	88.8	1.5	100.1	1.2
Cercopithecus	(CP)	11	49.3	1.8	86.5	4.6	95.5	4.6
Miopithecus	(MP)	7	45.8	2.9	85.2	2.4	93.8	2.4
Papio	(PP)	10	54.6	4.0	91.6	4.1	99.3	2.9
Macacus	(MA)	12	50.5	4.0	88.5	5.3	98.8	3.6
Semnopithecus	(SP)	5	37.0	6.2	81.9	5.8	99.7	5.4
Colobus	(CO)	10	22.3	5.8	84.6	5.0	100.7	5.8
Presbytis	(PB)	10	37.3	3.6	85.7	1.9	98.7	0.8
Rhinopithecus	(RH)	2	34.3	1.0	85.6	0.3	100.1	1.1
Trachypithecus	(TR)	2	33.4	0.1	85.6	2.8	100.8	0.4
Hylobates	(HY)	8	45.0	1.9	96.8	3.4	93.0	1.9
*Symphalangus**	(SY)	3	42.3	1.0	94.1	1.2	92.9	1.1
Pongo	(PO)	9	36.6	2.7	94.4	3.0	95.2	4.2
Pan	(PA)	14	41.0	2.4	89.4	2.9	93.5	2.6
Gorilla	(GO)	6	45.9	3.8	93.3	4.3	96.9	2.0
Homo	(HO)	3	65.6	3.2	97.8	3.4	93.3	2.2

* *Symphalangus* = *Hylobates symphalangus*

(= hypermesaxony) characterizes the humans, hylobatids, and tarsiers: *Tarsius* differs from all other primates by having the relatively shortest fourth ray in relation to the third one (89%). *Daubentonia* and *Arctocebus* have the longest fourth rays among primates (116% and 127%, respectively). The evolutionary significance of these findings is discussed at the end of this section.

Figure 8 also shows that there is an inverse correlation between the lengths of the second and fourth digital rays in relation to the third. The longer the fourth ray, the shorter the second one (*Perodicticus* and *Arctocebus*), and conversely (*Hylobates*, *Homo*). Most mesaxonic hands present at once two characteristics: (1) the ratio (m4 + d4)/(m3 + d3) is minimum, and (2) the two ratios (m4 + d4)/(m3 + d3) and (m2 + d2)/(m3 + d3) are almost equal. The second characteristic corresponds to the dashed diagonal line at the lower right square in *Figure 8*. Tarsiers, hylobatids and humans are the extreme examples of this mesaxonic type (hypermesaxony). In all nonhominoid simiiforms (*Cebus* excepted), *Pan*, *Tarsius*, and most strepsirhines, the length of the second ray in relation to the third one varies from about 80% to 93%. It is shorter in galagines,

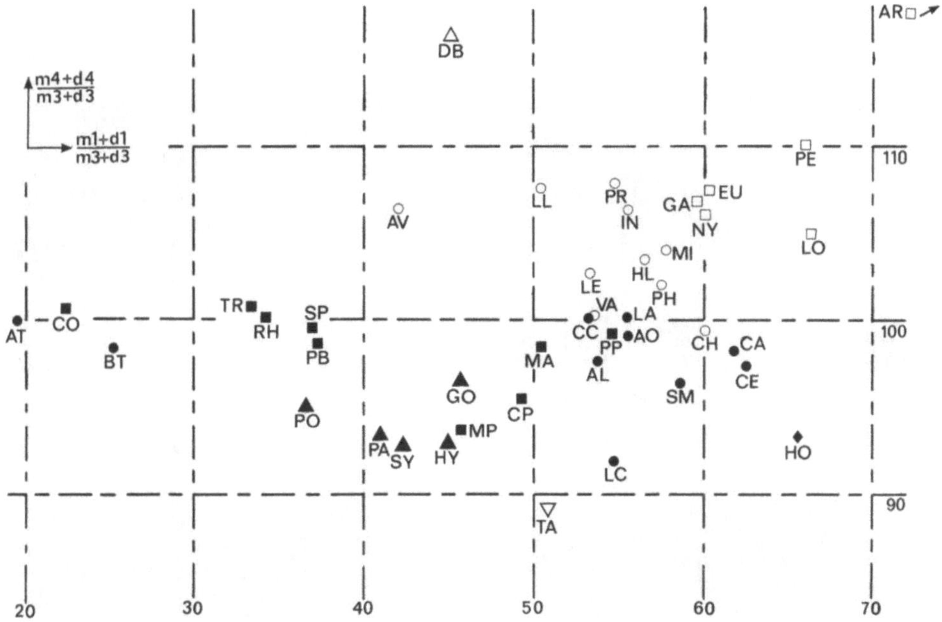

Figure 9 - Lengths of the digital rays 1 and 4 in relation to the digital ray 3 (in percentage). Same symbols as *Figure 3.*

indriids, *Daubentonia*, and lorisines (the African lorisines having the most reduced second digital ray among primates: 46% of digital ray 3 in *Perodicticus*). On the contrary, the second ray is longer than the average in apes (*Pan* excepted) and *Cebus*. It is even longer than the fourth ray in the hylobatids and *Homo*, the latter having the longest second ray among all primates (96% of the third ray).

As regards the relationship between the thumb ray and the other digital rays, *Figure 9* shows that in many strepsirhines, simiiforms, and tarsiers, the length of the first ray amounts to more than 50% of that of the third ray. Among genera with a shorter thumb ray, we find: the colobines, (less than 37%, with a minimum of about 22% in *Colobus*), the atelines *Ateles* and *Brachyteles* (25% and 19%, respectively), as well as the lesser and great apes (from 46% in *Gorilla*, to as low as 36% in *Pongo*). Among genera with the longest thumb rays, we find *Homo*, *Loris*, and the two African lorisines (as much as 93% in *Arctocebus*). It is to be noted that most catarrhine monkeys have relatively shorter thumbs (less than 50%) than most platyrrhines and strepsirhines. Among the notable exceptions are: (1) the terrestrial *Papio*, with a markedly longer thumb than the arboreal cercopithecines; (2) the atelines, which differ from other cebids by their greatly reduced thumbs. As discussed by STRAUS (1942), NAPIER & NAPIER (1967) and JOUFFROY & LESSERTISSEUR (1977), the *Ateles* hand is very similar to *Colobus* in this respect, although both taxonomy and locomotor behavior completely distinguish the two genera. Even though several researchers have interpreted thenar reduction in *Colobus* as related to suspensory behaviour (e.g., «brachiator» [STRAUS 1942, 1949]; «semibrachiator» [NAPIER & NAPIER 1967; discussion in STERN & OXNARD, 1973, MITTERMEIER & FLEAGLE, 1976]),

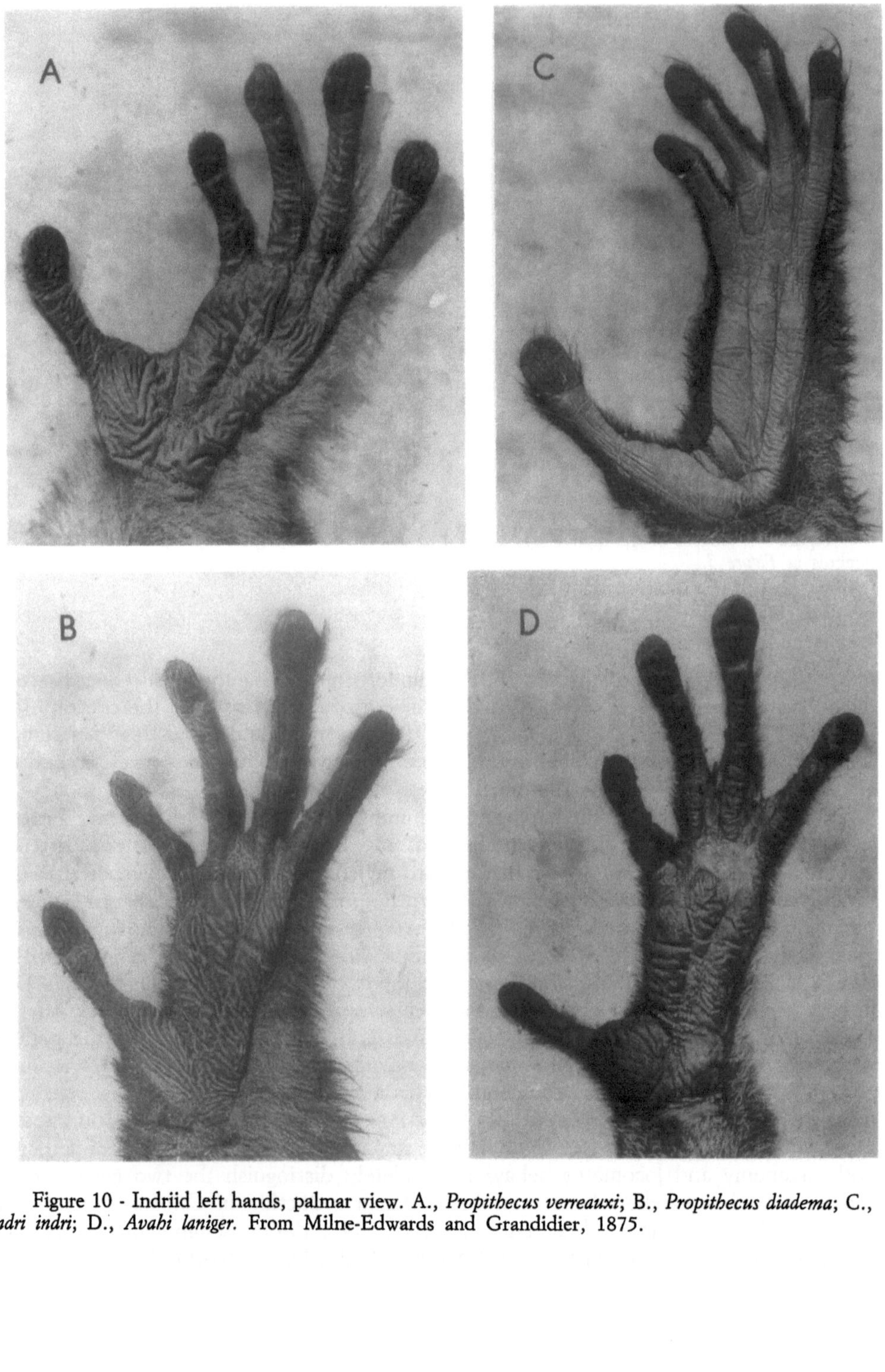

Figure 10 - Indriid left hands, palmar view. A., *Propithecus verreauxi*; B., *Propithecus diadema*; C., *Indri indri*; D., *Avahi laniger*. From Milne-Edwards and Grandidier, 1875.

TABLE 6 - *Lengths of the digital rays 1 (m1 + d1), 2 (m2 + d2), and 4 (m4 + d4) in relation to the digital ray 3 in strepsirhines and Tarsius. Mean and standard deviation.*

Genera		N.	(m1 + d1)/(m3 + d3) x̄	s.d.	(m2 + d2)/(m3 + d3) x̄	s.d.	(m4 + d4)/(m3 + d3) x̄	s.d.
Varecia	(VA)	9	53.6	2.8	88.7	2.9	100.3	12.1
Lemur	(LE)	39	53.4	2.7	87.0	2.7	102.7	3.4
Lepilemur	(LL)	12	50.4	2.5	80.9	3.2	107.6	1.9
Hapalemur	(HL)	8	56.6	1.8	79.2	3.0	103.5	2.5
Microcebus	(MI)	17	57.9	5.9	87.8	4.0	104.0	4.8
Cheirogaleus	(CH)	9	60.1	2.7	90.3	2.2	99.4	3.8
Phaner	(PH)	2	57.5	1.4	86.0	0.7	102.0	0.7
Propithecus	(PR)	6	54.7	2.5	78.7	1.3	107.8	3.8
Indri	(IN)	7	55.6	2.7	82.6	2.9	106.3	1.5
Avahi	(AV)	7	42.1	2.7	80.3	4.1	106.4	6.3
Daubentonia	(DB)	10	45.2	2.0	76.7	2.2	116.3	5.0
Loris	(LO)	9	66.3	2.2	73.0	4.3	104.9	3.0
Nycticebus	(NY)	22	60.1	9.1	66.9	4.7	106.0	4.1
Perodicticus	(PE)	8	66.0	4.0	46.1	4.1	110.0	5.7
Arctocebus	(AR)	18	93.4	5.7	53.1	3.8	127.2	5.0
Galago	(GA)	53	59.6	5.8	78.4	6.4	107.0	5.8
Euoticus	(EU)	20	60.3	2.5	82.9	2.2	107.4	3.7
Tarsius	(TA)	11	50.9	3.8	86.9	4.6	89.2	5.8

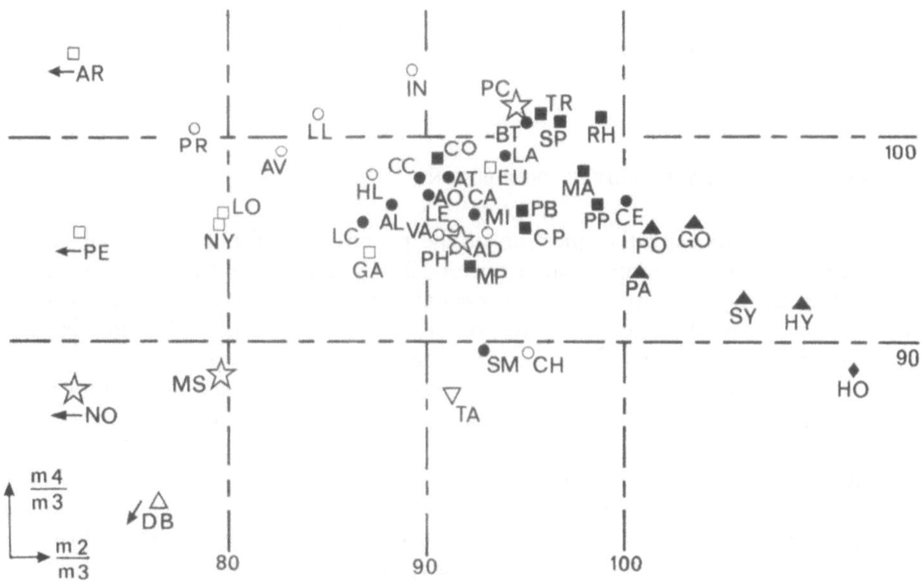

Figure 11 - Lengths of the metacarpals 2 and 4 in relation to the metacarpal 3 (in percentage). Same symbols as *Figure 3*.

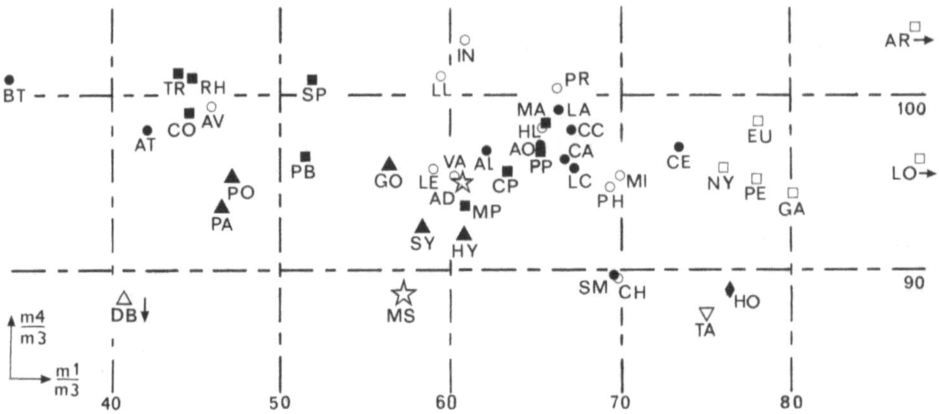

Figure 12 - Lengths of the metacarpals 1 and 4 in relation to the metacarpal 3 (in percentage). Same symbols as *Figure 3*.

field observations have demonstrated that free-ranging *Colobus*, unlike *Ateles*, does not brachiate (according to MITTERMEIER & FLEAGLE's definition, 1976) and rarely practices arm suspension (MORBECK, 1979); (3) *Avahi* is unexpectedly very different from the two other indriids with regard to its very short thumb ray (*Figures 9, 10*); and (4) the relative shortness of *Daubentonia*'s thumb is one among the numerous features that make this genus «unique» (OXNARD 1981, 1983) among primates.

There is no precise correlation between the relative lengths of the first and fourth rays, except among lorisines, the elongation of their fourth digital ray being associated with that of the thumb ray. In the paraxonic type (colobines and atelines), defined by the equal length of the third and fourth rays, we find a series of genera showing diverse degrees of thumb ray atrophy (shortest ray in *Ateles*: less than 20% of the third ray length).

ASSESSMENT OF PRIMITIVE DIGITAL PROPORTIONS

The above results raise the question of the primitive condition for thumb proportions. *Figure 9* suggests a proportion between 50% and 60% of the third ray length, corresponding to the majority of genera, assuming that there was not an evolutionary trend common to their lineages. Likewise, concerning the length of digit 2 in relation to digit 3, we can postulate that the ancestral primate proportion amounted to about 80-90%. A great taxonomic diversity, including genera from several families of catarrhines, platyrrhines and lemuriforms, as well as tarsiers, appears in *Figure 8* around a proportion of 85%.

As for the variability of digit 4 length in relation to digit 3, it is contained between 97% and 101% for the majority of platyrrhines and cercopithecids (*Table 5*), expressing paraxony. Nearly all simiiforms as well as tarsiers are located below the 100% line, whereas almost all lemuriforms are located above it. The principle of parsimony would suggest that ectaxony is primitive for lemuriforms and mesaxony for haplorhines; but what would be then the primitive condition for both groups together? (i.e. what is the

TABLE 7 - *Lengths of the metacarpals 1 (m1), 2 (m2), and 4 (m4) in relation to the metacarpal 3 (m3) in simiiforms. Mean and standard deviation.*

Genera		N.	m1/m3 x̄	m1/m3 s.d.	m2/m3 x̄	m2/m3 s.d.	m4/m3 x̄	m4/m3 s.d.
Callithrix	(CA)	6	66.8	1.6	92.5	6.2	96.3	5.4
Leontocebus	(LC)	7	67.4	9.8	86.9	4.2	95.9	3.8
Cebus	(CE)	11	73.5	5.2	100.2	10.4	97.0	3.1
Saimiri	(SM)	8	69.7	4.7	93.0	4.8	89.7	5.0
Aotus	(AO)	8	65.5	4.1	90.2	5.8	97.2	5.5
Callicebus	(CC)	3	67.2	2.3	89.8	3.6	98.0	1.8
Alouatta	(AL)	8	62.3	2.6	88.3	3.6	96.7	7.1
Ateles	(AT)	9	42.3	4.5	91.1	2.9	98.0	3.5
Brachyteles	(BT)	1	34.0	—	95.2	—	100.8	—
Lagothrix	(LA)	6	66.6	5.7	94.1	3.7	99.1	2.6
Cercopithecus	(CP)	11	63.6	2.5	95.1	4.0	95.7	3.1
Miopithecus	(MP)	7	61.2	6.5	92.2	5.8	93.8	3.8
Papio	(PP)	10	65.5	3.5	98.7	3.8	96.8	2.4
Macacus	(MA)	12	65.7	5.7	98.0	5.4	98.3	5.6
Semnopithecus	(SP)	5	52.0	4.1	96.6	9.1	100.9	4.0
Colobus	(CO)	10	44.9	5.0	90.6	7.2	99.0	1.9
Presbytis	(PB)	10	51.6	4.0	94.9	1.3	96.4	1.8
Rhinopithecus	(RH)	2	44.8	4.3	98.9	3.8	101.0	2.0
Trachypithecus	(TR)	2	44.1	0.8	95.7	2.5	101.2	3.0
Hylobates	(HY)	8	61.1	5.0	109.1	4.4	92.0	2.1
*Symphalangus**	(SY)	3	58.5	2.7	106.1	0.6	92.3	2.4
Pongo	(PO)	9	47.4	3.6	101.5	2.5	95.7	2.7
Pan	(PA)	14	46.8	3.4	100.8	1.8	93.5	2.7
Gorilla	(GO)	6	56.6	6.8	103.6	7.2	95.9	3.0
Homo	(HO)	13	76.3	7.9	111.8	13.0	88.9	10.0

* *Symphalangus* = *Hylobates syndactylus*

euprimate morphotype?). Because mesaxony is the primitive mammalian pattern and ectaxony a derived condition, it is tempting to assume that mesaxony was also primitive for euprimates and retained in haplorhines. However, the locomotor mode of the euprimate morphotype is a difficult and controversial question (see discussion in Godinot & Beard, 1991 and this volume). As there is a transitional zone between lemuriforms and simiiforms around the 100% line (*Figure 8* and 9), we will hypothesize here that this transitional zone represents the euprimate morphotype (alternatively, it might be a morphology convergently attained).

The accentuated mesaxony («hypermesaxony») of *Tarsius*, with its extremely predominant third digital ray, as well as hylobatid and hominid mesaxony, with their unusually short second digital ray, probably result instead from secondary specializations.

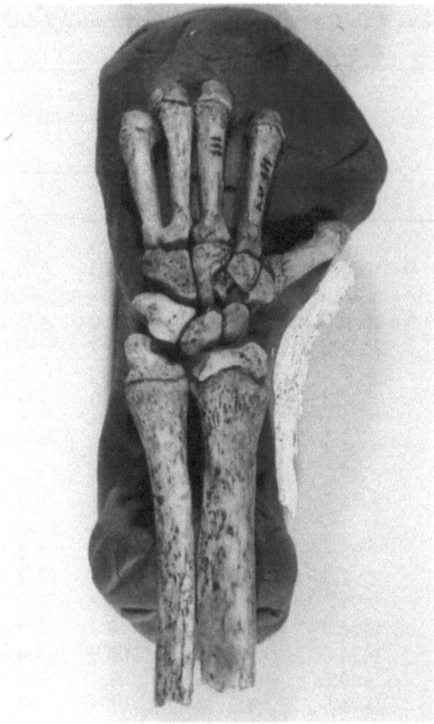

Figure 13 - Left *Adapis* hand, dorsal view.

Lengths of the metacarpals in relation to each other

Unlike fossil primate phalanges, only a few of which have been discovered, and in incomplete sets, the five metacarpals are known in four fossil genera (see below). Analyzing the relationship between the length of each metacarpal and the length of the digital ray it belongs to would aid in interpreting fossil hand morphology and function.

Figure 11 shows that in most primates, the fourth metacarpal is slightly shorter than the third ($90\% < m4/m3 < 100\%$),even in those genera whose fourth digit ray as a whole is definitely longer than the third one (e.g. *Loris*, *Nycticebus*, *Perodicticus*, *Euoticus*, and *Galago*). Due to the extreme shortness of its fourth metacarpal, and despite the great length of its fourth finger, *Daubentonia* is the most striking example of such a disparity between metacarpal and digital ray relative length ($(m4+d4)/(m3+d3) = 116\%$; as opposed to $m4/m3 = 68\%$), which expresses the unusually long phalanges of its fourth finger. Very few primates have a fourth metacarpal markedly longer than the third one, as in *Arctocebus* and *Indri*. In addition to *Daubentonia*, *Tarsius* and *Homo* possess the shortest m4 in relation to m3.

The wide diversity of m2 length in relation to m3 is also shown in *Figure 11*. In most nonhominoid simiiforms and nonlorisine strepsirhines, m2 amounts to 86-100% of m3. It is relatively much longer in the hylobatids and *Homo* and much shorter in the lorisines. Thus, marked specializations of the hand, such as the pincer-like function in lorisines, the hook-like function in hylobatids, and the prehensile function in humans, entail noticeable changes in metacarpal proportions.

TABLE 8 - *Lengths of the metacarpals 1 (m1), 2 (m2), and 4 (m4) in relation to the metacarpal 3 in strepsirhines and Tarsius. Mean and standard deviation.*

Genera		N.	m1/m3 x̄	m1/m3 s.d.	m2/m3 x̄	m2/m3 s.d.	m4/m3 x̄	m4/m3 s.d.
Varecia	(VA)	8	60.4	5.4	90.7	4.1	95.2	1.6
Lemur	(LE)	40	59.1	4.1	91.3	4.7	95.8	3.9
Lepilemur	(LL)	12	59.8	3.3	84.6	4.8	101.1	3.4
Hapalemur	(HL)	9	65.7	5.6	87.4	4.3	98.1	4.3
Microcebus	(MI)	19	70.1	7.6	93.2	6.9	95.4	6.0
Cheirogaleus	(CH)	10	69.9	5.5	95.1	5.8	89.7	6.9
Phaner	(PH)	3	69.5	0.5	91.7	0.5	94.7	1.7
Propithecus	(PR)	9	66.3	7.2	78.3	9.5	100.4	5.5
Indri	(IN)	7	61.2	3.7	89.4	4.5	103.1	2.7
Avahi	(AV)	7	46.0	3.4	82.7	1.8	99.3	7.0
Daubentonia	(DB)	9	40.4	3.5	52.7	2.4	68.4	3.8
Loris	(LO)	9	90.8	6.6	79.7	9.7	96.2	6.5
Nycticebus	(NY)	20	76.0	6.2	79.6	5.6	95.8	5.7
Perodicticus	(PE)	6	78.2	7.9	66.2	8.9	95.2	4.7
Arctocebus	(AR)	14	103.5	12.5	63.8	9.4	103.9	7.4
Galago	(GA)	59	80.3	8.4	87.2	7.0	94.3	6.7
Euoticus	(EU)	20	78.3	5.6	93.3	4.2	98.5	9.1
Tarsius	(TA)	12	75.2	6.6	91.4	4.7	87.8	6.4

As regards the relative length of the first metacarpal (*Figure 12*), we see that in most primates the length of m1 amounts to 50 to 75% of m3 length. The colobines, as well as *Pongo*, *Pan*, *Daubentonia*, and *Avahi* (distinguishing itself from the other indriids), have the shortest first metacarpals (i.e. less than 52% of m3 length). Surprisingly, *Brachyteles*, rather than *Ateles* or *Colobus*, has the shortest first metacarpal among all primates. As expected, because of their pincer-like hands, the lorisines have the longest first metacarpals, with the slender ones, *Arctocebus* and *Loris*, showing the highest m1/m3 ratio among primates (respectively 103% and 91%). Comparing the relative lengths of the digital rays (*Figures 8* and *9*) to the relative lengths of the metacarpals (*Figures 11, 12*) shows that there is not a strict relationship between the relative lengths of digital rays 1, 2 or 4 in relation to the third, and the relative lengths of the metacarpals 1, 2 or 4 in relation to the third. However, an exceptionally short or long metacarpal frequently goes together with a respectively short or long digit ray. As noted above, the fourth ray of *Daubentonia* (extremely long digit ray and extremely short metacarpus) is the most illustrative example of the exceptions to this general rule. The relative lengths of the metacarpals in relation to one another can be a helpful, though debatable, clue in interpreting fossils, as discussed below. The primitive condition is discussed in the Eocene adapiforms section.

Despite the paucity of complete forelimbs belonging to a single fossil primate individual, it was possible to obtain reliable length estimations for several fossil primates representing various families and epochs : namely, a Paleocene plesiadapiform (*Plesiadapis*

Fossil hands

TABLE 9 - *Estimated forelimb and hand proportions of fossil primates.*

		Forelimb			Hand		
		Hu/FL	Ra/FL	Hd/FL	c/Hd	m/Hd	d/Hd
Megaladapis edwardsi	(ME)	35.2	32.1	32.7			
Proconsul africanus	(PC)	38.3	35.7	26.0	18.5	33.5	48.0
Notharctus sp.	(NO)	32.5	32.0	35.5	15.2	20.7	64.1
Plesiadapis insignis	(PL)	36.0	31.5	32.5			

TABLE 10 - *Lengths of the metacarpals 1 (m1), 2 (m2), and 4 (m4) in relation to the metacarpal 3 (m3) in fossil primates.*

		m1/m3	m2/m3	m4/m3
Mesopithecus penteleci	(MS)	57.2	79.7	88.7
Proconsul africanus	(PC)		94.7	101.3
Notharctus tyrannus	(NO)		65.8	87.8
Adapis parisiensis	(AD)	60.4	91.8	95.0

insignis), two Eocene adapiforms (*Adapis parisiensis* and *Notharctus*), a Miocene cercopithecoid (*Mesopithecus pentelici*), a Miocene hominoid (*Proconsul africanus*), and a Malagasy subfossil (*Megaladapis edwardsi*). However, some of these estimations cannot be compared precisely to the measurements made on living primates. The corresponding forelimb and hand ratios were calculated by GODINOT & BEARD (1991 and this volume). The ratios expressing the lengths of the metacarpals in relation to one another were calculated: (1) from data in the literature for *Proconsul* and *Notharctus*; here, metacarpal lengths are the maximum lengths, i.e., to the apex of the distal joint surface; and (2) from GODINOT & BEARDS's (this volume) measurements for *Adapis* and *Mesopithecus*, i.e., length to the proximal dorsal border of the distal joint surface, which is closer to the length measured in this study on the mounted skeletons of extant primates. In both cases, fossil measurements are slightly overestimated as compared to those of extant genera. Because metacarpals were compared to one another, such a difference in measurements does not modify significantly the corresponding ratios.

On the basis of the above results, we will discuss each of the fossils, from the most recent to the oldest ones.

A SUBFOSSIL MALAGASY LEMURIFORM: *Megaladapis*

One carpal, a third metacarpal and several phalanges of *Magaladapis edwardsi* were described by LORENZ VON LIBURNAU (1905). Four metacarpals were carefully studied by

LAMBERTON (1934). The hand length used to plot *Megaladapis* in *Figure 3* is only an estimate (GODINOT & BEARD, this volume). *Figure 3* shows that the hand of *Megaladapis*, whose large size has been emphasized previously (LORENZ von LIBURNAU, 1905; LAMBERTON, 1934; ZAPFE, 1963), is relatively long in relation to forelimb length, when compared to the primate average. However, it does not differ from that of extant lemuriforms. It is slightly shorter than the hand of indriids, a feature which is probably not related to body size, because the largest indriid, *Indri*, has the relatively longest hand. Concomitantly, the *Megaladapis* radius is very short and its humerus is moderately long, when compared to other primates; however, the humerus is markedly longer than that of strepsirhines (three lorisines excepted: *Loris*, *Nycticebus* and *Arctocebus*).

It would be hazardous to infer the mode of locomotion for *Megaladapis* only from its hand length in relation to the forelimb, because the nearest genera in *Figure 3* are the much smaller, clawed callitrichids and the fossil plesiadapid. It has been suggested that *Megaladapis* practiced a koala-type of «modified vertical-clinging-and-leaping» (WALKER, 1974), or «vertical climbing» (JUNGERS, 1977). By relating forelimb length to that of the trunk, JUNGERS (1978) pointed out that *Megaladapis* presented a «moderate forelimb elongation». This applies to the humerus, but not to the radius. Indeed, the shortness of the *Megaladapis* radius in relation to the forelimb might be primitive, as such as shortness also characterizes the Paleogene fossils *Plesiadapis* and *Notharctus*.

A MIOCENE HOMINOID: *Proconsul africanus*

Most elements of a subadult left forelimb, including the almost complete hand, allocated to the Miocene hominoid *Proconsul africanus* and associated with skull parts and teeth, have been comparatively described by NAPIER & DAVIS (1959). Recent expeditions to Rusinga Island have yielded the partial skeleton of at least 9 individuals, including 49 carpals, so that the wrist anatomy is now completely known, in addition to several metacarpals and phalanges (BEARD *et al.*, 1986, BEARD *et al.*, this volume). For our purpose, it is sufficient to recall that *Proconsul* forelimb features have been interpreted controversially in terms of locomotor adaptations, being considered morphologically similar to either platyrrhine (and some catarrhine) palmigrade quadrupedal monkeys, or to chimpanzee-like arm swinger/ knuckle-walkers.

As regards the hand length related to forelimb length, *Figure 4* shows that the *Proconsul* hand is shorter than that of *Pan*, *Pongo*, and the arboreal cercopithecids. By comparison with these extant primates, the most salient feature is the greater length of the humerus (it is nevertheless close to that of *Cercopithecus*), which probably constitutes a primitive characteristic, as it is shared by many platyrrhines. *Figure 4* shows that the *Proconsul* radius length in relation to forearm length is a standard one for cercopithecids, pongids, and platyrrhines. The moderate length of the radius and the resulting low brachial index were interpreted by NAPIER & DAVIS (1959) as generalized features.

The relative shortness of the *Proconsul* hand in comparison to that of *Pan, Pongo* and the arboreal cercopithecids was also taken as a generalized feature by NAPIER & DAVIS (1959). According to this character, *Proconsul* is located in *Figure 4* at the lower limit of the generalized arboreal quadruped area. We noted above that among extant simiiforms a short hand characterizes terrestrial quadrupeds. Could the shortness of the *Proconsul* hand therefore be an indication of some degree of terrestriality? New data on the different species of *Cercopithecus* and other semi-terrestrial primate genera are required, before suggesting such a hypothesis. In any case, the *Proconsul* hand is notably short, which appears to be a specialized, rather than a generalized feature.

As for the intrinsic proportions of the hand, i.e. the relative lengths of its carpus, metacarpus and digital components (*Figure 6*), *Proconsul* is located within the cercopithecid area, next to *Macaca*. It should be noted that *Proconsul* stands completely apart from the platyrrhines, due to the great length of its carpus and the shortness of its digits. NAPIER & DAVIS (1959) suggested that these proportions are adaptive features. The intrinsic proportions of the *Proconsul* hand do not suggest any specialization for arm-swinging, or with chimpanzee-like knuckle-walking. Rather, we observe that *Proconsul* is located near the arboreal quadrupeds, on the same side of the cercopithecid area as the terrestrial *Papio*. This corroborates the emphasis placed by several authors (BEARD *et al.*, 1986; BEARD *et al.*, this volume) on the many cercopithecid-like characters of *Proconsul* limb bones. The proximity of *Proconsul* and *Macaca* in *Figure 6* should be noted; it substantiates the results of MCHENRY & CORRUCCINI (1983), who pointed out numerous similarities in the carpus of these two genera.

Concerning the relative lengths of metacarpals 2,3 and 4 in relation to one another (*Figure 11*), *Proconsul africanus* is located close to *Brachyteles* and the two colobines *Trachypithecus* and *Semnopithecus*. In *Figure 11*, *Proconsul* appears to be similar to living cercopithecids and to differ significantly from extant hominoids by its lower m2/m3 ratio and a higher m4/m3 ratio, i.e. by its shorter second metacarpal and longer fourth one. This probably reflects the primitive condition of hominoids; it may also be related to the quadrupedal locomotion of *Proconsul* (see BEARD *et al.*, this volume). In any case, the relative elongation of the second metacarpal, which is typical of all living hominoids, was not yet established in the Miocene genus *Proconsul*.

A MIOCENE CERCOPITHECID: *Mesopithecus pentelici*

The only ratios that we have estimated for the fossil colobine *Mesopithecus pentelici* are the lengths of its metacarpals in relation to one another (*Table 10*). Unfortunately, the third metacarpal of the specimen under study was fractured during fossilization, resulting in an overestimation of its length (see Figure 14 in GODINOT & BEARD, this volume). The three metacarpal ratios m1/m3, m2/m3, and m4/m3 are therefore underevaluated, making their interpretation somewhat uncertain. Nevertheless, the ratio m4/m3 must have been within the range of the extant cercopithecids, or very close to it (*Figure 11*). The ratio m2/m3 may have been rather low, possibly lower than in *Colobus* and *Miopithecus*. If so, it would represent the lowest value among cercopithecids, raising the question of whether it may represent the primitive condition for this family ? Particular attention should be paid to the ratio m1/m3, because the extant colobines are characterized by varying stages of first digital ray shortening. Even if underestimated, the *Mesopithecus* first metacarpal is still notably longer, in relation to m3, than that of the extant colobines, and is quite similar to that of the cercopithecines. *Mesopithecus* is evidence of a primitive colobine stage, in which the first metacarpal reduction had not yet begun.

TWO EOCENE ADAPIFORMS: *Notharctus and Adapis*

The hand of the Middle Eocene adapid *Notharctus* has been well-known since GREGORY'S (1920) monograph. Several hand bones, associated with the humerus and radius, are known from two individuals (see GODINOT & BEARD, 1991 and this volume). Data used in this paper are drawn from GREGORY's (1920) study.

The location of *Notharctus* in *Figure 3* shows that the great length of its hand in relation to the forelimb is correlated with a relative shortness of its humerus and radius. Today, a similar long hand and short humerus occur only in *Daubentonia*, as well as in vertical-clingers-and-leapers. In *Notharctus*, they may be related to vertical-clinging-and-leaping locomotion (WALKER, 1974; ROSE & WALKER, 1985; see discussion in GODINOT & BEARD, this volume). However, *Notharctus* differs from the extant vertical clingers and leapers by the shortness of its radius. This feature is also found in callitrichids and some cebids, and it might well be a primitive character. *Figure 3* supplements what was deduced from the relatively high brachial index of *Notharctus* by ROSE & WALKER (1985; radius/humerus = 0.94). This index indicates a long radius as related to the humerus, which was commented upon by NAPIER & DAVIS (1959): «in *Notharctus*, a relative elongation of the forearm is already apparent». When related to the entire forelimb, the *Notharctus* radius is short.

Concerning the intrinsic proportion of the hand, the location of *Notharctus* in *Figure 5* is approximate because it is obtained from GREGORY's (1920) reconstruction. For that reason, had there been only minor differences with other genera, they would have been difficult to interpret, whereas the distant location of *Notharctus* in *Figure 3* is meaningful. The length of the carpus in relation to the entire hand length is average in *Notharctus*, compared to extant primates. On the contrary, the metacarpus is relatively very short and the fourth digit extremely long. These characters do not seem to be correlated with vertical-clinging-and-leaping (though GREGORY, 1920, suggested it) because the similar-sized indriids have much shorter digits. These characters are very likely primitive, or they might also be related to some distinctive features shared with *Daubentonia*.

In *Notharctus tyrannus*, the indices of m2 and m4 lengths in relation to m3 (*Table 10*) have been evaluated from the maximal lengths measured by GREGORY (1920). It appears (*Figure 11*) that m4 is slightly shorter in relation to m3 than in most extant primates, but close to *Tarsius*, *Cheirogaleus*, and *Samiri*. This raises the question of whether such hand intrinsic proportions are primitive. It suggests that a marked elongation of the third metacarpal still existed, a likely character of the euprimate morphotype that is not apparent from the distribution of this trait in living genera. The marked relative shortness of m2 is a significant character. This feature was noticed by GREGORY (1920) and interpreted by him as primitive because it was shared with *Daubentonia*; it is more likely a specialization. The relative shortness of the *Notharctus* second metacarpal does indicate a strong reduction of the second digit ray, very likely related to a functional specialization, as in *Daubentonia*. As far as we know, this is not understood in the latter genus. Its hand function is very distinct from that of the two other extant primate genera having an equally great reduction of their second metacarpals, *Perodicticus* and *Arctocebus*.

The hand of the Upper Eocene adapid *Adapis parisiensis* (*Figure 13*) is known from a single specimen with a complete set of carpus and metacarpal elements (GODINOT & JOUFFROY, 1984). Consequently, metacarpal lengths in relation to one another were the only ratios that we could calculate for this study (*Table 10*). In both *Figures 11* and *12*, the location of *Adapis* is central, near *Lemur* and *Varecia*. *Adapis* is notable as being the unique Palaeogene fossil in which m1 length can be compared to m3. The ratio m1/m3 is at the lower range of living lemuriforms (*Avahi* excepted), and also at the lower range of non-ateline platyrrhines. Yet, it seems premature to hypothesize that these proportions are primitive for the Euprimates, due to the complexity of thumb and prehension evolution in the order Primates. In view of the difficulty in understanding *Adapis* locomotion, it must be noted that this genus does not show any m2 reduction or m1 elongation, which is a marked dissimilarity with extant lorisines (see dicussion in GODINOT & BEARD, 1991 and this volume). In both *Figures 11* and *12*, *Adapis* is near *Lemur*, *Varecia*, and *Miopithecus*, and slightly further removed from *Alouatta* and *Cercopithecus*, all of them being middle-sized arboreal quadrupeds. Might this be a tenuous indication of a similar way of life?

A PALEOCENE PLESIADAPIFORM: *Plesiadapis insignis*.

Plesiadapis remains the only Paleocene species for which the postcranial skeleton is known. The forelimb proportions given here are those of *P.insignis* from Menat (France), calculated from the measurements by GINGERICH (1976). Even if approximate, the *Plesiadapis* location in *Figures 3* and *4* is noteworthy. Its hand is relatively long in relation to the entire forelimb, when compared to that of all extant primates, but it is average-sized as compared to lemuriforms, especially the vertical-clingers-and-leapers. The relatively great length of the *Plesiadapis* hand among primates is associated with an average-sized humerus and, as in *Daubentonia* and *Callicebus*, an extremely short radius. Such shortness of the radius is very likely a primitive character.

The only simiiform having as long a hand is *Leontocebus*. The proximity of *Plesiadapis* and the two callitrichids in *Figure 3* is probably related to the three genera being small clawed primates. This suggests that *P. insignis* was arboreal as well. This conclusion, though based on limited data, is similar to the conclusions of GODINOT & BEARD (1991) on the locomotion of *P. tricuspidens*.

Because the archaic plesiadapiforms are sometimes considered as nonprimate mammals, it is interesting to compare the relative hand length and intrinsic forelimb proportions of *Plesiadapis* to those of some nonprimate mammalian taxa. *Plesiadapis* relative hand length is definitely different from that of the nonprimate mammals shown in *Figure 3* : it is longer than that of the scandentian *Tupaia*, the carnivoran *Potos* (kinkajou) and the marsupial *Didelphis*, but is much shorter than that of the rodent *Sciurus*. The *Plesiadapis* humerus is relatively as long as that of *Tupaia* and *Didelphis*, and it is relatively longer than that of *Sciurus*. Its radius is relatively as short as that of the kinkajou, and close to that of *Sciurus*; it is shorter than that of *Tupaia* and *Didelphis*.

To summarize, there is no reason to exclude *Plesiadapis* from the primate order on the basis of its relative hand length and intrinsic forelimb proportions, because various nonprimate mammals are spread out among primates in *Figure 3*.

Conclusion

A number of general conclusions emerge from the above results and discussion. First, there is a contrast between the significance of the extrinsic hand proportions (in relation to forelimb length) and the intrinsic ones (the lengths of the various elements in relation to one another); according to our results, the hand length in relation to forelimb length is not closely related to taxonomy but rather to locomotor modes, whereas the converse appears true for the hand intrinsic proportions. The fact that the intrageneric variability is much smaller for the former (*Tables 1* and *2*) than for the latter (*Tables 3* to *8*) suggests that the biomechanical constraints related to the lengths of forelimb bone lever arms are strong and do not allow marked variations from the optimum type, whereas the biomechanical constraints related to grasping and supporting hand function are smaller and compatible with some biometrical diversity. Nonetheless, the reason for the small standard deviations in *Tables 1* and *2* as compared to *Tables 3* to *8* may also be technical; the absolute precision in measuring is the same for long bones as for short ones, and thus relatively more accurate data are obtained with the humerus and the radius than with hand elements.

As regards the precision of our measurements, it was satisfying to note that the standard deviations do not increase according to the size of the samples. Concerning the relative humerus length percentage (*Tables 1, 2*), the genera represented by great numbers of individuals (e.g., 38 *Lemur*, 22 *Nycticebus*, 20 *Euoticus*) show very small standard deviations (0.5/0.8). *Galago* (55 individuals) makes a notable exception, because its humerus standard deviation (1.4) is the greatest within our entire primate sample. The reason is that four species, or even subgenera, of galagines are included within the genus *Galago* : *Galago crassicaudatus*, *G. alleni*, *G. senegalensis* and *Galagoides demidovii*. The biometrical diversity of their hindlimb elements has previously been reported (JOUFFROY *et al.*, 1984; JOUFFROY & GÜNTHER, 1985; BERGE & JOUFFROY, 1986).

An interesting insight into the evolution of the primate hand is brought about by noting that the hands of a great number of simiiforms and lemuriforms are nearly paraxonic. Rather than the classical general dichotomy into ectaxony (the Strepsirhini) and mesaxony (the Haplorhini), it is possible that the primitve mammalian mesaxonic pattern evolved in several directions within the order Primates: (a) very early, toward para-mesaxony and then true paraxony in many strepsirhines and catarrhines; (b) from paraxony to ectaxony in strepsirhines, reaching the pincer-like morphology of lorisine hands; and (c) hypermesaxony developed secondarily in *Homo* and the hylobatids (*Figure 8*), associated with an increase in length of the second ray in *Homo*. Thus, human mesaxony probably is not a primitive character. Due to the morpho-functional peculiarities of its unmatched long hand, we cannot go as far as to assume that *Tarsius*' hand hypermesaxony is primitive.

The relationship between the relative length of metacarpals in relation to each other and the relative lengths of the corresponding digital rays is important for interpreting fragmentary fossil hands. Even though a precise relationship does not exist, we have seen that a very short metacarpal (in relation to the third) corresponds most of the time to a short digital ray (in relation to the third one), e.g. the first ray in *Daubentonia*, *Colobus*, *Ateles* and *Avahi*, and the second ray in *Perodicticus* and *Arctocebus*.The only major exception is the fourth ray in *Daubentonia*, with its extremely short metacarpal (m4/m3 68%, i.e. the lowest ratio among primates) and extremely long digital ray (116% of the digit ray 3, i.e. the greatest among primates). This peculiarity also explains the separate location of *Daubentonia* in *Figure 5*, this genus differing from other primates by its unmatched short metacarpal (19% of the total hand length) and long digit (71%), because the lengths of the three main elements of the hand have been measured through the axis of the fourth digital ray (the longest one in *Daubentonia)*. In general, a marked elongation of the third metacarpal is probably a retention of the euprimate morphotype.

As pointed out by JOUFFROY & LESSERTISSEUR (1977), the different parts of digital rays are unequally affected in primates by regressive processes: it starts generally with the terminal phalanx, then the proximal ones and, ultimately, the metacarpal. The thumb of *Ateles* is an example of an extreme stage of digital ray reduction, insofar as the terminal phalanx is always absent, the first one is generally absent or extremely reduced, and the first metacarpal is also reduced, to become the relatively shortest one among primates.

Radiographs of a *Perodicticus* foetus and an *Ateles* newborn have shown that the length reduction, or atrophy, is associated with a delay of ossification and even, in the thumb of *Ateles*, chondrification (JOUFFROY & LESSERTISSEUR, 1977). According to NAUCK's (1967) terminology, primate digital ray atrophy results from a «distal cataplasy» phenomenon.

Studying primate hand biometry raises the question of whether there is some «homotypy» relationship between hand and foot lengths in relation to the respective entire limb lengths, as suggested by MOLLISSON (1911). In other words, are the short-handed (or long-handed) primates endowed with short (or long) feet (harmonic homotypy, or «homopody», LESSERTISSEUR & SABAN, 1967), or is there no such correspondence (dysharmonic homotypy, or heteropody)? Comparison of our results with previous studies (LESSERTISSEUR & JOUFFROY, 1973, 1978; JOUFFROY & LESSERTISSEUR, 1977, 1978, 1979) provides some answers. As examples of harmonic homotypy, we find : (1) primates with very short hands and feet, like the slender lorisines *Loris* and *Arctocebus* and the hylobatids, and (2) primates with very long hands and feet, like *Tarsius* and *Callithrix*. Several examples of dysharmonic homotypy are also found. *Pongo* is unique among primates in having the relatively longest foot (38% of the hindlimb), whereas its hand length is moderate. Such an extremely long foot is very likely related to the «hindlimb-hanging» ability of this genus. The analogy between brachiator hand and *Pongo* foot specializations for hanging is strengthened by the unique and striking atrophy of the *Pongo* hallux (JOUFFROY & LESSERTISSEUR, 1977). Indriids have very long hands and very short feet. Unlike other vertical-clingers-and-leapers, such as galagines and tarsiers, indriids do not show any elongation of their entire foot or tarsus (LESSERTISSEUR & JOUFFROY, 1971, 1978; JOUFFROY & LESSERTISSEUR, 1978, 1979; OXNARD *et al.*, 1981, 1983; JOUFFROY et *al.*, 1982). However, in each case, the biomechanical constraints of vertical-clinging-and-leaping require powerful extremities, ensuring a safe grip when landing on vertical trunks (PETERS & PREUSCHOFT, 1984). Hand elongation makes it easier to grasp thick branches. Finally, *Daubentonia* is exceptional among primates in having extremely long hands, whereas its foot relative length is quite similar to that of many other lemuriforms (e.g. *Lepilemur* and *Cheirogaleus*) (JOUFFROY & LESSERTISSEUR, 1978, 1979). The extrinsic proportions of its hand, as well as the intrinsic ones (see above), confirm the «uniqueness» of *Daubentonia* (OXNARD, 1981).

These examples show that no general homotypy relationship between hand and foot relative lengths can be identified within the order Primates.

From a methodological point of view, one can discuss the scope of the biometrical approach for comparative investigation of primate hand functional morphology. Standard deviation ranges show that most genera can be characterized by one or the other of their hand length proportions. However, because different morphologies may correspond to similar metrical features, whereas the converse is impossible, emphasis should be placed mainly on biometrical dissimilarities. Thus, the differences that we have found between genera taxonomically as close as *Lemur* and *Varecia*, *Miopithecus* and *Cercopithecus*, and *Avahi* and *Indri* demonstrate clearly that species of the same family, or subfamily, can differ from each other by biometrical characteristics that are probably related to differences in locomotor modes and/or body size.

The last question to be evoked in this discussion concerns the contribution of this study to our knowledge of euprimate hand evolution. In the course of our investigation, we first studied extant primates, and then the fossils. The former have showed some gradistic trends, from which we have tried to infer some of the morphotypic euprimate characteristics. These, in turn, imply hypotheses concerning the historical evolutionary trends. In considering the scanty evidence that can be drawn from the fossils, we can summarize what appear to us as possible hypotheses of broad evolutionary trends in the course of euprimate hand and forelimb evolution: (1) relative elongation of the radius and (2) relative shortening of the hand within the forelimb; (3) relative elongation of the longest metacarpal; and (4) relative shortening of the longest digit within the hand. These four hypothetical trends are corroborated by the Eocene adapid *Notharctus*. The first one is also corroborated by the Paleocene *Plesiadapis*, even though it is not a euprimate. Because hand length in relation to forelimb length is more correlated with locomotor modes than with systematics, it is not useful by itself for distinguishing primates from nonprimates.

From an evolutionary point of view, some of the fossils that we have studied show how difficult it may be to reconstruct the phylogeny of a group by studying only its extant members, without integrating fossil data. The parsimony hypothesis applied to the fact that all extant colobines have shortened m1 (as related to m3) would prompt one to infer that ancestral colobines had this shared derived character. However, the Miocene colobine *Mesopithecus* did not have such a first metacarpal reduction. From the relative lengthening of m2 in all extant hominoids, one could infer likewise that this shared derived character should also be a characteristic of their common ancestor. But the Miocene hominoid *Proconsul* has a non-elongated m2. Unless these two fossils are completely outside the two monophyletic groups including all extant forms, which is unlikely, they clearly suggest that the derived characters mentioned above are the result of evolutionary trends having occurred in parallel in several lineages, leading to the extant members of each group. That the above-mentioned characters are derived is clear from intergroup comparisons. When such characters are found in all members of a group, they are usually considered as «shared derived» characters of the group.

The two examples cited above, which are not isolated cases, show that the evolutionary history of animals frequently involved evolutive trends common to many, if not all, members of the group to which they belonged. It has frequently been pointed out that evolution is not a parsimonious process. Deciphering it requires a continual ebb and flow process associating the morpho-functional, biomechanical, biological and ecological data provided by investigations on extant animals and those drawn from the study of fossils, which ultimately make up the sole historical evidence, the irrefutable «archives» of evolution.

ACKNOWLEDGMENTS — We express our gratitude towards all the curators of the primate collections that we have been allowed to study in the Museums and Institutes listed in the «Material» section. Their helpful welcome permitted us to compile the set of 5.300 measurements that make up the basis of this study.

We especially wish to thank Mr A. Collier, who entrusted us with his beautiful fossil hand of *Adapis*.

Many thanks go to H. Ishida, who gave us access to his computational statistical program at Osaka University and provided us with fruitful comments, to B. Jouffroy, who voluntarily carried out an important part of the statistical processing, and to K.D. Rose who invited one of the authors (M.G.) to work in the Department of Cell Biology and Anatomy at the Johns Hopkins University School of Medicine (Grant from the C.N.R.S. on the N.S.F./C.N.R.S. exchange program).

We are greatly indebted to M.C. Dane, B. Demes, W.L. Jungers, P. Luckett, H. Preuschoft, and M. H. Stack for stimulating comments and stylistic help.

The technical assistance of A. Gordon, B. Orth and E. Pellé for the illustrations is acknowledged.

The investigation has been supported mainly by the «Centre National de la Recherche Scientifique» (C.N.R.S., «A.T.P. Evolution» , U.R.A. 1137, and U.R.A.327), with the benefit of additional support from the Muséum national d'Histoire Naturelle (Paris), the Osaka University, and the Japanese Society for promotion of Science (J.S.P.S).

References

ARISTOTLE, 4th century B.C. *Historia animalium*. Engl. transl. A.L. Peck, 1979, Harvard University Press, Cambridge.

BEARD K.C., TEAFORD M.F. & WALKER A., 1986. *New wrist bones of Proconsul africanus and P. nyanzae from Rusinga Island (Kenya)*. Folia Primatol., 47: 97-118.

BERGE C. & JOUFFROY F.K., 1986. *Morpho-functional study of Tarsius foot as compared to the galagines': What does an «elongate calcaneus» mean?* In: Current Perspectives in Primate Biology (D.M. Taub & F.A. King eds), pp. 146-158. Van Nostrand Reinhold, New York.

BERGE C. & KAZMIERCZAK B., 1986. *Effects of size and locomotor adaptations on the hominid pelvis: Evaluation of australopithecine bipedality with a new multivariate method*. Folia Primatol., 46: 185-204.

BIEGERT J. & MAURER R., 1972. *Rumpfskelettlänge, Allometrien und Körperproportionen bei catarrhinen Primaten*. Folia Primatol., 17: 142-156.

BISHOP A., 1964. *Use of the hand in lower primates*. In: Evolutionary and Genetic Biology of Primates (J. Buettner-Janusch ed.), pp. 133-225, Academic Press, New York & London.

BLUMENBACH J.F., 1790. *Beiträge zur Naturgeschichte*. DIETERICH, GÖTTINGEN.

BUFFON G.L. LECLERC, COMTE DE, 1790 (ed. nov.). *Histoire Naturelle, Générale et Particulière*. T. XII, Imprimerie royale, Paris.

CARTMILL M., 1974. *Pads and claws in arboreal locomotion*. In: Primate Locomotion (F.A. Jenkins, Jr. ed.), pp. 45-83. Academic Press, New York & London.

ETTER H.-U. F., 1974. *Morphologisch- und metrisch- vergleichende Untersuchung am Handskelett rezenter Primaten*. Gegenb. Morph. Jahrb., 120: 1-21, 22-48, 172-184, 457-484.

ETTER H.-U. F., 1978. *Lorisiform hands and their phylogenetic implications: a preliminary report*. In: Recent Advances in Primatology, Vol. I, sect. 3 (D.J. Chivers & K.A. Joysey eds.), pp. 161-170, Academic Press, New York & London.

FORD S.M.. 1987. *Callitrichids as phyletic dwarfs and the place of Calllithricidae in Platyrrhini*. In: Primate Evolution and Human Origins (R.L. Ciochon & J.G. Fleage eds.), pp. 136-140. Aldine de Gruyter, New York.

GINGERICH P.D., 1976. *Cranial anatomy and evolution of early Tertiary Plesiadapidae (Mammalia, Primates)*. Univ. Michigan Papers on Paleontol., 15, 1-141.

GODINOT M. & JOUFFROY F.K., 1984. *La main d'Adapis (Primate, Adapidé)*. In: Actes du symposium Paléontologique Georges Cuvier (E. Buffetaut, J.M. Mazin & E. Salmon eds.), pp. 221-242. Le Serpentaire, Montbéliard.

GODINOT M. - BEARD K.C., 1991. *A review and an evolutionary inquiry emphasizing early forms*. Human Evolution 6: 307-354.

GREGORY W.K., 1920. *On the structure and relations of Notharctus, an American Eocene primate*. Mem. Amer. Mus. Nat. Hist., n.s., 3: 51-243.

HAINES R.W., 1944. *Mechanism of rotation at the first carpo-metacarpal joint*. J. Anat. London, 78: 44-46.

HERSHKOVITZ P., 1977. *New World Monkeys (Platyrrhini)*. Vol. 1. University of Chicago Press, Chicago.

HOFFSTETTER R., 1982. *Les primates simiiformes (= Anthropoidea). Compréhension, phylogénie, histoire biogéographique*. Ann. Paléont., 68: 241-290.

JOUFFROY F.K., 1962. *La musculature des membres chez les lémuriens de Madagascar. Etude descriptive et comparative*. Mammalia, suppl. 2: 1-326.

JOUFFROY F.K., 1971. *Musculature des membres*. In: Traité de Zoologie (P.P. Grassé ed.), XVI (Mammifères), 3: pp. 1-475, Masson, Paris.

JOUFFROY F.K., 1975. *Osteology and myology of the lemuriform postcranial skeleton*. In: Lemur Biology (I. Tattersall & R.W. Sussman eds.), pp. 149-192. Plenum, New York.

JOUFFROY F.K. & GÜNTHER M., 1985. *Interdependence of morphology and behavior in the locomotion of galagines*. In: Primate Morphophysiology, Locomotor Analyses, and Human Bipedalism (S. Kondo ed.), pp. 201-234. University of Tokyo Press, Tokyo.

JOUFFROY F.K. & LESSERTISSEUR J., 1959. *La main des lémuriens malgaches comparée à celle des autres primates*. Mém. Inst. Sc. Madagascar, A, 13: 195-219.

JOUFFROY F.K. & LESSERTISSEUR J., 1960. *Les spécialisations anatomiques de la main chez les singes à progression suspendue*. Mammalia, 24: 93-151.

JOUFFROY F.K. & LESSERTISSEUR J., 1977. *Processus de réduction des doigts (main et pied) chez les primates*.

In: Mécanismes de la Rudimentation des Organes chez les Embryons de Vertébrés (A. Raynaud ed.). Coll. Intern. CNRS n. 266. pp. 381-392. Editions du CNRS, Paris.

JOUFFROY F.K. & LESSERTISSEUR J., 1978. *Etude écomorphologique des proportions des membres des primates et spécialement des prosimiens.* Ann. Sc. Nat., Zool., Paris, 20: 99-128.

JOUFFROY F.K. & LESSERTISSEUR J., 1979. *Relationships between limb morphology and locomotor adaptations among prosimians: an osteometric study.* In: Environment, Behavior, and Morphology: Dynamic Interactions in Primates (M.E. Morbeck, H. Preuschoft & N. Gomberg eds.), pp. 143-181. Fischer, New York & Stuttgart.

JOUFFROY F.K., BERGE C. & NIEMITZ C., 1984. *Comparative study of the lower extremity in the genus Tarsius.* In: Biology of Tarsiers (C. Niemitz ed.), pp. 167-190. Fischer, Stuttgart & New York.

JOUFFROY F.K., OXNARD C.E. & GERMAN R.Z., 1982. *Interpretation phylogénétique des proportions des membres des tarsiers, par comparaison avec les autres prosimiens sauteurs. Analyse multivariée.* C.R. Acad. Sc. Paris, 295: 315-320.

JUNGERS W.L., 1977. *Hindlimb and pelvic adaptations to vertical climbing and clinging in Megaladapis, a giant subfossil prosimian from Madagascar.* Yrb. Phys. Anthropol., 20: 508-524.

JUNGERS W.L., 1978. *The functional significance of skeletal allometry in Megaladapis in comparison to living prosimians.* Am. J. Phys. Anthrop., 9: 303-314.

KAPANDJI I.A., 1966. *Physiologie Articulaire.* Fascicule 1, Maloine, Paris.

KAPLAN E.B., 1965. *Functional and Surgical Anatomy of the Hand.* Lippincott, Philadelphia.

KAZMIERCZAK J.B., 1985. *Analyse logarithmique. Deux exemples d'application.* Rev. Statist. Appl., 33: 13-24.

KUCZYNNSKI K., 1974. *Carpometacarpal joint of the human thumb.* J. Anat. 118: 119-126.

LAMBERTON C., 1934. *Contribution à la connaissance de la faune subfossile de Madagascar: lémuriens et ratites.* Mém. Acad. Malgache, 17: 1-168.

LANDSMEER J.M.F., 1955. *Anatomical and functional investigations on the articulation of the fingers.* Acta Anat., suppl. 24.

LANDSMEER J.M.F., 1976. *Atlas of Anatomy of the Hand.* Churchill Livingstone, London.

LE GROS CLARK W.E., 1936. *The problem of the claw in primates.* Proc. Zool. Soc. London, 1-24.

LESSERTISSEUR J. & JOUFFROY F.K., 1973. *Tendances locomotrices des primates traduites par les proportions du pied.* Folia Primatol., 20: 125-160.

LESSERTISSEUR J. & JOUFFROY F.K., 1978. *La morpho-écologie du système locomoteur des primates dans ses rapports avec leur phylogenèse.* In: XIXe Congressus Morphologicus Symposia (E. Klika ed.), pp. 79-87. Univerzita Karlova, Praha.

LESSERTISSEUR J. & SABAN R., 1967. *Squelette appendiculaire.* In: Traité de Zoologie (P.P. Grassé) XVI (Mammifères), 1, pp. 709-1078. Masson, Paris.

LORENZ VON LIBURNAU L.R., 1905. *Megaladapis edwardsi G. Grandidier.* Denkschr. Kais. Akad. Wiss Wien, 77: 451-490.

LUCAE J.C.G., 1865. *Die Hand und der Fuss. Ein Beitrag zur vergleichenden Osteologie der Menschen, Affen, Beutelthiere.* Abh. Senckenb. Naturf. Ges., 5: 275-332.

MCCONAILL M.A. & BASMAJIAN J.V., 1977. *Muscles and Movement.* Krieger Publishing Company, New York.

MCHENRY H.M. & CORRUCCINI R.S., 1983. *The wrist of Proconsul africanus and the origin of hominoid postcranial adaptations.* In: New Interpretations of Ape and Human Ancestry (R.L. Ciochon & R.S. Corruccini eds.), pp. 353-367. Plenum Press, New York.

MIDLO C., 1934. *Form of the hand and foot in primates.* Amer. J. Phys. Anthrop., 19: 337-389.

MILNE-EDWARDS A. & GRANDIDIER A., 1875. *Histoire Naturelle des Mammifères.* In: Histoire Physique, Naturelle et Politique de Madagascar (A. Grandidier ed.), T. IV. Atlas I. Imprimerie nationale, Paris.

MITTERMEIER R.A. & FLEAGLE J.G., 1976. *The locomotor and postural repertoires of Ateles geoffroyi and Colobus guereza and a reevaluation of the locomotor category semibrachiation.* Am. J. Phys. Anthrop., 45: 235-256.

MIVART ST G., 1867. *On the appendicular skeleton of the primates.* Phil. Trans. Roy. Soc. London, 157: 299-429.

MOLLISON T., 1911. *Die Körperproportionen der Primaten.* Gegenb. Morph. Jahrbuch, 42: 79-304.

MORBECK M.E., 1979. *Forelimb use and positional adaptation in Colobus guereza: Integration of behavioral, ecological, and anatomical data.* In: Environment, Behavior and Morphology: Dynamic Interactions

in Primates. (M.E. Morbeck, H. Preuschoft & N. Gomberg eds.), pp. 95-118. Fischer, New York & Stuttgart.
NAPIER J.R, 1955. *The form and function of the carpo-metacarpal joint of the thumb.* J. Anat. London, 89: 362-369.
NAPIER J.R., 1956. *The prehensile movements of the human hand.* J. Bone Jt Surg., 38 B: 902-913.
NAPIER J.R., 1960. *Studies of the hands of living primates.* Proc. Zool. Soc. London, 134: 647-657.
NAPIER J.R., 1961. *Prehensility and opposability in the hands of primates.* Symp. Zool. Soc. London, 5: 115-132.
NAPIER J.R., 1966. *Functional aspects of the anatomy of the hand.* In: The Hand (R.G. Pulvertaft ed.), pp. 1-31. Butterworths, London.
NAPIER J.R. & DAVIS P.R., 1959. *The forelimb skeleton and associated remains of Proconsul africanus.* Fossil Mammals of Africa, n. 17. Br. Mus. (Nat. Hist.) London, 41-69.
NAPIER J.R. & NAPIER P.H., 1967. *A Handbook of the Living Primates.* Academic Press, London.
NAPIER J.R. & WALKER A.C., 1967. *Vertical-clinging-and-leaping, a newly recognised category of locomotor behaviour among primates.* Folia Primatol., 6: 204-219.
NAUCK E.T., 1967. *Extremitätenskelett der Tetrapoden.* In: Handbuch der Vergleichenden Anatomie der Wirbeltiere (L. Bolk, E. Göppert, E. Kallius & W. Lubosch eds.), 5: 71-248.
OXNARD C.E., 1981. *The uniqueness of Daubentonia.* Am. J. Phys. Anthrop., 54: 1-22.
OXNARD C.E., 1983. *The Order of Man.* Hong Kong University Press, Hong Kong.
OXNARD C.E., JOUFFROY F.K. & LESSERTISSEUR J., 1981. *The morphometrics of limb proportions in leaping prosimians.* Amer. J. Phys. Anthrop., 54: 421-430.
PETERS A. & PREUSCHOFT H., 1984. *External biomechanics of leaping in Tarsius and its morphological and kinematic consequences.* In: Biology of Tarsiers. (C. Niemitz ed.), pp. 227-255. Gustav Fischer, Stuttgart & New York.
PETTER J.-J., 1962. *Recherches sur l'écologie et l'éthologie des lémuriens malgaches.* Mém. Mus. Hist. Nat., 27: 1-146.
PETTER J.-J., SCHILLING A. & PARIENTE G., 1971. *Observations éco-éthologiques sur deux lémuriens malgaches nocturnes: Phaner furcifer et Microcebus coquereli.* La Terre et La Vie, 25: 287-327.
PIERON A.P., 1973. *The mechanism of the first carpo-metacarpal joint.* Acta Orthop. Scand. suppl. 148.
PREUSCHOFT H. & DEMES B., 1984a. *Biomechanics of brachiation.* In: The Lesser Apes. H. Preuschoft, D.J. Chivers, WY. Brockelman & N. Creel eds., pp. 96-118. Edinburgh University Press, Edinburgh.
PREUSCHOFT H. & DEMES B., 1984b. *Influence of size and proportions on the biomechanics of brachiation.* In: Size and Scaling in Primate Biology (W.L. Jungers ed.), pp. 383-399. Plenum Press, New York & London.
ROLLINSON J. & MARTIN R.D., 1981. *Comparative aspects of primate locomotion with special reference to arboreal cercopithecines.* In: Vertebrate Locomotion (M.H. Day ed.), pp. 377-427. Academic Press, London.
ROSE K.D. & WALKER A., 1985. *The skeleton of early Eocene Cantius, oldest lemuriform primate.* Am. J. Phys. Anthrop., 66: 73-89.
ROSENBERGER A.L., 1977. *Xenothrix and ceboid phylogeny.* J. Hum. Evol., 6: 561-581.
SCHUHL P.M., 1954. *L'Oeuvre de Platon.* Hachette, Paris.
SCHULTZ A.H., 1930. *The skeleton of the trunk and limbs of higher primates.* Human Biol., 2: 303-438.
SCHULTZ A.H., 1962. *Muscle function in the fingers.* J. Bone Jt Surg., 44 B: 899-1022.
SCHULTZ A.H., 1972. *Les Primates.* Editions Rencontre, Lausanne.
STACK H.G. 1962. *Muscle function in the finger.* J. Bone Jt Surg., 44B: 899-1022.
STERN J.T., Jr. & OXNARD C.E., 1973. *Primate locomotion: some links with evolution and morphology.* Primatologia, 4: 1-93.
STERN J.T., Jr., WELLS J.P., JUNGERS W.L., VANGOR A.K. & FLEAGLE J.G., 1980. *An electromyographic study of the pectoralis major in Atelines and Hylobates, with special reference to the evolution of a pars clavicularis.* Am. J. Phys. Anthrop., 52: 13-25.
STRAUS W.L., Jr., 1942. *Rudimentary digits in primates.* Quart. Rev. Biol., 17: 228-243.
STRAUS W.L., Jr., 1949. *The riddle of man's ancestry.* Quart. Rev. Biol., 24: 200-223.
SZALAY F.S. & DAGOSTO M., 1980. *Locomotor adaptations as reflected on the humerus of Paleogene primates.* Folia Primatol., 34: 1-45.
TATTERSALL I., 1982. *The Primates of Madagascar.* Columbia University Press, New York.

Tuttle R.H., 1970. *Postural, propulsive, and prehensile capabilities in the cheiridia of chimpanzees and other great apes.* In: The Chimpanzee (G.H. Bourne, ed.), pp. 167-253. Karger, Basel & New York.

Walker A.C., 1967. *Locomotor adaptations in Recent and Fossil Madagascar Lemurs.* Ph. D. Thesis University of London, London.

Walker A.C., 1974. *Locomotor adaptations in past and present prosimian primates.* In: Primate Locomotion (F.A. Jenkins, Jr, ed.), pp. 349-381. Academic Press, New York.

Wood Jones F., 1941. *The Hand.* Baillière, Tindall, and Cox, London.

Zapfe H., 1963. *Lebensbild von Megaladapis edwardsi Grandidier.* Folia Primatol., 1: 178-187.

Ziemer L.K., 1978. *Functional morphology of forelimb joints in the woolly monkey Lagothrix lagothricha.* Contributions to Primatology, 14: 1-125.

A. Ikeda
S. Matsumoto

New results concerning the vascularization of primate hands
Part I: The palmar arterial arches in Cercopithecidae, Pongidae, Hominidae and other primates

A three-dimensional angiographic analysis of the deep palmar and related arteries in Prosimians, Cercopithecidae, Hylobatidae, Pongidae and Hominidae revealed the most complex pattern in Old World monkeys:
In Cercopithecidae, three palmar arches are formed by the perforating branches of the second dorsal metacarpal artery: two proximal (catella volaris proximalis and arcus volaris profundus) and one distal (catella volaris distalis). In Hylobatidae and Pongidae, the arches are formed by the perforating branches of the first dorsal metacarpal artery, and they are composed of only the catella volaris proximalis and catella volaris distalis. In Hominidae, the arches are formed by the perforating branches of the first dorsal metacarpal artery, and they consist of the arcus volaris profundus and an incomplete catella volaris distalis. In Prosimians, the deep palmar arterial arch (arcus volaris profundus) is formed by the dorsal branch of the median artery in conjunction with the deep branch of the ulnar artery. In Lorisiformes, the catella volaris proximalis and distalis were absent. Our detailed studies of the arterial patterns in the hands of primates show various types reflecting phylogenetic relationships as well as morphological differentiation and rearrangements and comparative data for discussion.

Key words: angiography, arterial arches, catella volaris, arcus volaris, palmar arteries, stereoscopic x-ray.

Introduction

The comparative anatomy of the arterial patterns in the hand of primates has long attracted the attention of anatomists. Nishi (1939) and Koch (1939) observed three palmar arterial arches in rhesus monkeys. All are derived from the perforating branches of the second dorsal metacarpal artery: the arcus volaris profundus, catella volaris proximalis and catella volaris distalis. In textbooks of human anatomy, however, only one deep and one superficial arterial arch are described. The relation between both was the focus of this study.

It was difficult, or even impossible, however, to study the architecture of the fine arterial vessels in the small hands of non-human primates by the dissection technique usual in gross anatomy. A precondition for the better identification of the three-dimensional arterial network is an improved technique.

This is now available (see below) so that arterioles as small as 60 microns in the inner diameter could be identified. Here we present the design and results of the comparative analysis of arteriographs. Special emphasis is put on the deep arterial arches and its descendents which supply the interosseous and lumbrical muscles as well as the metacarpal bones.

In primates, like other land-living vertebrates, flexion and extension of all fingers are controlled by the forearm muscles (LANDSMEER, this vol.). For movements which are not simple opening and closing of the «whole hand grip» the intrinsic hand muscles come into action: The interosseous muscles combine metacarpo-phalangeal flexion with interphalangeal extension, whereby the latter seems to be the predominant action. The lumbrical muscles are silent during full flexion, but are very active in extension of the proximal or distal interphalangeal joints, and when these joints are maintained in extension combined with metacarpo-phalangeal flexion. We assume that the results of our investigation contribute to better understanding of the functional development of finger movements in primates, including man.

Materials and Methods

The study is based on hands of the following taxa:

Lemuridae
Lemur catta, Lemur macaco.

Lorisidae
Galago crassicaudatus, Galago senegalensis, Loris tardigradus, Nycticebus coucang, Perodicticus potto.

Cercopithecidae
Cercocebus torquatus lunulatus, Cercocebus torquatus torquatus, Cercopithecus cephus, Cercopithecus mona, Cercopithecus mitis albogularis, Erythrocebus patas, Macaca fascicularis, Macaca fuscata fuscata, Macaca fuscata yakui, Macaca maura, Macaca mulatta, Macaca nemestrina, Macaca nigra, Macaca radiata, Macaca sylvanus, Mandrillus leucophaeus, Papio anubis, Papio hamadryas, Presbytis cristata, Presbytis pileatus.

Hylobatidae
Hylobates l. agilis.

Pongidae
Pan troglodytes, Gorilla gorilla.

Hominidae
Homo sapiens.

The dead non-human primates were obtained from the Japan Monkey Center at Inuyama. After pathological dissection, the cadavers were put in cold storage (−20°C) until used. The human corpses (of both sexes, and between 40 and 80 years of age), were subject to investigation 28 to 48 hours after death.

The embalming technique used to prepare the materials in this study has been described detailedly in previous papers (IKEDA, UGAWA *et al.*, 1988; IKEDA, MIYAKE *et al.*, 1988; IKEDA, MATSUMOTO *et al.*, 1988; UGAWA & IKEDA, 1985).

Our technique is composed of pre-embalming treatment with a blood clot disperser (pH-A solution, Champion, USA), drainage of blood, and arterial embalming with an embalming machine via the axillary artery. The embalming fluid consists of 95% ethyl

Abbreviations used in all figures:

AVP arcus volaris profundus
CVD catella volaris distalis
CVP catella volaris proximalis
CPD common palmar digital artery
DD distal deep branch of the superficial palmar arch
DM dorsal metacarpal artery
DP distal perforating artery
DR dorsal radial artery of the thumb
DU dorsal ulnar artery of the thumb
IM intermetacarpal artery
MA median artery
MB muscular branch of the dorsal ulnar artery
PD proximal deep branch of the superficial palmar arch
PM palmar metacarpal artery
PP proximal deep branch of the superficial palmar arch
R radial artery
SPA superficial palmar arch
U ulnar artery
UM ulnar marginal artery

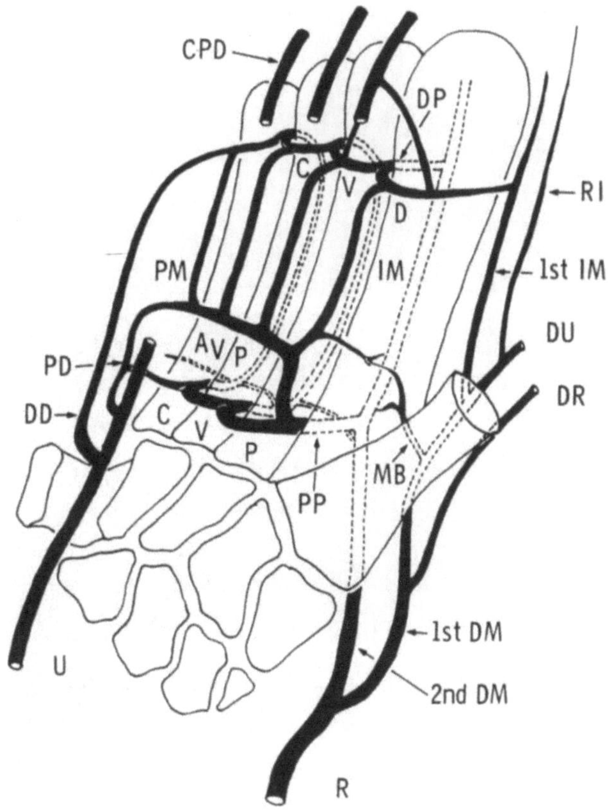

Figure 1 - Arterial pattern of the palmar side in Cercopithecidae (except the superficial palmar arterial arch).

alcohol and formalin as the fixative, ethylene glycol as the preservative, liquified phenol as the mould preventative, and water.

Following embalming, radioopaque substances were injected into the axillary artery using a 5 ml syringe with hand pressure. Stereoscopic angiographs were taken under exposure conditions of 80 kV, 20 mA, and 1.2 mm focus diameter for 7-12 s with Kodak X-Omat TL film at an 80 cm focal spot to film distance. The focal spot position was shifted by 60 mm to produce a stereoscopic pair of radiographs.

The stereoscopic observations of angiographs were made with a Nikon stereoscope (model II). Subsequently, the specimens were dissected for comparison with the radiographs.

Observations

Cercopithecidae (Figures 1, 2, 3, 4)

The typical arrangement of the palmar arterial arches in Cercopithecidae was observed in Japanese monkeys (*Macaca fuscata*).

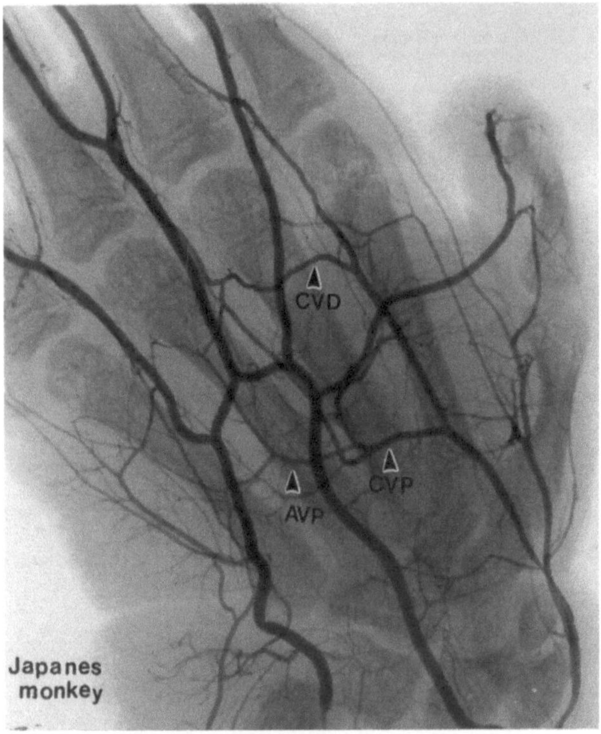

Figure 2 - Arteriograph of the hand in *Macaca fuscata*.

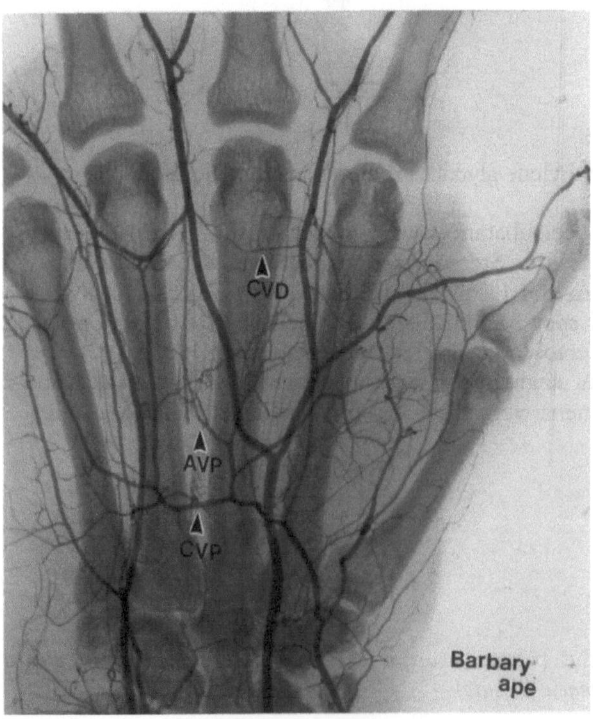

Figure 3 - Arteriograph of the hand in *Macaca sylvanus*.
Abbreviations as in Figure 1.

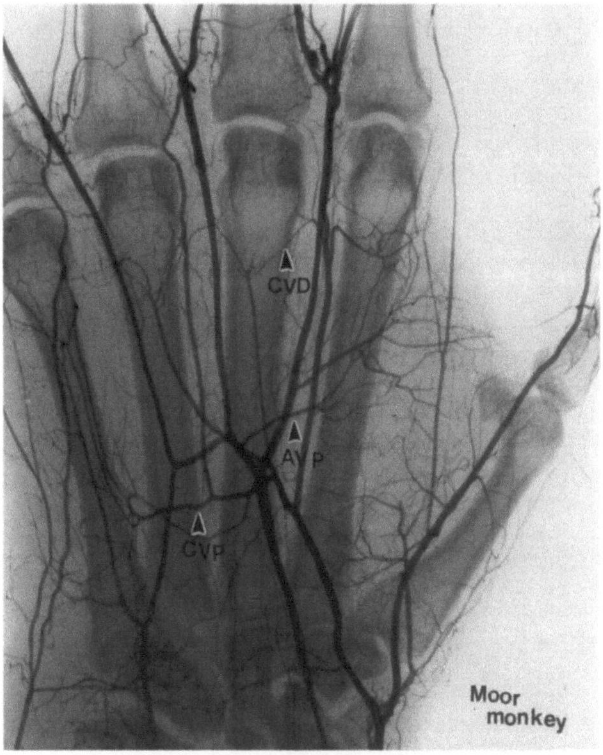

Figure 4 - Arteriograph of the hand in *Macaca maura*.
Abbreviations as in Figure 1.

The second dorsal metacarpal artery descends in the second intermetacarpal space and is the main source of the deep palmar arterial arches. The proximal perforating branch of the second dorsal metacarpal artery enters the palm at the base of the second intermetacarpal space to form the «catella volaris proximalis». This arterial chain runs in an ulnar direction on the palmar side of the third, fourth and fifth metacarpal bones and joins the deep branch of the ulnar artery.

One branch leaves the catella at the third metacarpal bone, and, perforating the palmar interosseous fascia, divides radially and ulnarly to form the arcus volaris profundus (= the deep palmar arch in man) along the deep branch of the ulnar nerve. It is accompanied by a perforating branch of the ulnar nerve. A distal perforating branch of the second dorsal metacarpal artery enters the palm at the distal end of the second intermetacarpal space and then forms the catella volaris distalis.

In the third and fourth intermetacarpal spaces, the intermetacarpal arteries arise from the catella volaris proximalis, and, passing down the interosseous spaces, join the catella volaris distalis. The arcus volaris profundus gives origin to four palmar metacarpal arteries which descend along the palmar sides of the second to fifth metacarpal bones to join the catella volaris distalis. The catella volaris proximalis does not branch into the palmar metacarpal arteries and the arcus volaris profundus does not give off the intermetacarpal arteries.

There is a definite regularity in the formation of the intermetacarpal arteries between

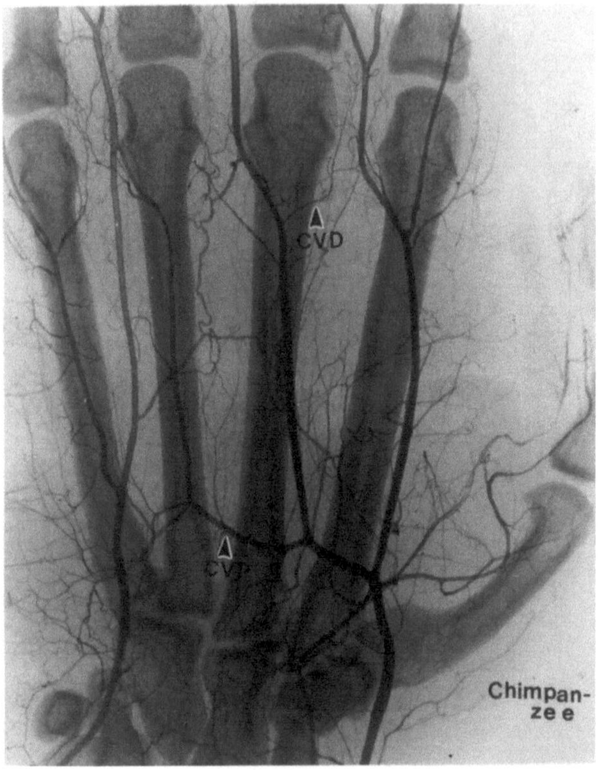

Figure 5 - Arteriograph of the hand in *Pan troglodytes*.
Abbreviations as in Figure 1.

the catella volaris proximalis and the catella volaris distalis and of the palmar metacarpal arteries between the arcus volaris profundus and the catella volaris distalis.

In a previous paper (UGAWA & IKEDA, 1985), the variations of the deep palmar arches in the Japanese monkey were classified into three types. The most common, general and complete form described above is called «type 2» (68%). In type 1 (20%), the arcus volaris profundus is incomplete, and is formed by branches coming from the ulnar and radial arteries separately. In type 3 (12%), the catella volaris proximalis is incomplete. The anastomoses among the catellae, arches and other branches are sometimes very slight or absent.

Hylobatidae and Pongidae (Fig. 5)

The radial artery reaches the proximal end of the first interosseous space on the dorsal side and enters the palm at the base of the first intermetacarpal space. It then turns ulnarwards along the palmar side of the second to fifth metacarpal bones and continues with a slight undulating convexity across the interosseous muscles.

The deep branch of the ulnar nerve is located distally and superficially to this arterial chain. This chain seems to correspond to the catella volaris proximalis in Cercopithecidae. It joins the deep branch of the ulnar artery. The first dorsal metacarpal artery runs to the distal end of the second metacarpal bone, where it branches off to form a weak transverse chain on the palmar side of the metacarpals, namely, the catella volaris distalis.

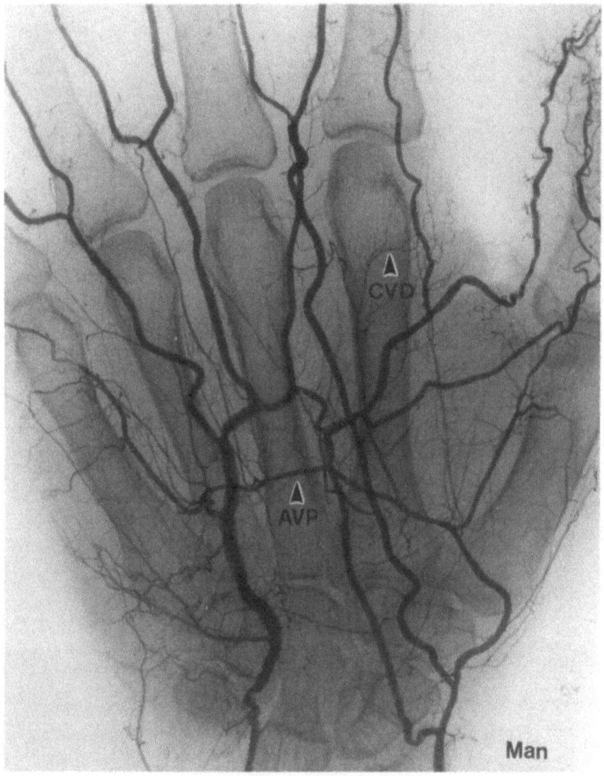

Figure 6 - Arteriograph of the hand in *Homo sapiens*.
Abbreviations as in Figure 1.

Both, the palmar metacarpal and intermetacarpal arteries arise from the catella volaris proximalis. The former join the catella volaris distalis, while the latter are thin, ending in the interosseous muscles. The second dorsal metacarpal artery was not developed in the specimens examined.

Hominidae (Fig. 6)

In most cases the radial artery (first dorsal metacarpal artery) enters the palm at the base of the first intermetacarpal space, and forms the deep palmar arterial arch in conjunction with the deep branch of the ulnar artery. The deep palmar arterial arch occupies the most proximal part of the cup of the palm and lies below the long flexor tendons under cover of the palmar interosseous fascia. The arch is accompanied by the deep branch of the ulnar nerve. This arterial arch corresponds to the arcus volaris profundus in Cercopithecidae.

The cạtella volaris proximalis is absent in man.

The first dorsal metacarpal artery gives off a perforating branch at the distal end of the second metacarpal bone into the palm, though the so-called catella volaris distalis is formed incompletely in man. The palmar metacarpal and intermetacarpal arteries arise together from the arcus volaris proximalis.

Of interest is the fact that in addition to a branch of the radial artery (see above), another perforating branch often arises from the second dorsal metacarpal artery into the

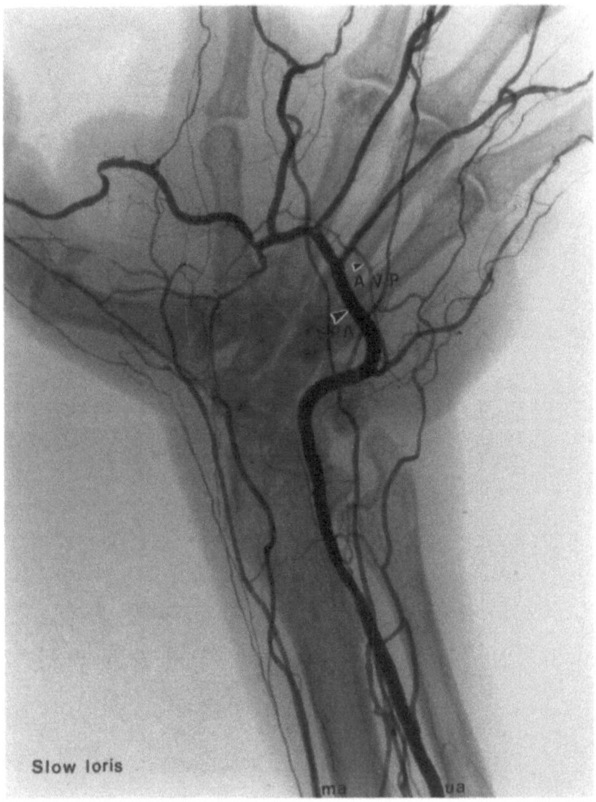

Figure 7 - Arteriograph of the hand in *Nycticebus coucang*. Abbreviations as in Figure 1.

palm at the base of the metacarpal space. These findings occur in 52% of the 220 human hands examined (IKEDA, UGAWA *et al.*, 1988). This branch joins the deep palmar arch as the third arterial source.

As in Pongidae and Hylobatidae there is only one proximal arterial arch in the depth of the palm. This single arch is usually treated as the arcus volaris profundus, as in prosimians, New World monkeys and colobines (MANNER-SMITH, 1911; NISHI, 1939; ZUCKERKANDL, 1907).

Lorisiformes (Fig. 7)

In the lorisids, the ulnar artery departs from the brachial artery and unites distally with the median artery to form the superficial carpal arch. The rete mirabile, which corresponds to the superficial brachial artery in man (IKEDA, MIYAKE *et al.*, 1988), traverses the medial surface of the arm, running diagonally from the ulnar to the radial side of the forearm. It ends as the median artery and a slender twig reaches the radial half of the volar surface of the hand on the thenar eminence.

The deep palmar arterial arch is formed by the dorsal branch of the median artery in conjunction with the deep branch of the ulnar artery. The dorsal branch of the median artery supplies the first dorsal metacarpal space. It then turns in an ulnar direction and

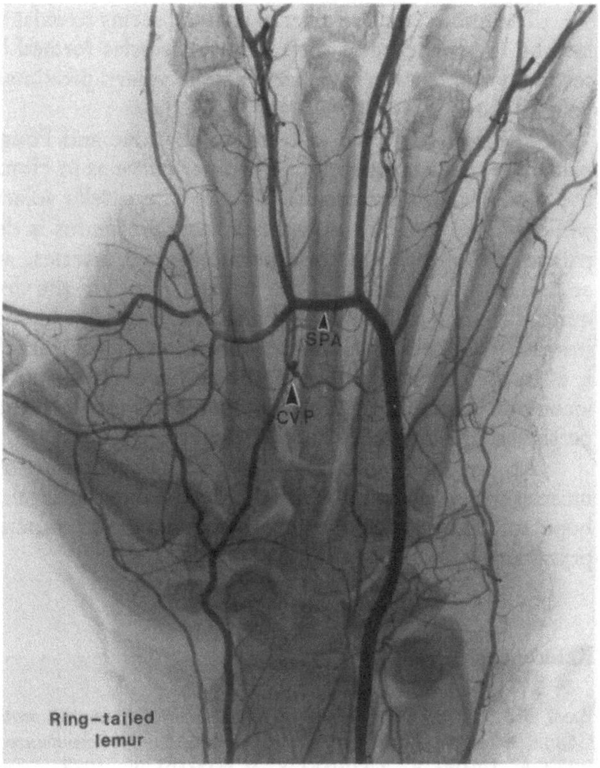

Figure 8 - Arteriograph of the hand in *Lemur catta*
Abbreviations as in Figure 1.

forms the deep palmar arterial arch, which in man is accompanied by the deep branch of the ulnar nerve. The catella volaris proximalis and distalis are absent in Lorisiformes.

Lemuriformes (Fig. 8)

Two types of the deep palmar arterial arches are observed in Lemuriformes:

1) Only the catella volaris proximalis and catella volaris distalis exist, and the arches are formed by the perforating branches of the second dorsal metacarpal artery;

2) the arches consist of the arcus volaris profundus and an incomplete catella volaris distalis and they are formed by the perforating branches of the second dorsal metacarpal artery.

Discussion

The simple pattern of the arterial arches at the deep proximal palm in Prosimians (Lorisiformes and Lemuriformes) seems to be a pattern quite different from that which we found in Hominoidea. However, according to our previous paper (IKEDA, MIYAKE *et al.*, 1988), the arterial pattern in Lorisiformes corresponds to human embryonic stages of 18-21 mm length. It possibly relates to the fact that the development of the vessels during embryonic growth is similar in both groups.

The most complete arterial pattern seems to exist in Cercopithecidae. It is characterized by the existence of three arterial arcades formed by the perforating branches of the second dorsal metacarpal artery: catella volaris proximalis, arcus volaris profundus, catella volaris distalis.

The deep proximal arch in Hylobatidae and Pongidae examined here is not accompanied by the deep branch of the ulnar nerve as in Hominidae, but instead corresponds to the catella volaris proximalis. That is, the catella volaris proximalis persists in Pongidae, while, on the contrary, an arcus volaris profundus is developed in Hominidae. Both, the palmar metacarpal and the intermetacarpal arteries, arise from the remaining proximal arch. The simplification of the arterial arches in the deep proximal palm in Hominidae is quite different from that in Pongidae. The development of an arcus volaris profundus instead of a catella in Hominidae must have occurred after separation from the Pongidae, i. e. in the course of rapid modification of the hand caused by complete freedom from locomotor tasks. It possibly is related to the fact that the palmar metacarpal arteries persist as the branches to the fingers.

At present, we start with micro-angiography of the interosseous and lumbrical muscles which are supplied by the deep palmar arterial arches. By using this technique we hope to further conclude on phylogenetic and functional relationships among various primate groups.

References

BANG F.B., 1936. *Observations on limb arteries of woolly monkey*. Anat. Rec. 66: 387-395.

GLIDEN E.M. & GARIS C.F., 1936. *Arteries of the chimpanzee*. Am. J. Anat. 58: 501-529.

IKEDA A., UGAWA A., KAZIHARA Y. & HAMADA N., 1988. *Arterial patterns in the hand based on a three-dimensional analysis of 220 cadaver hands*. J. Hands Surg, 13: 501-509.

IKEDA A., MIYAKE S., IKEDA K. & HAYAMA S., 1988. *Three-dimensional analysis of the arterial pattern of the upper limbs in Lorisiformes (Prosimiae, Primates)*. Acta Anat. 132: 297-303.

IKEDA A., MATSUMOTO S. & TANAKA H., 1988. *Comparative anatomical study of arteriographs of hand of primates*. Acta Anat. 133: 30-34.

KOCH ?., 1939. *Das Verhalten der Arterienbögen in der Vola manus and Planta pedis bei Affen*. Z. anat. Entw.Gesch. 110: 81-97.

LINEBACK P., 1933. *The vascular system*. In: Hartman & Straus (ed.) The Anatomy of the Rhesus Monkey, chapt. 10, Hafner, Baltimore.

MANNERS-SMITH T., 1911. *The limb arteries of primates*. J. Anat. Physiol. 45: 23-64.

MURAKAMI T., 1969. *On the position and course of the deep palmar arteries, with special reference to the so-called palmar metacarpal arteries*. Okajimas Folia Anat. Jap., 46: 177-199.

NISHI S., 1939. *Über die Arterien der Affenhand. Ein Beitrag zur vergleichenden Angiologie*. Jap. J. Med. Sci., 1, Anat. 7: 81-97.

SONNTAG C.F., 1924. *The morphology and evolution of the apes and man*. Bale. Sous & Danielsson, London.

UGAWA A. & IKEDA A., 1985. *Study of arterial patterns of the hand in primates. Deep palmar arterial arches and their correlating arteries in Macaca fuscata*. Acta Anat. 123: 82-89.

UMEDA & IKEDA, 1988. *Scanning electron microscopic study of capillary loops in the dermal papillae. Skin of the hand of Japanese monkey* (Macaca fuscata). Acta Anat. 132: 270-275.

ZUCKERKANDL E., 1907. *Zur Anatomie und Morphologie der Extremitätenarterien*. Sber. Akad. Wiss. Wien, Abt. III, 66.

A. Ikeda
N. Umeda

New results concerning the vascularization of primate hands Part II: The capillaries in the dermal ridges of fingers (and palms) in man and monkeys

A study of the microvasculature of the skin of the hand of Japanese macaques and humans by means of scanning electron microscopy of corrosion casts revealed the following:
1) The typical structure of the capillary loops in monkeys was observed in the balls of the fingers. Capillary loops are formed out of one, two or three capillary vessels. Each capillary vessel arises in the papilla and divides into several branches, and they all pass a hairpin turn to become descending limbs and then extrapapillary venules. The postcapillary venules fuse with the primary and secondary venous arcades.
2) The microvascular patterns in different areas of the hand are fundamentally similar, but size, length and number of the capillary loops differ.
3) In the thenar eminence, a well-developed interpapillar capillary network was observed between the lines of capillary loops underlying the epidermal ridges.
4) The microvascular pattern of the capillary loops in humans is principally the same as in monkeys.
5) In an 85-year-old man a somewhat deviating capillary pattern, probably due to age changes, was observed.

Key words: capillary loops, corrosion technique, dermatoglyphics, microvascularisation, subcutaneous arteries.

Introduction

The microcirculation system of the skin has been studied by light microscopy of sectioned or injected preparations and by vital dissecting microscopy (Spalteholz, 1957; Horstmann, 1957; Saunders, 1961; Odland, 1961; Winkelmann, Scheen & Conventry, 1961; Okajima, 1979; Imayama, 1981). The vascular system of the skin of the human finger and of its nail bed has also been investigated by the same methods (Popoff, 1934; Fleischhauer & Horstmann, 1955; Flint, 1955; Maruyama, 1955; Samman, 1977; Okajima, 1979). Morphology, length and density of the capillary loops in the hand have been described, but these techniques have been unable to provide a three-dimensional picture of the architecture of the fine vessels of the skin. A systematic ultrastructural and three-dimensional examination of the capillary system has not yet been done.

A first approach to the three-dimensional architecture of the skin was made by Imayama (1981) and Inoue (1978). Imayama, however, used the foot of a rat. Rats do not possess the specialized dermatoglyphic skin so characteristic of primates. Their capillary loops have a very simple structure. Although Inoue used an amputated human finger, the particular structure of the capillary loops was not demonstrated.

In our preliminary report (Umeda & Ikeda, 1988), the microvasculature of the skin of the hand in Japanese monkeys was examined by means of scanning electron microscopy of corrosion casts. In this study, we will describe the microvascular patterns and structures in different areas of the skin of the hand in primates, including man.

Material and Methods

The details of the technique used to prepare the materials in this study has been described previously (UMEDA & IKEDA, 1988). In this study, resin casts of the fine vasculature of fingers and palms of Japanese macaques (3-5 years old) and of a 85-year-old man were studied by scanning electron microscopy.

For this purpose, Mercox (Dainihon, Tokyo, Japan; principally composed of an acrylate monomer) was injected through the axillary artery until the axillary vein was filled with the injected resin. After this the resin was polymerized for two to three hours. The upper limbs were removed and macerated completely in a 20% aqueous potassium hydroxyde solution for three to four days, and then washed in running water. In the next step, these macerated arms were washed in a microwave box and rinsed. After thourough rinsing, the casts were air dried. The dried casts of the vessels were mounted, coated with gold-palladium, and examined with a scanning electron microscope (Model S-570, Hitachi).

Identification of the arteries and venules was made by determining the subpapillary venules, which exhibited prominent endothelial nuclear impressions, and was confirmed by examination of digital arteries.

Observations

Japanese macaques (Figures 1, 2, 3, 4)

The arteries that supply the skin are located in the subcutaneous layer. They branch into arterioles and capillaries which reach upward and form a network (i. e. rete cutaneum) parallel to the surface on the boundary line between the dermis and the hypodermis. From one side of this network, branches are given off that nourish the subcutaneous stratum, which includes fat cells, sweat glands, and the deeper portions of hair follicles. At the boundary between the capillary and reticular layers the branches form a denser, subpapillary network or rete subpapillare. This gives off capillaries to the papillae.

The palmar skin has a double capillary loop line of the papillae in each epidermal ridge which regularly follows the lines of the finger- and palmprints. The capillary loops are formed by the capillary vessels which arise from the subpapillar arterial capillary network and follow and coil around the subpapillary venules. The capillary loops of the skin can be divided into two segments: intrapapillary and extrapapillary *(Figure 4)*.

In the intrapapillary segment, each papilla has a loop of capillary vessels with ascending limbs, a hairpin turn, and descending limbs. The ascending limbs of the winding capillary loop passes either directly to a papilla or subdivides once, twice, or three times before giving rise to a final ascending limb that enters the papilla. After the loop has passed the hairpin turn, the descending limbs are 1.5 times wider than the ascending limbs in the intrapapillary portion, and form single or plural (1-5) descending capillary vessels in each dermal papilla.

Figure 1 - Ball of the finger (taken from an angle of 45° and following the epidermal ridge). The bar represents 100 μm. The arrow indicates the typical structure of the capillary loops in Japanese macaques.

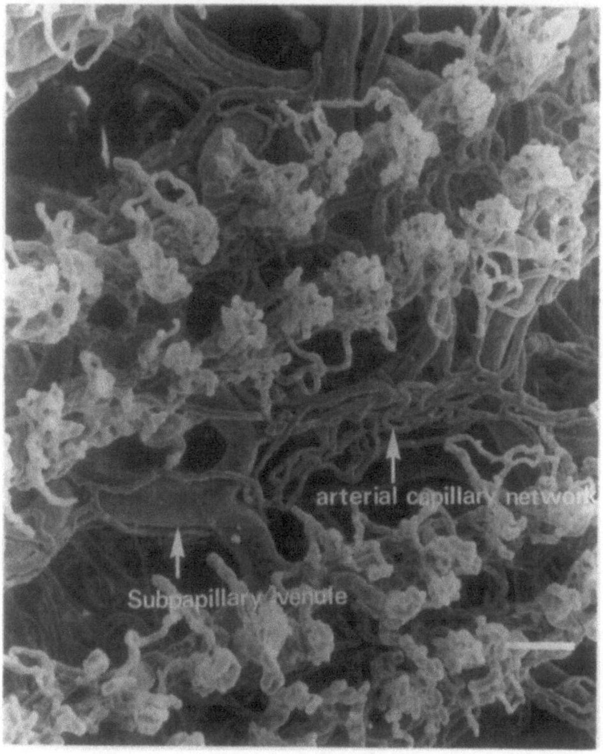

Figure 2 - Top view of the ball of a finger in a Japanese macaque. The bar represents 100 μm.

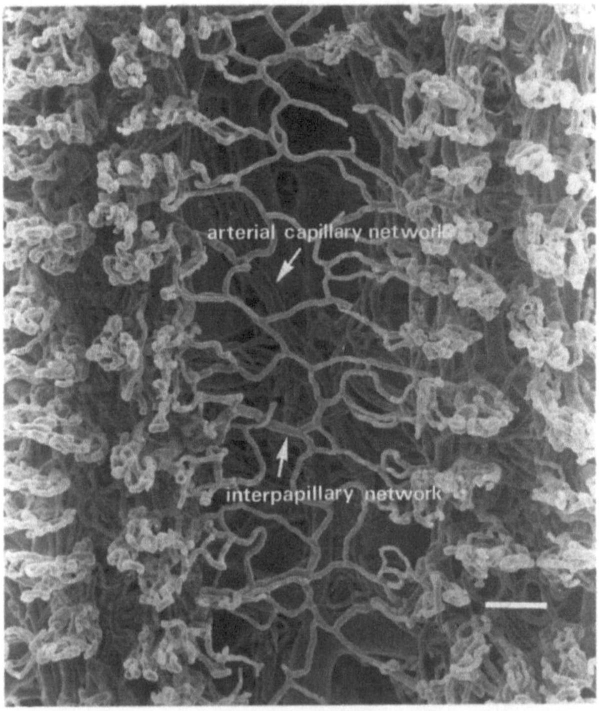

Figure 3 - Top view of the thenar eminence in a Japanese macaque. The bar represents 100 μm.

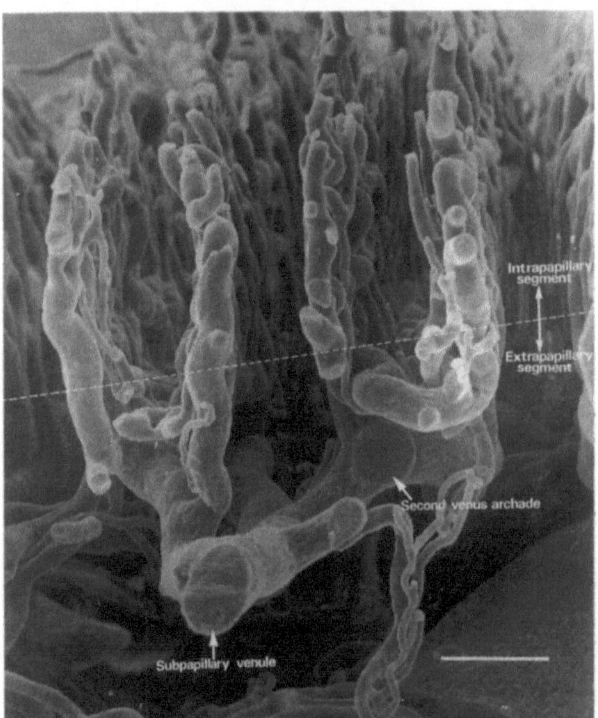

Figure 4 - Transverse section through the ball of the finger in a Japanese macaque. The bar represents 10 μm.

In the extrapapillary segment, the descending limbs from the intrapapillary portion are connected with the postcapillary venules in the postpapillary portion and with the horizontal network. The postcapillary venules fuse with the venules of some neighboring capillary loops and form the primary venous arcade which regularly follows the dermal ridges.

The subpapillar arterial capillary network *(Figure 3)* is formed by arterial vessels arising from the deep portion of the subpapillary layer. The capillary loops are formed by vessels arising from that network. They often divide into several branches at the intrapapillary segment (see above). Each of these branches reaches upward, and then becomes a descending limb. These descending limbs similarly anastomose with each other and reach into the subpapillary venules to form the primary and secondary venous arcades which regularly follow the line of the fingerprint. The subpapillary layer contains most of the dermal blood vessels, and is composed primarily of terminal arterioles, arterial capillaries and postcapillary venules.

The fundamental pattern of the microvasculature has the same structure in different areas of the hand, but size, length and number of the capillary loops differ. In the thenar and hypothenar eminences, the capillary loops were a little lower and their grooves wider than in the balls of the fingers. A well-developed interpapillar capillary network was observed at the epidermal furrow located between the capillary loops. This network existed in the superficial layer of the subpapillar arterial capillary network *(Figure 3)*. The new capillary loops frequently arose from the interpapillar network and then proliferated either alone or in small groups.

Human (Fig. 5)

Each papilla has loops of capillary vessels with ascending limbs, a hairpin turn and descending limbs. After the loops passed a hairpin turn, the descending limbs are 1.3 times wider than the ascending limbs. Three patterns of capillary loops were observed in this area:

1) A single loop pattern, straight or coiled in structure.
2) A complex loop pattern, in which the capillary loops are formed out of not just one capillary vessel, but out of two or three vessels. Each capillary vessel arises and divides into several branches in the papilla and they become descending limbs.
3) A pattern midway between the types 1 and 2.

We believe that these differences in the papillae are due to aging changes.

Proliferation of the capillary loops, and the resulting formation of more than the ususal two lines of loops in one dermal ridge was observed. Although proliferation of dermal papillae on the dermal ridges was evident, details of the process of development and aging are not clear at this time. In addition to this common phenomenon, dermal papillae also occasionally proliferate on the furrows to various degrees (OKAJIMA, 1979).

In this study, the capillary loops presented complicated features that changed with age. In some of the more complex loop patterns, the capillary loops lost their complexity and each loop assumed a simpler structure: The root of the descending loop, however, was crowd and became a single descending limb *(Figure 5)*.

In monkeys and in the human subject, the new capillary loops frequently arose from the interpapillar network which was observed in the epidermal furrows located between the capillary loops *(Figures 3, 5)*.

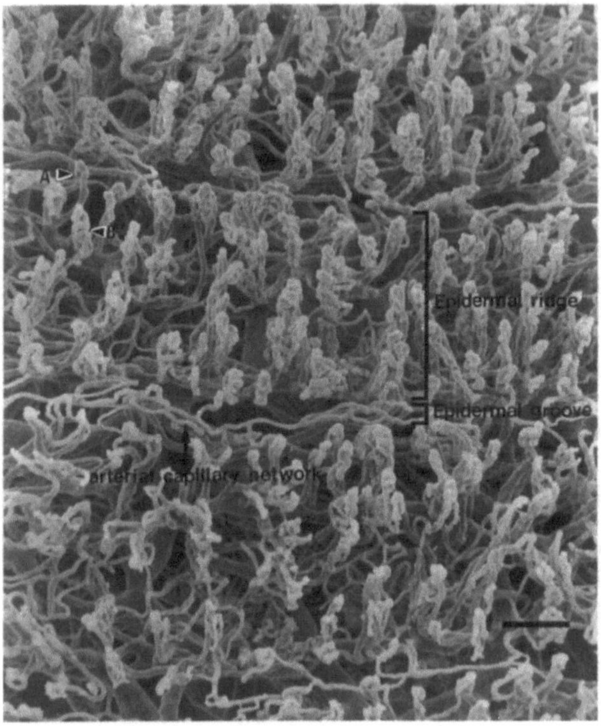

Figure 5 - The ball of the finger in a human (taken from an angle of 45° and following the epidermal ridges). The bar represents 100 Ðm. arrow A = «straight structure», arrow B = «coiled structure», arrow C = arterial capillary network.

These findings provide clear evidence regarding the proliferation process of the dermal papillae in the ridges and their occasional proliferation in the furrows in various gradiation, which was observed by OKAJIMA (1979).

Discussion

IMAYAMA (1981) stated that the development of the capillary loops might explain why the skin can be vertically mobile without injury to the vessels and why normal blood flow is maintained in spite of external pressure applied vertically to the skin surface. Our observations support this opinion.

At present, we are beginning to study the changes in the capillary loops with age, and we believe further morphofunctional conclusions can be drawn in the future.

References

FLEISCHHAUER K. & HORSTMANN E., 1955. *Der Papillarkörper und die Kapillaren des Paronychium.* Z. Zellforsch. mikrosk. Anat. 42: 213-228.

FLINT M.H., 1955. *Some observations on the vascular supply of the nail bed and terminal segments of the fingers.* Br. J. Plast. Surg. 8: 186-195.

HORSTMANN E., 1957. *Blutgefäße der Haut.* In: v. Möllendorff (ed.), Handbuch der mikroskopischen Anatomie des Menschen. VIII (Springer, Berlin), pp. 198-207.

IMAYAMA S., 1981. *Scanning and transmission electron microscope study on the terminal blood vessels of the rat skin.* J. Invest. Dermatol. 76: 151-157.

INOUE H., 1978. *Three-dimensional observations of microvasculature of human finger skin.* Hand 10: 144-149.

MARUYAMA M., 1955. *A study on the morphology and capillary system of the dermal papilla in different parts of the body surface in the Japanese population (in Japanese).* J. Kurume Igaku 18: 798-811 .

ODLAND G.F., 1961. *The fine structure of cutaneous capillaries.* In: Montagna, Ellis (ed.), Advances in Biology of Skin, vol. II: Blood vessels and circulation (Pergamon Press, Oxford), pp. 57-70.

OKAJIMA M., 1979. *Dermal and epidermal structure of the volar skin.* Birth Defects 15: 179-198.

POPOFF N.W., 1934. *The digital vascular system, with reference to state of glomus in inflammation, arteriosclerotic gangrene, diabetic gangrene, thromboangitis obliterans and supernumerary digits in man.* Archs Path. 18: 295-330.

SAMMAN P.D., 1977. *The nails in desease.* Heinemann Medical Books, London, pp. 7-12.

SAUNDERS R.L., 1961. *X-ray projection microscopy of the skin.* In: Montagna, Ellis (ed.), Advances in Biology of Skin, vol. II: Blood vessels and circulation, pp. 38-56 (Pergamon Press, Oxford).

SPALTEHOLZ W., 1927. *Blutgefäße der Haut.* In: Judassohn, (ed.) Handbuch der Haut-und Geschlechts-krankheiten (Springer, Berlin), pp. 373-433.

UMEDA N. & IKEDA A., 1988. *Scanning electron microscopic study of the capillary loops in the dermal papillae.* Skin of the Hand of Japanese Monkey (*Macaca fuscata*). Acta Anat. 132: 270-275.

WINKELMANN R.K., SCHEEN S. R., PYKA R.A. & CONVENTRY M.B., 1961. *Cutaneous vascular patterns in studies with injection preparation and alkaline phosphate reaction.* In: Montagna, Ellis (ed.), Advances in Biology of Skin, vol. II: Blood vessels and circulation, Pergamon Press, Oxford, pp. 1-9.

W. Maier

Adaptations in the hands of cercopithecoids and callitrichids

The differences in the 'biological roles' and in correlated structural adaptations of the hands of geladas and baboons are demonstrated. The hands of geladas show a suite of peculiarities which appear to be connected with their locomotor as well as with their 'grass-plucking' habits. In a second part, the long standing problem of nails and claws in primates is reviewed. An abnormal extra-pollex of an old female *Cebuella* is described and discussed. The relatively flat nail of this small extra-finger is distinctly broader than the 'claw' of the adjacent thumb. This find is interpreted as an 'atavistic recapitulation' and thus as a further indication of the secondary derived nature of callithrichied claws.

Key words: atavism, biological roles, claws, extra digit, form-function relationship, nails.

Introduction

The hands are very important 'tools' for all Primates, and they are used in a variety of biological roles. They are primarily employed in different sorts of locomotor functions, but manipulation of food items is of great importance as well. Thus hands support the body weight during standing; they transmit positive or negative impulses to the substrate during acceleration or deceleration; the body can hang by the hands during brachiation; hands can grasp branches or fruits and pull them towards the mouth; and hands can handle and break food items before ingestion. These last mentioned acts will mostly be done under visual control, and this visuomotor-controlled manipulation of objects is suspected to have played an important role for the development of a primate-specific sort of perception of the outside world; in the german language we still call this form of understanding 'begreifen'.

Different species of primates live in different habitats and they have different life-strategies (Charles-Dominique, 1977). Even in pre-Darwinian times it has been noted that certain structural features of each species seem somehow to be correlated with their various biological roles (Cuvier); this mutual fitting has been called adaptation. It becomes clear from this viewpoint that the term adaptation can be applied to many different sorts of correlative situations, and that it may have many different meanings (Bock & v. Wahlert, 1965). It should be specified in each case, what sort of correlation is addressed. Used in such an instrumental way, the term 'adaptation' would seem to be not only useful but indispensable for many sorts of biological research.

It is postulated, therefore, that different features of primate hands are linked with species-specific patterns of activities. This premise will be examined for two cases.

Adaptations in the hands of the gelada baboon

Problems of functional adaptations cannot be properly formulated without reference to the phylogenetic and systematic framework of the studied taxa. Adaptations are realized through natural selection in the evolutionary process, either by improving the efficiency of an existing system or by accomodating it to changing conditions of life. Extant adaptations are derived from older adaptation states by gradual modification. Related species are likely to share similar form-function relationships.

There can be little doubt that the genera *Papio* and *Theropitecus* belong to the Old World monkey group of the Cercopithecinae (KUHN, 1966; MAIER, 1970; SZALAY & DELSON, 1979). However, there exists no unanimous consensus as to the precise phylogenetic relationships of the various genera. The fossil record demonstrates that the papionine and the theropithecine tribes are clearly distinguishable in the upper Pliocene of Africa (MAIER, 1972a; DELSON, 1973; SIMONS & DELSON, 1978). STRASSER & DELSON (1987) group *Theropithecus* with *Cercocebus* and *Papio* into a common subtribe Papionina «until further study clarifies the relations of fossil and recent baboons and mangabeys» (p.97). While the theropithecines were abundant in the Plio-Pleistocene of Africa, they have survived with only one species in the montane regions of Ethiopia. The baboons sensu stricto seem to have sprung from the more primitive, macaque-like genus *Parapapio* and show some radiation during the Pleistocene, resulting in a recent distribution throughout the African faunal region. Both geladas and baboons have been extensively studied in their natural habitats during the last 25 years, and we are able to define fairly well their ecological niches, their social structures, and their behaviour (KUMMER, 1968; ALTMANN & ALTMANN, 1970; DUNBAR & DUNBAR, 1974; DUNBAR, 1977; CLUTTON-BROCK, 1977).

True baboons occupy variable habitats throughout Africa; they make their living both in semi-deserts and in the tropical rain forest. They are often found on trees, where most populations also rest overnight. They take a wide array of different foods, and ingest them without a specified manual technique. They prefer all sorts of plant-parts but have been proven to show even carnivorous habits; most taxa are classified as frugivorous (CLUTTON-BROCK, 1977). The body size of the various species and subspecies changes considerably, but all have a very distinctive sexual dimorphism. Group size and social organisation also show considerable differences between the baboon-taxa. Their biological versatility in some way reminds one of the Eurasian macaques (LINDBURGH, 1980). This is not the place to further elaborate on the enormous bulk of ecological and behavioural literature on this group of primates.

Gelada baboons are socially organized in bands of one-male and all-male groups (CROOK, 1966; CROOK & ALDRICH-BLAKE, 1968). They rest at the steep crevices of the gorges of High Simen. They ascend in the morning to the grassy plateaus and the bands may gather into foraging units, which can form herds of several hundred animals. They start feeding on grass blades, seeds or rhizomes sitting on their naked, greyish pads. They continue with their grazing activities for many hours a day, slowly shuffling forwards on their haunches. Fossil geladas seem to have been adapted to this sitting and feeding technique by a shortening of the hindlimbs and an elongation of the arms. Geladas have distinctly smaller day ranges than the other baboons, and they rarely climb low bushes and trees. During all seasons of the year more than 90% of their diet consists of grass, and they certainly are the most specialized terrestrial grazers of all primates. DUNBAR (1977) has described their feeding technique in detail: «Grass blades are collected by alternate rotary motions of the hands, snipping off the leaves between the thumb and the index finger. The blades are gathered into the palm of the hand, and when a handful has been collected the bundle is transferred to the mouth. Grass seeds are collected either by pulling the stems through the teeth one at a time, or by stripping standing stems between the thumb and the index finger ... When feeding on grass rhizomes, the hands are held flexed at right angles to the forearm and the earth is removed with rapid alternate digging

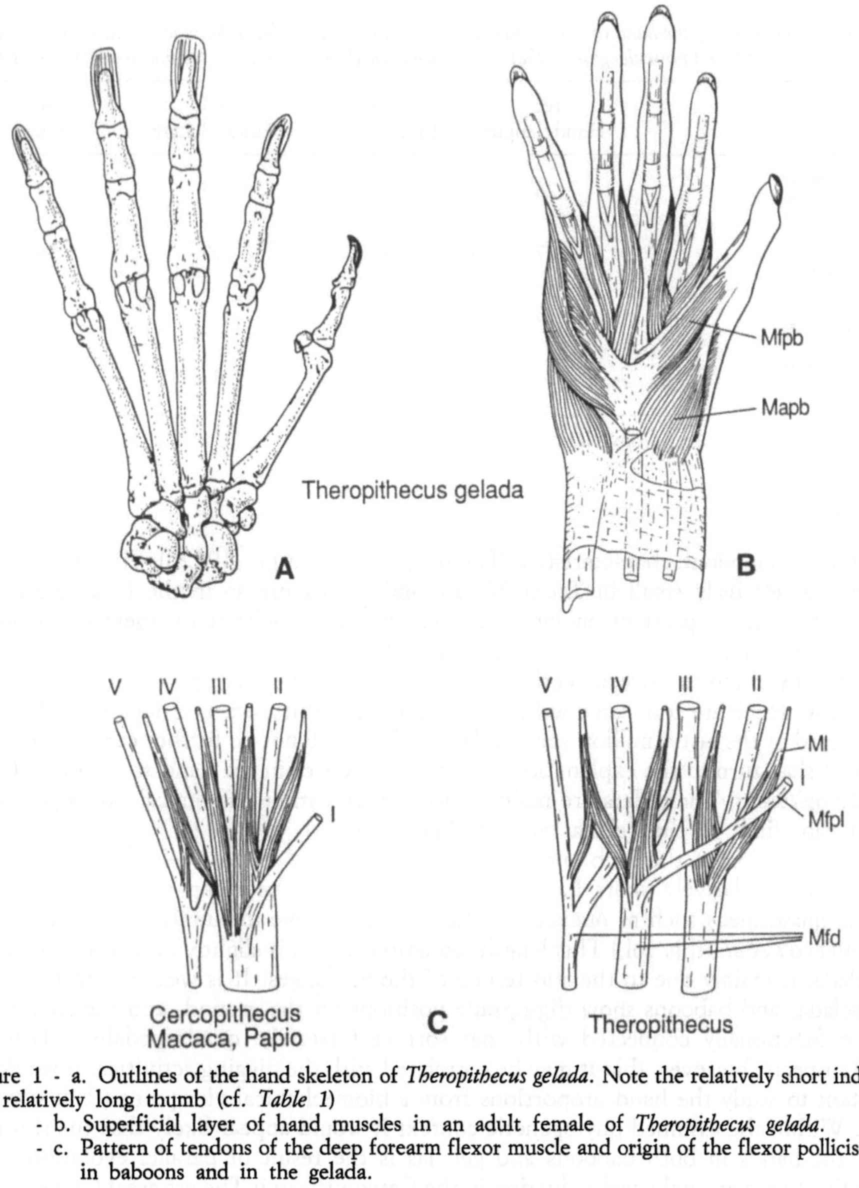

Figure 1 - a. Outlines of the hand skeleton of *Theropithecus gelada*. Note the relatively short index and the relatively long thumb (cf. *Table 1*)
- b. Superficial layer of hand muscles in an adult female of *Theropithecus gelada*.
- c. Pattern of tendons of the deep forearm flexor muscle and origin of the flexor pollicis longus in baboons and in the gelada.

(Mfpb: M. flexor pollicis brevis; Mapb: M. abductor pollicis brevis; Ml: M. lumbricalis; Mfpl: M. flexor pollicis longus; Mfd: M. flexor digitorum profundus)

movements... These feeding actions are highly skilled, and permit the gelada to feed on grasses at a much faster rate than other baboons» (p. 258). It is to be expected that this specialized feeding technique is somehow reflected by the involved body structures.

For example, baboons and geladas are clearly distinguishable in the structure of their dentitions. Whereas the front teeth of baboons are very prominent, their postcanine teeth

TABLE 1 - *Proportions of the hand of some cercopithecids. The handlength is related to length of the forelimb, the index- and pollexlengths to that of the hand length (for measuring technique cf* MAIER, *1971).*

	rel. Hand length	rel. Index length	rel. Pollex length	Opp. Index
Cercopithecus (n=13-19) *	26	—	38	53
Macaca (n=10-18) *	27	—	39	54
Papio (n=10-17) *	23	74	44	60
Theropithecus (n=3-4) *	21	68	47	70

* Values partly taken from NAPIER & NAPIER (1967).

show rather bunodont characteristics (DELSON, 1973; MAIER, 1972b). In contrast, the incisors are relatively small in *Theropithecus*, and even more so in the fossil species of *Simopithecus*; the hypsodont molars show complicated foldings of their bilophodont enamel loops: «tough small food items requiring little incisal preparation, but prolonged trituration by complex cheeek teeth» (SIMONS & DELSON, 1978; p. 106). These dental differences, which are connected with specific feeding habits, are well documented by the fossil record of the last 4 million years at least. They shall not be further considered here; instead, I shall turn to the explanation of some differences in the hand structure of *Papio* and *Theropithecus*. These data are mainly based on an earlier anatomical investigation of the anterior limb of the gelada baboon (MAIER, 1971).

An inspection of the limb and hand proportions (*Table 1*) reveals that geladas and baboons have relatively short hands as compared with the rump or the arm length; arboreal quadrupeds such as *Macaca* and *Cercopithecus* have distinctly longer hands (see also JOUFFROY *et al.*, this vol.) This length reduction, which is slightly more pronounced in the gelada, is mainly due to the shortening of the phalanges. It is well documented that both geladas and baboons show digitigrade positions on the ground, and the shortening may be functionally connected with that sort of terrestrial quadrupedalism; DUNBAR (1977) suspects, however, that it may be correlated with the digging activities. It would be important to study the hand proportions from a biomechanical viewpoint (PREUSCHOFT, 1979). Within the assumed phylogenetic context it would appear likely that the shortening of the hands in both baboons and geladas is the result of parallel evolution. It is interesting to see an analogical reduction in the Cercopithecini: The arboreal *Cercopithecus* shows long hands and fingers, whereas the terrestrial quadruped *Erythrocebus* shows considerable length reduction. In arboreal monkeys long hands would appear favourable for grasping branches of various sizes.

However, the hands of baboons and geladas also show structural differences in detail which are likely to be connected with the different feeding strategies. It becomes obvious at first sight, that the index fingers of the geladas are very short, and that the thumbs are relatively long. This results in a very high 'opposability-index' of 70, which is higher than in man. NAPIER & NAPIER (1967) assumed that this index reflects the manipulatory abilities between these two fingers, but it is clear that suitable neuromotor and behaviour

TABLE 2 - *Relative muscle weights of the hypothenar and thenar in three species of Cercopithecids; the weight of dried muscles are related to the total weight of the intrinsic hand muscles.*

	Cercopithecus aethiops (n = 3)	*Macaca mulatta* (n = 1)	*Theropithecus gelata* (n = 4)
Thenar	27.5	28.5	34.9
Hypothenar	27.5	22.2	23.0

components are needed as well. However, suitable length relations are a necessary prerequisite for achieving some sort of 'precision grip' between thumb and index.

Investigation of the intrinsic hand muscles shows that in *Cercopithecus* the thenar and hypothenar muscle weights are almost identical; in the Rhesus monkey I observed a distinct increase of the thenar muscles, and in *Theropithecus* the thenar muscles are even more preponderant; unfortunately, no baboons have been studied so far (*Table* 2). These data seem to indicate a distinct increase of the thumb musculature, which suggests an increased functional activity of the first digit. In *Macaca*, *Cercopithecus,* and *Papio*, the common radial tendon of the deep flexor muscle of the forearm gives origin to the tendon of the flexor pollicis; powerful flexion of the thumb is therefore closely linked with that of the other fingers. In four specimens of *Theropithecus*, the tendon of the deep flexor is subdivided into three portions, the medial of which gives origin to the thumb tendon (*Figure 1*). In analogy with the situation in man, this arrangement is likely to give more freedom and independence of movement to the thumb. JOLLY (1970) has used the 'T-complex' (*Theropithecus*-complex) of structural and behavioural adaptations as a model for explaining some features of human evolution.

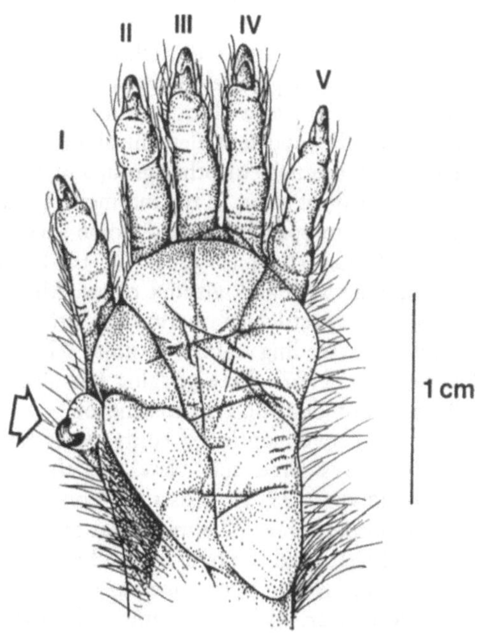

Figure 2 - Drawing of the left hand of an old female of *Cebuella pygmaea* (SAI 8672) showing a small abnormal extra digit at the base of the pollex. This digit has no claw but a nail-like structure.

These data seem to suggest that in fairly closely related taxa minor structural changes of the hand can be correlated with differences in feeding behaviour. Hands of a very similar basic structure, which seem to be subject to slightly different average activity patterns, tend to show gradual change. It would seem to be legitimate to state that they become 'adapted' to modified mechanical requirements linked with these patterns (PREUSCHOFT, 1979). It is admitted that these conclusions are based mainly on rather vague analogies, but it would seem worthwhile to pursue this sort of investigation.

The 'claws' of the Callitrichidae

The existence of flattened nails, at least at the hallux, has for a long time considered to be a diagnostic key character of the order Primates (WEBER, 1928; CLARK, 1962). Among simian primates, the South American Callitrichidae (tamarins and marmosets) are distinguished by sharp, recurved claws on all fingers and toes except for the hallux. The clawed thumb (pollex) does not exhibit much opposability, but is aligned almost in one row with the other fingers. There exists an old discussion in primatology whether the 'claws' of callitrichids are a primitive retention or a secondary adaptation of the nails for climbing at smooth tree-stems.

CLARK (1962) has described the morphological and evolutionary problems that are connected with the occurrence of claw-like structures in callitrichids: «They are very similar in general appearance to typical claws, such as those of the modern tree-shrews, but in the presence of a more conspicuous groove on their under surface they show what might be taken for an incipient tendency to open out and thus mark the earliest morphological stage in the transformation of the specialized claw into the less elaborate nail which is characteristic of most modern Primates. If it is proper to speak of the so-called retention of claws in the marmosets, this is of considerable interest in the problem of pithecoid ancestry. It remains possible, of course, that the marmoset claw is a secondary modification of a true nail such as is found in the Cebidae, and that it merely simulates a true claw in its general shape and functions. But microscopic examination shows that there still persists a thin stratum of the deep layer of horny substance which is the mechanically important element of a true claw, a stratum which seems to have disappeared entirely in the nails of other groups of monkeys ... From the functional point of view the development of claws in the marmosets is associated with their special arboreal habits, for they can run spirally up and down large tree trunks in the manner of squirrels...» (p. 203/204). CARTMILL (1972; 1974; 1985) has further elaborated the biomechanical aspects of climbing with and without claws. Meanwhile, there also exists good evidence that the climbing capacities of marmosets on big trunks may be correlated with the specific sapivorous feeding habits (COIMBRA-FILHO & MITTERMEIER, 1977; MAIER, ALONSO & LANGGUTH, 1982). Tree gouging has also had influence on the formation of a scoop-like lower front dentition in marmosets.

The above-mentioned controversy has not been settled in substance. WEBER (1928, p. 761) considered the 'claws' of the callitrichids as extremely compressed 'tegulae'. HERSHKOVITZ (1977) insists in considering the claws of callitrichids as a primitive plesiomorphic trait. The investigations of THORNDIKE (1968) indicate that both the claws of the fingers and the nails of the hallux of callitrichids retain a primitive terminal matrix and a deep layer of horn, much as in *Daubentonia* (CLARK, 1936). *Cebus* shows remnants of this deep layer, which is absent in all other investigated prosimians and simians, however.

The functional meaning of this layer for stabilizing a claw-like curvature remains an unsolved problem. The cladograms of ROSENBERGER (1981) would imply a secondary nature of the 'claws' without further structural analysis. Other features also suggest that this group is a specialized side-branch rather than a primitive stem-group (MAIER, 1980; ROSENBERGER, 1981). Of course, such a statement does not by itself preclude that a specific structural complex has retained certain primitive characteristics (plesiomorphies) of remote ancestors. Paleontology will probably not be able to tell us the whole story of claw and nail evolution in primates, which is essentially a softpart phenomenon.

We report here the abnormal occurrence of an additional digit at the left hand of an old female of *Cebuella pygmaea* from the Zoological Garden Frankfurt/Main (SAI 8672). The small extra-digit has a length of less then 4 mm and a breadth of about 1.7 mm, and it is situated at the basis of the thumb (*Figure* 2). The digit consists of two bony elements (phalanges); the basal one does not directly articulate with any of the carpal or metacarpal bones, but springs from a ligament fixed to the basis of the first metacarpal and the trapezio-metacarpal-joint. The trapezium forms a prominent medial process which serves as insertion for the long abductor-tendon of the thumb (BEATTIE, 1927). In contrast to many other studied skeletons of callitrichids, no sesamoid bone is inserted into the tendon of the abductor; the problem whether this sesam bone is homologous with a prepollex shall not be discussed here. The most intriguing feature of this pathological extra-finger, however, is its relatively flat nail, which is in contrast with the claw of the true thumb. Although this abnormal finger is much smaller than the thumb, its short nail has a breadth of about 1.3 mm, whereas the claw of the thumb only measures 0.8 mm.

It is certainly problematical to interpret the features of such an abnormal finger adequately. It seems more likely, however, that such a nail, if it is distinct from equivalent structures of the same individual, represents an atavistic degenerate ('Rückschlag') rather than a progressive neomorph (ZIMMERMANN, 1959). Therefore, I am inclined to consider this pathological nail as a further piece of evidence for the assumption that the claws of callitrichid monkeys are in fact secondarily modified nails. Therefore, callitrichid 'claws' should be considered as a structural 'adaptation' of a group of small, sapivorous primates which enables them to climb smooth tree-stems to reach their specific food sources. *Euoticus elegantulus*, a sapivorous galagid prosimian, has solved the same locomotory problem by forming 'claws' in a somewhat different way (CHARLES-DOMINIQUE, 1977).

References

ALEXANDER, R. McN., this volume.

ALTMANN S.A. & ALTMANN S., 1970. *Baboon ecology: African field research.* Chicago.

BEATTIE J., 1927. *The anatomy of the common marmoset.* Proc. Zool. Soc. London 1927, 593-718.

CARTMILL M., 1972. *Arboreal adaptations and the origin of the order Primates.* In: Functional and Evolutionary Biology of Primates (ed. R.H. Tuttle), Chicago, 97-122.

CARTMILL M., 1974. *Pads and claws in arboreal locomotion.* In: Primate Locomotion (ed. F.A. Jenkins), New York, 45-83.

CARTMILL M., 1985. *Climbing.* In: Functional Vertebrate Morphology (eds. M. Hildebrand, D.M. Bramble, K.F. Liem, D.B. Wake), Cambridge, Mass, 73-88.

CLARK W.E. & LE GROS, 1936. *The problem of the claw in Primates.* Proc. Zool. Soc. London 1936, 1-24.

CLARK W.E. & LE GROS, 1962. *The Antecedents of Man. An introduction to the evolution of Primates.* Edinburgh.

CLUTTON-BROCK T.H. ed., 1977. *Primate Ecology. Studies of feeding and ranging behaviour in lemurs, monkeys and apes.* London.

Coimbra-Filho A.F. & Mittermeier R.A., 1977. *Tree-gouging, exudate eating and the 'short-tusk' condition in Callithrix and Cebuella.* In: The biology and conservation of the Callitrichidae (ed. D.G. Kleiman), Washington, 105-115.

Delson E., 1973. *Fossil colobine monkeys of the circum-Mediterranean region and the evolutionary history of the Cercopithecidae (Primates, Mammalia).* Ph. D. Thesis, Columbia University, New York.

Dunbar R.I.M. & Dunbar I.P., 1974. *Ecological relations and niche separation between sympatric terrestrial primates in Ethiopia.* Folia primatol., 21: 36-60.

Dunbar R.I.M., 1977. *Feeding ecology of gelada baboons: a preliminary report.* In: Primate Ecology (ed. T.H. Clutton-Brock), London, 251-273.

Hershkovitz P., 1977. *Living New World monkeys.* Chicago.

Kuhn H.J., 1967. *Zur Systematik der Cercopithecidae.* In: Neue Ergebnisse der Primatologie (eds. D. Starck, R. Schneider, H.J. Kuhn), 25-46.

Kummer H., 1968. *Social organization of hamadryas baboons.* Chicago.

Maier W., 1970. *Neue Ergebnisse der Systematik und Stammesgeschichte der Cercopithecoidea.* Z. Säugetierk., 193-214.

Maier W., 1971. *Vergleichend- und funktionell-anatomische Untersuchungen an der Vorderextremität von Theropithecus gelada (Rüppell 1835).* Abh. Senckenb. Naturf. Ges., 527, 1-284.

Maier W., 1972a. *The first complete skull of Simopithecus darti from Makapansgat, South Africa, and its systematic position.* J. Human Evol., 1: 395-405.

Maier W., 1972b. *Anpassungstyp und systematische Stellung von Theropithecus gelada Rüppell, 1835.* Z. Morph. Anthrop., 63: 370-384.

Maier W., 1980. *Nasal structures in Old and New World Primates.* In: Evolutionary Biology of the New World Monkeys and Continental Drift (eds. R.L. Ciochon & A.B. Chiarelli), New York, 219-241.

Maier W., 1984. *Tooth morphology and dietary specialization.* In: Food acquisition and processing in Primates (eds. D.J. Chivers, B.A. Wood & A. Bilsborough), New York, 303-330.

Maier W., Alonso C. & Langguth A., 1982. *Field observations on Callithrix jacchus jacchus L.* Z. Säugetierk., 47: 334-346.

Napier J.R. & Napier P.H., 1967. *A Handbook of Living Primates.* London.

Nieschalk, U. & Demes, B., this volume.

Preuschoft H., 1979. *Motor behavior and shape of the locomotor apparatus.* In: Environment, Behavior, and Morphology: Dynamic Interactions in Primates (eds. M.E. Morbeck, H. Preuschoft, N. Gomberg), Stuttgart, 263-275.

Preuschoft, H., 1992: *Biomechanics and functional morphology.* Karger Gazette 54: 7.

Rosenberger A.L., 1981. *Systematics: The higher taxa.* In: Ecology and Behavior of Neotropical Primates (eds. A.F. Coimbra-Filho & R.A. Mittermeier), Rio de Janeiro, 9-27.

Simons E.L. & Delson E., 1978. *Cercopithecidae and Parapithecidae.* In: Evolution of African Mammals (eds. V.J. Maglio & H.B.S. Cooke), Cambridge, Mass., 100-119.

Strasser E. & Delson E., 1987. *Cladistic analysis of cercopithecoid relationships.* J. Human Evol., 16: 81-99.

Szalay F.S. & Delson E., 1979. *Evolutionary History of the Primates.* London.

Thorndike E.E., 1968. *Microscopic study of the marmoset claw and nail.* Amer. J. Phys. Anthrop., 28: 247-262.

Weber M., 1928. Die *Säugetiere.* Einführung in die Anatomie und Systematik der recenten und fossilen Mammalia. vol. II. Jena.

Zimmermann W., 1959. *Methoden der Phylogenetik.* In: Evolution der Organismen (ed. G. Heberer), Stuttgart, 25-102.

R. McN. Alexander

Joints and muscles of hands and paws

The joints of the human hand allow about 24 degrees of freedom of relative movement, about the same as in the fore feet of most mammals. Only ungulates and whales have substantially fewer degrees of freedom. Humans, cercopithecoid monkeys and primitive mammals such as *Didelphis* all seem to have 40-50 muscles, or independently-usable parts of muscles, in their hands or fore feet. Dogs, ungulates and whales have fewer. The extrinsic muscles of the fore foot are larger fractions of body mass in cercopithecoid monkeys than in *Didelphis* or in fissipedes. Their muscle fibres are longer (relative to ulna length) in humans and monkeys than in cursorial fissipedes, but only flexor digitorum superficialis has markedly longer fibres than in *Didelphis* and *Meles* (a non-cursorial fissipede). Ratios of muscle fibre length to moment arm indicate capacity for producing angular movements. They are larger for various hand muscles of humans, a monkey (*Macaca*) and a wallaby (*Macropus*), than for dogs and sheep. The difference is attributed to conflicting requirements for prehension and running.

Key words: degrees of freedom, mobility, moments arms, muscle fibre length, muscle mass.

Introduction

The hands of primates perform mechanical functions in locomotion and prehension. This paper considers their mechanical design, looking for major points of difference from the fore feet of other mammals. It examines the mobility allowed by the joints and the number and dimensions of the muscles that work them, in a variety of mammals. It avoids discussing details of individual joints and muscles, looking instead for major differences in overall design.

Mobility

The first question we will ask is whether Primate hands are capable of an exceptionally wide variety of movements. Engineers specify the number of kinds of movement that a mechanism can make by stating the number of degrees of freedom of relative movement of its parts. A hinge joint allows one degree of freedom. The interphalangeal joints of the human hand are examples: they can be flexed and extended, but do not allow appreciable movement of any other kind. A universal joint allows two degrees of freedom. In this case, the metacarpophalangeal joints of digits 2 to 5 will serve as examples. Any movement that they make can be described as a combination of just two kinds of movement: flexion/extension and adduction/abduction (Youm, Gillespie, Flatt & Sprague, 1978). A ball and socket joint such as the human gleno-humeral joint allows three degrees of freedom.

Figure 1 is a model of the human hand. Each of the joints that is shown allows one degree of freedom, except those labelled 2, which allow two. The wrist is shown as a single joint allowing two degrees of freedom as implied by the analysis of Youm, McMurtry,

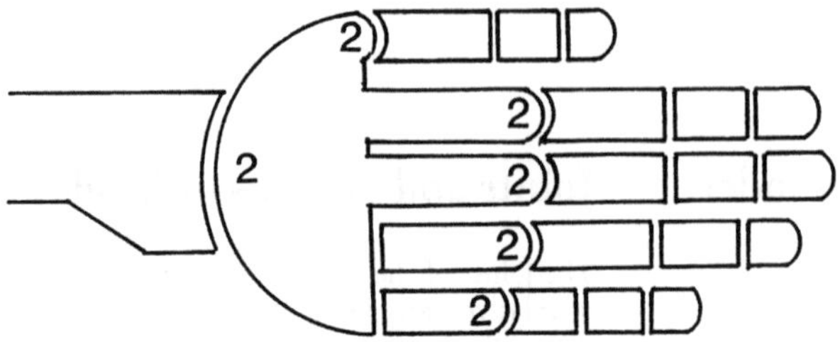

Figure 1 - This diagram of the human hand is explained in the text.

FLATT & GILLESPIE (1978). (The radiocarpal and intercarpal joints are not shown separately, because they do not seem capable of moving independently of each other. This is the only discrepancy between this analysis and that of MORECKI, EKIEL & FIDELUS, 1984.) The carpometacarpal joint of the thumb is given two degrees of freedom following KUCZYNSKI (1974). The carpometacarpal joints of digits 2 and 3 are omitted because they allow very little movement (YOUM, MCMURTRY *et al.*, 1978). The model is a simple mechanism without loops so its mobility can be assessed by adding together the degrees of freedom allowed by individual joints: the total is 24.

Most mammalian fore feet have as many joints as the human hand and should probably be represented by models allowing about the same number of degrees of freedom. The exceptions include ungulates (which have fewer digits) and whales (which have no freely movable joints in the hand).

How many degrees of freedom are required for hand function? An unrestrained object can move along any of three mutually perpendicular axes, or rotate about any of these axes, so it has six degrees of freedom. If it is to be held firmly, all six freedoms must be removed. Assume that fingers make point contact with objects and that there is friction at the points of contact. In that case, two fingers gripping an object can remove five of its degrees of freedom, preventing it from doing anything except rotating about the axis joining the two finger tips. A third finger can remove the last degree of freedom. Each finger needs joints that will give it three degrees of freedom, so that its tip can be moved relative to the palm of the hand, to any required position. MASON & SALISBURY (1985) used this argument to justify the design of the Stanford/JPL robot hand which has three fingers, each with three hinge joints.

The human hand is far more dexterous than any robot so far invented. Try knotting together the ends of a piece of string. You will probably find that at one stage of the process, each hand is grasping the string in three different places. Similarly BOESCH and BOESCH (this volume) have shown that chimpanzees breaking open nuts may hold the nut they are breaking, and several other nuts, simultaneously in the same hand. The more degrees of freedom a hand has, the more complex are the functions that it is in principle capable of performing. However, the remarkable manipulative ability of human and ape hands does not depend on unusually large numbers of degrees of freedom. It has been achieved by more effective exploitation of freedoms already present in primitive mammals.

The usefulness of a degree of freedom depends on the angular range of the movement it allows. Humans and other primates do not seem to have any remarkable advantage over other mammals, in this respect. Most of the joints in the fore feet of non-primates such as *Didelphis*, *Macropus* and *Meles* seem capable of moving through about the same ranges of angles as homologous joints of the cercopithecoid *Macaca*. This statement is based on manipulation of fresh feet of dead animals, and X-radiography of them in extreme positions.

The degrees of freedom of a hand cannot be fully exploited unless there are enough muscles, suitably arranged. A minimum of $n + 1$ muscles are needed to operate a hand that has n degrees of freedom. Thus a hinge joint requires two muscles, a flexor and an extensor. A universal joint requires three: for example, a flexor, an adductor and a muscle that tends both to extend and to abduct the joint. Extension can be performed by the extensor-abductor, with the adductor also active to prevent abduction. Similarly, abduction can be performed by the extensor-abductor with the flexor also active. Again, SPOOR (1983) has shown that a digit with three hinge joints (three degrees of freedom) can be operated by four muscles. In general, a system with n degrees of freedom can be operated by $n + 1$ muscles arranged as follows. Call one direction of movement for each degree of freedom the *A* direction, and the other the *B* direction. Allocate one muscle to each degree of freedom, arranging it so that it tends to produce the *A* movement. Add one more muscle crossing all the joints in such a way as to tend to produce the *B* movement for every degree of freedom.

The disadvantage of such a system is that many muscles may have to be activated to produce certain combinations of moments about the joints. Partially antagonistic muscles may each have to exert large forces, to obtain a relatively small moment. The more muscles that are activated and the larger the forces required of them, the more metabolic energy is used (See MCMAHON, 1984 p. 214, where the significance of the tension-time integral is explained). If there are $2n$ muscles, suitably arranged, antagonistic muscles need never be activated simultaneously. It therefore seems desirable to have 48 muscles to operate the 24 degrees of freedom of the human hand.

How many muscles are there? If we simply count the muscles given individual names in anatomical textbooks, we will get a misleadingly small total. WILLIAMS & WARWICK (1980) list 35 muscles of the human hand (15 extrinsic and 20 intrinsic). Three of the extrinsic muscles have four tendons each, one to each finger. These tendons emerge more or less separately from the muscle bellies, which can apparently pull on them individually. There are, however, connections between the tendons that may restrict the extent to which they can move independently. It seems best to regard each of these muscles as being effectively four muscles. Thus the total number of muscles available to work the joints shown in Fig. 1 can be reckoned rather tentatively as 44. This is only slightly less than twice the number of degrees of freedom.

Similar counts give 48 hand muscles for a cercopithecoid (*Macaca*; muscles described by HARTMAN & STRAUS, 1933) and 44 for a primitive marsupial (*Didelphis*, STEIN, 1981). Humans, other primates and primitive mammals all seem to have similar numbers of hand muscles. Even dogs, with their non-prehensile feet and reduced first digits, have only slightly fewer (about 36, muscles described by GETTY, 1975.). Ungulate mammals have many fewer hand muscles, and some cetaceans have none. (My colleagues and I found no hand muscles in a dissection of *Lagenorhynchus acutus*.)

So far, we have found no remarkable peculiarities of primate hands. They have about the same number of degrees of freedom and the same number of muscles as the hands of primitive mammals such as *Didelphis*.

Muscle dimensions

ALEXANDER, JAYES, MALOIY & WATHUTA (1981) determined the masses of major groups of limb muscles, and the mean lengths of their muscle fibres, for many species of mammals. They found that the flexor muscles of the wrist and hand tend to be larger in primates than in other mammals of equal body mass, and to have longer muscle fibres. Some of their data, and additional data, are shown in *Tables 1, 2*. Further data are given by Rauwerdink (this volume).

Table 1 shows hand muscle masses as fractions of body mass. This seems a sensible way to make comparisons between mammals of different sizes, because the masses of homologous limb muscles tend to be about proportional to body mass (ALEXANDER *et al.*, 1981). A group of Cercopithecoidea are compared to some other mammals. *Didelphis* has been chosen as a primitive mammal. The Fissipedia are mammals with fairly unspecialized fore feet, that span the size range of the Cercopithecoidea. *Meles*, which uses its fore feet for digging, has been separated from the other Fissipedia. The reason for separating it will become clearer, when *Table 2* is discussed.

Table 1 shows that *Didelphis* and the «other Fissipedia» have fairly similar hand muscle masses, relative to body mass, with one striking exception: *Didelphis* has a very small flexor digitorum superficialis. *Meles* has some peculiarities, and an exceptionally large total for a fissipede. The Cercopithecoidea have all their mean masses larger than for *Didelphis* and «other Fissipedia», many of them about twice as large. Large hand muscles seem to be a distinctive feature of Cercopithecoidea, in comparison with non-primates.

Table 2 shows muscle fibre lengths as percentages of ulna length. This seems a convenient way of making comparisons between mammals of different sizes. The allometric equations of ALEXANDER *et al.* (1979, 1981) show that muscle fibre length and ulna length generally scale similarly with body mass.

Notice in *Table 2* that the values for *Meles* are all larger than the means for other Fissipedia. They are also larger, in every case, than any of the individual values for other species of Fissipedia. *Meles* has much longer muscle fibres than is usual in its order.

Compare the muscle fibre lengths of «other Fissipedia» with those of homologous muscles of other mammals in *Table 2*. In most cases, they are remarkably short. This may indicate cursorial adaptation, absent in *Meles*.

TABLE 1 - *Masses of extrinsic muscles of one hand (g/kg body mass) for various mammals. The Cercopithecoidea are the four specimens of different species dissected by Alexander et al. (1981) and a 15 kg Macaca nemestrina. The Didelphis is a 3.5 kg D.marsupialis and the Meles is a 10 kg M.meles. The other Fissipedia are the Genetta, Ichneumia, Canis mesomelas, Crocuta and Panthera dissected by Alexander et al. (1981) and a 21 kg Canis familiaris. Means are given with 95% confidence limits.*

	Cercopithecoidea	*Didelphis*	*Meles*	other Fissipedia
F. carpi ulnaris	1.65±0.40	1.04	1.12	0.86±0.15
F. carpi radialis	0.87±0.39	0.37	0.29	0.33±0.15
F. digitorum superficialis	1.54±0.72	0.06	0.82	0.81±0.21
F. digitorum profundus	3.90±1.79	2.67	3.33	1.72±1.48
E. carpi ulnaris	0.55±0.14	0.37	0.80	0.46±0.18
E. carpi radialis	1.86±0.76	0.95	0.67	0.92±0.31
Extensores digitorum	0.96±0.22	0.63	1.05	0.57±0.30
Total of above muscles	11.33	6.09	8.08	5.67

TABLE 2 - *Muscle fibre lengths as percentages of ulna length. If fibre lengths vary within a muscle, a weighted harmonic mean is taken (Alexander et al., 1981). Muscle fibre lengths for Homo are from Brand, Beach & Thompson (1981), with an assumed ulna length of 250 mm. Sources of other data as in Table 1. Means are given with 95% confidence limits.*

	Homo	Cercopithecoidea	*Didelphis*	*Meles*	other Fissipedia
F. carpi ulnaris	17	13±3	28	9	4±1
F. carpi radialis	21	14±4	18	28	7±2
F. digitorum superficialis	28	20±3	11	9	6±2
F. digitorum profundus	26	23±3	41	24	13±5
E. carpi ulnaris	18	12±4	16	11	4±2
E. carpi radialis	30	23±6	33	51	24±8
Extensores digitorum	23	15±5	22	29	12±3

Ignore the «other Fissipedia» as possibly specialized for running, and compare the primates in Table 2 with *Didelphis* (which may have some adaptations for climbing) and *Meles* (which may be specialized for digging). The only clear difference between the primates on the one hand, and *Didelphis* and *Meles* on the other, is that primates have very long muscle fibres in flexor digitorum superficialis. This is the only feature in *Table 2* that sets the primates clearly apart from all the other mammals, with their various specializations.

Muscle fibres lengthen and shorten a little as the animal moves. The lengths given in *Table 2* refer to particular sets of joint positions which we attempted to make similar in every case. Fortunately, the differences that have been pointed out are too large to be due to discrepancies of position. The data of DIMERY (1985) indicate that the minimum lengths of muscle fibres, during locomotion, are seldom less than 70% of the maximum lengths. The range of errors due to discrepancies of position is probably much less than 30%.

Long muscle fibres can lengthen and shorten more than short ones, and can therefore produce larger angular movements of joints. However, the angular movement also depends on the moment arm of the muscle about the joint. The important quantity is the ratio (fibre length) / (moment arm). This point was discussed more fully by ALEXANDER (1981).

My colleagues and I have measured the moment arms of some hand muscles, in various mammals. We detached the muscles from their origins and inserted pins through the tendons. We kept the tendons taut by means of rubber bands and observed how the pins moved relative to the bones, as we moved the joints. The method was modified from ALEXANDER, MALOIY, KER, JAYES & WARUI (1982), using X-radiography instead of photography. The results were used to calculate the ratios given in *Table 3*. The wallaby *Macropus* has reduced fore limbs and was therefore omitted from the other tables, but is nevertheless included here. The sheep *Ovis* is also included, as an example of a highly specialized cursorial foot.

Each of the muscles included in *Table 3* crosses more than one joint, for each of which a ratio could have been calculated. The examples given in the table will nevertheless suffice to make a point. Remember that large ratios indicate ability to move joints through large angles. The ratios for *Macaca* and *Macropus* are closely similar, and the two available ratios for *Homo* are not much different from them. The ratios for the flexor carpi ulnaris

TABLE 3 - *Ratios of muscle fibre length to moment arm. The fibre lengths of wrist muscles have been divided by their moment arms about the wrist, and the fibre lengths of digital flexors have been divided by their moment arms about the metacarpophalangeal joint of digit 3. Data for Homo are from Brand et al. (1981) and Youm , Gillespie et al. (1978). Other data are from specimens used for Tables 1, 2, together with a 21 kg wallaby (Macropus rufogriseus) and a 62 kg sheep (Ovis aries).*

Muscle	Joint	*Homo*	*Macaca*	*Macropus*	*Canis*	*Ovis*
F. carpi ulnaris	wrist	—	4.1	3.5	0.8	0.9
F. carpi radialis	wrist	—	4.1	4.5	—	4.0
E. carpi radialis	wrist	—	6.3	6.4	—	4.3
F. digitorum superficialis	mcp	5.1	7.2	—	1.2	0.5
F. digitorum profundus	mcp	6.0	8.2	7.1	2.3	1.3

and digital flexors of *Canis* and *Ovis* are very much smaller. These are muscles that are active in the stance phase of running.

Running requires large forces which have to be exerted with the limbs in particular positions, which are about the same in every stride. Prehension has less predictable requirements. Objects of different sizes may have to be grasped, in various positions relative to the body. Thus prehension requires exertion of forces in a wide range of hand positions. Table 3 shows strikingly similar values for the prehensile hand of *Macaca* and the much less dexterous hand of *Macropus*. It seems possible that muscle fibres long enough to move joints through large angles are a primitive characteristic, not a specialization of primates. The much shorter fibres of *Canis* and *Ovis* (giving smaller ratios of fibre length to moment arm) are a specialization for running, as Ker's paper in this volume shows.

Conclusion

ALEXANDER *et al.* (1981) found that primates have exceptionally large hand flexor muscles, for their body masses, with exceptionally long fibres. The more detailed analysis in this paper confirms that many of the extrinsic muscles of their hands are exceptionally large. However, only flexor digitorum superficialis has exceptionally long fibres when *Didelphis* and *Meles* are included in the comparison, as well as cursorial fissipedes. The mammals compared with primates by ALEXANDER *et al.* (1981) were artiodactyls (which were recognized as having exceptionally short muscle fibres), fissipedes, *Lepus* and *Myocastor*. The artiodactyls, many of the fissipedes and *Lepus* were cursorial mammals that may have had short muscle fibres as a specialization for running. It seems possible that fairly long fibres are primitive, in the extrinsic muscles of the fore feet of mammals, and that the primates are less remarkable, in this respect, than suggested by ALEXANDER *et al.* (1981).

This paper has discovered no other remarkable specializations of the primate hand, but several well-known and presumably important features have not been discussed. These are the replacement of claws by nails in most primates, the absence of thick paw pads, and specializations of the thumb for grasping.

The new data presented in this paper were obtained in joint work with Drs. R.F. KER and M.B. BENNETT.

References

ALEXANDER R. McN., 1981. *Mechanics of skeleton and tendons.* In: V.B. Brooks (ed.) Handbook of Physiology. The Nervous System. Bethesda: American Physiological Society. 2: 17-42.
ALEXANDER R. McN., JAYES A.S., MALOIY G.M.O. & WATHUTA E.M., 1979. *Allometry of the limb bones of mammals from shrews (Sorex) to elephant (Loxodonta).* J. Zool., Lond. 189: 305-314.
ALEXANDER R. McN., JAYES A.S., MALOIY G.M.O. & WATHUTA E.M., 1981. *Allometry of the leg muscles of mammals.* J. Zool., Lond. 194: 539-552.
ALEXANDER R. McN., MALOIY G.M.O., KER R.F., JAYES A.S. & WARUI C.N., 1982. *The role of tendon elasticity in the locomotion of the camel (Camelus dromedarius).* J. Zool., Lond. 198: 293-313.
BRAND P.W., BEACH R.B. & THOMPSON D.E., 1981. *Relative tension and potential excursion of muscles in the forearm and hand.* J. Hand Surg. 6: 209-219.
DIMERY N.J., 1985. *Muscle and sarcomere lengths in the hind limb of the rabbit (Oryctolagus cuniculus) during a galloping stride.* J. Zool., Lond. A 205: 373-383.
GETTY R., 1975. *Sisson and Grossman's* (ed.) *The anatomy of the domestic animals* 5. Philadelphia: Saunders.
HARTMAN C.G. & STRAUS W.L., 1933. *The Anatomy of the Rhesus monkey (Macaca mulatta).* London: Baillière, Tindall & Cox.
KUCZYNSKI K., 1974. *Carpometacarpal joint of the human thumb.* J. Anat. 118: 119-126.
MASON M.T. & SALISBURY J.K., 1985. *Robot hands and the mechanics of manipulation.* Cambridge, Massachusetts: MIT Press.
McMAHON T.A., 1984. *Muscles, reflexes and locomotion.* Princeton: Princeton University Press.
MORECKI A., EKIEL J. & FIDELUS K., 1984. *Cybernetic Systems of Limb Movements in Man, Animals and Robots.* Chichester: Horwood.
SPOOR C.W., 1983. *Balancing a force on the fingertip of a two-dimensional finger model without intrinsic muscles.* J. Biomechan. 16: 497-504.
STEIN B.R., 1981. *Comparative limb myology of two opossums, Didelphis and Chironectes.* J. Morph. 169: 113-140.
WILLIAMS P.L. & WARWICK R. 1980. *Gray's* (ed.) *Anatomy* 36. Edinburgh: Churchill Livingstone.
YOUM Y., GILLESPIE T.E., FLATT A.E. & SPRAGUE B.L., 1978. *Kinematic investigation of normal MCP joint.* J. Biomechan. 11: 109-118.
YOUM Y., McMURTRY R.Y., FLATT A.E. & GILLESPIE T.E., 1978. *Kinematics of the wrist, I. An experimental study of radial-ulnar deviation and flexion-extension.* J. Bone Jt Surg., 60 A: 423-431.

G.P. Rauwerdink

Muscle fibre and tendon lengths in primate extremities

The dimensions of the distal leg muscles in ten different primate species were examined, in order to determine whether locomotor specialisation is reflected in the fibre and tendon lenghts of the muscles. A non-primate was included in the study for comparison. The locomotor specialisations which the investigated species displayed were *leaping, quadrupedal walking or running on the ground* and *brachiation*.
Muscles were removed and the tendon lengths measured, then fibres were isolated from several locations in the muscle and their lengths measured. Additionally the weight of the fibre mass was measured after drying at room temperature.
Statistical tests revealed no differences in length between the fibres within a muscle. The flexors of the digits were found to be the heaviest muscles of the distal part of the extremeties. Some differences between the species were found in fibre and tendon lengths but these were not conclusive. The combined results did not show a correlation between muscle dimensions and locomotor specialisation. It was concluded that every primate species has the potential to specialize in any locomotor mode.

Key words: locomotion, muscle dimensions, brachiation, climbing, walking.

Introduction

The importance of gripping for primates does not need to be reasserted. Their fingers and toes are outstandingly long and mobile. The versatility of movements characteristic of the hands and feet in the whole group should leave its imprints on the muscular system: in fact, several authors have shown that the distal limb muscles, which agitate the fingers and toes, are very well developed. This can be demonstrated through the relative weights of these muscles better than by descriptions (Oxnard 1963, Preuschoft 1963, Jolly & Gorton 1977).

In parallel to this, Alexander *et al.* (1981) examined the dimensions of the distal leg muscles in different species of mammals. He demonstrated that primates tend to have longer muscle fires than other, more cursorial mammals.

This paper is intended to continue this approach. The dimensions of the distal leg musles of ten primate species are examined in order to determine whether locomotory specialisation is reflected in the fibre and tendon lengths of the muscles. This hypothesis is based on the following considerations:

When flexion occurs in a joint, the point of insertion of the flexor tendon makes an excursion. The stronger the flexion, the longer the excursion, so, the shorter the muscle must become.

Since shortening of the muscle is achieved by the shortening of the muscle fibres to maximally 30% of their length at rest (Huxley & Peachy 1961, Wilkie 1976), the length of the muscle fibres is restrictive for shortening of the muscles. Thus for pronounced, active excursions, long muscle fibres are needed. Great shortening distances are also needed for the muscles of hands and feet that cross several joints. This is especially important in primates with grasping hands and feet, and in an arboreal environment. They flex their digits to acute angles and need a powerful grip.

As various primate species have specialized in different locomotor modes, in which the mode of gripping and the preferred substrate for locomotion (branches, thin or thick boughs, ground) may influence the excursions of the joints, the muscle fibre length might be adapted to a locomotor mode.

In addition, the tendon length of the same muscles was considered. According to ALEXANDER & BENNET-CLARK (1977), DIMERY, ALEXANDER & KER (1986), elastic energy can be stored in tendons. The excursions produced by elastic stretching and subsequent resilience depend on the length of the tendons, as long as Young's modulus is a constant value (which seems to be the case for all tendinous and ligamentous structures). A precondition for making use of elastic storage of energy seems to be that the animal performs regularly swinging, oscillatory movements. However, having long tendons which extend when the muscle contracts may hinder the effectiveness of joint excursions (KER *et al.* 1988).

Materials

One specimen of each of the following species was studied:

Strepsirhini:

Lemuridae;	— *Lemur catta* (Ring-tailed lemur)
Lorisidae;	— *Loris tardigradus* (Slender loris)
	— *Galago senegalensis* (Lesser bush baby)

Haplorhini:

Ceboidea;	— *Cebus capucinus* (White-faced capuchin)
	— *Lagothrix lagotricha* (Woolly monkey,)
Cercopithecoidea;	— *Macaca mulatta* (Rhesus monkey)
	— *Papio hamadryas* (Hamadryas baboon)
	— *Cercopithecus aethiops* (Grivet)

Hominoidea;

Hylobatidae;	— *Hylobates lar* (White-hand gibbon)
Pongidae;	— *Pan troglodytes* (Chimpanzee)

For comparison a chinchilla (Rodentia, *Hystricognathi*: *Chinchillidae*: Chinchilla chinchilla) was included in the study.

All were adult animals, except for Lagothrix, and fixed in formalin for extended periods, some already partly dissected. Therefore body masses were not available. Except when mentioned otherwise, sexes are unknown.

Methods

Muscles were dissected out with complete tendons. The tendon length was then measured from the insertion to the place of attachment of the first muscle fibres, this being the «free tendon length». Out of each muscle, several bundles of fibres were isolated from different places:

— one near the beginning of the free tendon
— one at half the length of the muscle belly
— one from the most proximal part of the muscle belly

The lengths of the isolated fibre bundles were then measured. When variations between the lengths of these three bundles were great, intermediate fibre bundles were also measured. After measuring, the mean fibre length was calculated. Afterwards all tendon material was removed from the muscle and the resulting muscle fibre mass dried between tissue paper until no moisture left under pressure. Then the «wet weight» was measured and the muscle left to dry at room temperature until mass constant, the «dry weight» of the muscle. Muscle mass was then expressed as a percentage of the total muscle mass of the segment.

«Weighted harmonic mean fibre length» for the various muscle groups was calculated after ALEXANDER *et al.* (1981):

$$\frac{\Sigma m_j}{\Sigma(m_j/l_j)} = L,$$

where m_j = mass of muscle j
l_j = mean length of fibres in muscle j

In order to make comparisons between animals of different body size, the muscle fibre lengths measured were multiplied by a size factor. To obtain this factor, the value 100 was divided by the actual length of the tibia or ulna of the individual under investigation.

Differences between species, between individual muscles and within muscles were then statistically tested with the parametric t-test, confidence limits 95%.

For the tests of significance the data were processed as follows:

Muscle mass

For each individual muscle the «relative muscle mass» was calculated

$$\frac{\Sigma \text{ (relative muscle mass of each species)}}{\text{number of species}}$$

These means were compared with those of other muscles to find the differences between individual muscles.

To find differences between one particular muscle and all others of the same segment, the «average muscle mass of the segment» was calculated

$$\frac{\Sigma \text{ (relative muscle masses of all muscles)}}{\text{number of muscles}}$$

This mean was then compared with the relative mass of a particular muscle.

Muscle fibre length

1) For finding differences between fibre lengths within a muscle over all species for a given muscle was calculated;

$$\frac{\Sigma \text{ (minimal absolute fibre length of a given muscle)}}{\text{number of species}}$$

This was compared to

$$\frac{\Sigma \text{ (maximal absolute fibre length of the same muscle)}}{\text{number of species}}$$

2) For finding differences between individual muscles the «mean fibre length of the muscle» was calculated

$$\frac{\Sigma \text{ (mean fibre length of this muscle)}}{\text{number of species}}$$

This mean was compared with that of the other muscles.

3) For finding differences between one muscle and all other muscles of the segment, the «mean fibre length of the segment» was calculated

$$\frac{\Sigma \text{ (mean fibre length of each muscle)}}{\text{number of muscles}}$$

This was compared to the mean fibre length of individual muscles.

4) For finding differences between species the «mean fibre length of a species» was calculated

$$\frac{\Sigma \text{ (mean fibre length of each muscle of the segment)}}{\text{number of muscles of the segment}}$$

This mean was compared with that of other species. The segments of the same species were also compared to find differences between front- and hindlimbs.

Tendon length

The same calculations as for fibre length were performed.
Regressions were calculated using the least square fit of log transformed data.

Identification of muscles

Except when indicated otherwise, origin and insertion of the muscles are comparable to those in humans. It should be remarked that the aponeurosis dorsalis of the digits begins at the second phalanx and not at the third as is the case in humans. For a more detailed description of the finger extensors and flexors see Landsmeer (this volume).

When measuring tendon lengths the complete length of tendon plus aponeurosis was considered. The muscles investigated can be seen from figures, 1, 2, 6 and 8.

The m. peroneus quartus branches off from the tendon of the brevis and inserts into the dorsal aponeurosis of the fifth digit.

For calculations of the weigthted harmonic mean fibre length (after ALEXANDER *et al.* 1981), the following groups were formed:

I Fore flexors = flex. carpi uln. + flex. carpi rad. + flex. digg. sup. + flex. digg. prof. + palm. long.
II Wrist and finger extensors = ext. carpi uln. + ext. carpi rad. brev. + ext. carpi rad. long. + ext. digg. sup. + ext. digg. prof. + ext. dig. min. + ext. poll. long.
III Deep hind flexors = flex. digg. fib. + flex. digg. tib.
IV Ankle extensors = gastr. + sol. + plant.

Results

Muscle masses

Figures 1 and *2* show that the flexor digitorum profundus of the forearm, the mm. triceps surae and the mm. flexores digitorum of the leg are the heaviest muscles of their respective segments. Each muscle makes up more than 20% of the total muscle mass of the segment. Compared to the average muscle mass of the segment, these differences are significant. Only in Hylobates are the superficial finger flexors slightly better developed than the deep flexors; the latter however, being still more than 20% of the segment muscle mass. Significant differences between species have not been found.

Fibre lengths in primates

Figures 3 to *5* show Alexander's data on primates, fissipedes and bovids (ALEXANDER *et al.*, 1981) and the data collected on the primates and chinchilla in the present study.

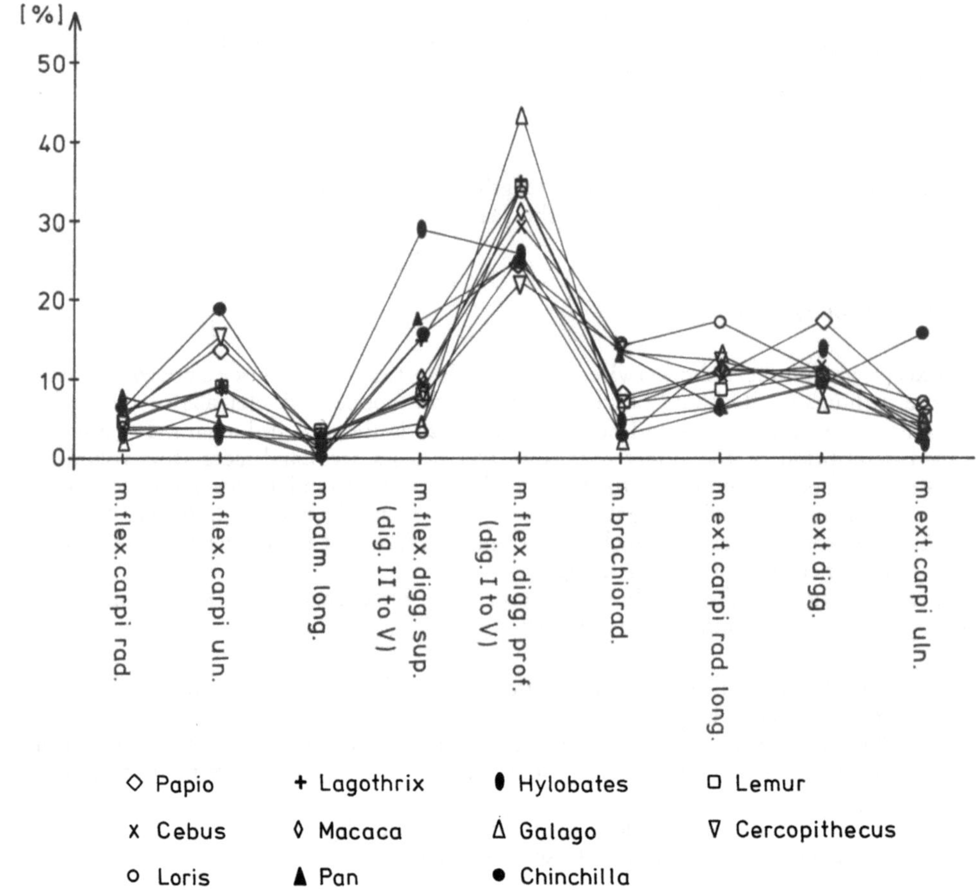

Figure 1 - Graphs showing the relative muscle mass of muscles in the forearm. Points for individual muscles are shown.

Between the present data on primates and Alexander's there is a systematic discrepancy: fibre lengths measured in the present study are longer.

Although lengths of the fibres, taken from very different places in the muscle vary (see *Table 1*), even the largest of these differences were insignificant.

Except for *Loris* and *Lagothrix*, who have longer muscle fibres in the hindlimb muscles, no significant differences could be demonstrated in fibre length or tendon length between front- and hindlimb of the other species (see *Table 2*).

Muscle fibre- and tendon lengths, forearm

Figures 6 and 7 show, for all species, muscle fibre and tendon lengths. Data have been standardized for an ulna length of 100mm.

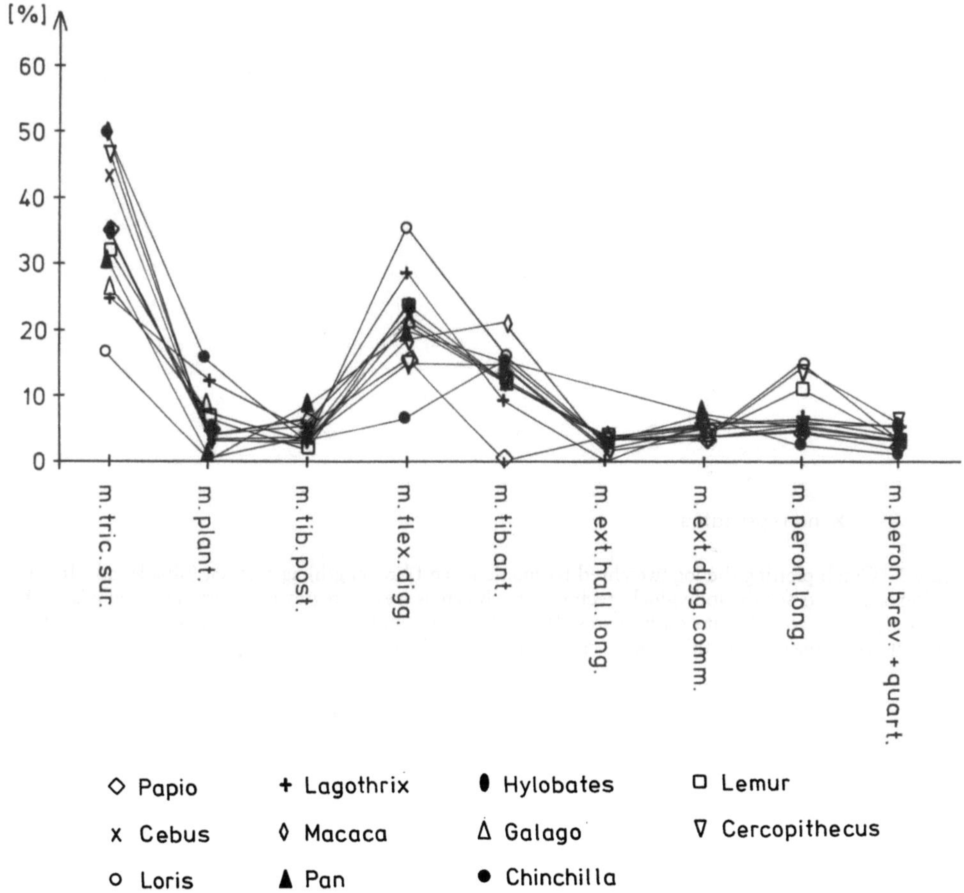

Figure 2 - Graphs showing the relative muscle mass of muscles in the leg. Points for individual muscles are shown.

TABLE 1 - *Minimal and maximal fibre lengths of a particular muscle corrected by size factor. Data represent a mean value over the species Papio, Lagothrix, Hylobates, Lemur and Cebus.*

Muscle	Minimum length measured	Maximum length measured
flex. carpi rad.	23.38 ± 3.72	27.54 ± 5.26
flex. carpi uln.	14.86 ± 5.87	16.20 ± 10.92
palm. long.	19.63 ± 4.86	26.40 ± 5.24
flex. digg. sup.	23.50 ± 7.55	31.49 ± 7.96
flex. digg. prof.	26.46 ± 5.91	35.20 ± 9.67

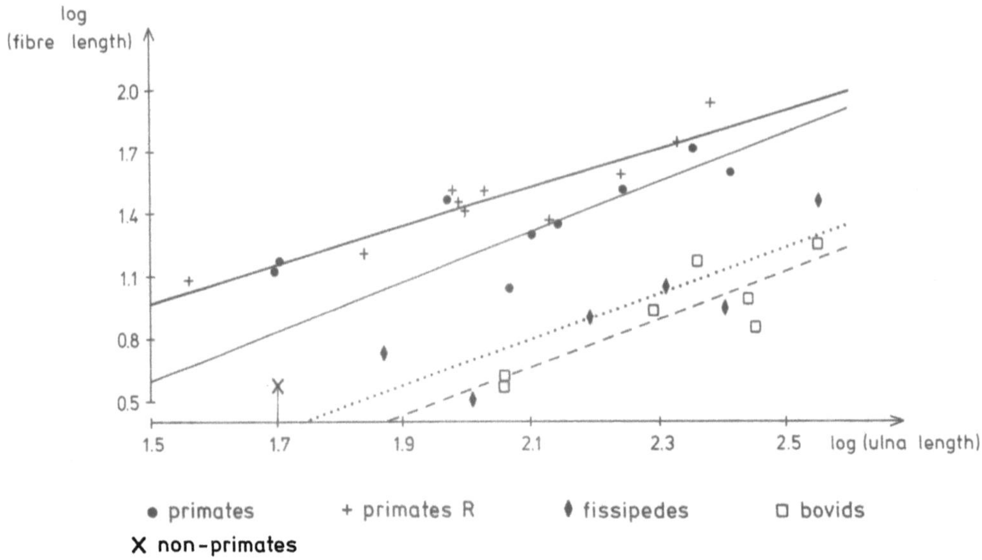

Figure 3 - Graph plotting the log (weighted harmonic mean fibre length) against log (ulna length) for the fore flexors. Points for the individual specimen are shown as well as regression lines. The symbols •, ♦, □ represent Alexander's data on primates, fissipedes and bovids respectively. The symbols + and × represent data collected in this study on primates and chinchilla respectively.

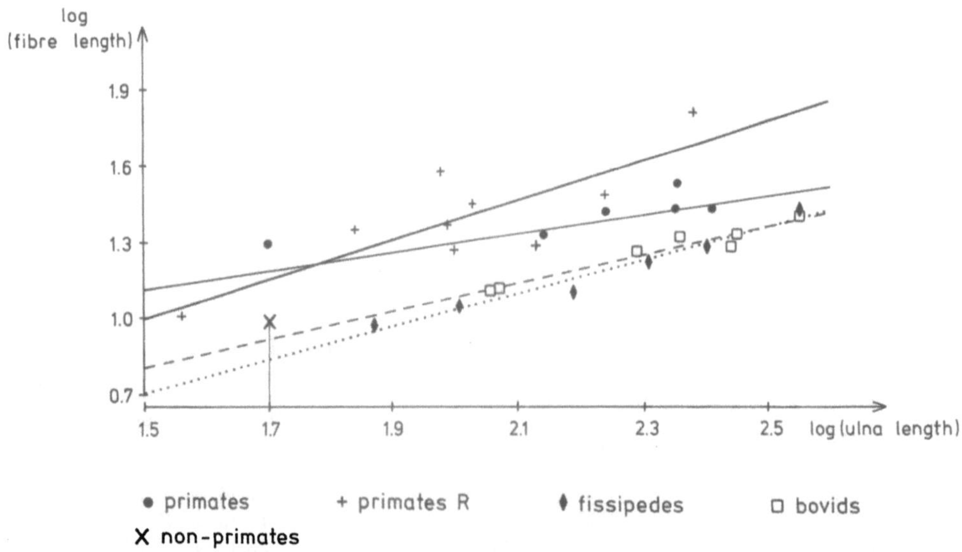

Figure 4 - Graphs plotting the log (weighted harmonic mean fibre length) against log (ulna length) for the deep hind flexors. Further details as in *Figure 3*.

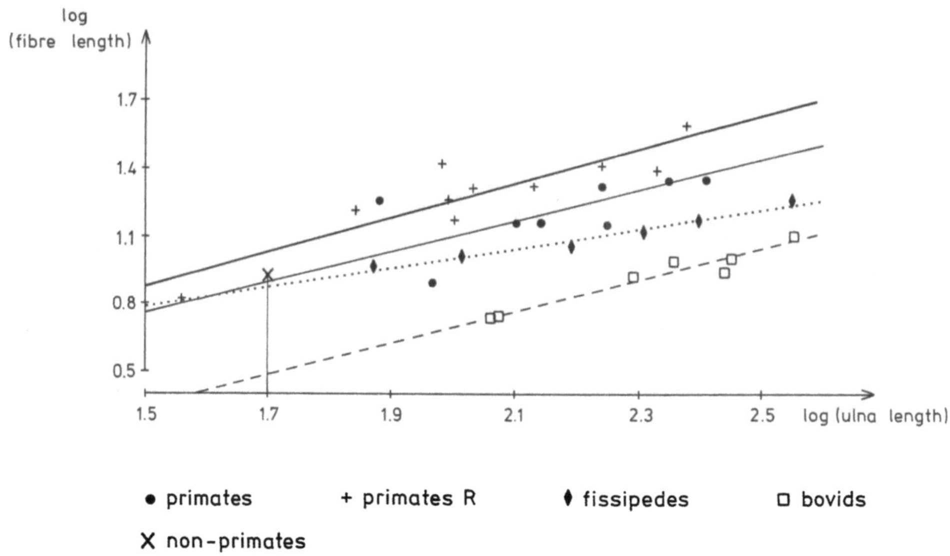

Figure 5 - Graphs plotting the log (weighted harmonic mean fibre length) against log (ulna length) for the ankle extensors. Further details as in *Figure 3*.

TABLE 2 - *Mean fibre lengths and tendon lengths of the segments for all species.*

Species	Fibre length forelimb	Tendon length forelimb	Fibre length hindlimb	Tendon length hindlimb
Papio	20.5 ± 4.2	47.2 ± 17.9	20.3 ± 8.9	45.7 ± 21.3
Lagothrix	30.9 ± 7.3	37.0 ± 22.8	43.1 ± 10.3	27.3 ± 23.7
Hylobates	21.3 ± 4.6	44.9 ± 23.7	27.1 ± 10.9	46.6 ± 18.5
Cebus	29.9 ± 4.9	44.9 ± 20.8	27.2 ± 7.6	32.5 ± 15.5
Lemur	26.8 ± 7.3	45.3 ± 23.5	21.4 ± 12.9	45.6 ± 22
Galago	30.9 ± 5.7	60.9 ± 32.1	24.8 ± 12.9	71.7 ± 32.2
Macaca	29.0 ± 7.0	50.6 ± 21.4	31.2 ± 11.8	47.1 ± 27.7
Cercopi	21.4 ± 10	49.8 ± 21.5	18.4 ± 8.11	43.9 ± 26.7
Loris	24.4 ± 3.6	47.7 ± 26.8	32.1 ± 7.0	44.0 ± 25.2
Pan	24.5 ± 6.2	50.6 ± 24.8	27.2 ± 9.9	45.4 ± 28.3
Chinchil	16.1 ± 10.6	58.4 ± 15.9	20.1 ± 9.5	58.3 ± 21.4

It could be shown, after statistical analysis, that the m. flexor carpi ulnaris, the m. extensor digitorum profundus and the m. abductor pollicis longus have shorter muscle fibres (ca. 20mm) than the «mean fibre length of the segment» (ca. 26mm). The m. flexor digitorum profundus and the m. extensor carpi radialis longus were shown to have longer fibres (ca. 31mm) than this mean length.

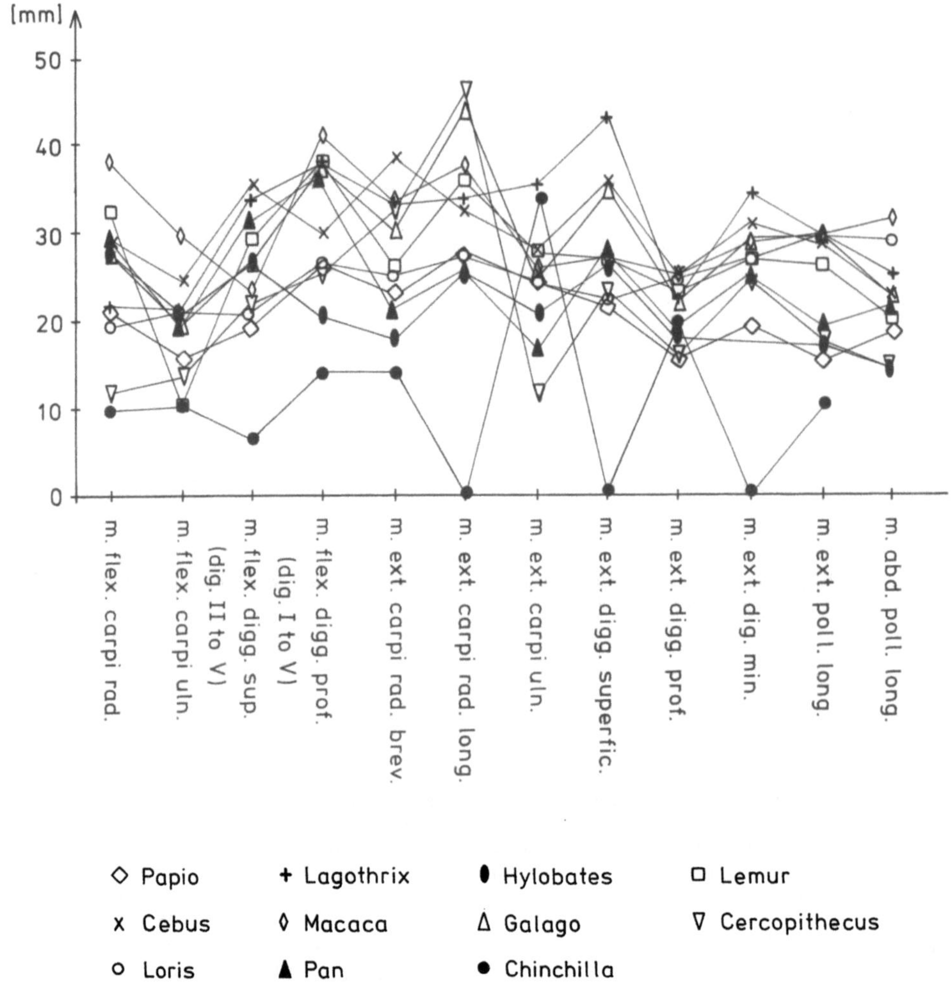

Figure 6 - Graphs showing the fibre lengths of muscles in the forearm at ulna length 100mm.

Tendon length is shorter in the hand flexors and the m. extensor carpi ulnaris than the «mean tendon length of the segment» (49mm), the superficial finger extensor has an especially long tendon (84mm).

Differences between the species could be found only for muscle fibre length. *Papio, Hylobates* and *Cercopithecus* having the shorter «mean fibre length of the species» (ca. 20mm), *Loris* and *Pan* having intermediate lengths (ca. 24mm) and the other species having the longest fibres (ca. 30mm). The species thus form three groups.

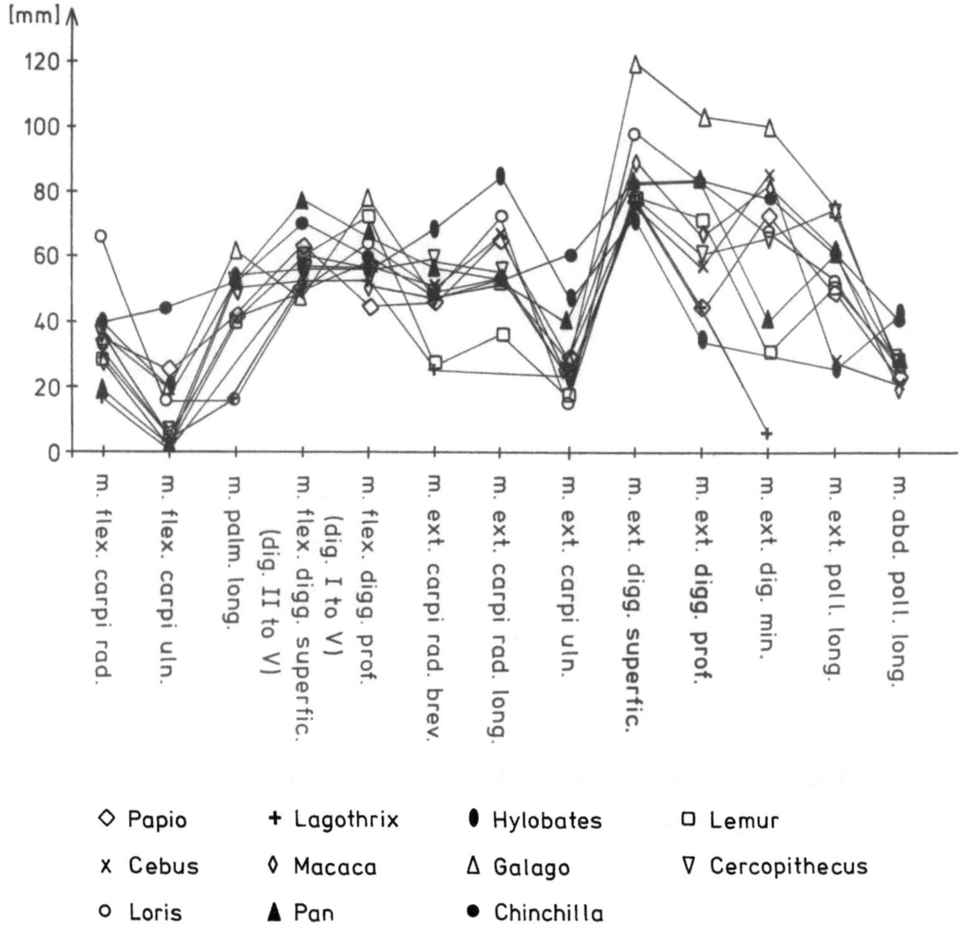

Figure 7 - Graphs showing the tendon lengths of muscles in the forearm at ulna length of 100mm.

Muscle fibre- and tendon length, leg

Figures 8 and 9 show, for all species, tendon- and muscle fibre lengths. Data have been standardized for a tibia length of 100mm.

The m. tibialis posterior could be shown to have shorter fibres (ca. 15mm) than the «mean fibre length of the segment» (27mm). The m. tibialis anterior and the m. ext. hallucis longus have longer fibres (40-46mm). The tendons the m. soleus and the m. tibialis anterior have shorter tendons (7 and 27mm resp.) than the «mean tendon length of the segment» (45mm). The toeflexors and common extensor have longer tendons (> 65mm).

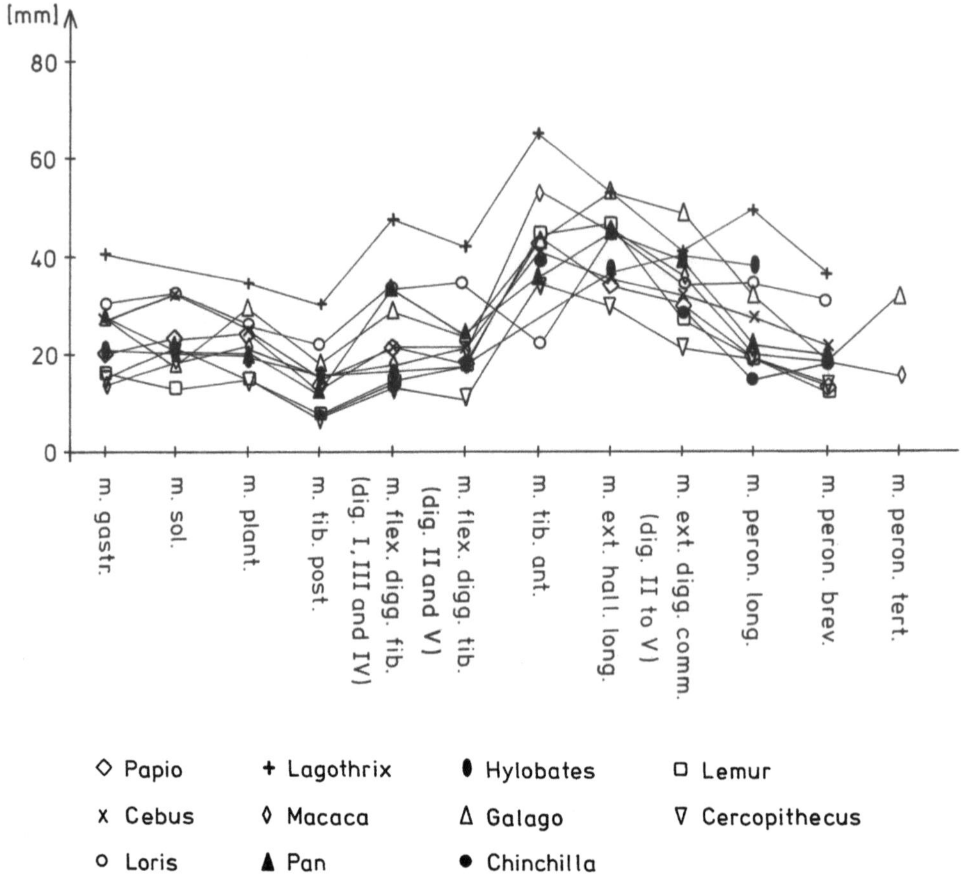

Figure 8 - Graphs showing the fibre lengths of muscles in the leg at tibia length 100mm.

Differences between the species could be found for fibre and tendon lengths. *Lagothrix* having the longest fibres (43mm) and *Cercopithecus* the shortest (18mm). The other 8 species having intermediate fibre lengths of ca. 27mm. *Galago* shows the longest tendons (72mm). Between the other species no significant differences occurred in either fibre- or tendon length.

Data on Chinchilla

It could be shown that *Chinchilla* has significantly shorter (mean fibre length of the species = ca. 19mm) muscle fibres in muscles of the forearm than «mean fibre length of the

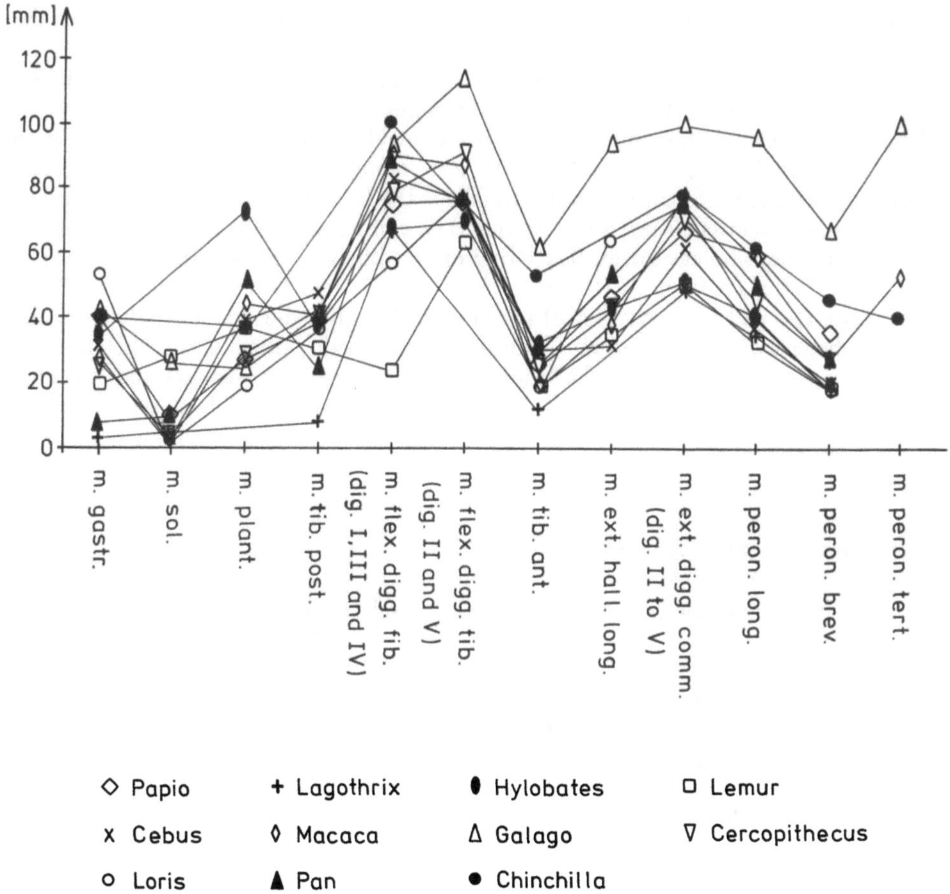

Figure 9 - Graphs showing the tendon lengths of muscles in the leg at tibia length 100mm.

segment» in primates (=26mm). In the leg no differences in fibre lengths with the primate species were found.

In both forearm and leg there are no differences in tendon length between *Chinchilla* and the primates.

Discussion

Muscle masses

The relative weights of muscles can give valuable information about their function, because in muscle mass the cross sectional area of the muscle is included. This area is proportional to the force a muscle can develop. The heavier the muscle, the greater the force it can develop.

In the forearm and leg we find the deep flexors of the digits well developed: finger- and toeflexion is an important function in primates.

The locomotor mode was found to be reflected in the relative muscle mass of certain muscles: in slow climbing species such as *Loris tardigradus*, *Nycticebus coucang* and *Perodicticus potto* the m. flexor digitorum tibialis is the best developed muscle of the leg. In leaping species, such as *Galago senegalensis* and *G. crassicaudatus* the m. gastrocnemius is the heaviest muscle, whereas the m. flexor digitorum tibialis weighs about half as much as the m. flexor digitorum fibularis (JOLLY & GORTON 1977 and GEBO & DAGOSTO 1988).

Though GEBO and DAGOSTO (1988) conclude differently, their data clearly show the Lorisines to have a well developed m. flexor digitorum tibialis, which shows this muscle's importance in slow climbing. For the leaping species; Galagines, Tarsiers and, to some extend, Lemurids, they also find the m. gastrocnemius to be the best developed.

Data presented in this study are consistent with those mentioned above (see *Figure 10*).

Weighted harmonic mean fibre length

It could be shown that primates have longer muscle fibres in the forearm than non-primates, this in accordance with ALEXANDER *et al.* (1981). The discrepancy between Alexander's and the present data is due to a difference in methods: in this study whole muscle fibres are dissected out, whereas Alexander took measurements within the muscle.

The longer muscle fibres are needed to allow the great joint excursions that can be performed in primate hands and feet.

Since the flexors of the digits are the heaviest muscles of the segments and their muscle fibres are not longer than the average of all muscles, they also are the strongest, and their strength provides the force necessary for balancing the great rotating moments of the external forces.

In the leg *Chinchilla* does not show shorter muscle fibres than the primates. This seems to be what ALEXANDER *et al.* (1981) found for the «hoppers», which he therefore treated separately in his analysis. It may be due to the fact that *Chinchilla* is a specialized «jumper» with elongated tarsus. This elongation probably resembles the elongation we find in primate feet, though here it serves a different purpose.

Muscle fibre lengths

The fact that no significant differences were found between fibres taken from the same muscle, implies that there are no marked differences in fibre length within a muscle.

One might expect that differences could occur between primate species who display specialized modes of locomotion (or live in different habitats).

These locomotor modes could be, firstly, «*leaping*» where hands and feet may have very different functions. Here the hindlimb in the small animal has elongated elements to produce a long acceleration distance (SCHULTZ 1963, DEMES & GÜNTHER, 1989). Joint excursions are great, requiring long muscle fibres. Additionally, a specialized jumper will accumulate the greater mass of the limb proximally, thus causing long tendons (HALL-CRAGGS, 1977; MC ARDLE, 1981; PETERS & PREUSCHOFT, 1984). These features could not

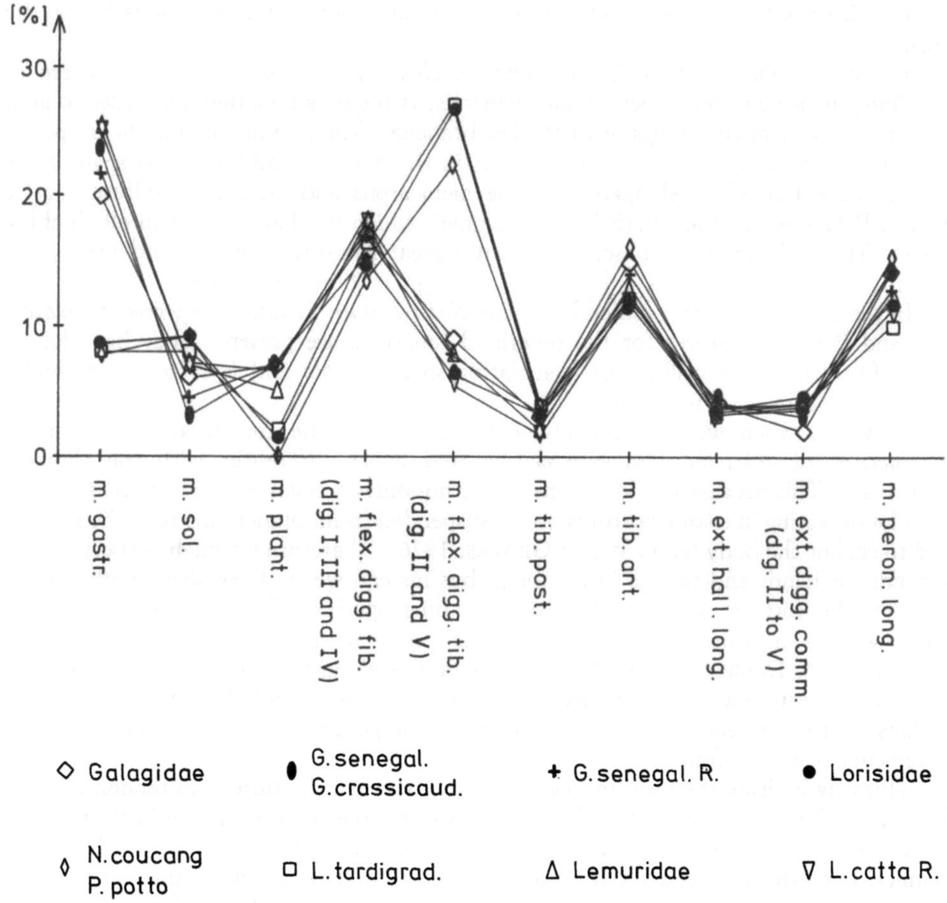

Figure 10 - Graphs showing the relative muscle mass of muscles in the leg. Data on «Galagidae». «Lorisidae», and «Lemuridae» after GEBO & DAGOSTO (1988). Data on *G. senegalensis* + *crassicaudatus*, *Nycticebus coucang* + *Perodycticus potto* after JOLLY & GORTON (1977). Data marked «R» were obtained in this study.

both be found in this study. The specialized jumper *Galago* shows longer tendons in the leg than other species, but the muscle fibre length was no different from most species. As for *Chinchilla*, this species has no longer tendons in the leg muscles than the primates. Neither were muscle fibres shorter than in primates.

Secondly there is «*quadrupedal walking or running on the ground*», where hands and feet are subject to similar demands. In order to increase speed in a pendulum model of walking, parts of the limbs (such as forearm, carpus or metacarpus, or leg, tarsus or metatarsus) are elongated (SCHULTZ, 1956; JOUFFROY *et al.* this volume), which may lead to great rotating moments, needing strong muscles. Joint excursions, however, need not be very large, requiring no long muscle fibres. In the «walker» *Papio* however, the only obvious difference with the other species found are shorter fibres in muscles of the forearm.

Thirdly, one might distinguish «*quadrupedal climbing*» as a locomotor mode, where, as in walking, hands and feet meet similar demands. It requires long digits in order to grab hold of a number of thin twigs or of thicker branches. The size of the objects gripped is inversely proportional to the joint excursions in the digits. In addition, the span of the grip can be extended by elongation of the metacarpus and metatarsus (NIESCHALK & DEMES, PREUSCHOFT *et al.*, both in this volume), which is what we find in all climbing species. These elongated limb segments cause great rotating moments and need great flexibility in both wrist and ankle, so causing great joint excursions.

In the climbing species studied here: *Lagothrix*, *Lemur*, *Cebus*, *Cercopithecus*, *Macaca*, *Loris* and *Pan*, the features for the described functions are present. They have longer muscle fibres than the walking or brachiating species *Papio* and *Hylobates* and similar situations in fore and hind limbs.

However, when we consider *Hylobates*, displaying the fourth locomotor mode, «*brachiation*» or «*bimanual suspension*», we find these differences with the climbers inconsistent. This because this locomotor mode not only involves the real brachiation, but also a lot of resting in sitting postures on, or suspended from, branches, where the feet are used to anchor the body (ANDREWS & GROVES, 1976). Therefore we might expect similar demands for hands and feet as in climbing, but instead we find *Hylobates* with shorter fibres in the forearm muscles than the climbing species. This may be due to the characteristic elongation of the forearm (STERN & OXNARD, 1973; JUNGERS, 1985 and PREUSCHOFT & DEMES, 1985) which surpasses the elongation of the muscle fibres.

Another feature which could be expected at least in *Papio* and *Hylobates*, namely long tendons in the extremities for elastic energy storage during their regular rythmical movements, was not found.

This may indicate the need for a compromise between the fibre- and tendon lengths. When a tendon acts as a spring, the only necessary role for the muscle is to maintain tension, for which short fibres are mechanically adequate (KER *et al.* 1988, KER, this volume). But primates need active, mobile fingers and toes, which require long muscle fibres and minimum tendon extension, and therefore shorter tendons.

Considering all the above, we may arrive at the general conclusion that in primates the different locomotor modes are not reflected in the fibre and tendon lengths of the forearm and leg muscles. In muscle mass some different locomotor modes (leaping and slow climbing) can be distinguished.

This does not necessarily imply that a specific primate would be incapable of performing certain locomotor actions. It may well be that every primate species has the potential for displaying all locomotor modes and the ability to specialize in any of them.

References

ALEXANDER R. MCN. & BENNET-CLARK H.C., 1977. *Storage of elastic strain energy in muscle and other tissues*. Nature, 265: 114- 117.

ALEXANDER R. MCN., JAYES A.S., MALOIY G.M.O. & WATHUTA E.M., 1981. *The allometry of the leg muscles of mammals*. Journal of Zoology, 194: 539-552.

ANDREWS P. & GROVES C.P., 1976. *Gibbons and brachiation*. In: Gibbon and Siamang (D.M. Rumbaugh ed.) Karger AG Basel, 4: 167-218.

DEMES B. & GÜNTHER M.M., 1989. *Wie die Körpermasse den Springstil von Halbaffen und deren Proportionen bestimmt.* Z. Morph. Anthrop., 77: 209-225.

DIMERY N.J., ALEXANDER R. McN. & KER R.F., 1986. *Elastic extension of the leg tendons in the locomotion of horses (Equus caballus).* Journal of Zoology, 210: 415-425.

GEBO D.L. & DAGOSTO M., 1988. *Foot anatomy, climbing, and the origin of the Indriidae.* Journal of Human Evolution, 17: 135-154.

HALL-CRAGGS E.C.B., 1977. *Physiological and histochemical parameters in comparative locomotor studies.* In: Prosimian Anatomy, Biochemistry and Evolution (R.D. Martin, G.A. Doyle & A.C. Walker eds.) Duckworth & Co Ltd London, 11: 829-845.

HUXLEY A.F. & PEACHEY L.D., 1961. *The maximum length for contraction in vertebrate striated muscle.* Journal of Physiology, 156: 150-165.

JOLLY C.T. & GORTON A.T., 1977. *Proportions of the extrinsic foot muscles in some Lorisid prosimians.* In: Prosimian Anatomy, Biochemistry and Evolution (R.D. Martin, G.A. Doyle & A.C. Walker eds.) Duckworth & Co Ltd London, 11: 801-815.

JUNGERS W.L., 1985. *Body size and scaling of limb proportions in primates.* In: Size and Scaling in Primate Biology (W.L. Jungers ed.) Plenum Press New York/London, 345-382.

KER R.F., ALEXANDER R. McN. & BENNETT M.B., 1988. *Why are mammalian tendons so thick?* J. Zool. Lond. 216: 309-324.

MCARDLE J.E., 1981. *Functional morphology of the hip and thigh of Lorisiformes.* Contrib. Primatol., 17: 1-132.

OXNARD C.E., 1963. *Locomotor adaptations in the primate forelimb.* Symposium of the Zoological Society of London, London, 10: 165-182.

PETERS A. & PREUSCHOFT H., 1984. *External biomechanics of leaping in Tarsius and its morphological and kinematic consequences.* In: Biology of Tarsiers (C. Niemitz ed.) Gustav Fischer Verlag, Stuttgart.

PREUSCHOFT H., 1963. *Muskelgewichte bei Gorilla, Orang-Utan und Mensch.* Anthropologischer Anzeiger 26, 4: 308-317.

PREUSCHOFT H., 1965. *Muskeln und Gelenke der Vorderextremität des Gorillas* (Gorilla gorilla, *Savage et Wyman 1847).* Morphologisches Jahrbuch, 107 (2): 99-183.

PREUSCHOFT H. & DEMES B., 1985. *Biomechanic determinants of arm length and body mass in brachiators.* In: Fortschritte der Zoologie (Duncker & Fleischer ed.) Gustav Fisher Verlag, Stuttgart, 30: 39-43.

SCHULTZ A.H., 1956. *Postembryonic age changes.* In: Primatologia, Handbuch der Primatenkunde: (H. Hofer: A.H. Schultz & D. Starck ed.) S. Karger Basel/New York, 887-964.

SCHULTZ A.H., 1963. *The relative lengths of the foot skeleton and its main parts in primates.* Symposium of the Zoological Society of London, 10: 199-206 London 1963.

STERN J.T. & OXNARD C.E., 1973. *Primate locomotion: some links with evolution and morphology.* In: Primatologia, 4(II) (H. Hofer, A.H. Schultz & D. Starck ed.) S.Karger Verlag Basel.

WILKIE D.R., 1976. *Muscle.* Studies in Biology 11. Edward Arnold Ltd London.

U. Nieschalk
B. Demes

Biomechanical determinants of reduction of the second ray in Lorisinae

Members of the subfamily Lorisinae are characterized by a reduced second ray. When walking on branches, lorises grip the support with the second ray pointing in the direction of movement. This does not contribute to the grip of either the hand or the foot. The finger is hyperextended at the metacarpophalangeal joint, flexed at the proximal interphalangeal joint, and hyperextended at the distal interphalangeal joint. As a result, the palmar (volar) pad proximal to the second finger and the distal pad of the digit are pressed against the underlying support. The toe is sometimes placed on top of the third digit. The morphological and behavioral features have a mechanical basis. The orientation of the second ray in combination with the great stride length characteristic of these animals generates high stress in bones and tendons. Its reduction in length as well as its flexed posture during locomotion are means to avoid overstressing. Moments in the basal joints of fingers and metacarpals that must be counterbalanced by muscle action are also reduced. In the larger-bodied lorisine species bending stresses are higher than in the smaller ones: the reduction of the second ray is therefore more pronounced.

Key words: functional morphology, locomotion, prosimians, rudimentary digits, mechanical stress, postural behavior.

Introduction

The shape and structure of the hand and foot reflect an animal's behavior and ecology to such an extent that they provide an indicator of its lifestyle. In the Lorisinae, the hands are used for manipulatory activities like catching prey and handling food objects, as well as in locomotion. The animals walk and climb quadrupedally. The feet are used nearly exclusively in postural behavior and locomotor activities. Hands and feet of all lorisines are characterized by a reduced second ray (*Figure 1 a, b*). The functional meaning of this reduction is still unexplained. Because little reliable information exists about the use of hands and feet in lorisines, the various explanations necessarily remain speculative. Our new investigations are based on a study of gripping patterns in *Loris tardigradus* and *Nycticebus coucang*. Original data were collected by one of us (U.N.), and were supplemented by information from the literature. We have tried to characterize first, the way in which the second ray is used in contrast to the others, and, second, the situations in which it is exposed to particularly high stresses.

Morphology of hand and foot of the lorisines

Table 1 gives the length of the second ray of prosimian primates in percentage of the length of the longest ray. Ratios have been calculated from absolute lengths of the phalangeal and metacarpal bones published by Etter (1978). A relatively short second digit seems to be typical for prosimians in general. Reduction is most pronounced in the African lorisines, *Arctocebus calabarensis* and *Perodicticus potto*, followed by the Asian lorisines, *Loris tardigradus* and *Nycticebus coucang*, and by the Malagasy species of *Daubentonia*. Within the two geographical groups of lorisines, reduction is more pronounced in the larger species (*P. potto* and *N. coucang*).

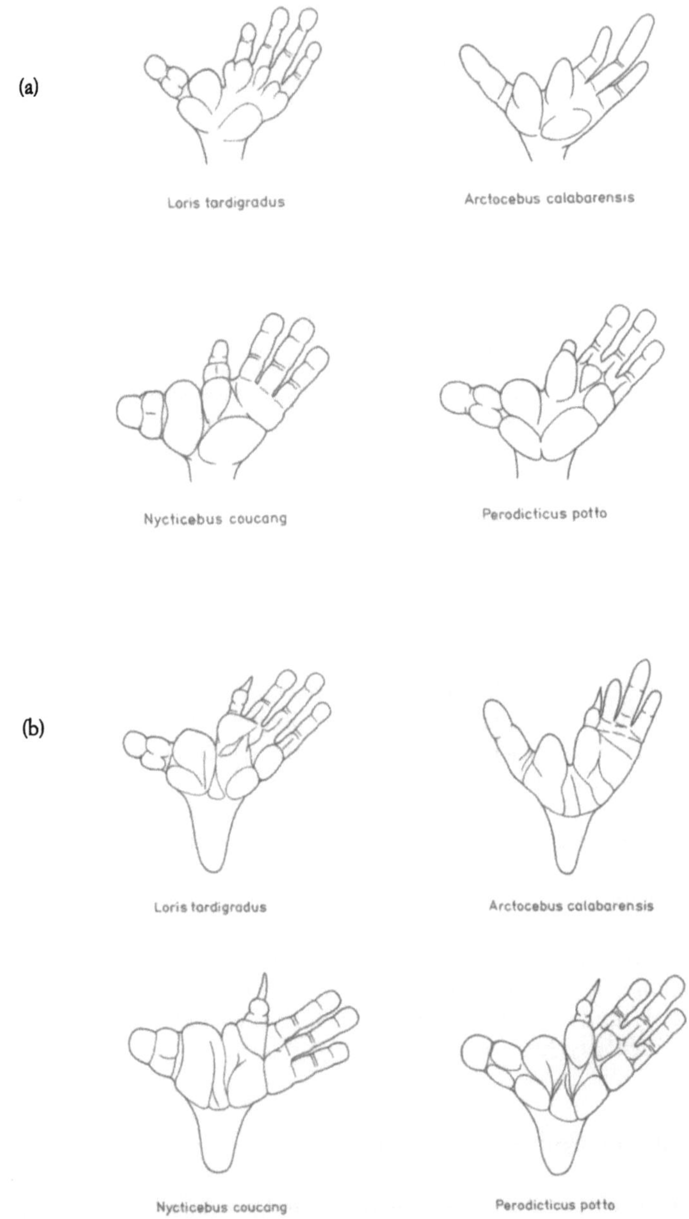

Figure 1 - Hands (a) and feet (b) of the Lorisinae. After BIEGERT, 1959.

TABLE 1 - *Lengths of the second and the longest ray in prosimian hands. Data from* ETTER, *1978; length of rays = sum lengths of metacarpal and phalangeal bones.*

Species	n	longest finger (mm)	second finger (mm)	ratio
Loris tardigradus	4	16.2	3.7	68.3
Arctocebus calabarensis	19	25.5	11.1	43.5
Perodicticus potto	25	47.4	19.4	40.9
Nycticebus coucang	7	38.2	24.1	63.0
Galago demidovii	4	19.9	13.9	69.8
Galago senegalensis	6	33.8	25.6	75.7
Galago (E.) elegantulus	13	36.2	28.0	77.3
Galago crassicaudatus	1	43.1	32.8	76.1
Propithecus verreauxi	3	82.2	59.6	72.5
Indri indri	1	136.0	103.2	75.8
Avahi laniger	1	61.7	45.7	74.0
Lepilemur leucopus	3	41.3	31.2	75.5
Hapalemur griseus	2	46.3	33.5	72.3
Lemur albifrons	1	59.8	50.4	84.2
Lemur catta	1	53.3	45.8	85.9
Microcebus murinus	2	16.9	13.3	78.6
Daubentonia	2	114.7	74.0	64.5
Tarsius syrichta	2	33.4	30.1	90.1
Tarsius bancanus	1	37.9	32.5	85.7

JOUFFROY & LESSERTISSEUR (1976) have calculated a similar index of length of the metacarpal plus phalangeal bones of the second vs. the fourth ray. Their results are principally in accordance with ETTER's, with the one exception that *Arctocebus* has a relatively shorter second ray than *Perodicticus*. As they have measured only one animal, this is probably an individual variant. New data from JOUFFROY *et al.* (this vol.) fully confirm ETTER's results. The index of length of the second metacarpal relative to the third has low values for all four lorisine species, indicating that the reduction of the second ray also involves its most proximal bone (*Table* 2).

TABLE 2 - *Length of the second ray in % of the third and second metacarpal in % of the third in Lorisinae. After* JOUFFROY *et al., this vol.*

Species	n	m2 + d2 / m3 + d3	n	m2/m3
Loris tardigradus	9	73.0 ± 4.3	9	79.7 ± 9.7
Nycticebus coucang	22	66.9 ± 4.7	20	79.6 ± 5.6
Perodicticus potto	8	46.1 ± 4.1	6	66.2 ± 8.9
Arctocebus calabarensis	18	53.1 ± 3.8	14	63.8 ± 9.4

m = length of metacarpal; d = length of digit.

Comparable measurements for the foot are not available in the literature, but some data are provided by MIDLO (1934), NAPIER (1967) and STRAUS (1942), and the figures published by BIEGERT (1959) and reproduced here in *Figure 1b* show essentially the same trends that exist in the hand.

The reduction in length of the second ray of the African species is accompanied by a reduction in the number of phalanges. *Perodicticus* variably possesses two or three phalangeal bones in the 2nd fingers, and *Arctocebus* regularly has only two (HILL, 1953; STRAUS, 1942; NAPIER & NAPIER, 1967). *Loris* and *Nycticebus* normally have the complete number of three bony elements. In all four species the second toe has the complete number of phalanges.

The reduction in length and number of bony elements in the second phalanx also affects muscle insertions (HILL, 1953). According to STRAUS (1942), the tendons of the superficial flexor to the index finger and the short flexor to the second toe are absent in *Nycticebus* and *Loris*, and in *Perodicticus* the extensor tendons of finger and toe are also lacking. Comparative data for *Arctocebus* are not available.

Behavioral data

The locomotion of lorisines has been characterized as «slow walking and climbing» (NAPIER & NAPIER, 1967; WALKER, 1969), or, more recently, «lorisiform crawling» (JOUFFROY, 1989). Among primates, this particular type of locomotion is unique. It enables them to move without making noises in an arboreal environment — an advantage in approaching prey as well as in avoiding becoming prey themselves. *Loris tardigradus* usually walks on top of branches rather than below them. Gaps are crossed by bridging. The animals are able to descend head-first and ascend branches of various angles of inclination. In the seminatural cage environment of our facilities, they prefer branches of 2 to 4 cm in diameter, horizontal over vertical branches, and continuous substrates over discontinuous ones. The gait never has an aerial phase between the steps. Walking speed is increased by increasing stride frequency and especially stride length. Forelimb stride length at high speeds is attained by increasing angular excursions, whereas lateral flexions of the trunk contribute to the long hindlimb strides. Average stride length is higher than in other primates of comparable body size. It is the same for the smaller bodied *Loris* and for the 3 times heavier *Nycticebus*, about 43 cm (DEMES, JUNGERS & NIESCHALK, 1990).

In manipulating objects by the hands, the second ray is used in concert with the others and has no special function. Whether reaching for an insect, for a piece of fruit, or for a conspecific, the grip is determined by the shape of the object. Opposition of two fingers, like in the so-called «precision grip» of the higher primates, was never observed. Except for cleaning, the foot is used exclusively in locomotion and posture. In autogrooming, the second toe bearing the toilet claw is moved up and down together with the other toes.

In postural behavior and locomotion, the hand grips between the thumb on one side, and the third to fifth finger on the other side of the branch. These lie oblique to the direction of the branch (*Figure 2a*), whereas the position of the second digit coincides with the direction of the branch. The index finger is usually flexed, with the metacarpo-phalangeal joint dorsiflexed. In the stance phase of walking, the arm is primarily rotated about the wrist joint, not between the hand and the substrate as in most terrestrial forms (compare PREUSCHOFT et al., this vol.). This guarantees a secure handhold. The hand is

(a)

(b)

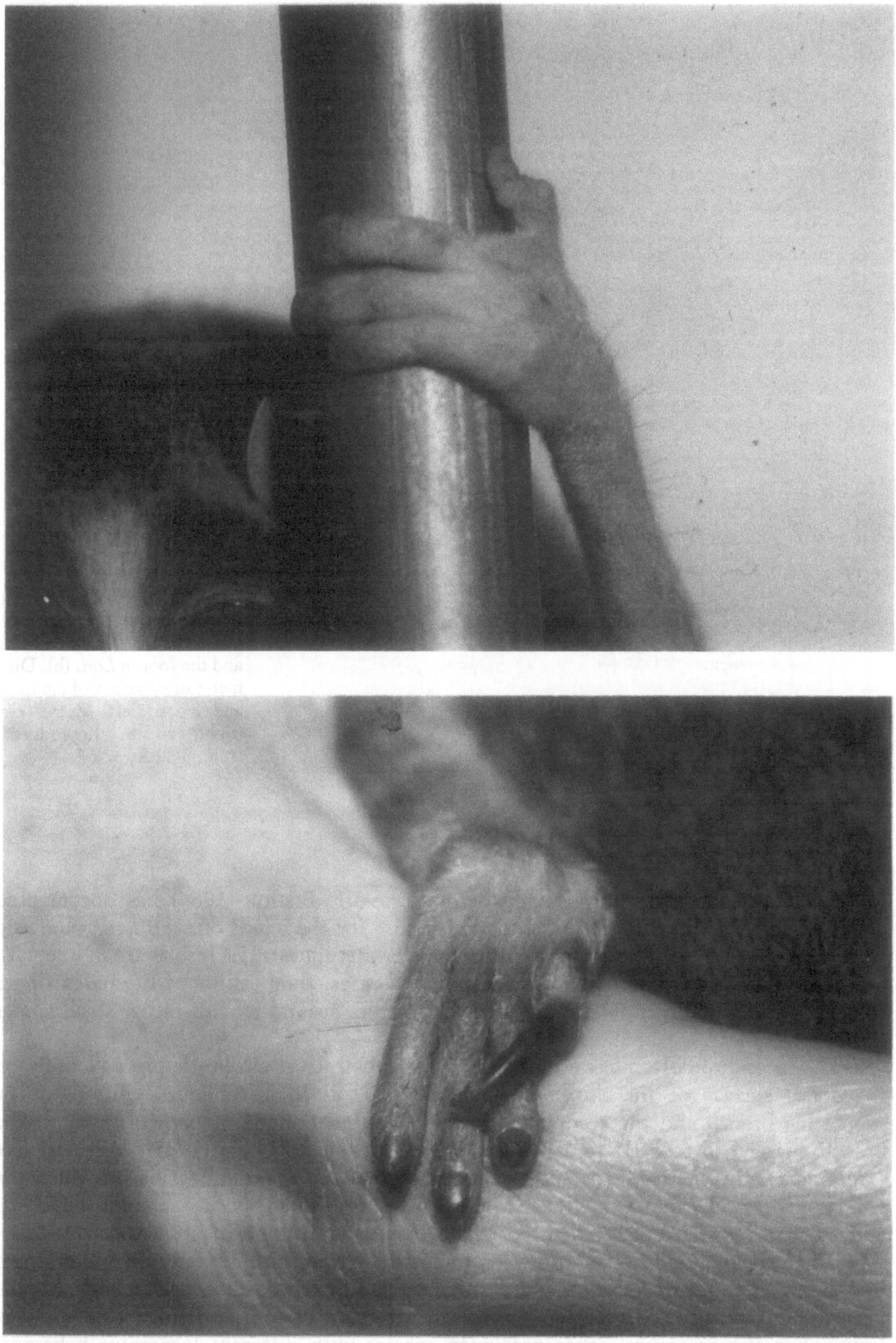

Figure 2 - Posture of hand (a) and foot (b) of *Loris tardigradus* when walking on a branch.

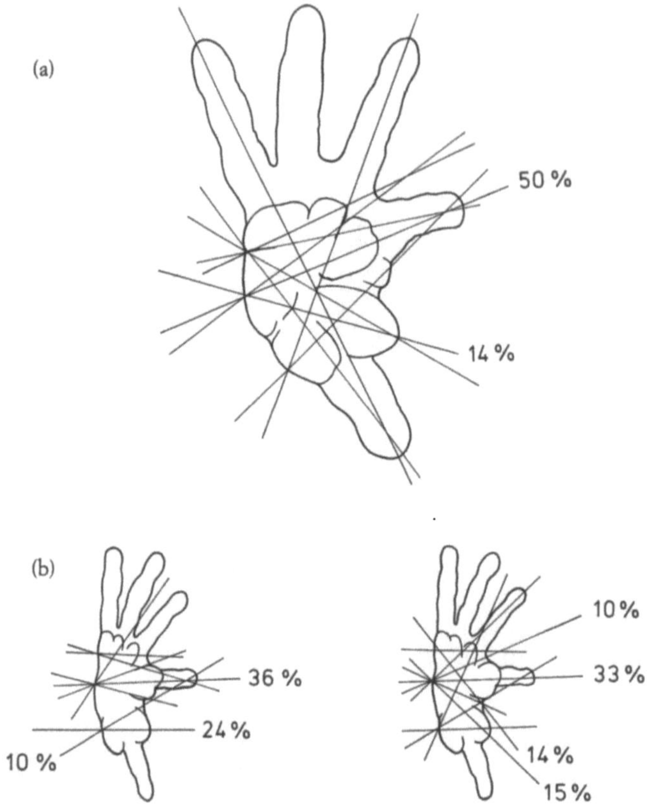

Figure 3 - Axes of movement in relation to the digits of the hand in *Nycticebus* (a) and the foot in *Loris* (b). Diameter of dowels = 4 cm in a, and 2 cm (left) and 4 cm (right) in b. Reproduced from BISHOP, 1964.

rolled off over the palm only shortly before take-off. BISHOP (1964) has documented quantitatively the direction of the «functional axis» for the hands of *Loris tardigradus* and *Nycticebus coucang* (see also RAWLINS, this vol., for macaques). His results are in line with our observations, showing that in the majority of cases, the position of the index finger coincides with the direction of movement or scatters around it with rather small angles (*Figure* 3).

The second toe often has no contact to the substrate at all, but is placed loosely on the dorsal surface of the third. The metatarsal II is in an oblique position to the supporting branch and the second toe itself lies either oblique or perpendicular to the branch, parallel to toes III to V (*Figure 2b*). According to GEBO (1985), lorisines, like *Daubentonia*, the cheirogaleids and tarsiids have a «I - V opposable grip», in which the hallux opposes the other four digits, whereas other prosimian species perform the «I - II adductor grip», with enhancement of digit II as one half of the pincer.

When walking on level ground or on top of thick branches, not only the index finger but all fingers and, to a lesser extent, all toes are in a flexed posture, with metapodiophalangeal joints dorsiflexed. We have observed a flexion of toes and fingers not only in lorises, but also in several other prosimian species with different locomotor behavior.

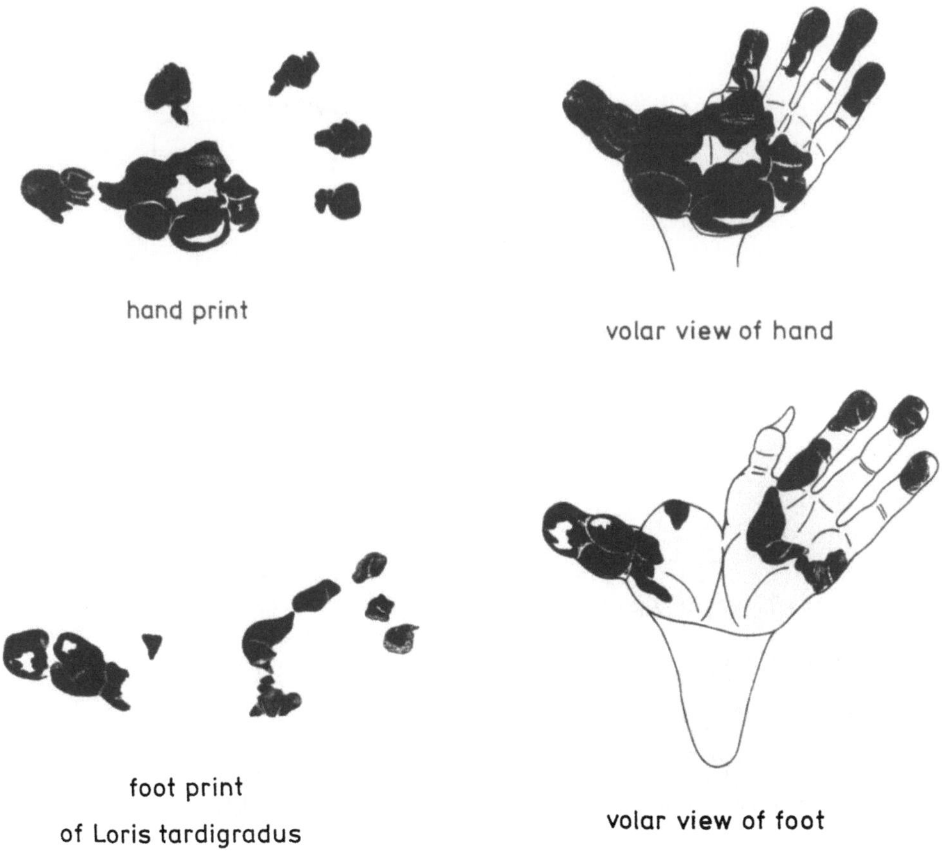

Figure 4 - Hand- (top) and footprints (bottom) of *Loris tardigradus*. Arrow indicates direction of movement. Contact areas are black, on right side the contours of the hand and the foot are outlined.

The position of the second ray described above during locomotion is also documented by hand- and footprints of *Loris tardigradus* (*Figure* 4). The palms and soles of an animal were stained with stamping ink: it was then made to walk along a branch of 2 cm in diameter that was wrapped with a sheet of paper. The index finger leaves an imprint in the direction of the branch, but the second toe does not leave a trace in most cases.

Functional analysis

In walking along branches, the lorisine grip between digits I and III — V enlarges the gap between the two counterparts of the pincer and enables the animals to grip around bigger branches. With the second digit in the direction of movement, load is concentrated at its terminal phalanx during roll-off. As has been detailed before, lorisines walk with long strides and great angular excursions, especially of the forelimb. At the push-off phase

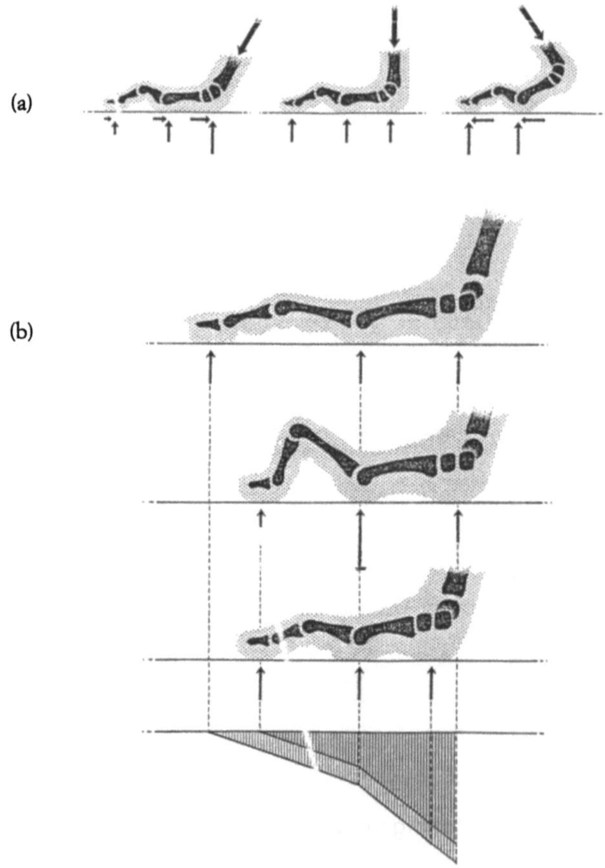

Figure 5 - Reaction forces and bending moments along the digit. (a) Force components at the joints at touch down, mid stance and take off phase of the gait. (b) Bending moments along the axis of a long (top), a flexed (middle), and a short (bottom) finger. If (functional) length is reduced, the moments become smaller, even if the same reaction forces are applied.

of the cycle, the weight force acts with a long lever arm at the basal joints of the second finger or toe (*Figure 5*). The lever arm of the flexor muscles is determined by the distance of their tendons from the center of rotation. This distance is small and remains nearly the same at various positions of the joints. The muscle forces necessary to balance the joint are high, and the phalanges are exposed to high bending stresses (PREUSCHOFT, 1969, 1973).

The reduction of the second ray reduces the length of the load arm and therefore also the rotating moment. The joints can be balanced with less muscle force, and the bending stresses are reduced (*Figure 5b*).

The lever length of the bending force is also reduced functionally by a flexion of the index finger. It is hyperextended in the metacarpo-phalangeal joint and flexed in the proximal interphalangeal joint (*Figure 2a* and *Figure 5*). The pad overlying the head of the second metacarpal is pressed firmly against the substrate and takes a greater share of the overall reaction force acting on the second ray. The contact area is enlarged. The reaction force at the tip of the phalanx is thereby reduced. The same mechanism is effective for the other fingers when the hand is not gripping around a branch, but placed on a level support. Consequences for friction and transmission of sensory informations are discussed by BUCK & BÄR (this vol.) and LANDSMEER (this vol.).

The second finger is exposed to higher stresses than the second toe because of the greater angular excursion at the forelimb, and the position of its long axis closer to the direction of the movement. In the case of the foot, high loads are simply avoided by removing the second toe from the direction of movement and placing it obliquely and in concert with the other toes (*Figure 2b*).

The special loading pattern of the second ray not only concerns the free digit, but also the metacarpal or metatarsal bones, respectively. However, the situation becomes less critical, as the load is distributed over the metacarpus or metatarsus as a whole. At the foot, the load is directed nearly perpendicularly to the long axis of the element, and it is pressed against and supported by the third, fourth, and fifth metatarsals (see PREUSCHOFT, 1969).

As mentioned before, the reduction of the second ray is more pronounced in the larger-bodied species (*Tables 1* and *2*). The ground force that causes a rotating moment at the finger or toe grows proportionally to body mass (with the cube of linear dimensions). The muscle forces that balance the rotating moments at the joints increase with the cross-sections of the muscles, that is only with the square of linear dimensions. The greater the body mass, the smaller becomes the ratio of available force to load. This may well explain the size-dependent difference in the degree of reduction of the second ray between *Nycticebus* and *Loris*.

The morphology of the wrist joint reflects the position of the hand of lorisids during locomotion. According to LEWIS (1985), the retreat of the ulna and distal displacement of the pisiform accommodates extreme ulnar deviation in the wrist. The axis of flexion and extension is not perpendicular to the long axis of the forearm, but deviates proximally to the ulnar side. Its direction is more or less perpendicular to the second ray and supports our hypothesis on the special loading of this ray in contrast to the others. The functional relationship between this orientation and the character complex of the ectaxonic hand is discussed by PREUSCHOFT *et al.* (this vol.).

The orientation of the hand and the foot in walking along branches, in combination with the great stride lengths of the animals, thus explains the reduction of the second ray. Other arboreal primates do not face the same loading patterns as they are less specialized for walking on thin branches and gripping around the support, but, instead, walk on top of thicker branches. As a consequence, the functional axes of hands and feet are different. Among prosimians, galagos come closest to the lorisine gripping pattern.

Reduction of phalangeal length, morphological as well as by flexion, is a common response to high loads. Chimpanzees, and more so gorillas, possess short toes. This trait is more pronounced in humans, where the lateral toes are also kept in a position like in prosimians. The first toe, which carries the greatest share of the weight force, is short and also reinforced.

ACKNOWLEDGEMENTS — We thank H. Preuschoft for his invitation to participate in the symposion. We are also indebted to H. Preuschoft and D. Chivers for comments on the manuscript, M. Wischniwetzki and H. Schulze for the graphics. This research was supported by DFG grants Pr 58/15-1.

References

BIEGERT J., 1959. *Die Ballen, Leisten, Furchen und Nägel von Hand und Fuß der Halbaffen.* Z. Morph. Anthrop., 49: 316-409.

BISHOP A., 1964. *Use of the hand in lower primates.* In: J. Buettner-Janusch (ed.), Evolution and Genetic Biology of Primates, Academic Press, London: 133-223.
BUCK C. & BÄR, H., this vol.
DEMES B., JUNGERS W.L. & NIESCHALK U., 1990. *Size- and speed-related aspects of quadrupedal walking in slender and slow lorises.* In: F.K. Jouffroy, H. Stack & C. Niemitz (eds.), Gravity in Primates, Il Sedicesimo, Florence, 175-197.
ETTER, H.U.F., 1978. *Lorisiform hands and their phylogenetic implications: a preliminary report.* In: D.J. Chivers & K.A. Joysey (eds.), Recent Advances in Primatology, Academic Press, London: 161-170.
GEBO D.L., 1985. *The nature of the primate grasping foot.* Am. J. Phys. Anthrop., 67: 269-277.
HILL W.C.O., 1953. *Primates*, vol. I. Edinburgh University Press, Edinburgh.
JOUFFROY F.K., 1989. *Quantitative and experimental approaches to primate locomotion: A review of recent advances.* In: R.K Seth & R. Seth (eds.), Perspectives in Primate Biology, Today and Tomorrow's, New Dehli.
JOUFFROY F.K. & LESSERTISSEUR J., 1976. *Processus de réduction des doigts (main et pied) chez les primates. Modalités, implications génétiques.* In: Coll. C.N.R.S. 266, Mécanismes de la rudimentation des organes: 381-392.
JOUFFROY F.K., GODINOT M. & NAKANO Y., this vol.
LANDSMEER J.M.F., this vol.
LEWIS O.J., 1985. *Derived morphology of the wrist articulation and theories of hominoid evolution. Part I. The lorisine joints.* J. Anat., 140: 447-460.
MIDLO C., 1934. *Form of the hand and foot in primates.* Am. J. Phys. Anthrop., 19: 337-386.
NAPIER J.R. & NAPIER P.H., 1967. *A Handbook of Living Primates.* Academic Press, London.
PREUSCHOFT H., 1969. *Statische Untersuchungen am Fuß der Primaten. I. Statik der Zehen und des Mittelfußes.* Z. Anat. Entwickl. -Gesch., 129: 285-345.
PREUSCHOFT H., 1973. *Functional anatomy of the upper extremity.* In: G.H. Bourne (ed.): The Chimpanzee, vol. 6, Karger, Basel: 34-120.
PREUSCHOFT H., GODINOT M., BEARD C., NIESCHALK U. & JOUFFROY F.K., this vol.
RAWLINS, this vol.
STRAUS W.L., 1942. *Rudimentary digits in primates.* Quart. Rev. Biol., 17: 228-243.
WALKER A., 1969. *The locomotion of the lorises with special reference to the potto.* East African Wildlife Journal, 7: 1-5.

J. Koebke

Functional morphology of the human carpus

Starting with the classical description of the wrist joint area it shows that for functional aspects this concept is not suitable. Instead of a description of the radiocarpal joint the wrist joint should be divided into a radial and ulnar compartment. Within the ulnar compartment the articular disk plays an important guiding and force-bearing role.
Based on densitometric analyses on the subchondral bone it can be concluded that in a healthy wrist joint the physiological stress in the scaphoid is higher than in the lunate bone. The subchondral bone of the distal radius also reflects the specific stress distribution of the radiolunate and radioscaphoid articulations. The lunate itself is stressed differentially in the radio-ulnar plane, demonstrated by its trabecular structure.
These results support the columnar concept of the carpus, describing a radial, a central and an ulnar column.
In addition to the wrist joint, of which the movements allow a precise orientation of the entire hand, the functional significance of the first and fifth carpometacarpal joints is pointed out. Both joints are of a sellar type and both are able of flexion/extension, abduction/adduction and axial rotation. Concerning the thumb saddle joint different theories of rotational movement are discussed.

Key words: densitometry, force distribution, longitudinal column, metacarpal mobility, ulnocarpal disc, wrist compartments, wrist mobility.

Introduction

Upon analyzing the movements of the hand, one can divide them schematically into movements of the wrist, of the fingers and of the thumb. In a strict sense, wrist movements consist of flexion and extension about one axis and radial abduction and ulnar abduction about a different axis. The functional study of the wrist is, however, inseparable from the rotational movements of the forearm, so that three principal axes exist, i.e. a longitudinal axis, a transverse axis and an dorsovolar axis. Due to the complexity of the wrists movements, however, these axes are not geometrically fixed.

The wrist is a transition zone from the architecture of the distal radius and ulna to that of the metacarpals. As an intercalated but not rigid block, it is in itself mobile while its individual bony elements can be displaced with respect to one another. The securing system of ligaments is responsable for restricting the motion of the wrist within its physiological range. Traumatic lesions of the capsuloligamentous apparatus lead to severe displacements of carpal bones. The columnar concept of the carpus is a key to understanding the nature of such displacements.

Radiocarpal Joint

The classical description of the carpal region in the skeleton of the human hand is quite clear. Proximally, the carpal region is limited by the cleft of the proximal wrist joint, and distally it extends to the bases of the metacarpal bones (ossa metacarpi).

This classical description is, however, inadequate for a functional analysis of the carpus under static and dynamic conditions. For example, the term «articulatio radiocarpalis» only describes the immediate contact between the radius, the scaphoid and the lunate. In this context, one should subdivide the wrist into radial and ulnar compartments, as clinicians have done for some time. Within the ulnar compartment, the head of the ulna indirectly articulates with the proximal carpus through the ulnocarpal disk. This triangular disk is part of an extensive fibrous complex which forms the sheath for the extensor carpi ulnaris and the ulnar collateral ligament (KAUER, 1975). The disk transmits static and dynamic pressure forces from the hand to the ulna and vice versa. Histologically, cartilage cells within the disk's fibrous tissue support the presumption that the disk transfers pressure forces (*Figure 1*). The magnitude of transferred forces is dependent on the positions of the hand and forearm (ROSSAK, 1965, 1970; FRYKMAN, 1967; KOEBKE, 1983; PALMER *et al.*, 1984). Fissuring and perforation of the disk are caused mechanically; a thin disk, as found in ulna-plus-variance, is perforated more readily than a thick disk, as found in ulna-minus-variance (STEINHÄUSER *et al.*, 1973; PALMER *et al.*, 1984).

The only stabilizing elements for the distal radio-ulnar joint are the deep head of the M.pronator quadratus, which stabilizes actively (KOEBKE *et al.*, 1984), the M.extensor carpi ulnaris with its tendon (KAUER, 1979; LENGSFELD *et al.*, 1988) and the ulnocarpal

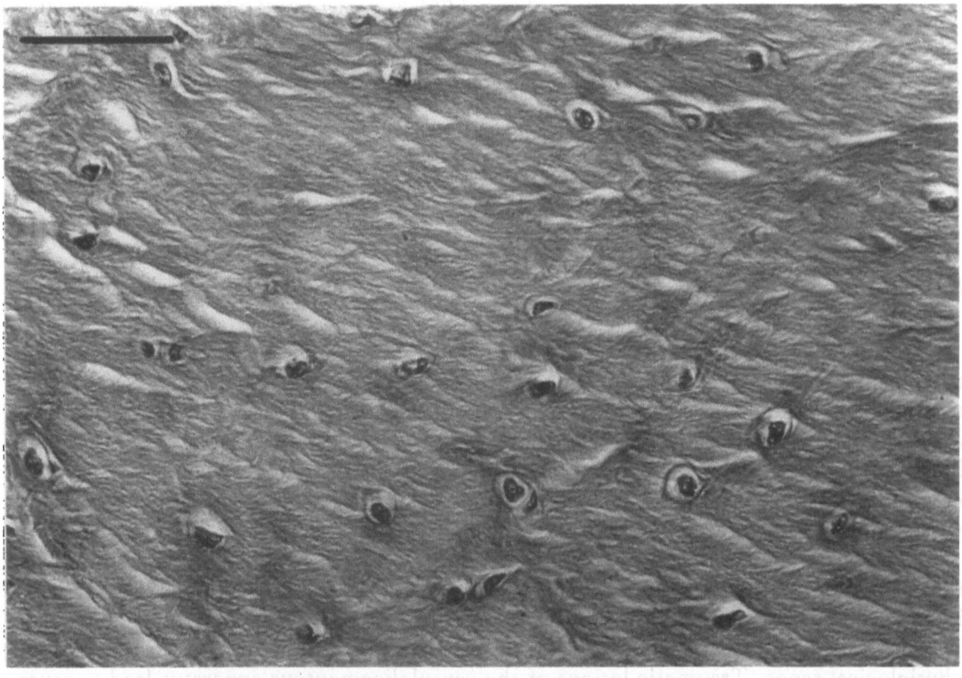

Figure 1 - Histological section through the central part of the ulnocarpal disk of a 62-year-old man; numerous cartilage cells are visible. Goldner, 10 μm; scale bar 300 μm.

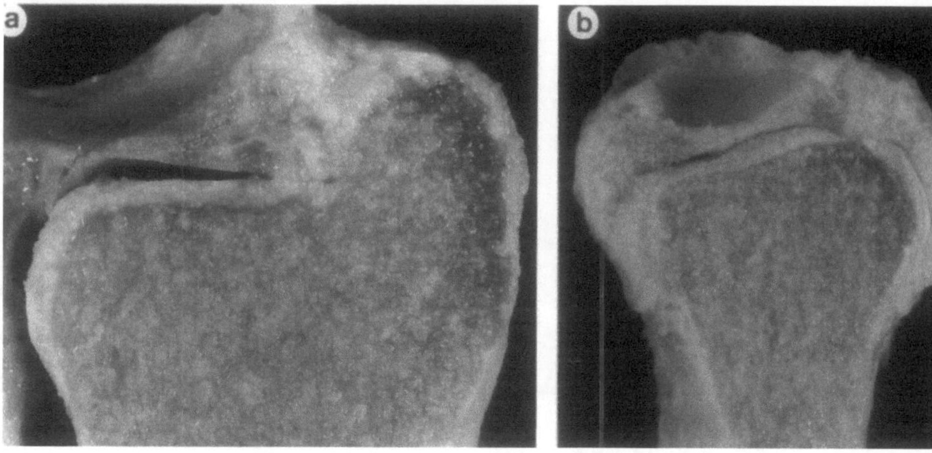

Figure 2 - a) Radioulnar section through the ulnar head and ulnocarpal disk. b) Dorsopalmar section through the ulnar head and ulnocarpal disk.

disk, which stabilizes passively. The disk has two insertions at the ulnar head. One is found in a basal cavity on the radial side of the styloid process (*Figure 2a*). The other insertion consists of the merger of the thickened palmar and dorsal peripheries of the disk with the capsular wall of the distal radio-ulnar joint (*Figure 2b*). Dorsal and volar ligaments, Ligamenta radioulnaria dorsale et palmare, as described by KAPANJI (1984) do not exist. As a consequence of this dual insertions, a fracture of the ulnar styloid process alone does not lead to a loss of the disk's guiding function (BRADE & KOEBKE, 1983).

Depending on the position of the hand in the plane of abduction, different parts of the lunate's proximal surface articulate with the ulnocarpal disk and the ulnar edge of the radius. This results in a characteristic stress pattern in the lunate, which is expressed by locally varying thickness and density of cortical and trabecular elements (*Figure 3*).

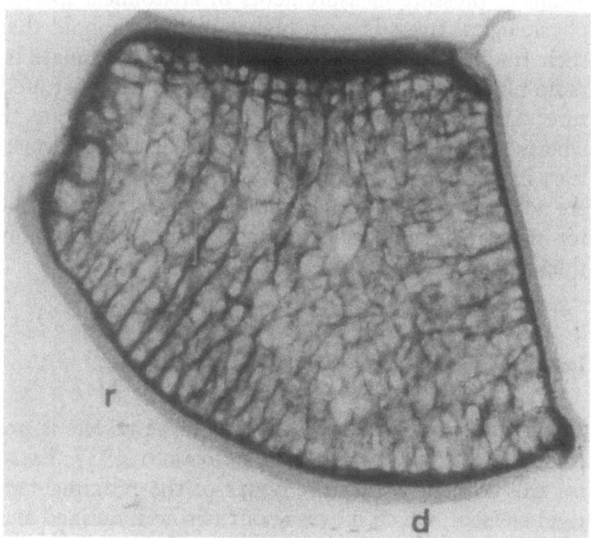

Figure 3 - Roentgenogram of a radioulnar section of a lunate. The «part» of the bone articulating with the radius (r) shows striking, compressive-stressed trabeculae. The «part» of the bone which articulates with the disk (d) shows a diffuse trabecular network.

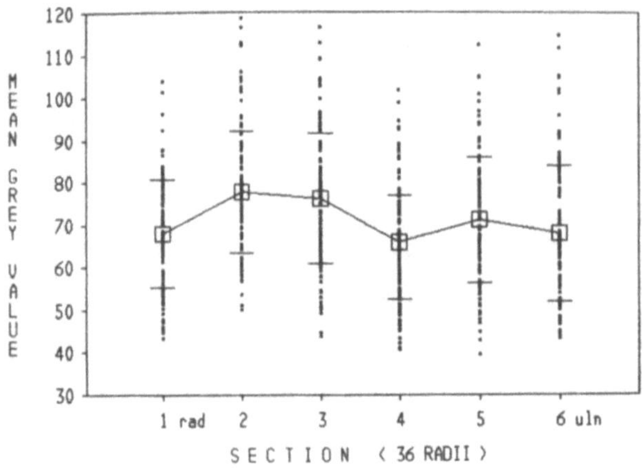

Figure 4 - Results of density analysis on the subchondral bone of the distal surface of the radius. The sections 1-3 represent the scaphoidal facet, the sections 4-6 the lunatal facet. Density (mean grey values) is about 15% higher in the scaphoidal facet (after MÖLLERS *et al.* 1986).

According to WOOD JONES (1944) and MAC CONAILL & BASMAJIAN (1969), the capitate and lunate bones, which morphologically represent the central axis of the carpus, also form the primary stress axis (see also PREUSCHOFT *et al.* in this volume). Forces are transferred from the hamate and the capitate to the lunate, so that, according to these authors' opinions, the lunate is the most stressed element of the proximal carpus.

For the analysis of the radial compartment, it bears review, that, generally, forces are transmitted from one bone to another via a joint cleft. This transmission occurs along the joint resultant which must be a normal to the joint surface (PREUSCHOFT, 1969). The quantity and distribution of bony substance within the subchondral cortical plate are an expression of the strain pattern induced by external forces (PAUWELS, 1973; KUMMER, 1985). Densitometric analyses on the subchondral bone of the distal radius reflects the specific stress-distribution in the radiolunate and radioscaphoid articulations. According to MÖLLERS *et al.* (1986) the scaphoidal facet of the distal radius must support more compressive stresses than the lunatal facet (*Figure 4).* KOEBBKE *et al.* (1989) support by means of pressure measurements in anatomical specimens this view. They show that the physiological stress is even higher in the scaphoid than in the lunate bone. Furthermore their findings lead to the suggestion that the lunate is itself stressed differentially in the radio-ulnar plane. Morphologically this is demonstrated by the lunatal trabecular architecture. The radial «half» of the lunate offers strong trabeculae running proximodistally, wheres the ulnar «half» shows a trabecular meschwork of minor density (see *Figure 3*). At least in pathological conditions of the wrist joint the pressure and the density distribution are adapted to the pathomorphology in each case, too. Concerning the ulna minus variant, for instance, it can be shown that pressure transmission via an ulnocarpal disk of abnormal thickness is obviously decreased.

Longitudinal Columns of the Wrist

A functional interpretation of these results is possible with the aid of the so-called columnar concept of the carpus (NAVARRO, 1937; TALEISNIK, 1976; FISK, 1981). The basis for this concept is that the bones of the proximal carpal row do not form a unique and rigid element, which moves about two well-defined axes. The specifications of these axes,

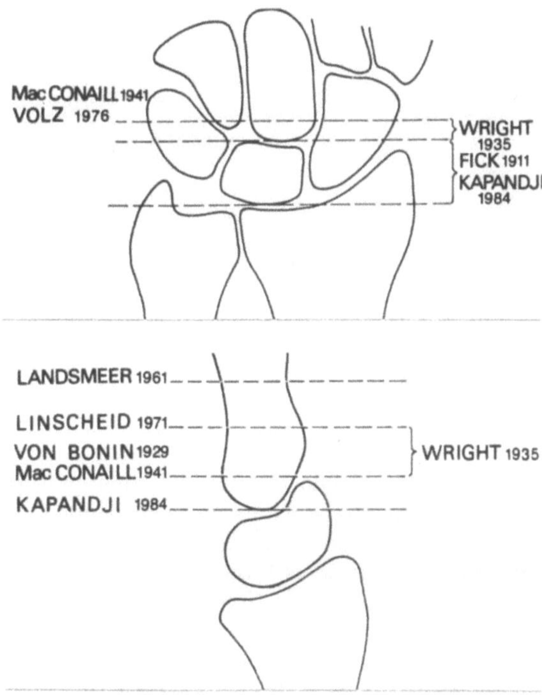

Figure 5 - Axes of flexion-extension (above) and ulnar or radial abduction (below). Several authors (right) postulate different axes for flexion-extension and ulnar or radial abduction respectively.

which have been given by several investigators vary greatly (*Figure 5*). It seems more appropriate to study the motion of each ossicle individually. In this context, the movements of the carpus can be compared to those made by three longitudinal chains (KAUER, 1964; MULDER, 1970; TALEISNIK, 1976). The radial chain is composed of the scaphoid, the trapezium and the trapezoid; the central chain is formed by the lunate and the capitate. The ulnar column consists of the triquetral and the hamate. Each chain or column of the carpus can be seen as a functional unit. The radial column, as the mobile column (NAVARRO, 1937) is the foundation for the thumb's skeletal elements. Movements between the first metacarpal and the trapezium, the trapezium and the scaphoid, and the scaphoid and the radius give this chain the greatest mobility of all. The central chain, as the flexion/extension column (TALEISNIK, 1976) is basally connected with the radial chain by the scapholunate ligament. This ligament is formed by fibers of unequal length (KAUER, 1974), so that the two bones can become mutually displaced during flexion and extension (*Figure 6*) as well as during radial and ulnar abduction. The most superficial fibers of the ligament are arranged dorsovolarly. Although the ligament does not hinder the mutual displacement of the lunate and the scaphoid, it seems important for the equilibration of forces in the radiocarpal joint. In cases where this ligament is torn, the radial chain is disrupted and the radioscaphoid articulation may become arthritic (*Figure 7*).

Both the lunate and scaphoid are intercalated bones, lying between the distal row of the carpus and the radius. The stability of the lunate and the scaphoid bones during carpal movements depends primarily on intact interosseous and carpal ligaments. Especially severance of the dorsal or palmar radiolunate and scapholunate ligaments leads to carpal instability and to a collapse of the radial or central columns.

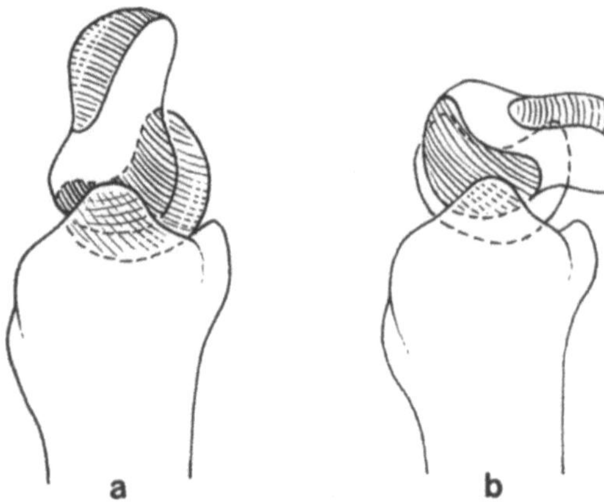

Figure 6 - Position of the lunate and the scaphoid in extension (a) and flexion (b) of the wrist, radial view (after KAUER, 1974).

The ulnar column (column of pronation/supination, TALEISNIK, 1976) and its ligamentous linkage to the radius are essential for stabilization of the central and radial carpal columns in the radio-ulnar plane. As shown by KUHLMANN (1977), the volar and dorsal radiotriquetral ligaments form an oblique sling, which functions as a harness around the ulnar side of the carpus (*Figure 8*).

This ligamentous harness counteracts a physiological shift of the whole carpus, which is effected by the articular surface of the distal end of the radius. This shift is directed ulnarly at an angle of 15-20 degrees.

The columnar carpus concept of TALEISNIK is critised by SENNWALD (1987). Based on anatomical studies he points out that there is no real ligament connecting the lunate to the capitate bone. From this a true central column for flexion and extension shall not exist. The stability of the two transverse rows of the carpus (the author favours this concept of the carpus) is for him given by two v-shaped ligamentous bonds, which will assure and guide the proximal and distal carpal rows. Nevertheless SENNWALD (1987) comes to the wrong conclusion that the lunate bone is the main element of the carpus for transmitting axial forces from the hand to the forearm.

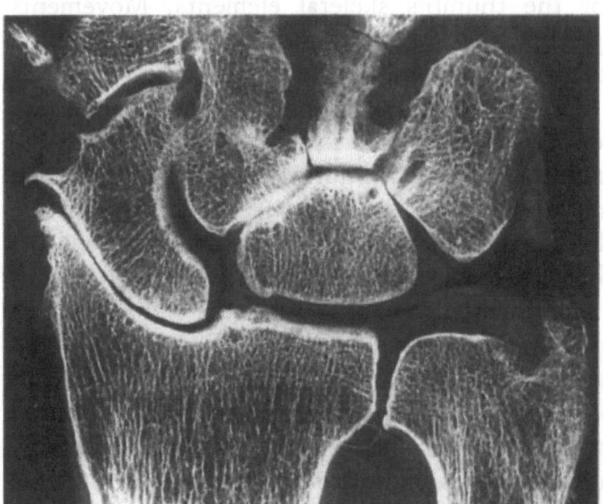

Figure 7 - Roentgenogram of a section of a wrist with arthritic radioscaphoid articulation. The scapholunate ligament is ruptured.

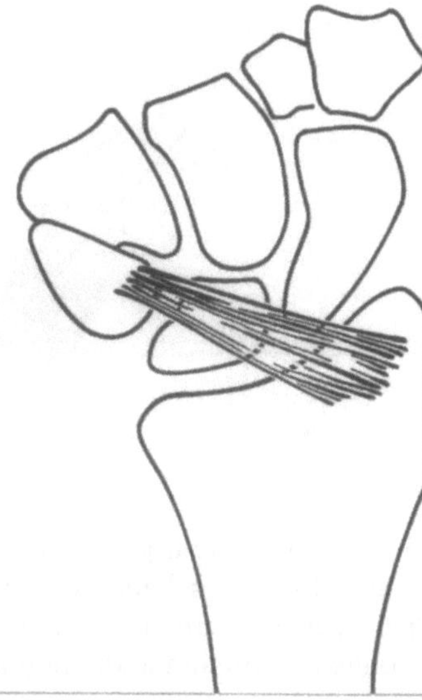

Figure 8 - Course of the palmar radiotriquetral ligament, schematically. Palmar and dorsal ligaments form a sling around the ulnar side of the carpus (after KUHLMANN, 1977).

First and Fifth Carpometacarpal Joints

While the distal radio-ulnar joint and the wrist joint are primarily important for the orientation of the entire hand, some of the carpometacarpal articulations play a fundamental role in manipulation i.e. movement of individual manual elements. For instance, while grasping a spherical object, the hollow of the palm is malleable. The concavity of the palm depends on changes in the curvature of the transverse metacarpal arch. This is accomplished by flexion and adduction movements of the first, fourth, and fifth metacarpal heads (*Figure 9*). In contrast to the intermediate metacarpals, which are united to the carpus by the intrinsic interlocking shapes of the bones themselves, the thumb metacarpal is totally independent. Similarly, the fifth metacarpal reveals an astonishing mobility. The latter has a range of flexion and extension of approximately 20 degrees. In the case of phalangization, the degree of mobility increases beyond this (TUBIANA & ROUX, 1973). The basal facet of the fifth metacarpal, which articulates with the hamate, extensively varies with respect to its outline and curvature. In most cases, however, it has a saddle-like curvature or is nearly flat (EL-BACHA, 1981). The articular facet of the hamate is always concave in the dorsopalmar plane, and it is often transversely convex. In principle, the

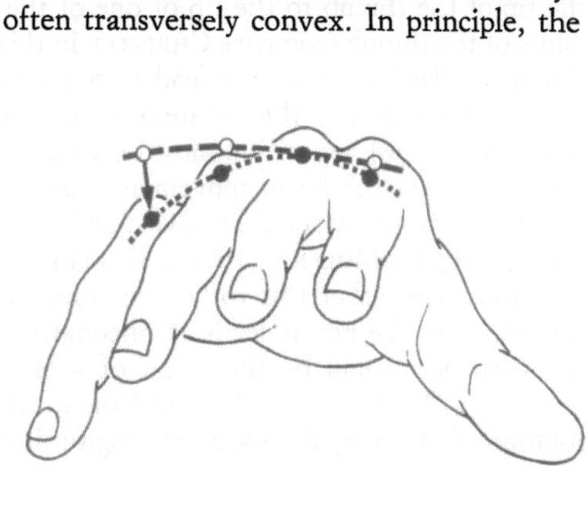

Figure 9 - The hollow of the palm adaptates by flexion of the fourth and fifth metacarpals (after KAPANDJI, 1984).

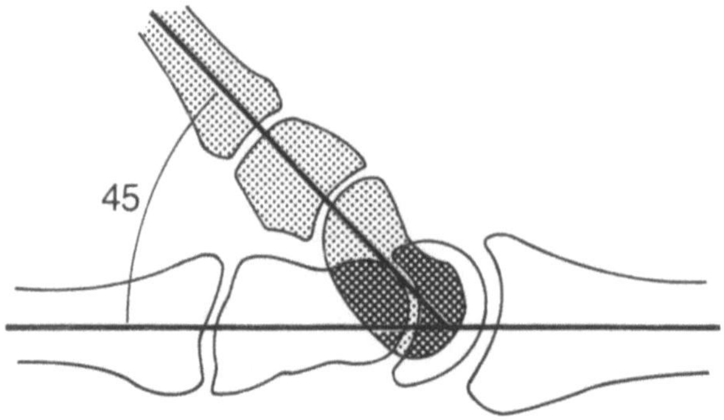

Figure 10 - Oblique position of the thumb ray, schematically (after TUBIANA, 1981).

fifth carpometacarpal joint may be regarded as a sellar joint. The capsule of the joint is thin and slack; dorsal and volar carpometacarpal ligaments are always present; intermetacarpal ligaments are often quite delicate. In addition to the principal movements of flexion/extension and abduction/adduction, slight rotation and thus opposition are possible.

This diversity of movements is very specific especially for the first carpometacarpal joint, the sellar joint of the thumb. It appears, at first sight, simple to define the morphology and the movements of the thumb's sellar joint. In contrast to investigations which try to describe the two articular facets as having a reciprocal curvature (FICK, 1854; HENKE, 1863; DU BOIS-REYMOND, 1896), it could be shown that the carpometacarpal thumb joint is really «a very unclean joint» (MEYER, 1973). BOPP (1966); KUCZYNSKI (1974); SCHMIDT & LIEB (1981) and PIERON (1981) found that the joint surface of the trapezium as well as that of the first metacarpal are actually asymmetrical.

In spite of this asymmetry, it is of didactic value to presume that the carpometacarpal thumb joint works as a saddle joint.

This means that there are two principal mechanical axes. It is noteworthy that the trapezium is not in one plane with the rest of the carpus. The oblique position is not limited to the trapezium, but involves the entire thumb skeleton, starting with the scaphoid (*Figure 10*). The thumb skeleton lies volar to the plane of the other metacarpals. In a neutral position, it forms an angle of approximately 45 degrees with the index finger (TUBIANA, 1981).

BUNNELL (1938) was the first to state, that the opposition movement, which brings the tip of the thumb to the tip of one of the other digits requires adjustment in all three joints of the thumb (compare CHRISTEL in this volume). Simple flexion or adduction of the thumb's sellar joint does not lead to opposition at all. Numerous, sometimes conflicting, analyses contribute to the debate over how opposition of the tumb is achieved. According to NAPIER (1956) opposition must be seen as a movement in the thumb's carpometacarpal joint consisting of basic movements with a rotational component. Such a rotational movement is mentioned in investigations by LANDSMEER (1962); EATON & LITTLER (1969); BOJSEN-MØLLER (1976), KOEBKE (1983) and KAPANDJI (1984). However, the question, of whether this rotational component is possible without losing congruency of the joint or whether it leads to incongruency of the articular surfaces, remains open. Incongruency would be the result of axial rotation of the first metacarpal bone. KUCZYNSKI (1974); PIERON (1981) and KOEBKE (1983) describe a loss of congruency following rotation of metacarpal I about its longitudinal axis. KUCZYNSKI (1974) observes a twisting

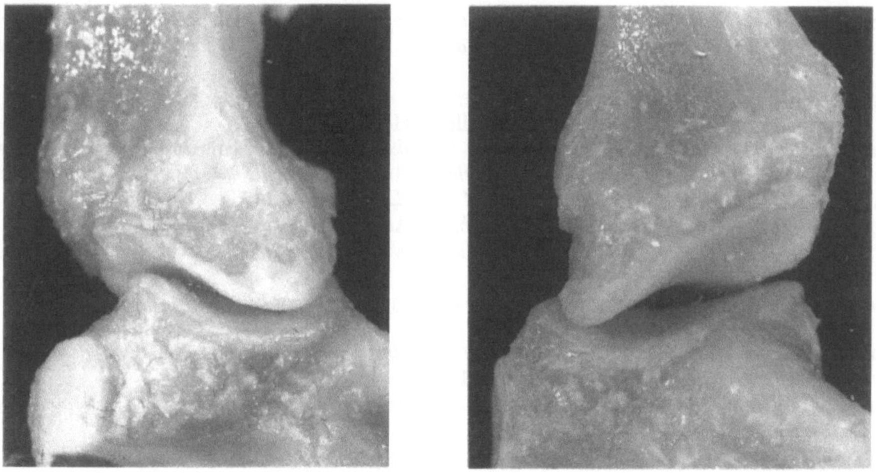

Figure 11 - Congruent thumb saddle joint (left); oppositional rotation leads to incongruence of the joint (after KOEBKE, 1983).

of the passively rotated metacarpal out of the saddle of the trapezium in a transfixed amputation specimen. The articular surfaces ride up on each other, and the joint contact area is reduced. KUCZYNSKI (1974) considers it likely, that the same phenomenon can be applied to active movement in final opposition or reposition of the living thumb.

Morphological investigations of the thumb's saddle joint articular surfaces reveal in many cases small local depressions of the cartilage (KOEBKE, 1983). These are an indication of rotary movement in this joint. The metacarpal is twisted out of the trapezium's saddle during rotation (*Figure 11*). Articular surface contact is diminished, and the joint pressure is concentrated locally.

In summary, the physiological rotation, which is part of thumb opposition leads to a pronounced incongruity in the saddle joint. The small lesions of the articular cartilage are early signs of arthritic change.

References

BOJSEN-MØLLER F., 1976. *Osteoligamentous guidance of the movements of the human thumb.* Am. J. Anat., 147: 71-80.

BOPP H.M., 1966. *Anatomische Untersuchungen der Daumensattelgelenksarthrose.* Diss. Köln.

BRADE A. & KOEBKE J., 1983. *Verletzungen des Discus articularis und des Radioulnargelenkes nach Handgelenkstraumen beim Sport.* In: Sportverletzungen und Sportschäden, Thieme Stuttgart, New York.

BUNNELL S., 1938. *Opposition of the thumb.* J. Bone Jt. Surg, 20: 269-284.

COONEY W.P., LUCCA M.J., CHAO E.Y.S. & LINSCHEID R.L., 1981. *The kinesiology of the thumb trapeziometacarpal joint.* J. Bone Jt. Surg., 63-A: 1371-1381.

DU BOIS-REYMOND R., 1896. *Die Gelenksbewegungen bei der Opposition.* Anat. Anz., 11: 464-467.

EATON R. & LITTLER J., 1969. *A study of the basal joint of the thumb. Treatment of its disabilities by fusion.* J. Bone Jt. Surg., 51-A: 661-668.

EL-BRACHA A., 1981. *The carpometacarpal joints (excluding the trapeziometacarpal).* In: The Hand; vol. 1, ed. by Tubiana, R., Saunders Philadelphia-London-Toronto.

FICK A., 1854. *Die Gelenke mit sattelförmigen Flächen.* Z. rat Med.NF, 4: 314-321.

FISK G., 1981. *Biomechanics of the wrist joint.* In: The Hand; vol. 1 ed. by Tubiana, R., Saunders Philadelphia-London-Toronto.

FRYKMAN G., 1967. *Fracture of the distal radius including sequelae-shoulder-hand-finger syndrome, disturbance in the distal radio-ulnar joint and impairement of nerve function. A clinical and experimental study.* Acta Orthop. Scand., (Suppl) 108: 1-153.

HENKE J., 1863. *Handbuch der Anatomie und Mechanik der Gelenke mit Rücksicht auf Luxationen und Contracturen.* Winter'sche Verlagsbuchhandlung Heidelberg.

KAPANDJI I.A., 1984. *Funktionelle Anatomie der Gelenke.* Bd. I obere Extremität, Enke Stuttgart.

KAUER J.M.G., 1964. *Een analyse van de carpale flexie (An analysis of the carpal flexion).* Thesis Leiden.

KAUER J.M.G., 1974. *The interdepence of carpal articulation chains.* Acta Anat., 88: 481-502.

KAUER J.M.G., 1975. *The articular disc of the hand.* Acta Anat., 93: 590-605.

KAUER J.M.G., 1979. *The collateral ligament function in the wrist joint.* Acta Morphol. Neerl-Scand., 17: 252-253.

KOEBKE J., 1983. *A biomechanical and morphological analysis of human hand joints.* Adv. Anat. Embryol. Cell. Biol., 80: 1-85.

KOEBKE J., WERNER A. & PIENING H., 1984. *Der Musculus pronator quadratus - eine morphologische und funktionelle Analyse.* Anat. Anz., 157: 311-318.

KOEBKE J., FEHRMANN PH. & MOCKENHAUPT J., 1989. *Zur normalen und pathologischen Beanspruchung des Handgelenks.* Handchirurgie, 21: 186-204.

KUCZYNSKI K., 1974. *Carpometacarpal joint of the human thumb.* J. Anat., 118: 119-126.

KUHLMANN N., 1977. *Anatomia descriptiva y functional del carpo en relacion con la pathologia traumatica del mismo.* Rev. Esp. Cirug. la Mano, 5: 13-104.

KUMMER B., 1985. *Einführung in die Biomechanik des Hüftgelenks.* Springer Berlin-Heidelberg-New York-Tokyo.

LANDSMEER J., 1962. *Power grip and precision handling.* Ann. Rheum. Dis., 21: 164-170.

LENGSFELD M., STRAUSS J.M. & KOEBKE J., 1989. *Funktionelle Bedeutung des M.extensor carpi ulnaris für das distale Radioulnargelenk.* Handchirurgie, 20: 275-278.

MAC CONAILL M.A. & BASMAJIAN J.V., 1969. *Muscles and Movements a Basis for Human Kinesiology.* Williams and Wilkins, Baltimore.

MEYER H., 1873. *Die Statik und Mechanik des menschlichen Knochengerüsts.* Engelmann Leipzig.

MÖLLERS N., LEHMANN J. & KOEBKE J., 1986. *Die Verteilung des subchondralen Knochenmaterials an der distalen Gelenkfläche des Radius.* Anat. Anz., 161, 151.

MULDER TH. J., 1970. *Röntgencinematografic en Draadreconstructil van de pols.* Thesis Leiden.

NAPIER J.R., 1956. *The prehensile movements of the human hand.* J. Bone Jt. Surg., 38-B: 902-913.

NAVARRO A., 1937. According to: Tubiana R., 1981. *Architecture and functions of the hand.* In: The Hand; vol. 1 ed. by Tubiana R., Saunders Philadelphia-London-Toronto.

PALMER A.K. & WERNER F.W., 1984. *Biomechanics of the distal radioulnar joint.* Clin. Orthop., 187: 26-35.

PAUWELS E., 1973. *Kurzer Überblick über die mechanische Beanspruchung des Knochens und ihre Bedeutung für die funktionelle Anpassung.* Z. Orthop., 111: 681-705.

PIERON A., 1981. *The first carpometacarpal joint.* In: The Hand; vol. 1 ed. by Tubiana R., Saunders Philadelphia-London-Toronto.

PREUSCHOFT H., 1969. *Statische Untersuchungen am Fuß der Primaten I.* Z. Anat. Entwickl.-Gesch. 129, 285-345.

ROSSAK K., 1965. *Druckmessungen am Ellenbogengelenk.* Verh. Dtsch. Orthop. Ges., 52: 417-420.

ROSSAK K., 1970. *Experimentelle Druckmessungen am Unterarm und ihre klinischen Folgen.* Ergebn. Chir. Orthop., 54: 141-175.

SCHMIDT H.M. & LIEB B., 1981. Die Articulatio carpometacarpea pollicis des Menschen: *Größenmerkmale und Krümmungsprofile der Gelenkflächen.* Verh. Anat. Ges., 75: 665-667.

SENNWALD G., 1987. *Das Handgelenk.* Springer Berlin-Heidelberg-New York-London-Paris-Tokyo.

STEINHÄUSER J., ABELE H. & SCHETTLER G., 1973. *Anatomisch-morphologische Studien zur sogen. Minusvariante der Elle am Handgelenk.* Z. Orthop., 111: 36-40.

TALEISNIK J., 1976. *The ligaments of the wrist.* J. Hand. Surg., 1: 110-118.

TUBIANA R., 1981. *Architecture and functions of the hand.* In: The Hand; vol. 1 ed. by Tubiana R., Saunders Philadelphia-London-Toronto.

TUBIANA R. & ROUX J.P., 1974. *Phalangization of the first and fifth metacarpals.* J. Bone Jt. Surg., 56-A: 447-452.

WOOD JONES F., 1944. *The principles of anatomy as seen in the hand.* 2nd edn. Bailliere, London.

H. Preuschoft
M. Godinot
C. Beard
U. Nieschalk
F.K. Jouffroy

Biomechanical considerations to explain important morphological characters of primate hands.

Two morphologically distinct types of primate hands, the «ectaxonic» hand of most strepsirhines, and the «mesaxonic» of many, epecially ground-dwelling, monkeys and of apes and humans, are subject to a biomechanical analysis. A functional explanation for the existence of these two combinations of skeletal and myological characteristics is presented.

Each combination of traits fulfills similar, but not identical biological roles: either climbing and walking on thin branches (ectaxonic), or walking and running on thick branches, or on the ground, with occasional climbing and precision handling (mesaxonic). In locomotion primarily a power grip is used. Precision handling is a sort of hand usage which not only appears late in primate evolution, but is also influenced by size. Humans rarely use their hands for locomotion, and so the requirements for a power grip are less rigorous; they proceeded farther on the path towards specialisation of the hand as a precision tool than other primates. The hand shape characteristic of humans is a consequence of the lack of locomotor tasks, of the «mesaxonic» structure it evolved from, and of its size.

Key words: ectaxony, forces, mechanical function, mesaxony, power grip, precision grip, stress.

Introduction

Among primates, at least two different types of hands can be distinguished morphologically. This has been observed by several authors in the past, as mentioned and reviewed in several articles of this volume, and emphasized in Jouffroy *et al.*, Godinot & Beard, Beard *et al.*. The basic variants are characterised as follows:

(1) *Ectaxonic*:
- ray IV is the longest,
- all digits are long (> 56% of total hand length),
- long pollex,
- short metacarpals (< 29% of hand length),
- carpus with large ulnar elements,
- extensive contact between carpal bones and ulna.

Among living forms, most strepsirhines are ectaxonic. We take loris hands as an example.

(2) *Mesaxonic*:
- ray III is the longest,
- all digits are short (< 57% of total hand length),
- short pollex,
- long metacarpals (> 27%),
- long carpus, with radial elements more developed,
- carpo-ulnar contacts reduced.

The mesaxonic hands are found among hominoids and the more terrestrial cercopithecoids. For this study, the human hand is the focus of our attention, but sometimes reference is also made to chimpanzee or macaque hands.

Intermediates between both extremes exist to the extent that JOUFFROY *et al.* (this volume) also define a paraxonic hand in which rays III and IV are of about the same length. It is found in *Varecia, Cheirogaleus, Ateles, Cercocebus and Colobus*, that is in «unspecialized» strepsirhines and highly arboreal simians.

All previous considerations of the origin of these hand forms leave open the question: What do the well-known morphological characters mean functionally ? More specifically: what are the selective advantages of these characters, for which biological reasons have these characters been developed? In discussions about living forms not much attention is paid to these morphological features. In discussions about fossils, the question is often raised as to what conclusions can be drawn from these characters with regard to the mode of life — or at least of locomotion — of their former possessors?

Reliable, biomechanically well-founded hypotheses can be derived from theoretical considerations based on morphological and behavioral observations. These hypotheses can only be tested by examination of living forms — for which we know their behavior as well as their morphology.

For the purpose of this study, we distinguish between a «power grip», that is, any sort of prehension executed with high forces, and the «precision grip». What we call power grip occurs regularly in locomotion, but also in fighting, or catching of prey, sometimes perhaps in feeding, for instance in destroying stems to get hold of the marrow. The force which acts in a power grip on the hand may be as large as total body weight or even a multiple of it (when stopping the animal falling down from a tree). Precision gripping in our understanding includes all behaviors in which the hands are involved with small and mobile objects, where muscle force is reduced. It occurs for instance in feeding, grooming, or in lifting and exploring objects, or in more complex manipulations.

The use made of the «ectaxonic» type of hand is described for lorises as an example (NIESCHALK & DEMES, this volume). The use made of mesaxonic hands is detailed by RAWLINS with emphasis on locomotion, and by REDSHAW as well as CHRISTEL, MAIER, and BOESCH & BOESCH, all in this volume) with regard to manipulation.

The most important results are the following: The ectaxonic hand of lorises is mainly used in locomotion. Preferred substrates of locomotion are more or less horizontal branches; preferred direction is along branches (not across). The hand is clasped around the supporting branch like a «pincer», with the digits II -V on one side and the pollex on the other (*Figure 1*). This requires at touchdown a pronounced ulnar abduction (see *Figure 2* in NIESCHALK & DEMES, this volume). During the stance phase, the arm swings as an inverse pendulum about the carpal joint as a pivot.

The pollex in the ectaxonic pincer hand of a loris (*Figure 1*) needs a strong flexor tendon to exert the force necessary for pressing the thumb firmly against the branch on which the animal moves. In the anterior view shown in *Figure 1*, the total reaction force exerted by the fingers in the direction towards the thumb must be equal to the force

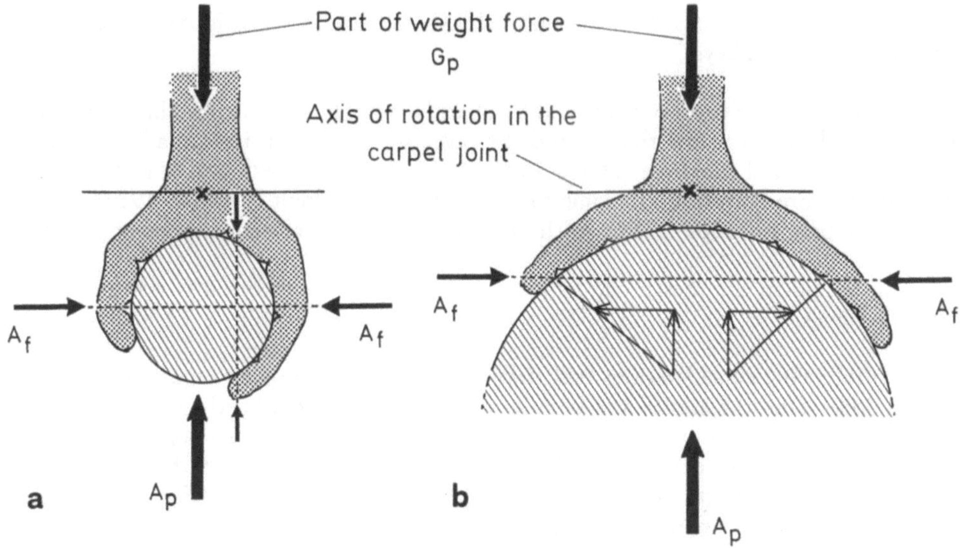

Figure 1 - Anterior views of a primate hand clasped around a branch.
a) on the left: thin branch
b) on the right: thick branch
A certain part of the body weight is carried by the hand (G_p). It evokes a reaction force A_p, of the same size and opposite direction. In addition, the fingers exert forces against the branch (A_f), the direction of which may vary. The horizontal and the vertical components of all forces A_f, produced by the fingers (and thumb) of one hand, add up to zero as long as the branch does not move under the grip.
In b), the forces G_p and A_f combine to the resultant as discussed in CARTMILL (1985).

exerted by the latter towards the fingers (otherwise the seized branch or other object would move medially or laterally irrespective of the weight force). Great length of the pollex is a means to produce a transverse force component that is opposed to and in line with the one exerted by the lateral fingers.

The mesaxonic hand of a macaque is usually placed on a more or less flat surface (ground, big branch). The metacarpals are either placed in or close to the direction of the walk, or touch the ground only with their distal joints. The hand, or only the digits, are «rolled off». In both cases, the pivot around which the forelimb rotates during the stance phase, is formed by the metacarpal heads, and not the carpus. In primates adapted to terrestrial locomotion (vervets, macaques, and more so baboons or patas) the carpus and metacarpus is integrated into the length of the «inverse pendulum» — which so increases stride length as to make locomotion faster (PREUSCHOFT & DEMES, 1984, 1985; WITTE, PREUSCHOFT & RECKNAGEL, 1990; RAWLINS, this volume; see also DEMES, 1991).

A similar situation occurs in knuckle-walking African apes. Here again, carpus and metacarpus are kept and exposed to loads in line with the forearm. Only the positions of the lateral fingers are different from those in ground-living monkeys, being flexed and loaded on their dorsal surfaces in the apes, while hyperextended and loaded on their volar surfaces in the monkeys.

In the «power grip» of pongids and hylobatids, a hook is formed by the lateral digits without involvement of the thumb, and the pivot in locomotion is normally the contact between fingers and substrate (PREUSCHOFT, 1964, 1973; PREUSCHOFT & DEMES, 1984, 1985, there more literature), in some cases also the carpal joint (JENKINS, 1974). Usually the metacarpals are kept in line with the forearm, so increasing pendulum length.

The thumb in the mesaxonic hands of monkeys and apes often is not exposed to external forces.

The following considerations will first propose a reasonable explanation which may help to understand the statics and kinetics of the forearm-wrist-hand-complex, and, as a consequence, the reason why the above-listed traits occur in combination. On this basis the differences between ectaxonic and mesaxonic hands are interpreted; then the evidence which at present supports the mechanical explanation is reviewed. Finally, some possible conclusions about the pathways of evolution will be sketched.

Biomechanics of both types of hands

Grasping between rays I and IV (and III) of an ectaxonic hand increases the width of span (*Figure 2*), and allows to get hold of thicker branches than grasping between rays I and II in a mesaxonic hand. The danger to slip off is also reduced (*Figure 1*, more details in CARTMILL, 1974, 1985).

Figure 3 shows the hand in prehension (as in *Figure 1b*) from above. The metacarpals and the digits in particular appear shortened because of the curved surface of the branch which acts as a substrate. Important for the further analysis is the fact that tendons run nearly parallel to the rays. For the sake of simplicity, the flexion of the digits in prehension is ignored in *Figures 4-7*.

Figure 4 shows schematic drawings of hand skeletons with their flexor tendons. If the flexor muscles are contracted in prehension, their pull acts more or less parallel to the forearm skeleton (M, see also PREUSCHOFT, 1973). At the level of the carpal tunnel, the tendons fan out. The forces they exert on the lateral digits can be combined to one resultant, named D. Another, smaller muscle force, named P, is exerted by the pollical flexor. The resultant of D and P is called R_f. Forces D and P on one side, and M on the other are «actio» and «reactio», therefore of equal magnitude. This is shown in *Figure 5*. That means they are in equilibrium. Because these forces do not pull in exactly opposite directions, they exert redirectional forces F as shown in *Figure 6*.

In the ulnarward abducted ectaxonic hand (*Figure 6b*) this force F_u pushes against the ulnar border of the carpal tunnel. In a radially abducted, or not abducted mesaxonic hand, the corresponding resultant F_r (which is largely produced by the pollical flexor tendon) pushes against the radial border of the carpal tunnel (*Figure 6a*). The structures which form the carpal tunnel are the tubercles of the trapezium and scaphoid on the radial and the

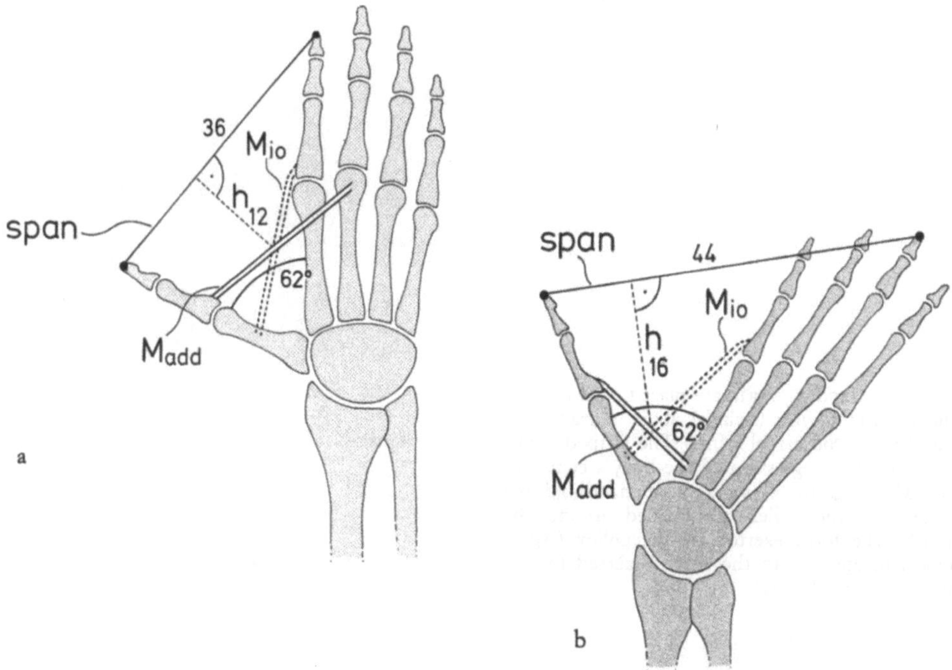

Figure 2 - a) schematic drawing of a mesaxonic (human) hand to show the most important muscles which adduct the first and the second rays, the span of the grip and the «depth» of the grip (h).
b) schematic drawing of an ectaxonic hand in the same position, and of nearly the same dimensions. Only the third finger is shortened, and the fourth elongated by the length of the terminal phalanx. The span width as well as depth are increased by much more than the length difference of the fingers.

hamulus and the pisiform on the ulnar side (*Figure 8*). These structures need to sustain the resultant forces F_u or F_r, respectively.

The same fact can be expressed in the following way: Ulnar abduction of the hand causes a redirecting of all digital flexor tendons (D in *Figures 4 and 5*). The tensile forces in these tendons create a resultant in ulnar direction (F_u) in *Figure 6b*. The point of force application is the ulnar eminence (pisiform and hamulus). This force is bigger than the radially-directed resultant F_r in *Figure 6a* that results from redirecting the long pollical flexor.

At a more proximal level, namely in the antebrachio-carpal joint, the forces F_r (*Figure 7a*) or F_u (*Figure 7b*) act together with the longitudinal pull of the flexor muscles M. The joint resultant (C_u) deviates ulno-proximally from the forearm axis. If this situation is characteristic for the animal (or occurs often), the contact between ulna and carpus must be expanded on the expense of the radial contact, and the ulnar carpal bones must be reinforced.

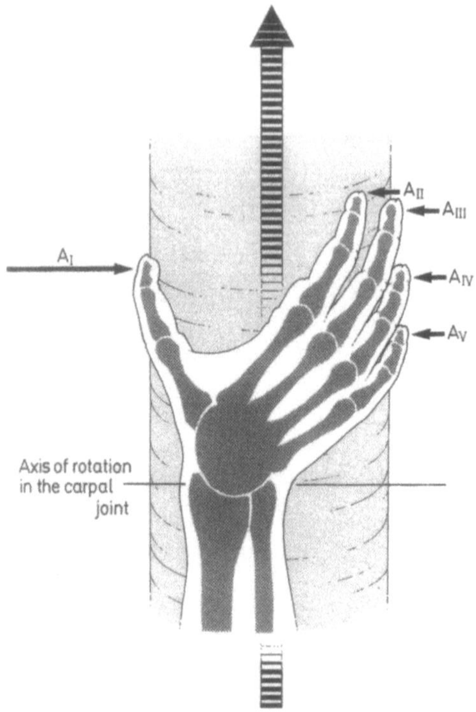

Figure 3 - Hand of a primate placed on a branch. The forearm rotates (swings) about a transverse axis through the carpal joints. The striped arrow illustrates the progression of the body's center of mass while walking. To secure a safe handhold, the fingers and the pollex are clasped around the branch. The force exerted by the pollex (A_I) is equal and opposed to the forces exerted by the lateral fingers (A_{II}-A_V).

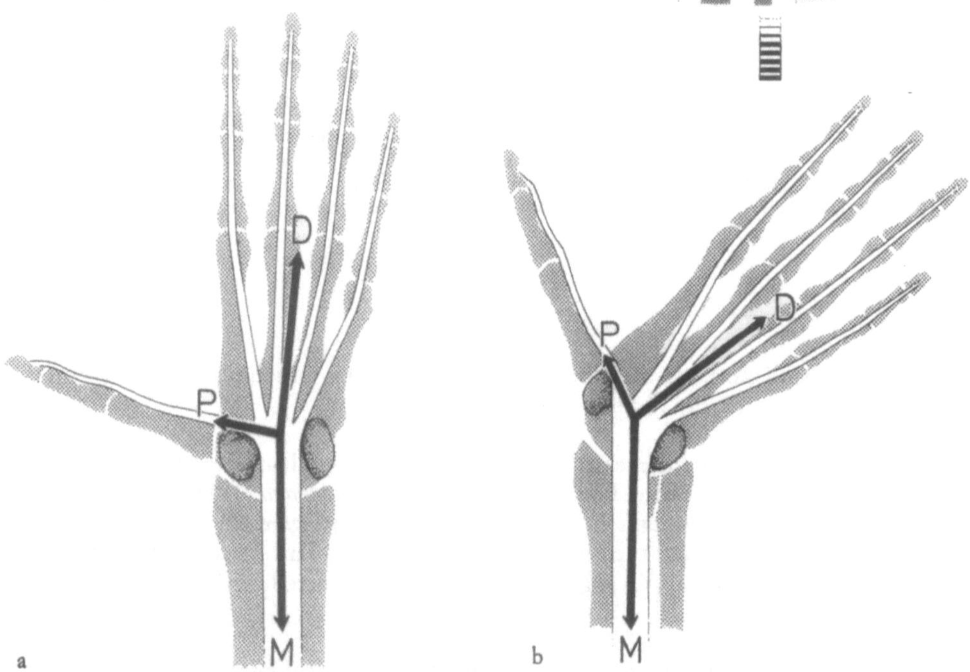

Figure 4 - The grasping hand shown in fig. 3 drawn in a more schematic way. If finger flexion is resisted by a reaction force from the substrate, several forces intersect in the carpal region: Pull of the flexor muscle M. Reaction force of the pollical flexor tendon P and of the digital flexor tendon D.
a) Neutral position as usual in hominids.
b) ulnar abduction as usual in lorisids.
Note the differences between a) and b) which only depends on the position of the carpal joints.

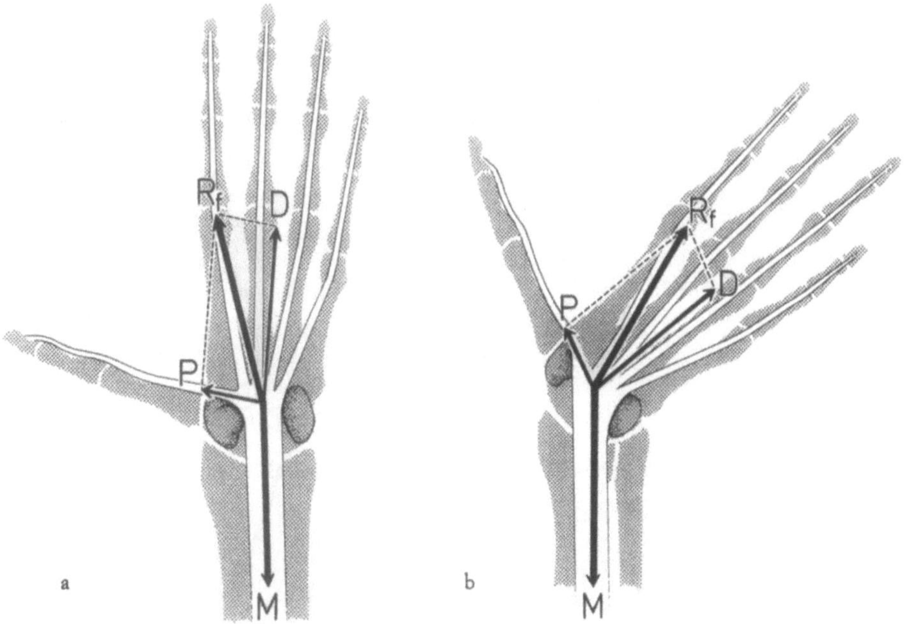

Figure 5 - The reaction forces acting on the digital flexor tendons D and on the pollical flexor tendon P combine to form a resultant R_f. Note the different direction of R_f in both drawings, a) hominids and b) lorisids.

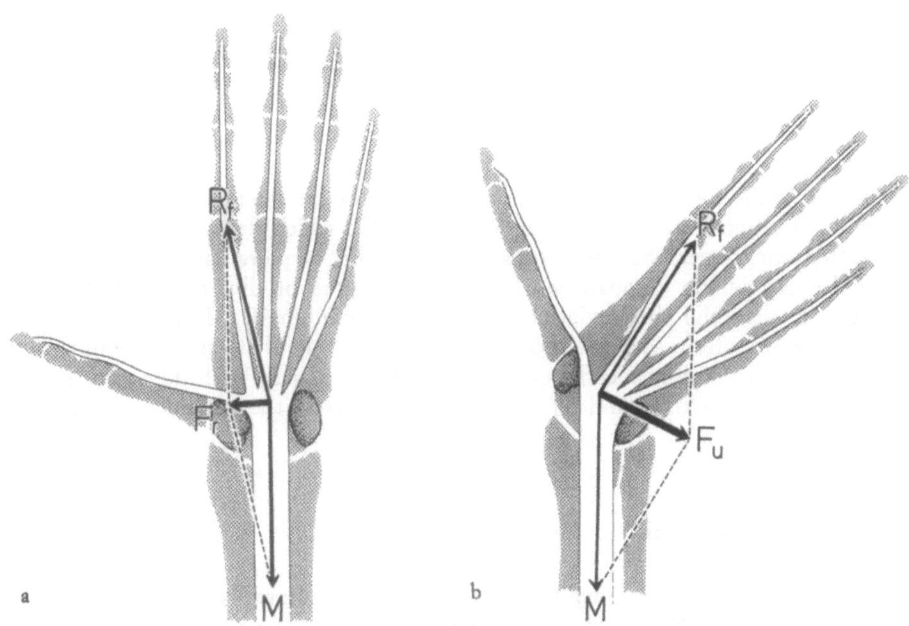

Figure 6 - The pull of the muscle M and its combined reaction forces R_f form a blunt angle. At the radial or ulnar margins of the carpal tunnel, a force F comes into action which is evoked by the change of direction of the tendon forces M and R_f. This force pushes against the respective carpal eminence, and requires a considerable strength of that structure.

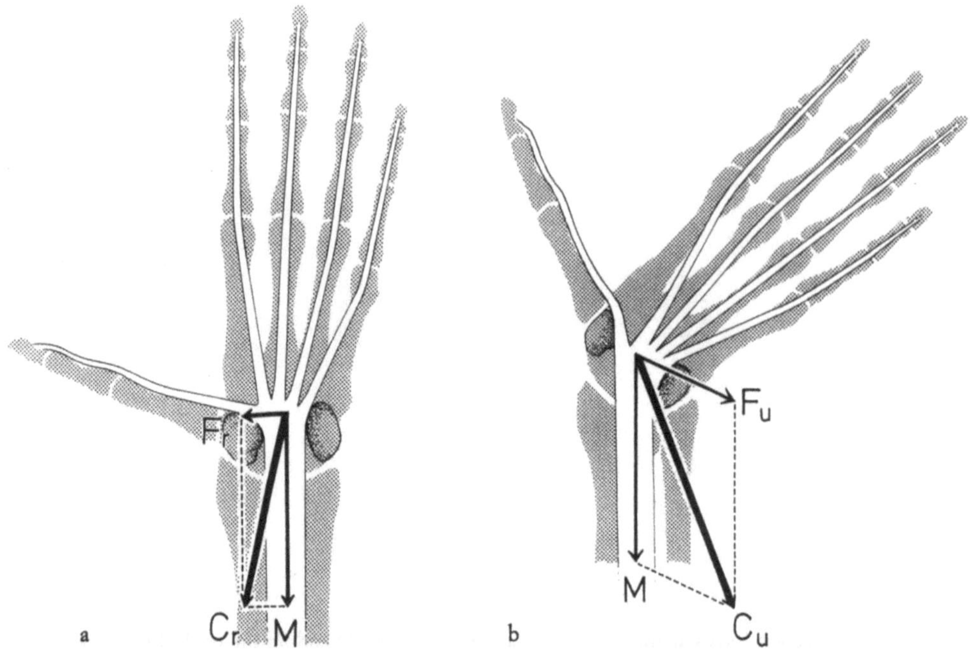

Figure 7 - The joint between the forearm bones and the proximal carpal bones is compressed by the force C. This is the resultant of the pull of the flexor muscle M and the force F (derived from redirection of the reactive forces of the tendons as shown in *Figure 6*). In the neutral position often assumed by humans (or apes, or macaques), the force C_r concentrates on the radius because of the hallux abduction, in ulnar abduction like in a loris hand, the force C_u requires an ulno-carpal contact.

In *Figure 7a* the forces which act on the carpal joint are the muscle force M and the resultant F_r. Both combine to form the joint force C_r, which deviates radio-proximally from the forearm bones. Stresses concentrate on the carpo-radial articulation while relieving the carpo-ulnar compartment. A greater degree of divergence of the pollex accentuates (moderately) the deviation of the joint resultant C_r. If this sort of force transmission occurs often, so that it becomes the typical way of stressing, the ulno-carpal contact becomes dispensible and the carpal bones that articulate with the radius must be strengthened. Both forces, C_r and C_u, deviate markedly from the longitudinal axes of the forearm bones.

Morphological consequences

The mechanical approach unveils the reason why the above listed features of ectaxonic and mesaxonic hands occur *in combination.* Lewis (1985) has pointed out that lorisids are characterized by their carpo-ulnar contact between hamate and the styloid process of the ulna. In hominoids on the other hand, there is a clearcut tendency to enlarge the carpo-radial contact on the expense of the carpo-ulnar. In addition, the radial tubercle as

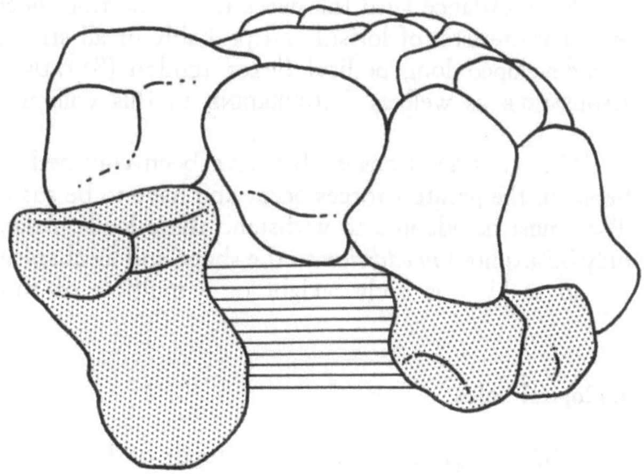

Figure 8 - Proximal view on the carpus of *Cebus* spec. to show the ulnar and radial eminences (stippled areas). The tendons pass through the carpal tunnel (hatched area).

the border of the carpal tunnel, and the hypomochlion for the redirectional force F_r is markedly developed in hominoids. The same combinations can be found not only in modern primates, but also in fossils (GODINOT & BEARD in this volume). The force C_r is nearly perpendicular to the joint surface of the radius. Its tilted position has been mentioned by KOEBKE in this volume. Joint forces, which do not act perpendicularly to the joint surfaces, cause uncontrolled shearing (see PREUSCHOFT, 1969). Low shearing components may be the reason for the development of otherwise unexplained ligaments (as described by KOEBKE in this volume).

In contrast to widespread opinions (textbooks of Anatomy), the transmitted *forces* influence the shape of the articulating surfaces as much as do their *movements.* Load concentrations on those rays which fall in line with the forearm axis lead to a reinforcement of the metacarpals and digits (PREUSCHOFT, 1969, 1973; see also NIESCHALK & DEMES, or KOEBCKE, this volume).

As LANDSMEER emphasizes in this volume (see also RAUWERDINK, this volume; STRAUS 1942; or PREUSCHOFT, 1964) the deep flexor of most, in particular lower, primates is not subdivided into bundles that give raise to the tendons running to individual fingers. Therefore, the force developed by the muscle is distributed on all rays. In *Figure* 4, each tendon is assumed to receive the same share of the total muscle force, which is not necessarily the case. The tendon of the longest finger probably receives the greatest part of the muscle force, because a reaction force applied to its tip creates greater extending moments than in a shorter finger. These moments of the load must be counterbalanced by increased pull of the flexor tendon. If so, the force F_u, derived from redirection of the tendons to the lateral fingers, becomes bigger and is shifted more in an ulnar direction. Extensive speculations about the share of tendon force passed on to individual digits are unnecessary because the influence exerted by the changes of direction of F on the joint force is not very great. The degree of divergence of the pollex also does not exert too great influence. The major tendencies, however, become clear from *Figures 4-7.*

In accordance with the necessity of exerting much force by the thumb (see above), one characteristic of lorisid and probably of all strepsirhine hands is the existence of a well-developed long pollical flexor tendon (STRAUS, 1941, 1942; PREUSCHOFT, 1964; LANDSMEER as well as RAUWERDINK in this volume).

Our considerations so far have been confined to the power grip, because in this behavior the greatest forces occur and have to be sustained by the anatomical structures. They must be adapted to withstand these loads. Adaptations for precision handling can only be acquired *in addition* to the shape and mechanical strength necessary for supporting a load as high as body weight or a multiple of this.

Biological roles

The biological roles of both types of hand diverge, though not entirely:

Primates have to find secure handholds on all supports available in their environment. Trees offer in the form of thin or thick branches supports measuring between 0.2 cm and 25 cm or even more in diameter. The «external» conditions given by their environment are at least similar for all primates, regardless of their sizes. On the other hand, small objects of importance for a primate (grains, insects, ticks, thorns etc.) are more difficult so snatch with a hand being 25 cm long and 15 cm wide (as in a large ape), than with a hand having 1/10 or less of these dimensions. This raises different problems for small and for large animals.

As soon as larger animals intend to get hold of small objects, like insects, or grains, they need to shift from «whole-hand-grip» to the use of only two fingers, that is to perform a precision grip. It may be that this size-dependent phenomenon accounts for the occurence of a «precision grip» only in the larger simiiform primates, but not among the smaller strepsirhines.

In precision handling, the «pincer» is reduced to a «pincette» (forceps), in which two neighboring digits, mostly the (possibly reduced) thumb and only one of the lateral fingers cooperate. It seems evident, that for such a purpose the second ray, adjacent to the pollex, becomes the best opponent of the thumb and has the best chances to be promoted by evolution.

Length of the fingers makes control over them difficult. Therefore forms which tend to precision handling will rarely develop very long digits.

Both hands are equally well suited for vertical clinging — the only condition is that the hand can find a safe handhold. A direct influence of leaping on hand shape cannot be expected, in particular since taking off as well as landing is performed for biomechanical reasons (PETERS & PREUSCHOFT, 1984; PREUSCHOFT, 1985; or DEMES & GÜNTHER, 1989a, b) mostly by the strong and often specialised hindlimbs.

The ectaxonic lorisid hand is well suited and probably adapted for grasping and moving on rather thin branches. In spite of this, it allows a wide span (NIESCHALK & DEMES, GODINOT & BEARD, both in this volume). This is especially important in relatively small animals. The mesaxonic hand, however, is advantageous for animals large enough to have no span problems, and which prefer walking on the ground or along thick branches, or which suspend themselves below branches.

Possible pathways of evolution

Terrestriality favors mesaxony, or at least paraxony, since loading leads to reinforcement of bone. A moderately mesaxonic frontlimb can be transformed into a prehensile hand in two ways: Either by a shift towards ectaxony, or by maintenance of the mesaxony and increasing mobility of the thumb. The success of one or the other direction depends on the the preferred substrates, the direction of movement in relation to them, size of the animal, and so forth. The pincer hand is one of the possible endpoints of adaptation to walking along branches in arboreal life.

If the course of evolution now changes again, back towards terrestriality, a persisting mesaxony would lead to a dog-like or ungulate structure, characterised by progressive shortening of all digits, combined with a strength (and length) increase in particular of the metacarpals. The necessity to use any sort of precision grip enhances the reinforcement of digit II as the one closest to I. This tendency is by no means in conflict with walking quadrupedally on flat ground or on thick boughs, nor with suspensory climbing by means of a hook-hand.

If, in contrast, an already developed ectaxonic hand is the starting point for adaptation toward terrestriality, a continued reduction of ray II and then a shortening of the rays I and III/IV is much more probable than the return to a mesaxonic form of the frontlimb.

References

CARTMILL 1974: *Pads and claws in arboreal locomotion.* In : F.A. Jenkins Jr. (ed.), Primate Locomotion, Academic Press, New York.

CARTMILL M., 1985 *Climbing.* — In: M. Hildebrand, D.M. Bramble, K.F. Liem & D.B. Wake (eds.), Functional Vertebrate Morphology, The Belknap Press of Harvard Univesity Press, Cambridge, Mass. and London, 73-88.

DEMES B., 1991: *Biomechanische Allometrie: Wie die Körpergröße Fortbewegung and Körperform von Primaten bestimmt.* - Courier Forsch. - Inst. Senckenberg, 141, Frankfurt a.M.

DEMES B & GÜNTHER. M.M., 1989a: *Wie die Körpermasse den Springstil von Halbaffen und deren Proportionen bestimmt.* - Z. Morph. Anthrop. 77: 209-225.

DEMES B. & GÜNTHER. M.M, 1989b: *Biomechanics and allometric scaling in primate locomotion and morphology.* - Folia Primatol. 53: 125-151.

JENKINS F., 1981: *Wrist rotation in primates: a critical adaptation for brachiators.* - In: M.H. Day (ed.), Vertebrate Locomotion. Symp. Zool. Soc. London, No. 48, Academic Press, London, 429-451.

PETERS A. & PREUSCHOFT H., 1984: *External biomechanics in leaping in* Tarsius *and its morphological and kinematic consequences.* - In: C. Niemitz (ed.), Biology of Tarsiers, G. Fischer, Suttgart, 227-255.

PREUSCHOFT H., 1964: *Die Nerven der Vorderextremität des Gorilla.* - Ant. Anz. 115: 313-334.

PREUSCHOFT H., 1969: *Statische Untersuchungen am Fuß der Primaten.* - I. *Phalangen und Metatarsalia.* - Z. Anat. Entw. Gesch. 129: 285-345.

PREUSCHOFT H., 1973: *Functional anatomy of the upper extremity.* - In: G.H. Bourne (ed.), The Chimpanzee, Vol. 6, Karger-Verlag, Basel, München und New York, 34-120.

PREUSCHOFT H., 1985: *On the quality and magnitude of mechanical stresses in the locomotor system during rapid movements.* - Z. Morph. Anthrop. 75 (3): 245-262.

PREUSCHOFT H. & DEMES B. 1984: *Biomechanics and brachiation.* - In: H. Preuschoft, W.Y. Brockelmann, D.J. Chivers & N. Creel (eds.), The Lesser Apes. Evolutionary and Behavioral Biology, Edinburgh University Press, 96-118.

PREUSCHOFT H. & DEMES B. 1985: *Biomechanic determinants of arm length and body mass in brachiators.* - Fortschr. d. Zoologie 30: 39-43.
STRAUS W.L. JR., 1941: *The phylogeny of the human forearm extensors.* - Hum. Biol. 13: 23-50, 203-238.
STRAUS W.L. JR., 1942: *Rudimentary digits in primates.* - Quart. Rev. Biol. 17: 228-243.
WITTE H., PREUSCHOFT H. & RECKNAGEL S., 1990: *Human body proportions explained on the basis of biomechanical principles.* - Z. Morph. Anthrop. 78: 407-423.

R.F. Ker

Elasticity of hand and forefoot tendons

The main finger muscles are in the forearm to reduce the volume and mass in the distal part of the limb; their tendons are therefore long and may stretch significantly. The stresses on and extensions of the extrinsic digital tendons were investigated for several mammals: an opossum, a macaque, a dog, a badger, a cow and a horse. For most of the tendons, including those of the macaque, the stresses are small compared to the breaking stress of tendon. The most notable exceptions are the digital flexor tendons of the ungulates, which are more highly stressed. This is associated with the specialisation of the leg for running.
Three consequences of tendon extension are considered. (i) If a muscle is to rotate a joint through a given angle, it will have to contract more to compensate for the extension of the tendon. This needs a larger muscle and wastes energy. (ii) The flexor tendons of forefeet act as springs for energy saving in running. The energy to stretch these tendons comes not from their own muscles but from mechanical energy changes associated with locomotion. (iii) The stretching of tendons is an important influence in the reflexes involved in the control of position and force. The extensibility of tendons may increase the precision of force control.

Key words: extrinsic muscles, length, mass reduction, stiffness, strength properties, stress, strain, tensile properties, tendon stretching.

Introduction

Many of the muscles operating the fingers have impressively long tendons, which allows the main volume and mass of the muscles to be situated in the forearm at a convenient distance from the digits. For example, the superficial digital flexor and the digital extensor both originate on the humerus and insert on the phalanges. The length of the muscle fibres represents only a small part of the distance between origin and insertion; the rest is tendon. Because they are long, the tendons may stretch significantly in use; my purpose is to consider the magnitude and functional consequences of any stretching. Examples will be taken from a range of mammals to compare the tendons of primates with those of other mammals and to explore for specialisation in tendon design.

I will not be considering the intrinsic hand muscles which, in primates and many other mammals, have short tendons relative to the length of the muscle fibres. The extrinsic muscles taken together are much larger than the intrinsic muscles: for the human, the extrinsic muscles constitute 87% of the total mass of hand muscle (from data in BRAND, BEACH & THOMPSON, 1981).

Tensile properties of tendons

All collagenous tendons show generally similar mechanical properties when subjected to longitudinal tension. A plot such as *Figure 1* may be obtained by clamping a tendon in a tensile testing machine and measuring load and extension during stretching (KER, 1981).

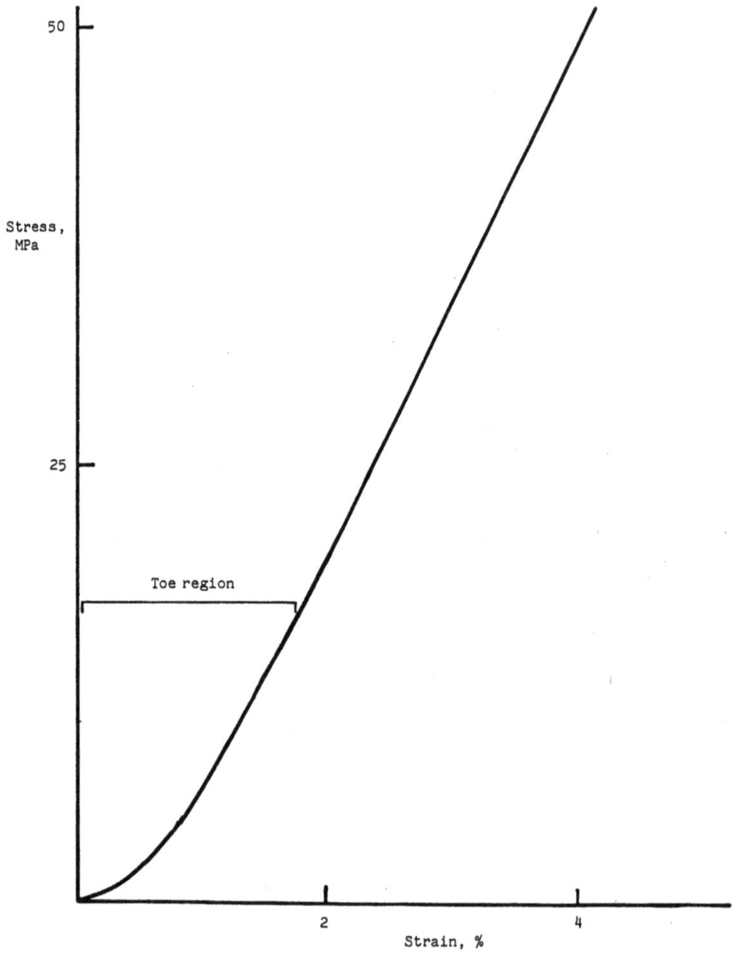

Figure 1 - The stress-strain curve for a typical mammalian tendon.

To allow for the dimensions of the particular specimen, the quantities plotted in *Figure 1* are *stress* (= load/cross-sectional area) against *strain* (= extension/length). The plot is linear at higher stresses (above about 20 MPa and close to linear from, say, 10 MPa upwards), but at low stresses it is curved. The curved portion is called the 'toe region' and is characteristic of collagenous tendons. The slope of the graph (called the *tangent modulus*) is a measure of the stiffness of the material. BENNETT, KER, DIMERY & ALEXANDER (1986) have compiled the available data for tangent moduli of mammalian tendon in the linear region and found that all the values lie between about 1 GPa and 2 GPa with a mean of 1.5 GPa. Some of the tendons covered by BENNETT *et al.* (1986) are from forefeet. In addition, I have made a measurement on a cow digital extensor tendon, which gave a tangent modulus of 1.4 GPa. Figure 1 was derived from this measurement, but the scales on the axes have been slightly altered so that the slope is 1.5 GPa. *Figure 1* can therefore be considered as representing a typical tendon.

TABLE 1 - *Mechanical properties of some biological materials used in tension. The values have been selected to give a general idea of properties; anything more precise would require a full definition of the specimens and of the test conditions. The tangent moduli (first column) are from the linear region of the stress-strain plot. For mammalian tendon this is at higher stresses (Figure 1): for the other materials, a (more-or-less) linear region starts at the origin. The second column gives estimates of the stress at which the function of each material is impaired: for wool and bone this is a yield stress, but for the other materials it is a fracture stress.*

Material	Tangent modulus GPa	Stress at which damage occurs MPa	References
Mammalian tendon	1.5	100	1
Wool	4	60	2
Silk (spider dragline)	4	600	3
Insect cuticle (tendon)	13	400	4
Mammalian long bone	20	140	2
Wood	10	100	2

References: (1) BENNETT *et al.* (1986); (2) WAINWRIGHT, BIGGS, CURREY & GOSLINE (1976); (3) DENNY (1980); (4) KER (1977).

To put this in perspective, *Table 1* gives the tangent moduli for several biological materials used for transmitting tension. The stiffness of tendon is relatively low. *Table 1* also gives estimates of strength in terms of the stress at which the material is damaged. Tendon is not notably strong, but, nonetheless, it can be stretched further than the other materials (except for spider dragline), because of its relatively low stiffness and because of the toe region. Achieving maximum stiffness and strength do not appear to have been dominant evolutionary pressures in the design of mammalian tendon. Tendons are *not* inextensible strings. Of course, in the evolutionary design of a particular mammal, tendons acquire a thickness appropriate for the compliance and weakness of the material.

Stretching the tendon of the human flexor pollicis longus

I will take the flexor of the thumb as an example because the necessary quantitative anatomical data are available (RACK & ROSS, 1984). Consider the thumb used as a hook supporting a load at its end. At the symposium, I hung my briefcase from the end of my thumb using a loop of cord. The angle of the terminal interphalangeal joint was about 150°, which was the angle used by Rack and Ross in their experiments. To maintain this position the flexor pollicis longus must be under tension. *Figure 2* shows the muscle and its tendon inserting on the loaded terminal phalanx. The muscle is unipennate with its origin on the radius bone. The tendon spreads as an aponeurosis over the muscle belly: about 115 mm length of tendon lies beyond the muscle and about an equal length is in the form of the aponeurosis. The fascicles of the tendon, therefore, range in length from 230 mm to 115 mm. In the calculations below, I will take the average length of the tendon as 170 mm.

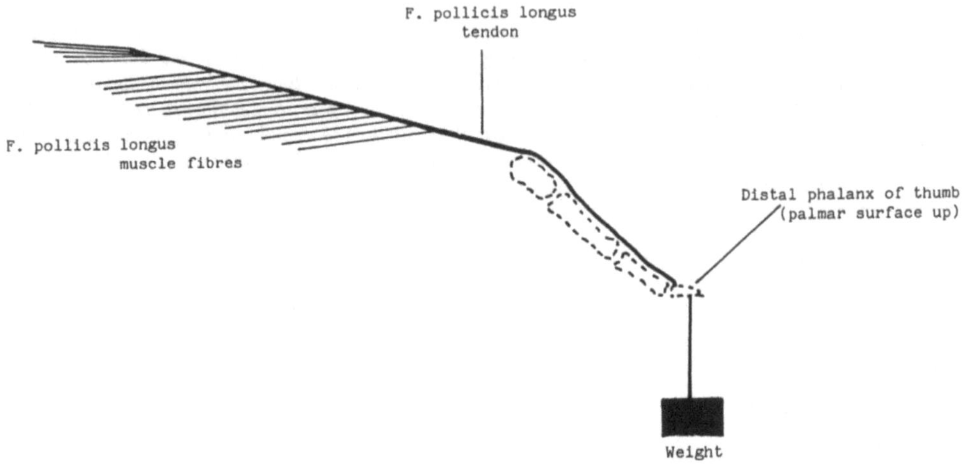

Figure 2 - The human flexor pollicis longus under tension (After RACK, 1985).

The tension, T, in the tendon is calculated by balancing moments about the axis of the terminal interphalangeal joint. The cord holding my briefcase (weight 68 N) was estimated to have a moment arm of 16 mm about this axis. The moment arm of the tendon about the axis is given by RACK & ROSS (1984) as 7.7 mm and, therefore,

$$7.7\,T = 16 \times 68$$

and

$$T = 140 \text{ N}$$

Taking the cross-sectional area to be 9.1 mm² (RACK & ROSS, 1984), the stress in the tendon is 140/9.1 = 15 MPa. This is the maximum value reached by Rack and Ross in their experiments — which is why I chose a briefcase weighing 68 N!

From *Figure 1*, the corresponding strain is about 1.7% and the 'average' extension is 2.9 mm. This is significant in comparison with the moment arm of 7.7 mm: a movement of the tendon of 2.9 mm corresponds to the joint turning through an angle of about 21°. The muscle, whose fibres have a cross-sectional area of about 380 mm² (calculated from data in BRAND *et al.*, 1981), also bears the load of 140 N and the stress in the muscle fibres is therefore 140/380 = 0.37 MPa. (This calculation ignores the non-zero angle of pennation, as will be explained in the next section.) This stress is at, or beyond, the expected maximum isometric value for a muscle and seems likely to be an overestimate. The data may not be entirely consistent as it comes from different sources, so the results of these calculations cannot be taken to be precise. The intention is to give an idea of the magnitudes involved and to show how the extension of the tendon comes about.

In this example, the tension in the tendon has been estimated from the applied load and the known geometry. For most tendons, information to assess the load on them in use is not available and to obtain a broad view of the stresses to which tendons are subjected it is necessary to proceed from the other end. Instead of calculating the muscle stress, the maximum probable muscle stress can be assumed and the maximum tendon stress found from that. This method requires only anatomical measurements and enables a wide range of tendons to be surveyed. The results of such measurements are presented in the next section.

Maximum stresses in digital flexor and extensor tendons

The work to be presented in this section was carried out jointly by the author, Professor Neill ALEXANDER and Dr Mike BENNETT.

A tendon is in series with its muscle fibres and so both tendon and muscle are subjected to the same tension. (Note: If the angle of pennation is θ, the ratio of the tensions is $\cos\theta$. For mammalian muscle, $\theta < 30°$ (ALEXANDER, 1981) and therefore $0.87 < \cos\theta < 1$. This factor will be neglected.) With equal tension, the ratio of stresses is equal to the reciprocal of the ratio of cross-sectional areas of the muscle fibres and of the tendon. I will assume below that the maximum stress in the muscle is 0.3 MPa, which is a commonly used value (WELLS, 1965; JAYES & ALEXANDER, 1982). (This is best considered as a value for the maximum isometric stress: the maximum possible stress is greater and is achieved when an active muscle is being rapidly stretched. This possibility should be born in mind when considering the absolute values for tendon stresses given below in *Table 2*). In summary:

$$\text{Maximum stress in the tendon} = 0.3 \times \frac{\text{area of muscle}}{\text{area of tendon}}. \qquad \text{(equ. 1)}$$

Materials

The mammals used were: an opossum *Didelphis virginialis* (mass 3.5 kg), representing 'primitive' mammals; a monkey *Macaca nemestrina* (mass 15 kg); a Dalmatian dog *Canis familiaris* (mass 21 kg); a badger *Meles meles* (mass 10 kg); and two ungulates, a horse *Equus caballus* (mass 570 kg) and an adult cow *Bos taurus* of unknown mass. The reasons for death were not related to this research. The carcasses, or sometimes the forelimbs only, were stored at −20 °C until required.

The results to be reported here come from measurements on all the extrinsic digital flexors and extensors of the forelimb. It would have been of interest to include the wrist muscles. However, many wrist muscles lack a usable length of tendon external to the muscle belly on which to measure the cross-sectional area. (Note. This does not mean that the tendon, as a whole, is necessarily short. The fibres of wrist flexor muscles are often very much shorter than the belly length and the tendon, in the form of aponeuroses, is correspondingly long.)

Methods and calculations

Fresh (i.e. unfixed) muscles were dissected out and the following measurements were made.

(i) Overall distance from origin to insertion, L_o.

(ii) Mass of the muscle, M_m.

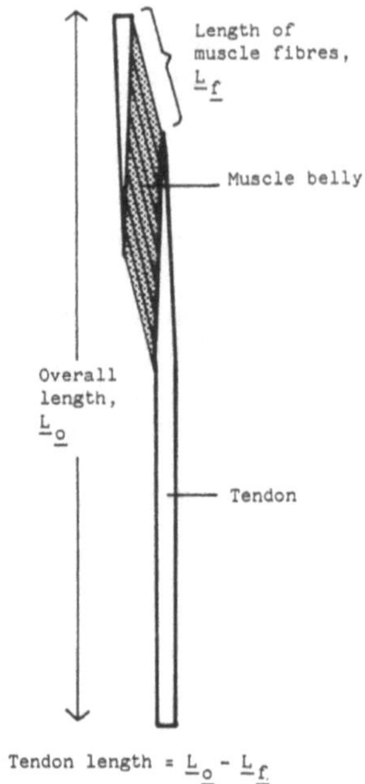

Figure 3 - Illustration of the estimation of the length of a tendon. In estimating the length of the tendon as the difference between the overall length and the length of the muscle fibres, the influence of the angle of pennation, θ, has been neglected because $\cos\theta \simeq 1$. (In this diagram $\theta = 15°$: $\cos 15° = .97$.) (After KER *et al.*, 1988).

(iii) Mass, M_t, of a measured length, L_t, of tendon. The chosen length was selected as having reasonably uniform cross-section.

(iv) Length of muscle fibres, L_f. This measurement required a cut through the belly of the muscle in the plane of the muscle fibres. Measurements were checked at a number of positions in the muscle. There is a risk of the muscle fibres becoming distorted and therefore the muscle fibre lengths were also measured on the other limb, which had been injected with fixative prior to dissection.

The length of tendon is $(L_o - L_f)$. *Figure 3* explains this calculation and, schematically, shows the way in which the cross-sectional area of the aponeuroses reduces deeper in the muscle. (RACK & WESTBURY, 1984, show, for the cat soleus, that the whole of the tendon behaves in the same way as the part which is external to the muscle belly.) When the muscle originates on bone, an 'average' tendon length was taken, as with the flexor pollicis longus in the previous section.

$$\text{The cross-sectional area of muscle fibres} = \frac{M_m}{1060 \times L_f}. \qquad \text{(equ. 2)}$$

In SI units, this assumes the density of muscle to be 1060 $kg.m^{-3}$ (MÉNDEZ & KEYS, 1960). For most of the muscles, the fibre length was fairly nearly constant throughout any one muscle head. The humeral head of the deep digital flexor of the cow and the horse has two parts, the fibre lengths in one part being ten times those in the other. In these cases, we had to cut between the two parts to weigh them separately, in order to obtain a proper value for the total cross-sectional area.

$$\text{The cross-sectional area of tendon} = \frac{M_t}{1120 \times L_t}. \qquad \text{(equ. 3)}$$

In SI units, this assumes the density of tendon to be 1120 kg. m^{-3} (KER, 1981).

Results

Table 2 compares the maximum stresses to which flexor and extensor tendons of the digits can be subjected. The tendons have been grouped into three assemblies, the superficial and deep digital flexors and the digital extensors. These broad groupings simplify comparisons, because each digital extensor or flexor tendon can be unambiguously placed in one of the assemblies, whichever animal it comes from. Table 2 gives the range of stresses found within each assembly for the respective mammals.

TABLE 2 - *Maximum stresses in MPa for the extrinsic flexor and extensor tendons of the digits. The values are derived from the ratio of the muscle to tendon cross-sectional areas and the assumption that the maximum muscle stress is 0.3 MPa. In several cases, a range of values is given to cover the results obtained for several different tendons within the muscle assembly listed.*

Tendon	*Didelphis*	*Macaca*	*Canis*	*Meles*	Ungulates
F. digitorum superficialis	9	14-30	24	33	41-75
F. digitorum profundus	22	20	27	9	35-40
Extensor digitorum	8-14	6-18	6-16	8-11	6-17

The number of measurements made for a given muscle assembly with a given animal depends on the details of the muscular anatomy. To make a separate measurement, a tendon must be unambiguously associated with its muscle belly. This can be done when the relevant volume of muscle is wrapped in its own connective tissue and it can, therefore, be separated from its neighbours during dissection using a blunt instrument. To illustrate this, I will describe the situation in the macaque.

The macaque has four superficial digital flexor tendons inserting on their respective fingers. Each of these is the tendon of insertion of a separable muscle belly and three of these muscle bellies also have tendons of origin on which measurement is possible. Thus, for the macaque superficial flexor tendon, seven separate stress values were found. These are (in MPa) 14, 15, 15, 17, 20, 27 and 30 giving a 'range' of 14-30.

In contrast, the deep flexor tendon of the macaque has only two separable muscle bellies. The tendon from digit V and its muscle belly can be separated from the rest. The other four tendons had to be considered together and relative to the whole of the attached muscle. No difficulty arose in applying equation (2) because the muscle fibre length was the same throughout. In applying equation (1), the 'area of tendon' was taken as the sum of the cross-sectional areas of all four tendons, giving an average value for the stress in the tendons. Two values were thus available for this particular assembly of tendons: they happened to be the same and, therefore, only a single value is given in *Table 2.*

The digital extensor assembly, covering all five digits, gave a total of 7 separate stress measurements. In MPa, these are 6, 8, 12, 13, 16, 18; a 'range' of 6-18.

The number of tendons and the extent to which their muscles can be separated differs among the species. A full description would be lengthy, so, to present general features of the results, it seems preferable to use the 3 broad and unambiguous groupings.

Discussion

The values for the maximum stresses to which tendons are likely to be subjected (*Table 2*) vary from 6 to 75 MPa. Most values are fairly low in this wide spread, so most tendons have a much larger cross-sectional area than is necessary to achieve sufficient strength.

It may be that tendons which are never highly stressed are less strong than those which are subjected to higher stresses. To investigate this possibility, we carried out tensile tests on one of the extensor digitorum communis tendons from the cow and one from the horse. The tendons chosen were of a size appropriate to the tensile testing equipment we have available (BENNETT *et al.*, 1986). The cow tendon broke at a stress of 77 MPa and the horse tendon at 104 MPa, which compares with maximum stresses in the tendons from their muscles of 8 MPa and 15 MPa respectively. Tensile tests tend to give a low value for the ultimate tensile stress because of stress concentrations in the clamped regions. For the experiment with the horse tendon 'frozen clamps', constructed according to the design of RIEMERSMA & SCHAMHARDT (1982), were used and it is possible the measured value is close to the true ultimate tensile stress of the material. Anyway, these tendons have far greater strength than is necessary. On the other hand, the superficial flexor of the horse, which is loaded by its muscle to 75 MPa according to our measurements, has only a modest safety factor.

All the extensor tendons bear only low stresses and are never stretched beyond the toe region, so that the stress-strain curve in the working range is markedly non-linear (*Figure 1*). Many of the other tendons are also never stressed much beyond the toe region. The digital flexor tendons of the ungulates are the main group of tendons which can be highly stressed by their muscles. This is presumably related to the specialisation of the leg for running, as will be explained below.

Extension of tendons

Even when the stresses are fairly small, the extensions can be significant relative to the displacement of tendons during the rotation of joints. The moment arms of most joints are not known; instead, extensions will be compared to the length of the muscle fibres (*Table 3*). Muscles may be able to contract by about 50% in use, though, at least when substantial stresses are being developed, the working range is likely to be less than this (DIMERY, 1985). With a muscle contraction of 50%, when the ratio of muscle fibre length to the possible extension of the tendon given in *Table 3* is 2, the muscle and tendon (when

TABLE 3 - *Ratio of muscle fibre length to maximum tendon extension for the flexors and extensors of the digits.*

	Didelphis	*Macaca*	*Canis*	*Meles*	Ungulates
F. digitorum superficialis	13	4-12	2	2	0.1-0.3
F. digitorum profundus	18	11	4	18	0.4-0.6 & 4-6*
Extensor digitorum	9-21	10-18	5-7	21-40	6-10

* Two ranges are given for the deep flexors of the ungulates because the muscle consists of two portions with fibre lengths differing by a factor of 10 times.

stretched) have equal effect on joint angles. Even with a ratio substantially larger than 2 the stretching of the tendon may be significant; for the human flexor pollicis longus the ratio is 13 and the stretching of the tendon was shown above to be equivalent to a significant change in joint angle. With a ratio much less than 2 (ungulate flexors), the possible change in length by the tendon is far greater than by the muscle; the muscle has, functionally, been replaced by its tendon. The advantage of this for running will be explained below.

Table 3 in the paper by ALEXANDER in this volume gives related information, from a slightly different viewpoint, on the significance of muscles in determining joint displacements. ALEXANDER'S *Table 3* also shows the relative length of the digital flexor muscles to decrease in the sequence primates — fissipedia — ungulates. The same specimens of macaque and dog are referred to in ALEXANDER'S *Table 3* and in *Table 3* here.

Consequences of tendon extension

EXTRA MUSCLE-MASS

Most tendons have a far greater cross-sectional area than is required for strength. Why are they so stout? If the function of the muscle is to change joint angle, stretching of the tendon represents a non-functional extra displacement. The muscle must be larger (with greater resting-length of the fibres) to take up this displacement. Consider again the data and results given above in the section headed 'Stretching the tendon of the human flexor pollicis longus'. If this tendon was halved in cross-sectional area, which would leave ample strength, a mass of 0.9 g would be saved. The maximum stress would be increased from 15 to 30 MPa and, from *Figure 1*, the strain would increase from 1.7% to 2.7% and the extension would increase by 1.7 mm. The muscle probably contracts through a maximum of 15 mm and therefore, with the thinner tendon, the muscle would have to be increased in size by a factor of about 16.7/15 = 1.11, to achieve the same displacement at the joint. The extra mass of muscle would be about 2.0 g. The opposite change of doubling the cross-sectional area of the tendon, would require 1.7 g more tendon and 1.3 g less of muscle. The tendon of the human flexor pollicis longus is probably about as stout as it should be to minimise mass.

The main reason for having a long tendon at all is to redistribute the mass (and volume) even though this is at the cost of the mass of the tendon. However, reduction of the total mass, for a given length of tendon, may be one of the reasons for having tendons

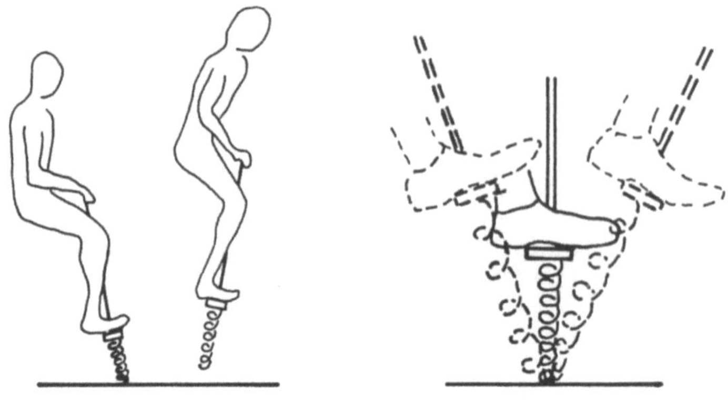

Figure 4 - (a) A child on a pogo stick. (b) The situation in 'mid-step', with the spring storing energy, is shown by the diagram with full lines: the positions before and after are shown by dashed lines. (After ALEXANDER, 1984.)

which are thicker than is needed on grounds of strength alone. This possibility is considered more fully by Ker, ALEXANDER & BENNETT (1988).

ENERGY LOSS

Stretching a tendon requires energy and this energy must be provided each time a load is applied. If you lift a suitcase, you waste energy stretching your digital flexor tendons; the thinner the tendons, the more they stretch and the more energy is wasted. The energy required to stretch a unit volume of tendon is given by the area under the stress-strain curve.

RUNNING

Large mammals use the extensibility of their tendons to save energy when running (Alexander, 1984). The principle is that of a child on a pogo stick (*Figure 4*). During the first half of the period in contact with the ground, the spring compresses storing energy which comes from a decrease in gravitational potential energy and of kinetic energy. During the second half of the period in contact with the ground the energy flows are reversed. The mammal uses its leg flexor tendons as the springs, which are stretched to store energy. Tendons are good springs and are capable of returning some 93% of the energy used to stretch them (KER, 1981). The energy used for running is much less than would be the case if all the energy changes were accomplished via muscles. In the horse, the legs have become highly evolved for running. The superficial flexor muscle of the hind leg has virtually vanished so that the tendon runs from bone to bone. In the foreleg, the process has not gone quite so far but, nonetheless, the muscle fibres of the superficial digital flexor are extremely short; about 4mm. The tendon can be stressed to about 75

MPa (*Table* 2) indicating a strain of 5.6% and an extension of 47 mm, which gives a ratio of muscle fibre length to tendon extension of about 0.1 (*Table 3*). DIMERY, ALEXANDER & KER (1986) estimated the extension in the superficial flexor tendon during running to be of this magnitude. Their method was based on the analysis of films of a running horse together with anatomical observations.

If the forelimb has more varied tasks than just running, as in the monkey, the opossum and even the dog, longer muscle fibres are required. The presence of relatively larger muscles does not prevent the tendons being used as energy stores to save energy during running. However, the net saving will not be quite as great because the muscles require energy for activation and to maintain tension even if they are remaining at constant length and therefore not contributing to the mechanical energy balance. It seems probable that this energy saving mechanism is used by humans (hindlimb) and other primates (possibly forelimb as well), though the details are not known. CAVAGNA, HEGLUND & TAYLOR (1977) provide evidence for significant elastic energy storage in the running of springhares, kangaroos, dogs and monkeys. This conclusion is based on a comparison of the metabolic energy used in running with the observed mechanical energies. Small animals do not seem to use tendon elasticity to save energy during running. BIEWENER, ALEXANDER & HEGLUND (1981) found this to be the case with the kangaroo rat: the tendons appear to be too stout to be useful energy stores.

The flexor muscles of the badger are interesting. The superficial flexor is highly stressed; the deep flexor is subject to very much less stress (*Table* 2). The forepaw has two contrasting major functions: (i) in locomotion for which a large muscle is not required but an energy store can be useful; (ii) for digging, which requires a large muscle, and for which stretching of tendon would be wasteful. It is tempting to suggest that the superficial flexor tendon may be primarily used for running and the deep flexor tendon for digging.

CONTROL OF THE DIGITS

RACK & ROSS (1984) and RACK (1985) investigated how the reflex control of the position of the digits is influenced by having an extensible tendon. They used the flexor pollicis longus of the human. The muscle spindles, which are sensors in the feedback loop, do not 'see' the correct position of the joint when the tendon is stretched. More subtly, having an elastic element (the tendon) in series with a viscoelastic element (the muscle) results in a phase shift of the muscle length relative to the joint position, when an oscillating load is applied. Investigations into reflexes often use oscillating loads and the phase shift has a very significant effect on the interpretation of the results.

RACK & ROSS (1984) point out the distinction between the control of position and of load. For control of *position*, the tendon should be stiff but, in controlling the *load*, limited stretchability of the tendon is advantageous This is like the brake pedal of a vehicle; if the travel of the pedal is too small it is difficult to brake smoothly and accurately. When the load is very low, a highly stretchable tendon would be particularly helpful and would improve the performance of delicate tasks. To carry the speculation a stage further, the precision of control would be the same at all loads if the slope of the graph of load against extension was always proportional to the load. This means load should vary exponentially with extension. *Figure 5* shows the toe region of *Figure 1* together with an exponential curve fitted at 2 points. Clearly, an exact fit is not to be expected since an exponential does not go through the origin. Apart from this, at low loads, the slope of the stress-strain curve is greater than that of the exponential: this is not

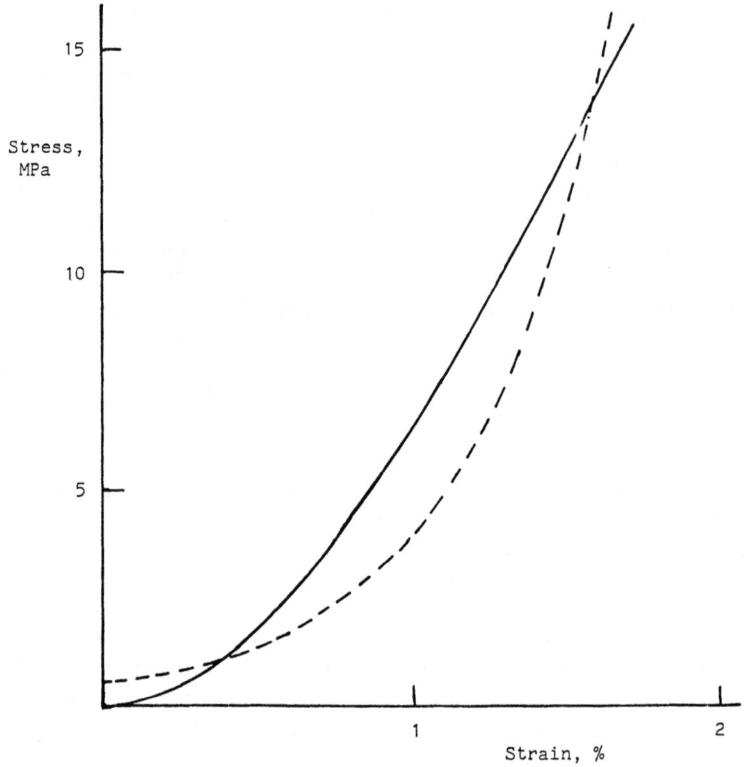

Figure 5 - *Full line*: The toe region of the stress-strain curve for mammalian tendon. *Dashed line*: An exponential curve fitted at two arbitrary points.

surprising because, in many other tasks, an excessively extensible region would be non-functional. Having tendons with a toe region may aid the precision of load control in delicate tasks; but this cannot be the only reason for having a toe region. All collagenous tendons (and ligaments) show a toe region, but not all are used in delicate tasks. RACK & ROSS (1984) suggest a possible mechanism for reducing the disadvantage of a stretchable tendon in the control of position. When a joint is to be positioned precisely, the flexors and extensors may be simultaneously taut, opposing each other; each is stressed to a higher point on the stress-strain curve at which the stiffness is greater.

Conclusion

Most digital tendons, including those of primates, are subjected to stresses which are far less than the breaking stress of tendon. Presumably, the need to limit the extension of the tendons is the main factor in determining their cross-sectional area. Nonetheless, the extension remains sufficient to be significant: it requires extra muscle mass and wastes energy; in the flexor tendons, which tend to be more highly stressed, it allows energy storage to aid running; it affects the action of muscle spindles in controlling position and force at the fingers.

The mechanical design of primate digital tendons is not specialised. The tendons are similar to those in a 'primitive' mammal (the opossum) with regard both to stresses and to extensions relative to the muscle fibre length. The primate hand can be put to a wide variety of uses; this is, in part, a consequence of the lack of specialisation of the design.

References

ALEXANDER R. McN., 1981. *Mechanics of skeleton and tendons.* In: V.B. Brooks (edit.) *Handbook of Physiology. The Nervous System.* 2: 17-42. American Physiological Society, Bethesda.

ALEXANDER R. McN., 1984. *Elastic energy stores in running vertebrates.* Am. Zool. 24: 85-94.

BENNETT M.B., KER R.F., DIMERY N.J. & ALEXANDER R. McN., 1986. *Mechanical properties of various mammalian tendons.* J. Zool., Lond. (A) 209: 537-548.

BIEWENER A., ALEXANDER R. McN. & HEGLUND N.C., 1981. *Elastic energy storage in the hopping of kangaroo rats (Dipodomys spectabilis).* J. Zool., Lond. 195: 369-383.

BRAND P.W., BEACH R.B. & THOMPSON D.E., 1981. *Relative tension and potential excursion in the muscles of the forearm and hand.* J. Hand Surg. 6: 209-219.

CAVAGNA G.A., HEGLUND N.C. & TAYLOR C.R., 1977. *Mechanical work in terrestrial locomotion: two basic mechanisms for minimizing energy expenditure.* Am. J. Physiol. 19(2): 249-256.

DENNY M.W., 1980. *Silks — their properties and functions.* In: J.F.V. Vincent and J.D. Currey (edit.). The Mechanical Properties of Biological Materials. 247-272. Cambridge University Press, Cambridge.

DIMERY N.J., 1985. *Muscle and sarcomere lengths in the hind limb of the rabbit (Oryctolagus cuniculus) during a galloping stride.* J. Zool., Lond. (A) 205: 373-383.

DIMERY N.J., ALEXANDER R. McN. & KER R.F., 1986. *Elastic extension of leg tendons in the locomotion of horses (Equus caballus).* J. Zool., Lond. (A) 210: 415-425.

JAYES A.S. & ALEXANDER R. McN., 1982. *Estimates of mechanical stresses in leg muscles of galloping greyhounds (Canis familiaris).* J. Zool., Lond. 198: 315-328.

KER R.F., 1977. *Some structural and mechanical properties of beetle and locust cuticle.* Ph. D. Thesis, University of Oxford.

KER R.F., 1981. *Dynamic tensile properties of the plantaris tendon of sheep (Ovis aries).* J. Exp. Biol. 93: 283-302.

KER R.F., ALEXANDER R. McN. & BENNETT M.B., 1988. *Why are mammalian tendons so thick?* J. Zool., Lond. 216: 309-324.

MÉNDEZ J. & KEYS A., 1960. *Density and composition of mammalian muscle.* Metabolism 9: 184-188.

RACK P.M.H. & ROSS H.F., 1984. *The tendon of flexor pollicis longus; its effects on the muscular control of force and position at the human thumb.* J. Physiol. 351: 99-110.

RACK P.M.H., 1985. *Stretch reflexes in man: the significance of tendon compliance.* In: W.J.P. Barnes and M.H. Gladden (edit.) Feedback and Motor Control in Invertebrates and Vertebrates. Croom Helm, London.

RIEMERSMA D.J. & SCHAMHARDT H.C., 1982. *The cryo-jaw, a clamp designed for in vitro rheology studies of horse digital flexor tendons.* J. Biomechan. 15: 619-620.

WAINWRIGHT S.A., BIGGS W.D., CURREY J.D. & GOSLINE J.M., 1976. *Mechanical design in organisms.* London: Edward Arnold Ltd.

WELLS J.B., 1965. *Comparison of mechanical properties between slow and fast mammalian muscles.* J. Physiol., Lond. 178: 252-269.

E. Doden

The relationship between the function and the inner cortical structure of metacarpal and phalangeal bones

The relationship between the function and the inner cortical structure of metacarpal, proximal and middle phalangeal bones was compared between gibbon and man. Cross sections, 1mm and 0.1mm thick, were taken from five shaft sites of the metacarpals and three shaft sites of the phalanges. Photon densitometric analyses were carried out on the 1mm thick cross sections, and the 0.1mm thinn cross sections were examined by microradiography and under circular polarized light.
Shape, density distribution and structural organisation, especially the orientation of the collagen fibres in the analysed bones, reflect more or less the functional adaptation to locomotion and gripping, as were demonstrated by calculations of Preuschoft (1973), who showed that the metacarpal and the phalangeal bones are subjected to bending stress in a dorsovolar plane mainly in a dorsoconcave direction and in a radioulnar plane as well in radioconcave as in ulnarconcave bending.

Key words: bone density, collagen fibres, osteons, cross sections, functional adaptation, bone structure, microradiography.

Introduction

In early primates the development from terrestrial quadrupedalism to arboreality has led to altered functions of the forelimbs. They changed from a supporting or locomotor organ to an organ of prehension. These alterations of function are reflected in morphological features of the skeleton, especially of the hand. Biomechanically the hand can be seen as a linkage system of intercalated bony segments (ALEXANDER, this volume; AN *et al.*, 1979). The motion of this system is made possible and is controlled by a complex network of muscles, tendons, ligaments and aponeurosis, which surrounds the metacarpals and the phalanges (LANDSMEER, this volume). In the quadrupedal mammals the movements of the hand as a locomotor organ consist mainly of anteversion and retroversion. This would be possibly obtained by transforming the joints into hinges. In order to increase the speed of the locomotion, the metacarpals are lengthened (vide NIESCHALK & DEMES, GODINOT & BEARD, JOUFFROY *et al.*, this volume), the number of digits is reduced and the hand is erected from plantigrady over digitigrady to unguligrady (STARCK, 1979).

However, the movements of a primate's hand are very variable. It can perform prehensile and non-prehensile movements (NAPIER, 1956; LANDSMEER, 1982; WYNN-PARRY, 1981; MATEV & BANKOV, 1982; but see also BOESCH and BOESCH , REDSHAW, CHRISTEL, this volume). The non-prehensile movements include for example pushing, punching and squeezing. The prehensile movements are generally divided into two basic patterns, the precision and the power grips.

Considering the bony segments during movement the question arises, in what way are the outer shape and the inner bone structure adapted to mechanical stress during locomotion and prehension respectively?

Materials and Methods

The investigation was made on metacarpal and proximal and middle phalangeal bones of humans (*Homo sapiens*) and gibbons (*Hylobates syndactylus*) (*Table 1*). In total 23 bones were analysed, 7 metacarpals, 3 proximal and 12 middle phalanges. For rough orientation the bones were X-rayed macroscopically. At five sites, 30%, 40%, 50%, 60% and 70% of the total length of each metacarpal bone and at three sites, 25%, 50%, and 75% of each phalanx bone three transverse sections were taken in succession using a precision saw with an annular blade (*Figure 1*). The first section was 1 mm thick and was used for density measurements. The following two sections were each 100 μm thick and were prepared for structural investigation.

Density

The density was expressed by γ-ray absorption using a ^{125}J Profile Scanner (Börner *et al.*, 1969). The measurements were taken with the beam path parallel to the shaft axis. From all measurements absorption curves were drawn, enlarged five times. the photon absorption was calculated by $\mu = (\ln(I_0/I)/d$, where μ is the absorption coefficient, I_0 the intensity of the unattenuated beam, I the attenuated intensity of the beam after the passage through the bone.

(1) Contour maps over the length of the shaft: The density was measured at eight locations on each cross section. From the single values obtained from the absorption curves at the eight locations and the five or three sites respectively, density contour maps were calculated by means of a program for the HP 9845B computer.

Table 1 - *Length of the investigated bones in mm.*

	Os metacarpale				
	I	II	III	IV	V
Homo	40.0	62.1	61.0	55.3	51.2
Hylobates			58.4	54.0	
			Phalanx proximalis		
			III	IV	
Homo			41.3		
Hylobates			42.8	39.7	
			Phalanx media		
			III	IV	
Homo			28.8*		
Hylobates			32.3	29.5	

* mean value of 10 middle phalanges.

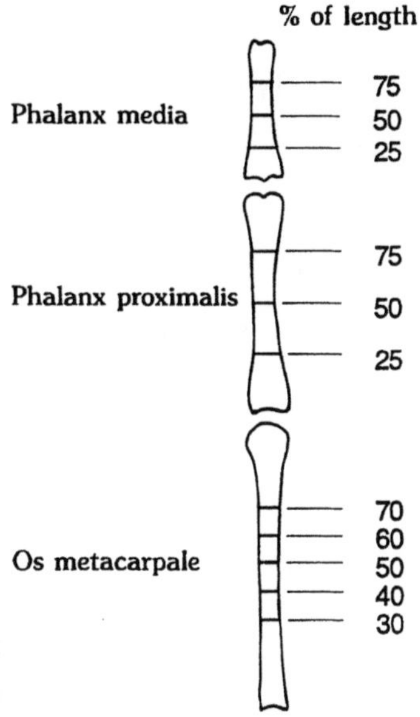

Figure 1 — Locations of the analysed cross sections.

(2) Contour maps over the cross section: The cross sections were scanned at 0.5 mm intervals in a dorsovolar direction from radial to ulnar. From the single absorption curves density contour maps were calculated for each cross section by means of a program for the HP 9825 computer (FÖRSTER, 1980). Three lines of equal absorption, isodenses, subdivide the cross section into several areas representing a density pattern with three different ranges.

MICROSTRUCTURE

Microradiographs were made from the 100 µm thinn cross sections. The structure and the micromineralization were evaluated on enlarged photo collages (x35) of the microradiographs.

The cross sections were embedded and the collagen fibre arrangements were analysed under circular polarized light.

Results and discussion

Metacarpal bones

DENSITY

The densitometric analysis shows on average higher density values in the gibbon than in the human (*Figure 2*). In both, density decreases from the proximal to the distal end of

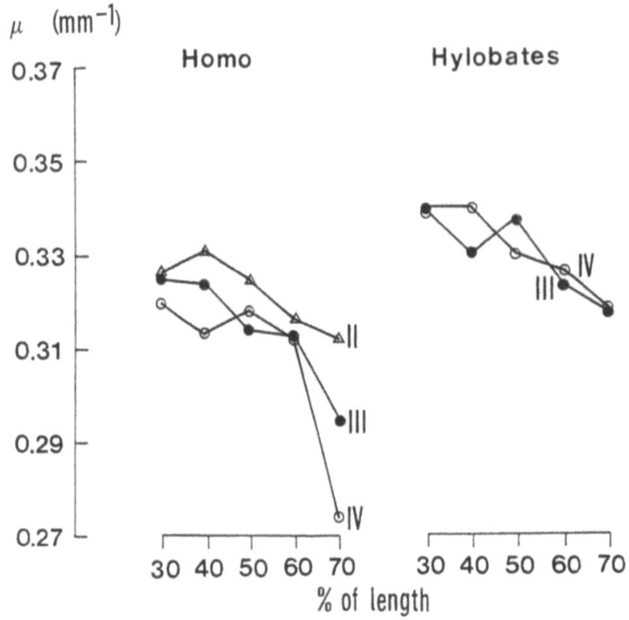

Figure 2 — Mean density as the mean absorption coefficient μ (mm^{-1}) of compact bone at five sites of the ossa metacarpalia II, III and IV of man, Homo, and of the ossa metacarpalia III and IV of the gibbon, Hylobates syndactylus.

the diaphysis. In the human the os metacarpale II exhibits highest density. On the assumption that there exists a positive correlation between density and stress, this means that the metacarpals of the human hand would be less stressed than the hand of gibbon, if «height of stress» is measured in fractions of body weight.

The density contour maps show the maximum density along the diaphysis of the metacarpal bones (*Figure 3*). Changes in the density pattern, as can be seen from different density values at the eight locations on the cross sections as well as from those at the five sites of the shaft, are likely due to different stress conditions.

In the gibbon (*Figure 3a*) the density decreases in proximodistal direction, especially obvious to see in the os metacarpale VI. High density surrounds the circumference of the metacarpal bones with a decrease of density only in the dorsal area of the os metacarpale III and in the voloulnar area of the os metacarpale IV. From this density patterns bending stresses may be deduced in rotating planes, as PREUSCHOFT (1973) showed in theoretical analyses.

In the human the corresponding contour maps of the metacarpal bones III and IV (*Figure 3b*) show a very heterogenous density distribution, making it diffiult to associate special functions. The question arises: Will the stresses deduced from the density distribution be reflected in the structural organisation of the bone, or will the structural organisation allow a better functional interpretation?

STRUCTURAL EVALUATION

In the gibbon (*Figure 4*) the cross sections of the metacarpal diaphysis III and IV are nearly circular in shape. The compacta is composed of an endosteal zone of less minera-

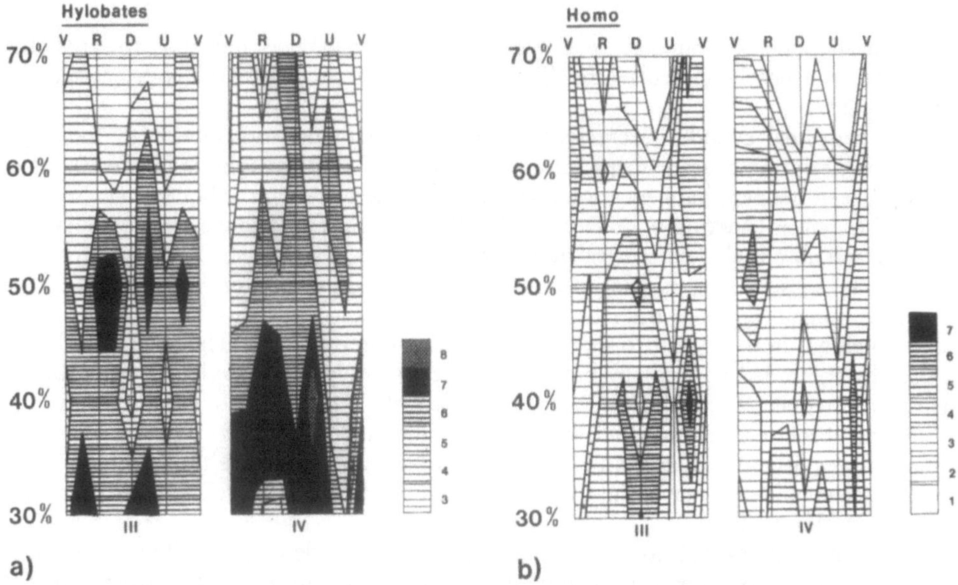

Figure 3 — Topographical distribution of the mean maximum absorption coefficient μ (mm^{-1}) as the mean maximum density in the third (III) and fourth (IV) metacarpal diaphysis a) of the gibbon, and b) of man, *Homo.* The compacta is to be concieved as cut open volarly and spread out in the drawing plane. D, dorsal; R, radial; V, volar; U, ulnar. The hatchings correspond to the density or absorption respectively, which increases from 1 with $\mu \leq 0.29 mm^{-1}$ to 8 with $0.35 mm^{1} \leq \mu \leq 0.36 mm^{-1}$, to once with $\mu = 0.01$ mm^{-1}.

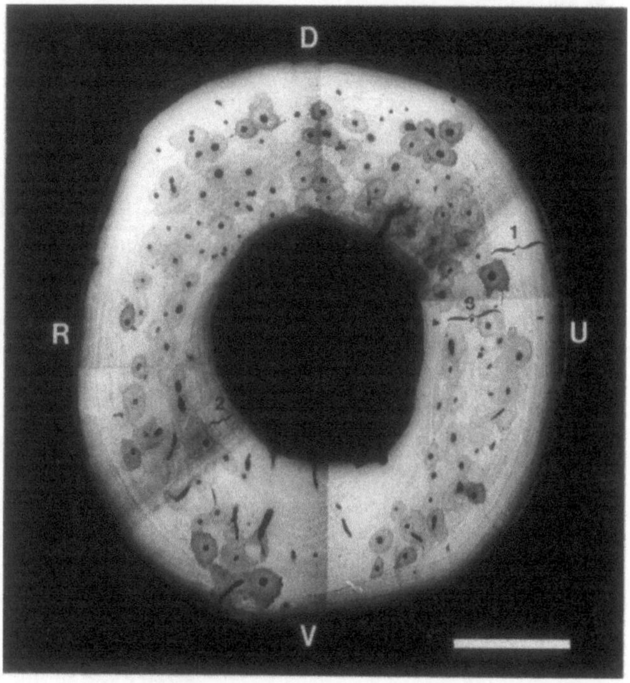

Figure 4 — Microradiographic collage of a cross section from the midshaft of the os metacarpale IV of the gibbon. 1, periosteal lamellar bone; 2, endosteal lamellar bone; 3, osteonal bone; D, dorsal; R, radial; V, volar; U; ulnar. Bar 1 mm.

lized lamellar bone, surrounded by a cylinder of osteonal bone and ends towards the periosteum in a nearly avascular and highly-mineralized circumferentially-orientated lamellar bone.

The periosteal cylinder decreases distally in extent and corresponds to the high density visible in the contour maps in the proximal half of the diaphysis (*Figure 3*). As a consequence of the peripheral location and the high mineralization the bone can be seen as «supporting bone» (AMTMANN & DODEN, 1981). In the osteonal bone the osteones are relatively and absolutely less mineralized than the interosteonal bone. They are probably more apt to subserve apposition and resorption related to calcium metabolism.

In the human (*Figure 5*) the cross sections of the metacarpals vary in shape along the shaft though not in the extent to show special strength in one plane. The compacta shows endosteally an eroded and spongy border and is completely penetrated by longitudinally orientated osteones. The osteones are in part relatively and absolutely larger than in the gibbon and absolutely less mineralized than the interosteonal bone. The dorsal compacta contains fewer osteones than the rest of the cross section. Therefore the highly mineralized interosteonal bone causes the increase of density and the bone seems to be more apt to stress in this region.

In the gibbon, as well as in the human, neither the density distribution nor the structural organisation shows a pattern of a clear relation to a specific stress pattern. Rather one can assume an intermitting bending stress in different planes, which corresponds better to the heterogeneous pattern of movements of a prehensile hand.

ASCENZI & BONUCCI (1967, 1968; ASCENZI *et al.*, 1990) demonstrated that bone with a longitudinal orientation of collagen fibres is adapted to high tensile strength, bone with

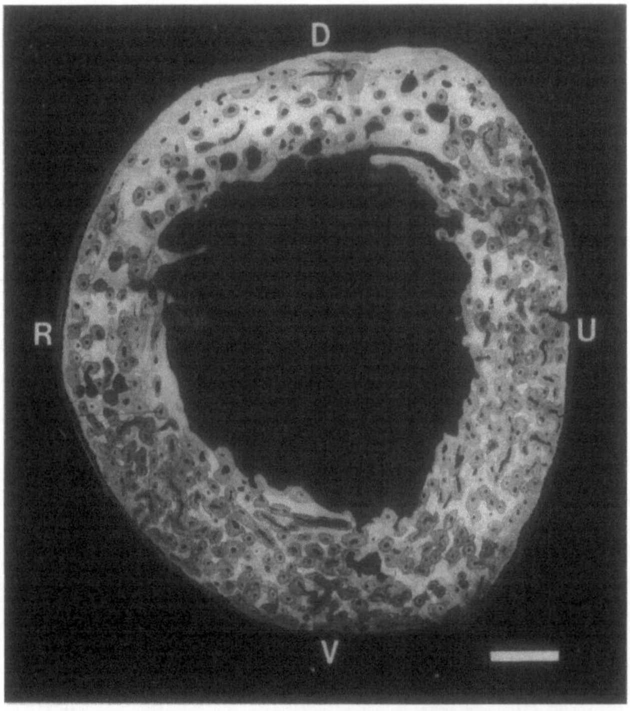

Figure 5 — Microradiographic collage of a cross section from the midshaft of the os metacarpale III of a 74 years old woman. D, dorsal; R, radial; V, volar; U, ulnar. Bar 1mm.

a more transverse fibre orientation is better for supporting high compressive stress, while bone with alternate fibre directions is optimal for withstanding bending stresses.

In the gibbon (*Figure 6*) the analyses of the metacarpal bones exhibit mainly longitudinally-orientated fibres with an endosteally located ring of transverse circumferentially orientated fibres. Also the periosteal longitudinal fibres are interlaced with circumferentially-orientated transverse fibres. The osteones with mainly longitidinal fibres are bordered by a ring of transverse fibres. This means the osteonal and the periosteal bone are better adapted to withstand tension stress combined with intermitting bending stress.

In the human (*Figure 7*) it seems that due to the mostly transverse fibres in the third metacarpal the adaptation to compressive stress prevails volarly and ulnarly, while radially and dorsally predominantly longitudinal fibres indicate tensile stress. This shows that the fiber course corresponds with a bone subjected to bending in a dorsovolar as well as in a radioulnar plane, which occurs mainly during gripping and holding loads.

Phalangeal bones

As expected, in the phalanges the density and the structural pattern is quite different from that in the metacarpus.

DENSITY

In the gibbon (*Figure 8a*) the density of the proximal and middle phalanx of the third and fourth finger is fairly equal. In the third proximal phalanx high density occurs

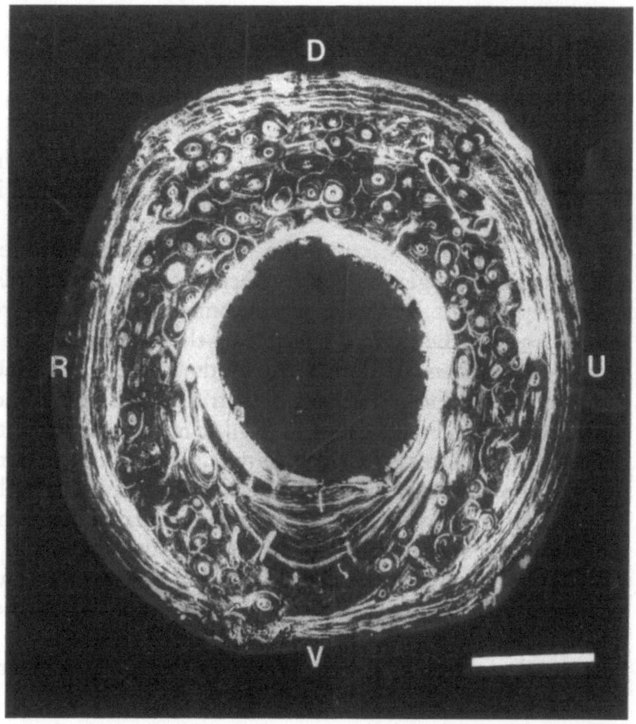

Figure 6 — Collagen fibre arrangement under circular polarized light of a cross section from the midshaft of the os metacarpale IV of the gibbon. Transversal fibres white; D, dorsal; R, radial; V, volar; U, ulnar. Bar 1 mm.

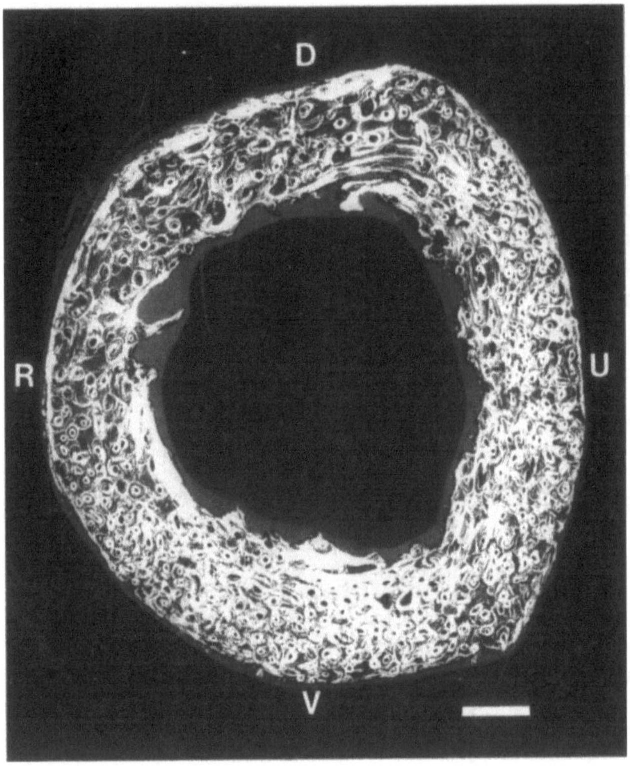

Figure 7 — Collagen fibre arrangement under circular polarized light of a cross section from the midshaft of the os metacarpale III of a 74 years old woman. Transversal fibres white; D, dorsal; R, radial; V, volar; U, ulnar. Bar 1 mm.

proximally, while in the middle phalanx it prevails distally in the ulnar half of the shaft. In the fourth finger high-density zones are especially located in the radial and ulnar sides of the compacta.

In the human (*Figure 8b*) the density is in general less than in the gibbon. In the analysed middle finger the density decreases towards the distal end of the proximal as well as middle phalanx. By means of contour maps over the cross sections NEUBAUER (1986) showed that in the middle phalanx of the middle finger (*Figure 9*), the high and mean density ranges I and II decrease towards the distal end more than towards the proximal end of the shaft. From this density distribution pattern, and because of the biomechanical considerations of PREUSCHOFT (1973), it may be deduced that the middle phalanx of the human finger is stressed by voloconvex bending in the dorsovolar plane with the highest stress in the midshaft and the highest compressive strength in the dorsal cortical region.

This is in line with the stress in the palmar pinch, where the object is held between the distal and middle phalanx of the middle finger and the thumb, which leads to a voloconvex bending of the middle phalanx. The bending will be intensified, if the grip is performed with maximum force and hyperextension in the distal phalangeal joint of the finger. The same sorts of stress has been predicted by PREUSCHOFT on the base of theoretical mechanics as early as 1969 for toe bones and for finger bones in 1973.

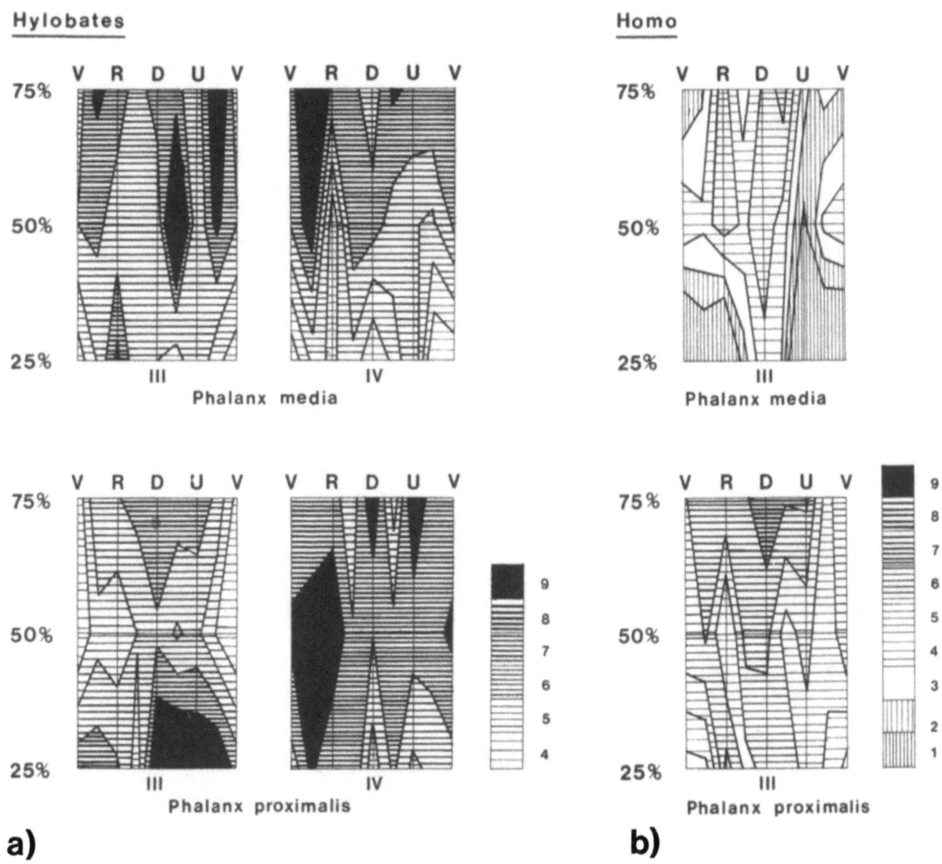

Figure 8 — Topographical distribution of the mean maximum absorption coefficient μ (mm^{-1}) as the mean maximum density a) in the third (III) and fourth (IV) proximal and middle phalangeal diaphyses of the gibbon, and b) in the third (III) proximal and middle palangeal bone of man. The compacta is to be concieved as cut open volarly and spread out in the drawing plane. D, dorsal; R, radial; V, volar; U, ulnar. The hatchings correspond to density or absorption respectively, which increases from 1 with $\mu \leq 0{,}27 mm^{-1}$ to 9 with $\mu > 0.34 mm^{-1}$, at a time with $\mu = 0.01 mm^{-1}$.

STRUCTURAL EVALUATION

In the gibbon (*Figure 10*) the structural pattern of the proximal and middle phalanx is similar to that in the metacarpals. But the shape is proximal more elliptic, in the midshaft the bone is more flattened in dorsovolar direction. Towards the distal end the cross section is sickle-shaped with the smallest dorsovolar diameter. However, the radioulnar diameter is almost constant over the shaft length. The wide circumferential, nearly avascular, bone lamellae are located periosteally and endosteally. In the circumferential lamellae highly mineralized lines occur predominantly in the dorsal periosteal cortex. The border of the marrow cavity is very smooth. In the middle phalanges III and IV two straight trabeculae pass the marrow cavity in a dorsovolar direction, moving from a more central position to a more peripheral one towards the distal joint. Osteonal bone is

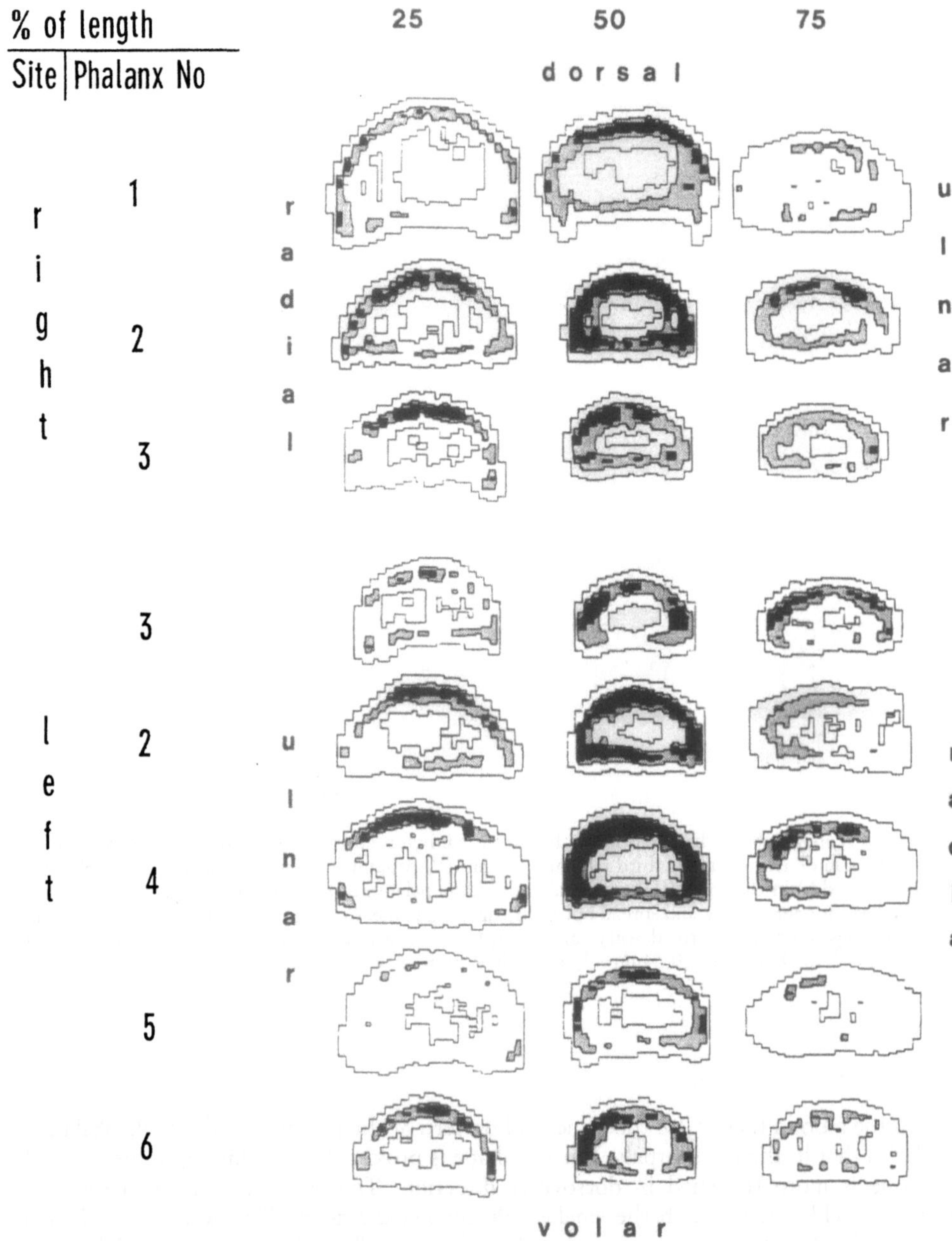

Figure 9 — Density contour maps of 1 mm thick bone slabs at three sites of three right and five left human middle phalangeal bones. Black, density range I - all points with $\mu \leq 0.22 mm^{-1}$; dotted, density range II - $0.16 mm^{-1} \leq \mu \leq 0.22 mm^{-1}$; white, density range III - $0.13\ mm^{-1} \leq \mu \leq 0.16 mm^{-1}$ (From NEUBAUER, 1986).

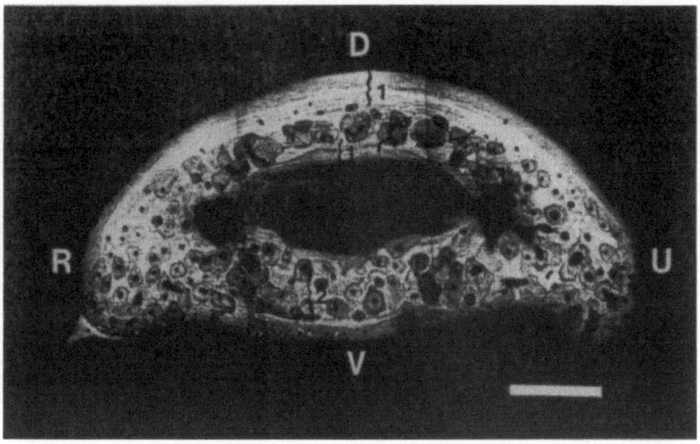

Figure 10 — Microradiographic collage of a cross section from the midshaft of the phalanx media IV of the gibbon. 1, lamellar bone; 2, osteonal bone; D, dorsal; R, radial; V, volar; U, ulnar. Bar 1 mm.

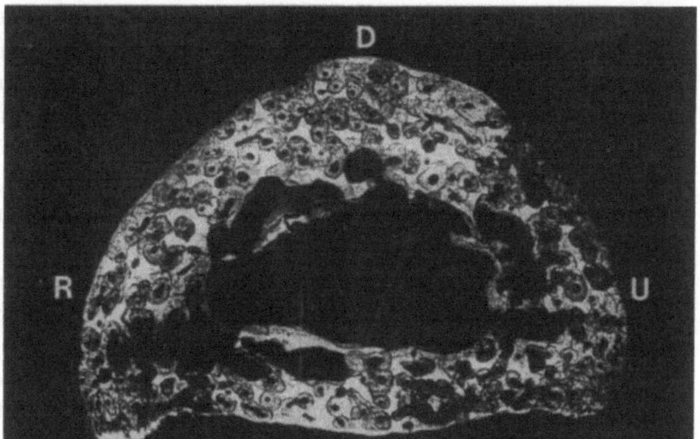

Figure 11 — Microradiographic collage of a cross section from the midshaft of the phalanx media III of a 74 years old woman. D, dorsal; R, radial; V, volar; U, ulnar. Bar 1mm.

arranged bandike over the whole circumference in the middle of the cortex, extending ulnarly and radially to the periosteal margin. The osteones are different in size and highly mineralized over the whole cross section. There exists only a small difference in mineralization between the osteones and the interosteonal bone.

In man (*Figure 11*) the phalanges are mostly composed of osteonal bone in the midshaft, whereas at the ends of the diaphysis the radial and ulnar cortex is opened up by large cavities and looks like spongy bone. The cortex is thickest dorsally and decreases on the radial and ulnar sides towards volar in thickness. Osteones are numerous in the dorsal cortex, running mostly parallel to the long axis of the shaft and showing low mineralization. The interosteonal bone is more highly mineralized and shows partly a waving pattern. Outer and inner circumferential lamellae are often interrupted by resorption lacunae. The marrow cavity has an intended border and often the compact bone changes into spongy bone. The thin trabeculae, running mainly in a dorsovolar direction and

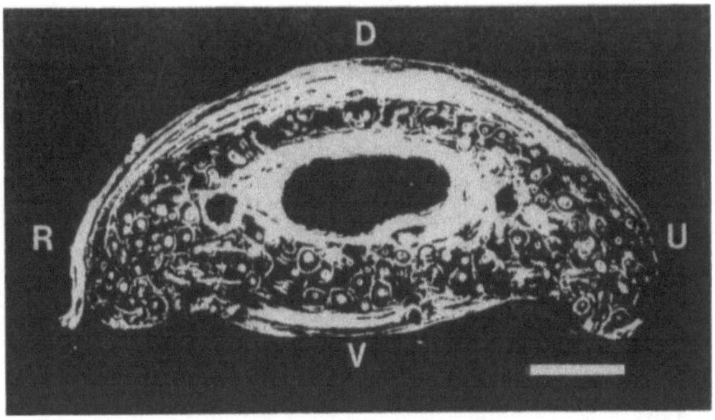

Figure 12 — Collagen fibre arrangement under circular polarized light of a cross section from the midshaft of the phalanx media IV of the gibbon. Transversal fibres white; D, dorsal; R, radial; V, volar; U, ulnar. Bar 1 mm.

sometimes passing through the marrow cavity, are centrally highly mineralized, but show low mineralization towards the ulnar and radial borders. The mean number of osteones is highest in the midshaft, and proximal and distal nearly equal. Within the cross section in the dorsal cortex the number of osteons is nearly twice that of the other regions. In other bones like femur, tibia and fibula a positive correlation was established between the compressive strength and the number of osteones per unit area (EVANS & BANG, 1967; EVANS & VINCENTELLY, 1974). The evidently higher number of osteons in the dorsal compact bone is a further indication of compressive stress in this region. Over the length of the shaft there seems to exist voloconvex bending, according to the stress distribution found by PREUSCHOFT (1973).

Apart from this the density was highest and most osteones occur dorsally. Due to the relatively low mineralization of the osteons, the interosteonal bone must be highly mineralized in order to compensate this defiency.

In all phalangeal cross sections of the gibbon (*Figure 12*) the collagen fibres are orientated circumferentially around the marrow cavity. In the periosteal bone lamellae the fibres run mostly transversely over the volar side of the shaft length, while dorsally the fibres run co-axially in the proximal sections and mainly transversely in the midshaft. In the osteones the collagen fibres run transversely only on the volar side of the proximal shaft, whereas in the other regions and the other shaft sites prevails a co-axially course. This pattern may be seen as an adaptation to comprenssive stress due to dorsoconcave bending during hanging and swinging.

In the human (*Figure 13*) the osteones with longitudinally-orientated collagen fibres prevail dorsally, while most of the volar osteones show lamellae with predominantly transversal spiral course of the collagen fibres. This would mean that dorsally tension and volarly compression will occur, which corresponds to the dorsoconvex bending in the dorsovolar plane. PREUSCHOFT (1973) used this type of bending to explain the shapes of the phalanges.

The finding is in contrast to the relationship between the density distribution pattern, the cortical thickness and the osteonal distribution and therefore in contrast to the voloconvex bending during palmar pinch. However, voloconvex bending in the dorsovolar plane occurs also during power grip, where the object is held between the fingers and the palm, and the finger joints are all flexed (PREUSCHOFT, 1973).

The orientation of the collagen fibers in the interosteonal bone is transverse in the proximal diaphysis on the volar side, whereas on the ulnar, dorsal and radial sides

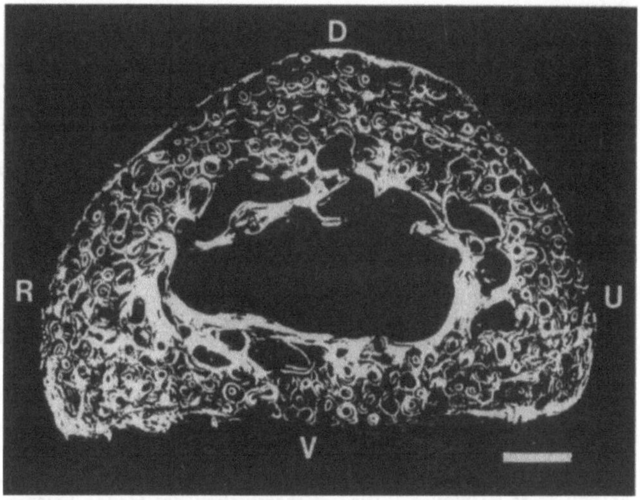

Figure 13 — Collagen fibre arrangement under circular polarized light of a cross section from the midshaft of the phalanx media III of a 74 years old woman. Transversal fibres white; D, dorsal; R, radial; V, volar; U, ulnar. Bar 1 mm.

longitudinally-running fibres prevail, which correspond to the tensile stresses occurring on the convex side of bending in both planes. In the midshaft and the distal diaphysis only dorsally a wide band of predominantly transverse fibres occurs in the periosteal margin. This is in line with proximal dorsoconvex bending near the proximal joint. From the midshaft towards the distal diaphysis voloconvex bending seems to occur in the proximal and middle phalanges.

General considerations

All analysed bones show transverse fibres, running circumferentially around the marrow cavity. The amount of these fibres increases with a decreasing area of the marrow cavity on the cross sections. This may correspond more with a centric compression, which leads probably to hydrostatic compression of the bone marrow during loading. Especially in narrow marrow cavities the hydrostatic compression of the bone marrow seems to be held off by transverse fibres. This supports the assumption of Otto (1984), that the long bones behave as a «pneu».

Summarizing the findings of the analysed bones of the hand, the combination of density distribution and the structural pattern, especially the arrangement of the collagen fibres seems to reflect a stress pattern which is adapted to withstand dorsoconcave bending in a dorsovolar plane and a bending in a radioulnar plane as well in radial as in ulnar direction, as Preuschoft has predicted from theoretical considerations (1973) and analysed experimentally by an enlarged model (1987).

Acknowledgement —I am deeply indepted to H. Preuschoft for sending me the skeletal material of *Hylobates* that was used in this investigation. Thanks also go to K. Westermann for her splendid technical assistance. I am most grateful to the editor H. Preuschoft for his great patience. This investigation was done while the author was an assistent at the Anatomical Institut of the Medizinische Hochschule Hannover.

References

AMTMANN E. & DODEN E., 1981. *Anpassung der Knochenstruktur an mechanische Beanspruchung.* Z. Morph. Anthrop., 72: 1-21.

AN K.N., CHAO E.Y., COONY W.P. & LINSCHEID S., 1979. *Normative model of human hand for biomechanical analysis.* J. Biomechanics, 12: 775-788.

ASCENZI A., BASCHIERI P. & BENVENUTI A., 1990. *The bending properties of single osteons.* J. Biomechanics, 23: 763-771.

ASCENZI A. & BONUCCI E., 1967. *The tensile properties of single osteons.* Anat. Rec., 158: 375-386.

ASCENZI A. & BONUCCI E., 1968. *The compressive properties of single osteons.* Anat. Rec., 161: 377-392.

BÖRNER W., GREHN S., MOLL E. & RAUH E. , 1969. *A ^{125}J Profile Scanner for measuring density in the fingers.* German Med. Monthly, 14: 301-305.

EVANS F.G. & BANG S., 1967. *Differences and relationship between the physical properties and the microscopic structure of human femoral, tibial and fibular cortical bone.* Am. J. Anat., 120: 79-88.

EVANS F.G. & VINCENTELLI R., 1974. *Relations of compressive properties of human cortical bone to histological structure and calcification.* J. Biomechanics; 7: 1-10.

FÖSTER E., 1980. *Zur Densitometrie von Knochenschnitten. Messungen am menschlichen Humerus.* Med. Diss., Hannover.

LANDSMEER J.M.F., 1962. *Power grip and precision handling.* Ann. Rheum. Dis., 21: 164-169.

MATEV I. & BANKOV S., 1982. *Rehabilitation der Hand.* Thieme, Stuttgart, pp. 76-87.

NAPIER J.R., 1956. *The prehensile movements of the human hand.* J. Bone Joint Surg., 38B: 902-913.

NEUBAUER M., 1986. *Zur funktionellen Morphologie der Phalanx media menschlicher Mittelfinger. Eine densitometrische, mikroradiographische und zirkularpolarisationsoptische Studie.* Med. Diss., Hannover.

OTTO F., 1984. *Pneu und Knochen. IL 35, Institut für leichte Flächentragwerke (IL).* Vorauflage, Stuttgart.

PREUSCHOFT H., 1969. *Statische Untersuchungen am Fuß der Primaten. Teil I: Phalangen und Metatarsalia.* Z. Anat. Entw. Gesch., 129: 285-345.

PREUSCHOFT H., 1973. *Functional anatomy of the upper extremity.* In: G. H. Bourne ed. The Chimpanzee, vol. VI, 34-115. Karger, Basel & New York.

PREUSCHOFT H., 1987. *Funktionsweise der Sehnen, Beanspruchung und Form der Phalangen und Metacarpalia.* Verh. Anat. Ges., 81: 279-292.

STARCK D., 1979. *Vergleichende Anatomie der Wirbeltiere auf evolutionsbiologischer Grundlage. Bd. 2. Das Skeletsystem. Allgemeines, Skeletsubstanzen, Skelet der Wirbeltiere einschließlich Lokomotionstypen.* Springer, Berlin-Heidelberg-New York, pp. 555-739.

WYNN-PARRY C.B., 1981. *Rehabilitation of the hand.* 4th ed., Butterworth & Co. Ltd., London.

C. Buck
H. Bär

Investigations on the biomechanical significance of dermatoglyphic ridges

To investigate the significance of the dermal ridges for the adhesion of the skin to a substrate an in vivo experiment with 46 subjects (each with 10 fingers) and 4 kinds of substrates was conducted. Subjects produced the friction «actively» by moving their fingertips perpendicularly or parallel to the orientation of their dermatoglyphic ridges on a force plate, covered with different materials.
A comparison between the measured coefficients of friction showed that frictional resistance was greater when the force was exerted perpendicularly, not in parallel to their orientation. The differences between the coefficients of friction were significant in 7 out of 8 experiments. With the same normal force, the coefficient of friction increases by ca. 20%, if the frictional force is acting perpendicularly to the dermal ridges, and not in parallel. These results are in accordance with existing information on the microscopic anatomy of the dermatoglyphic skin.

Key words: adhesion, dermal ridges, friction, force plate, human hand, slide friction, static friction.

Introduction and problem

As generally known, the hands of primates are equipped on their palmar surfaces with the so-called friction skin. The friction skin is characterized by dermatoglyphic patterns, formed by dermal ridges. The arrangements of the dermatoglyphic ridges are individually variable but, statistically, species specific, and inherited to a remarkable degree.

The fact that the ridges are found on the palmar and the plantar surfaces of hands and feet has led early to the conclusion that they fulfill a mechanical function in locomotion, in walking, climbing, and, of course, also in gripping. This function could be an «improvement» of the grip, associated with an increase of friction between skin and substrate. Whipple (1904) speculated that dermal ridges are arranged perpendicularly to forces which would otherwise cause a slipping of the skin on the substrate. Until now, this speculation has not been tested experimentally, although it gave reason to the name «friction skin» which is in common use.

Since the beginning of research on dermatoglyphic patterns more than 80 years ago, the biomechanical function of the dermal ridges remained uncertain.

Earlier work on the dermatoglyphics was primarily descriptive; for example, Whipple (1904) and Schlaginhaufen (1905) made extensive comparisons between the patterns on various primates' palmae and plantae.

Later the focus shifted to the hand's function as an organ of tactile sense. It has been demonstrated that the arrangement of sense organs within the dermal ridges contribute to the reception and discrimination of tactile stimuli (Andres, 1973).

It was not until the fifties that the skin as the area of contact to solid substrates, and, accordingly, its mechanical characteristics became a focus of interest. Naylor (1955) was the first to consider the frictional properties of the skin more detailedly. SULZBERGER (1966) investigated lesions of the skin caused by friction; his work was stimulated by the results obtained by BOWDEN and TABOR (1964), who demonstrated deviations from COULOMB's law of friction for all visco-elastic materials, to which the palmar and plantar surfaces of hands and feet belong. COULOMB's law states a direct proportionality between all forces that contribute to friction. COMAISH & BOTTOMS (1971) and EL SHIMI (1977) have also investigated the frictional propensities of the skin experimentally. Their experiments included the dermatoglyphic skin as well as the skin without dermal ridges from other parts of the body, but were not designed specifically to explain the functional meaning of the dermatoglyphics. In their experiments, the authors brought rotating cylinders covered by different materials into contact with the skin.

CARTMILL (1974, 1979) and NIEMITZ (1977) dealt with the function of the dermatoglyphic system in the order of primates. Both assumed a connection between the evolution of the dermatoglyphic patterns and arboreal locomotion. CARTMILL experimentated on *Microcebus murinus*, *Cheirogaleus medius*, *Lemur fulvus*, *Lemur macaco*, *Galago crassicaudatus*, as well as *H. sapiens*. For these species he determined the size of the coefficients of friction between a natural substrate (wood) and the entire palms and soles.

All this leaves open the question, whether the ridge patterns as a whole play any biological role, that is whether the orientation of the ridges relative to the direction of a force acting parallel to the skin surface has importance for survival. This question was discussed by NIEMITZ (1990). Our focus is a different one:

Supposed that WHIPPLE's assumption mentioned in the beginning is true, a force acting across the dermatoglyphic ridges should produce a higher friction than the same force acting in parallel to the ridges. The investigation presented here is aimed at testing this hypothesis experimentally.

Material and Methods

Experimental Design

The experiment was designed as *in vivo* investigation, because the natural conditions were thought to be represented most accurately, if the subjects had to produce the frictional force actively. In the investigations described in the literature up to now, subjects, or the areas of skin which were observed, have always been «passive» with respect to the generation of the forces.

In our experiment, subjects were told to slide their fingertips (areas Id — Vd, as defined by CHRISTEL, this volume) over the different test materials. The fingertips were chosen because their ridges have the same orientation regardless of the dermal pattern of the finger, namely from the radial to the ulnar side.

Experimental Device

The experimental apparatus was an especially designed force plate, which rests on three points. It was equipped with a frame, into which the plates of the 4 test substrates — glass, wood, plastic, and stone — could be inserted. The surfaces of the test materials were 30 cm long and 8 cm wide.

For «wood», smoothened and polished pinewood was selected. The wooden plate was

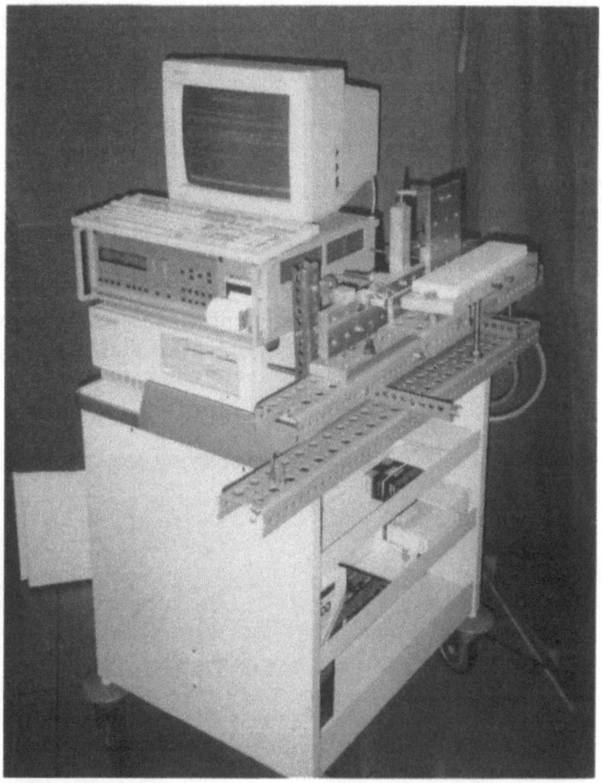

Figure 1a - Experimental setup for measuring the frictional resistance and the normal forces of finger tips on various materials

positioned in such a way, that the finger tips had to be moved parallel to its fibres. The «stone» material was polished marble. For «glass», ordinary window glass was chosen. The «plastic» material used was a common grey plastic.

The forces occurring in the three dimensions of space were measured as changes of the electrical resistance of strain gauges. All directions in space were described by the coordinates x, y, and z. The y-coordinate represented the frictional force $F_{(R)}$ in the direction opposite to the movement of the finger. The x-coordinate reflected the frictional force perpendicular to the direction of the movement of the fingers. Finally, the z-coordinates showed the normal force $F_{(N)}$. The measurements were transformed by an amplifier and a multi-channel digitizer and then registered by a computer (see *Figure 1a*).

Subjects

46 subjects aged between 18 and 69 years took part in the experiments; 22 of them were females and 25 males. The sample consisted of 43 right-handed and 3 left-handed persons.

Subjects were asked to classify their sweat secretion into one of three categories: weak, medium or strong. The following answers were obtained: 13 subjects rated their sweat secretion as «weak», 27 as «medium» and 6 as «strong». A question concerning occasional contact with a solvent like xylol or formalin was answered positively in two cases. None of the experimental subjects suffered from skin deseases or showed overt anatomical abnormalities at their fingertips. For documentation, prints of the dermal patterns were taken from the place where the measurements were made, namely the finger pulps.

Experimental procedure

The experiment consisted of two parts. In the first part (experiment 1) subjects were asked to slide their fingertips over the plate while exerting a normal force comfortable to them. In that experiment «slide friction» was thought to be produced.

In the second part (experiment 2), subjects were told to try to move away the force plate by exerting a strong pressure. The aim of that procedure was to produce «static friction» (see *Figure 1b*).

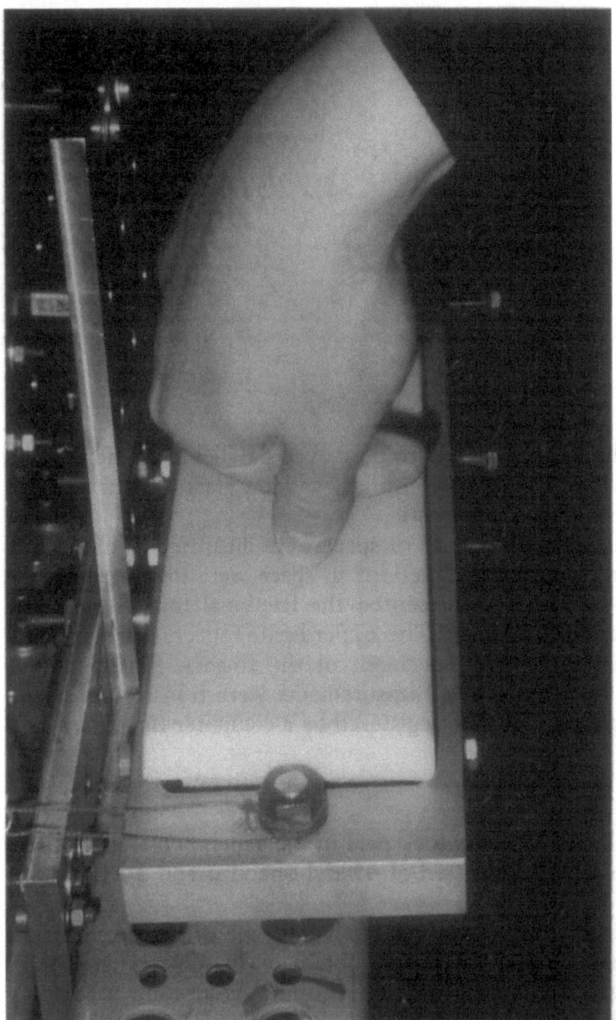

Figure 1b - Testing the frictional resistance of the thumb in the direction «pull» on stone.

All subjects involved had short fingernails, i. e. the nails were not longer than the fingertips. Errors due to the subject's clawing into the substrate could thus be avoided. The subject's hands were neither washed nor degreased to maintain «natural» conditions. The plates were cleaned with acetone and dried with a piece of cloth before each test run. Before the experiment started, subjects were given the instructions to follow:

Instructions for the test procedure

«Parallel»: slide the finger on the plate from right to left, or from left to right (parallel to the orientation of the dermal ridges).

«Perpendicular»: pull or push the finger vertical to the direction of the dermal ridges.

The direction of the movement called «parallel» in the instructions corresponds to the direction «radial/ulnar» (RU), the direction called «perpendicular» corresponds to the direction of the movement «pull-push» (SZ) which is the same as «distal/proximal» (DP).

Experiment 1, slide friction:

The finger is set on one edge of the plate with arbitrary pressure, moved to the opposite edge by an arbitrary speed, and then lifted from the plate. The fingernail must not touch the plate.

Order of the test runs:

right hand:	digits 1 to 5 parallel from left to right
	digits 1 to 5 parallel from right to left
left hand:	digits 1 to 5 parallel from left to right
	digits 1 to 5 parallel from right to left
right hand:	digits 1 to 5 perpendicular from left to right
	digits 1 to 5 perpendicular from right to left
left hand:	digits 1 to 5 perpendicular from left to right
	digits 1 to 5 perpendicular from right to left

Experiment 2, static friction:

The finger is pressed strongly onto one edge of the plate; then it tries to move the plate to the opposite side by a pushing or pulling movement, while maintaining the pressure. If the finger starts sliding, it is lifted from the plate. The finger nail must not touch the plate.

Order of the test runs:

right hand:	digits 1 to 5 parallel from left to right
	digits 1 to 5 parallel from right to left
left hand:	digits 1 to 5 parallel from left to right
	digits 1 to 5 parallel from right to left
right hand:	digits 1 to 5 perpendicular from left to right
	digits 1 to 5 perpendicular from right to left
left hand:	digits 1 to 5 perpendicular from left to right
	digits 1 to 5 perpendicular from right to left

Statistics

The test data were registered in the form of electrical tensions in mV, for each subject separately. After transforming the original values into the normal force $F_{(N)}$, the friction force $F_{(R)}$, and the coefficient of friction μ (the quotient of $F_{(R)}$ and $F_{(N)}$) these data were the input for a statistical analysis. The relationships between $F_{(R)}$ and $F_{(N)}$, respectively μ and $F_{(N)}$ were calculated as linear regressions and correlations for each experiment, for each direction of movement, and for every single finger. For the regression analysis the test data had to be logarithmic.

For each finger the values for $F_{(N)}$, $F_{(R)}$, and μ were calculated. According to the central criterion «direction of the finger's movement» these data were pooled. On the basis of these subsamples, statistical parameters like arithmetic mean, regression coefficient, etc. were calculated. «Direction of the finger's movement» was classified into two categories:

a) movement of the finger in direction «radial» or «ulnar», abbreviated R or U, respectively,

b) movement of the finger in direction «push» or «pull», abbreviated S or Z.

For each test material the following samples were accumulated:

1) all fingers, direction SZ
2) all fingers, direction RU
3) all fingers, direction S (push)
4) all fingers, direction Z (pull)
5) finger I, direction SZ
6) finger I, direction RU
7) finger II, direction SZ
8) finger II, direction RU
9) finger III, direction SZ
10) finger III, direction RU
11) finger IV, direction SZ
12) finger IV, direction RU
13) finger V, direction SZ
14) finger V, direction RU

The arithmetic means and the corresponding standard deviations were calculated from the data pools to be compared.

The t-test was used for testing the means of the forces $F_{(N)}$, $F_{(R)}$, as well as the coefficients of friction μ, for statistical significance. The preconditions for that were fulfilled, because:

1) the measurements to be compared were independent from each other,
2) the frequency distributions of the values were sufficiently symmetrical,
3) the standard deviation fell in the same order of magnitude.

Besides the t-test, the sign test according to DIXON & MOOD (SACHS, 1968) has been used for non-parametric testing of significance.

Results

Variation of measurements, regressions

The pairs of values for the friction $F_{(R)}$, and the normal force $F_{(N)}$ were entered into a linear coordinate system. Despite their great dispersion it is evident from the swarm of points that there is no linear proportionality between the frictional and the normal force.

However, if the values for $F_{(R)}$ and $F_{(N)}$ are transformed into their logarithms, and if these paired logs are plotted against each other, the linear interdependence between log $F_{(R)}$ and log $F_{(N)}$ becomes evident. Under that condition a linear regression can be calculated. The function of the linear regression follows

$$\log F_{(R)} = \log K + n \cdot \log F_{(N)} \qquad \text{(equ. 1)}$$

This function is identical to:

$$F_{(R)} = K \cdot F_{(N)}^{n} \qquad \text{(equ. 2)}$$

Accordingly, the relation between the coefficient of friction and the normal force $F_{(N)}$ is described by:

$$\log \mu = \log K + (n-1) \cdot \log F_{(N)} \qquad \text{(equ. 3)}$$

$$\text{or else: } \mu = K \cdot F_{(N)}^{n-1} \qquad \text{(equ. 4)}$$

As can be seen, the equations 2 and 4 represent an exponential relation between $F_{(R)}$, and $F_{(N)}$, or μ and $F_{(N)}$, respectively. This means that the coefficient of friction $F_{(R)}$ increases less than linearly with increasing normal force $F_{(N)}$, or that the coefficient of friction decreases with increasing normal force $F_{(N)}$. From the slopes of the regressions n (equ. 1) and n-1 (equ. 3) and the section of the ordinate log K, empirical values for the factors K and μ of the exponential function were calculated. This was done separately for each of the 4 substrates and for the directions SZ, and RU, in which the fingers were moved.

In experiment 1, values between 0.4 and 1.2 were found for the factor K. The values varied from 0.4 to 0.6 for the wooden substrate, and ranged from 0.7 to 1.2 for glass, plastic and stone. The values for the factor n vary from 0.7 to 1.0.

In experiment 2, the distribution of the values for K and n is much more homogeneous. Factor K varies between 0.5 and 0.6 for all substrates, for the exponent n we found values between 0.8 and 0.9.

Mean coefficients of friction

EXPERIMENT 1 («slide» friction with arbitrary normal force).

For all fingers, coefficients of friction are greater in the proximo-distal than in the radio-ulnar direction. This holds true for all four test substrates (see *Table 1*).

The greatest and highly significant (p = 1%) differences between the coefficients are found for wood and for stone. The difference is slightly smaller for plastic, while for glass

TABLE 1: *Experiment 1: Coefficients of slide friction μ; mean values and comparison between $\mu_{(SZ)}$ and $\mu_{(RU)}$.*

Material	μ(SZ)	μ(RU)	μ(SZ)-μ(RU)
glass	0.78	0.77	0.01
wood	0.46	0.41	0.05
plastic	0.57	0.53	0.04
stone	0.64	0.54	0.10

SZ: push/pull
RU: radial/ulnar

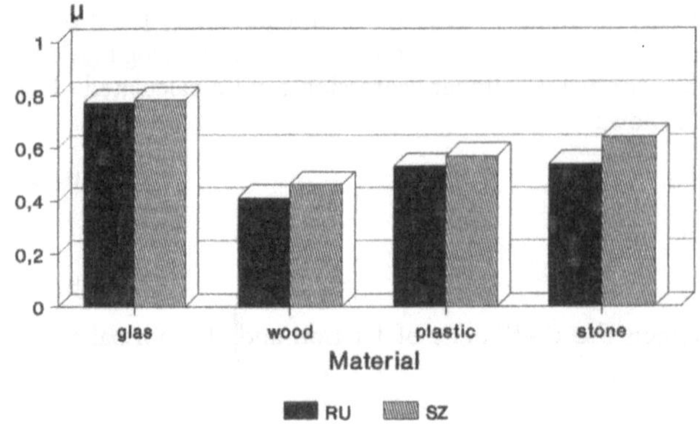

Figure 2 - Experiment 1, slide friction: Coefficients of friction μ. Comparison of mean values $\mu_{(SZ)}$ for push/pull and $\mu_{(RU)}$ for movements in radio-ulnar directions.

only small differences exist, which are not statistically significant (see *Figure 2*, and *Table 2*).

If the mean coefficients of friction $\mu_{(SZ)}$ and $\mu_{(RU)}$ are compared with regard to the normal force exerted, it can be shown that the size of the coefficient depends on the kind of substrate. The greatest mean coefficients are found for glass. Plastic and stone follow with similar values; μ-values for wood are least. Correspondingly, the greatest normal forces are found for wood, while plastic and stone behave similarly, and smaller values are observed for glass (*Figure 3*). The normal force exerted is always bigger for the direction SZ than for RU. Values of the mean coefficients of friction $\mu_{(SZ)}$ and $\mu_{(RU)}$ calculated for each finger individually (data pool: «μ-values of single fingers») coincide with the overall results. The coefficients for the proximo-distal direction (SZ) are equal to or bigger than those for the radio-ulnar direction (RU).

For «glass» only fingers I, II, and V show a difference. Values for the coefficients of fingers III and IV are of the same size. In contrast, for wood, plastic, and stone higher coefficients are found in all fingers moved in the proximo-distal direction SZ. It should be noted that the smallest differences between the coefficients for the directions SZ and RU are found for finger I (thumb) (see *Figures 4-7*).

TABLE 2: *Experiment 1: Coefficients of slide friction μ tested for significance of $\mu_{(RU)}$ versus $\mu_{(SZ)}$*

Material	Direction	Mean µ	SD	N	t-test
glass	RU	0.77	0.30	920	
	SZ	0.78	0.30	920	
wood	RU	0.41	0.20	900	**
	SZ	0.46	0.25	900	
plastic	RU	0.53	0.21	920	**
	SZ	0.57	0.22	920	
stone	RU	0.54	0.27	920	**
	SZ	0.64	0.32	920	

* difference is probable (p=5%)
** difference is significant (p=1%)
SZ: push/pull
RU: radial/ulnar

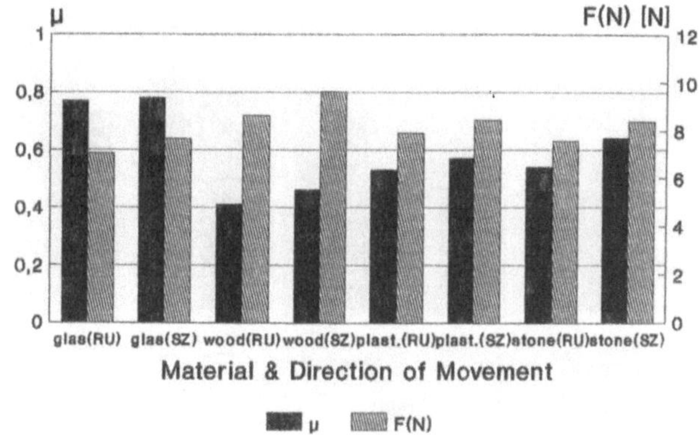

Figure 3 - Experiment 1, slide friction: Mean coefficients of friction $\mu_{(SZ)}$ and $\mu_{(RU)}$ as percentages of mean normal forces.

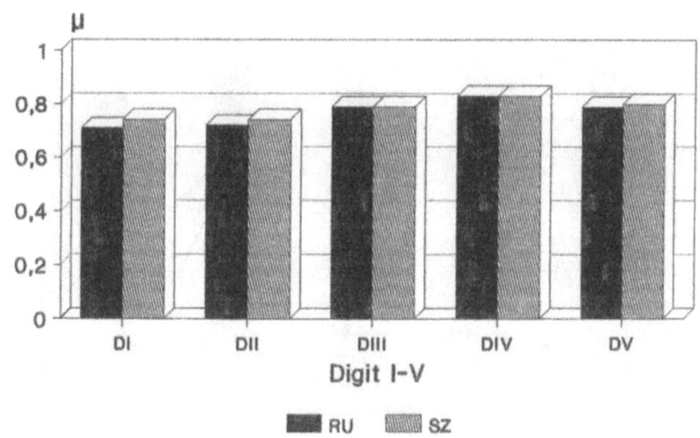

Figure 4 - Experiment 1: Coefficients of slide friction μ on glass. Comparison of mean values $\mu_{(SZ)}$ for push/pull movements and $\mu_{(RU)}$ for movements in radio-ulnar direction.

EXPERIMENT 2 («static» friction under strong pressure)

From experiment 2, a similar result is obtained as from experiment 1. Comparisons between mean coefficients of friction $\mu_{(SZ)}$ and $\mu_{(RU)}$ for all fingers on all substrates show a similar pattern in both experiments (see *Table 3*). For the direction SZ values of the coefficients are larger than those for the direction RU (see *Figure 8*). For glass, wood, and plastic this difference is highly significant (p = 1%; t-test). For stone a significant differ-

TABLE 3: *Experiment 2: Coefficients of static friction μ; mean values and comparison between $\mu_{(SZ)}$ and $\mu_{(RU)}$*

Material	μ(SZ)	μ(RU)	μ(SZ)-μ(RU)
glass	0.42	0.39	0.03
wood	0.41	0.38	0.03
plastic	0.36	0.33	0.03
stone	0.46	0.44	0.02

SZ: push/pull
RU: radial/ulnar

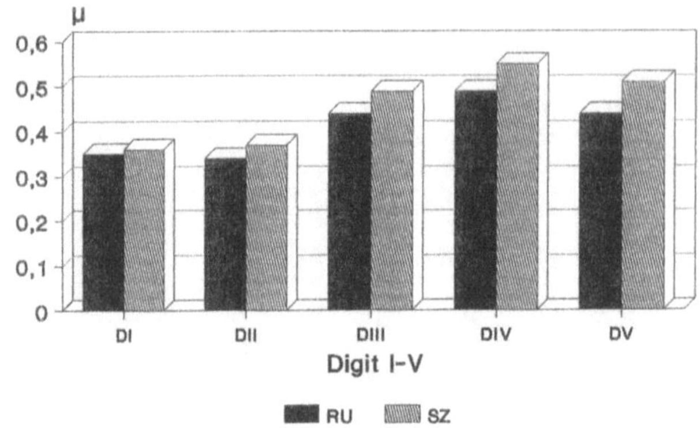

Figure 5 - Experiment 1: Coefficients of slide friction μ on wood. Comparison of mean values $\mu_{(SZ)}$ for push/pull and $\mu_{(RU)}$ for movements in radio-ulnar direction.

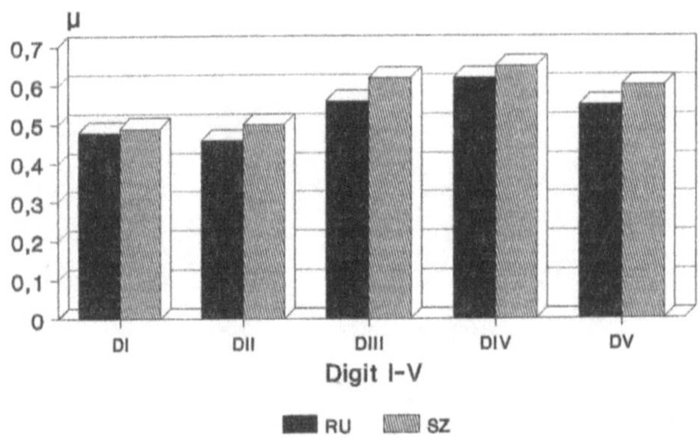

Figure 6 - Experiment 1: Coefficients of slide friction μ on plastic. Comparison of mean values $\mu_{(SZ)}$ for push/pull and $\mu_{(RU)}$ for movements in radio-ulnar direction.

TABLE 4: *Experiment 2: Coefficients of static friction μ tested for significance $\mu_{(RU)}$ versus $\mu_{(SZ)}$*

Material	Direction	Mean μ	SD	N	t-test
glass	RU	0.39	0.18	920	**
	SZ	0.42	0.20	920	
wood	RU	0.38	0.15	900	**
	SZ	0.41	0.16	900	
plastic	RU	0.33	0.11	920	**
	SZ	0.36	0.13	920	
stone	RU	0.44	0.17	900	*
	SZ	0.46	0.17	900	

* difference is probable (p=5%) SZ: push/pull
** difference is significant (p=1%) RU: radial/ulnar

ence (p=5%) was found (see *Table 4*). In relation to the exerted normal forces, the differences between the four materials are less pronounced than in experiment 1. The coefficients of friction vary from 0.3 to 0.5. Normal forces vary from 21.8 N to 24.8 N, and are thus increased by a factor 2.5 to 3, as compared to experiment 1. Forces acting in the proximo-distal direction SZ are always equal to or bigger than those acting in the radio-ulnar direction RU (see *Figure 9*).

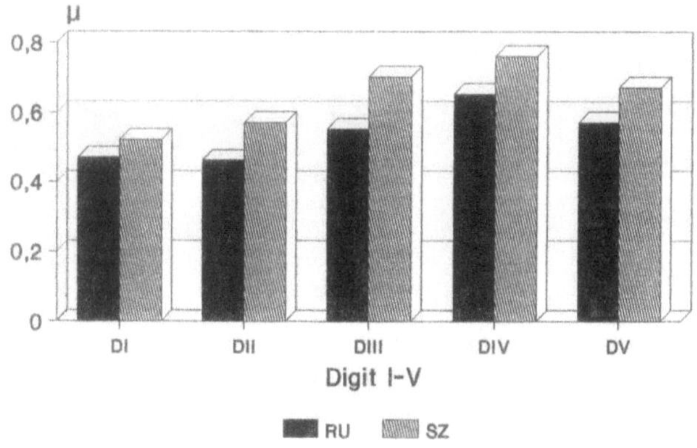

Figure 7 - Experiment 1: Coefficients of slide friction μ on stone. Comparisons of mean values $\mu_{(SZ)}$ for push/pull and $\mu_{(RU)}$ for movements in radio-ulnar directions.

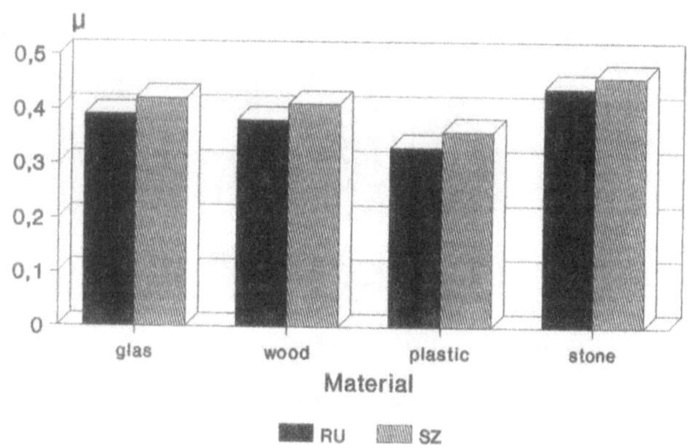

Figure 8 - Experiment 2: Coefficients of static friction μ on various materials. Comparison of mean values in radio-ulnar direction ($\mu_{(RU)}$) with push/pull ($\mu_{(SZ)}$).

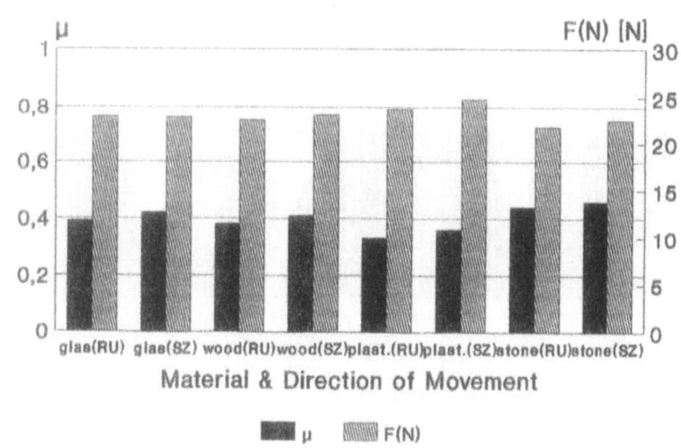

Figure 9 - Experiment 2, static friction: Mean coefficients of friction against push/pull $\mu_{(SZ)}$ and against movements in radio-ulnar directions $\mu_{(RU)}$ as percentages of the means of the respective normal forces $F_{(N)}$.

For all five fingers and all four materials the coefficients $\mu_{(SZ)}$ are larger than those of $\mu_{(RU)}$ (see *Figures 10-13*). For glass, wood, and stone the smallest differences are always found in finger I (thumb). Only on plastic the thumb does not deviate from the other fingers. More pronounced differences exist in fingers II — V.

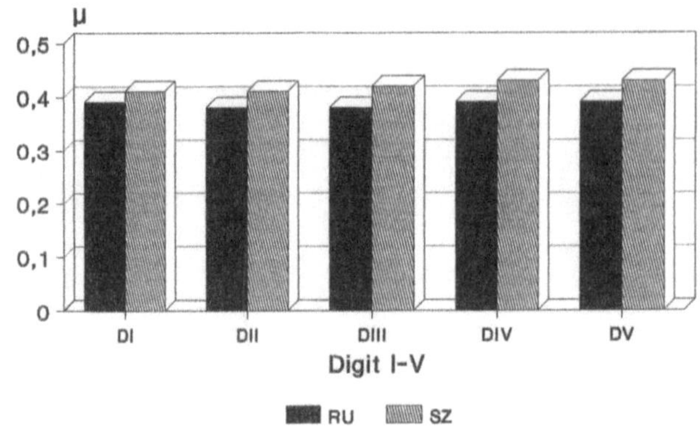

Figure 10 - Experiment 2: Coefficents of static friction on glass. Comparisons of mean values $\mu_{(SZ)}$ for push/pull and $\mu_{(RU)}$ for movements in radio-ulnar directions.

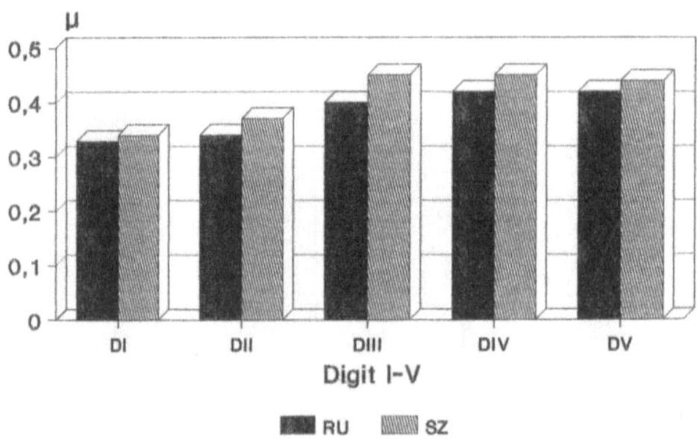

Figure 11 - Experiment 2: Coefficients of static friction on wood. Comparison of mean values $\mu_{(SZ)}$ for push/pull and $\mu_{(RU)}$ for movements in radio-ulnar direction.

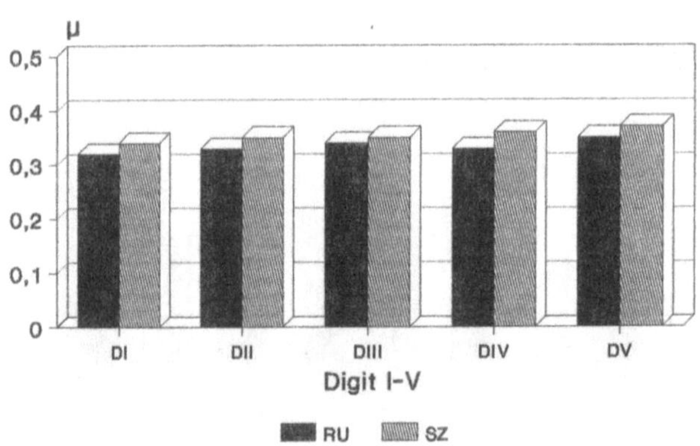

Figure 12 - Experiment 2: Coefficients of static friction on plastic. Comparison of mean values $\mu_{(SZ)}$ for push/pull and $\mu_{(RU)}$ for movements in radio-ulnar direction.

In the course of the investigation another question raised itself: are there differences in the pattern of friction between the two directions of finger movement «push» (distally) and «pull» (proximally)? A comparison of the mean coefficients of friction $\mu_{(S)}$ and $\mu_{(Z)}$ shows indeed higher values for pulling, $\mu_{(Z)}$, than for pushing, $\mu_{(S)}$. This holds true for both, experiment 1, and experiment 2.

In experiment 1, the difference between coefficients of friction are significant for glass (p = 1%; t-test; see *Table 5*). The corresponding mean normal forces $F_{(N)}$ in «pushing» are clearly higher than in «pulling» (see *Figure 14*). These results are highly significant for glass and stone (p = 1%; t-test), for wood and plastic they are probable (p = 5%; t- test; see *Table 6*).

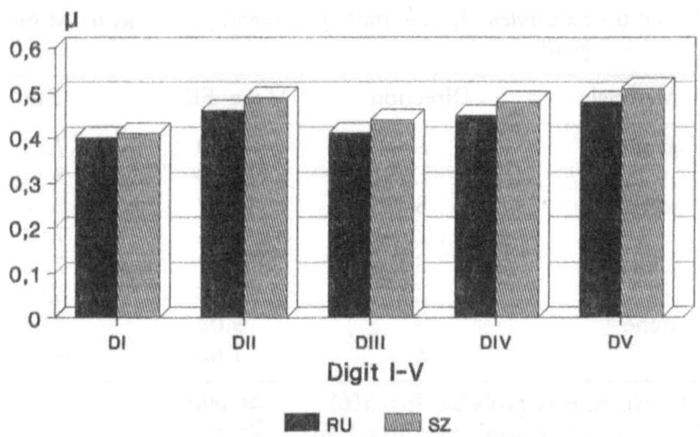

Figure 13 - Experiment 2: Coefficients of static friction on stone. Comparison of mean values $\mu_{(SZ)}$ for push/pull and $\mu_{(RU)}$ for movements in radio-ulnar direction.

TABLE 5 - *Experiment 1: Coefficients of slide friction* μ *tested for significance of* $\mu_{(S)}$ *versus* $\mu_{(Z)}$.

Material	Direction	Mean µ	SD	N	t-test
glass	S	0.75	0.27	460	*
	Z	0.80	0.32	460	
wood	S	0.42	0.23	450	**
	Z	0.49	0.27	450	
plastic	S	0.53	0.20	460	**
	Z	0.61	0.23	460	
stone	S	0.59	0.30	460	**
	Z	0.70	0.34	460	

* difference is probable (p = 5%)
** difference is significant (p = 1%)

S: push
Z: pull

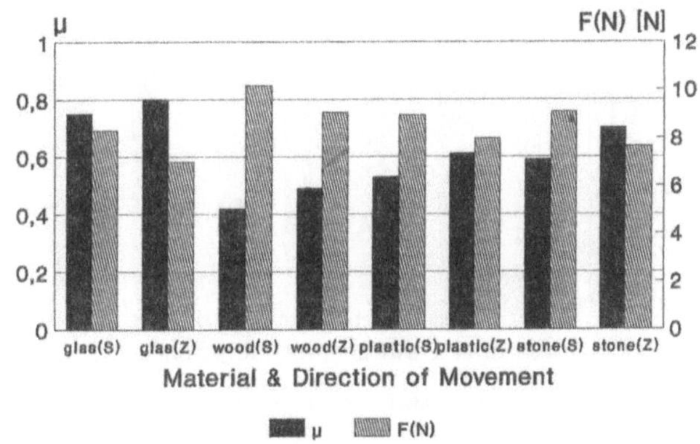

Figure 14 - Experiment 1: Mean coefficients of slide friction $\mu_{(S)}$ and $\mu_{(Z)}$ as percentages of the respective mean normal forces $F_{(N)}$.

In experiment 2, the analysis leads to nearly identical results: Mean coefficients of friction for pulling $\mu_{(Z)}$ are significantly greater for wood and plastic than the coefficients for pushing $\mu_{(S)}$ ($p = 1\,\%$; t-test). For the materials glass and stone no significant differences can be found (see *Table 7*).

TABLE 6 - *Experiment 1, slide friction: Normal force* $F_{(N)}$ *tested for significance of* $F_{(NS)}$ *(push) against* $F_{(NZ)}$ *(pull).*

Material	Direction	Mean F(N)	SD	N	t-test
glass	S	8.30	7.63	460	**
	Z	7.00	5.31	460	
wood	S	10.17	7.30	450	*
	Z	9.06	7.01	450	
plastic	S	8.92	7.10	460	*
	Z	7.95	6.67	460	
stone	S	9.08	7.63	460	**
	Z	7.62	6.29	460	

* difference is probable (p=5%) S: push
** difference is significant (p=1%) Z: pull

TABLE 7 - *Experiment 2: Coefficients of static friction* μ *tested for significance of* $\mu_{(S)}$ *versus* $\mu_{(Z)}$. *Symbols as in table 2.*

Material	Direction	Mean μ	SD	N	t-test
glass	S	0.42	0.19	460	
	Z	0.43	0.22	460	
wood	S	0.39	0.15	450	**
	Z	0.42	0.17	450	
plastic	S	0.34	0.12	460	**
	Z	0.37	0.14	460	
stone	S	0.45	0.16	450	
	Z	0.47	0.19	450	

S: push
Z: pull

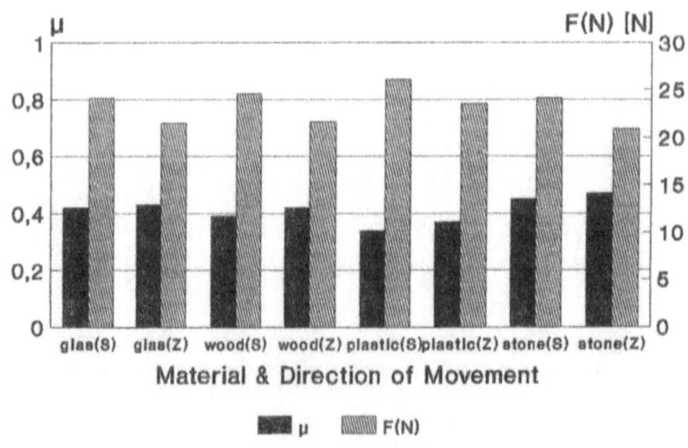

Figure 15 - Experiment 2: Mean coefficients of static friction $\mu_{(S)}$ and $\mu_{(Z)}$ as percentages of the respective normal forces $F_{(N)}$.

In experiment 2, the normal force $F_{(N)}$ acting in the direction of the fingers' movement is greater in the pushing movement than in pulling (see *Figure 15*). The differences between the mean normal forces $F_{(N)}$ are highly significant ($p=1\%$; t-test) for glass, wood, and stone, for plastic a probable difference has been found ($p=5\%$; t-test; see *Table 8*).

TABLE 8 - *Experiment 2, static friction: Normal force* $F_{(N)}$ *tested for significance of* $F_{(NS)}$ *against* $F_{(NZ)}$. *Symbols as in table 2.*

Material	Direction	Mean F(N)	SD	N	t-test
glass	S	24.09	15.43	460	**
	Z	21.47	13.70	460	
wood	S	24.54	16.57	450	**
	Z	21.63	12.89	450	
plastic	S	26.05	17.86	460	*
	Z	23.55	14.17	460	
stone	S	24.14	14.93	450	**
	Z	20.86	11.72	450	

S: push
Z: pull

Discussion

Relation between normal force and friction

The relation between normal force $F_{(N)}$ and the frictional force $F_{(R)}$ are best described by an exponential function of the type

$$F_{(R)} = K \cdot F_{(N)}^{n} \qquad \text{(equ. 2)}$$

The correlation coefficients set out in *Table 9* and *10* vary from $r=0.67$ to $r=0.86$. The correlation coefficients are great enough to show that the theoretical relation can be accepted as a good description of the data obtained experimentally. This is the same exponential function that has been deduced and considered to be optimal by BOWDEN & TABOR (1964) for describing the frictional resistance between visco-elastic plastic material and hard surfaces. It evidently also holds true for the frictional propensities of the skin on the fingertips equipped with dermal ridges.

The difference between our experimental results and those of BOWDEN & TABOR reflects the fact that we worked with living skin and the normal force had to be exerted actively by the subjects' muscle power. In contrast, BOWDEN & TABOR have put in contact solid materials, whereby the visco-elastic materials were exposed to a moving force. The exponential relation established by these authors has also been used in the experiments conducted later by COMAISH & BOTTOMS (1971), EL SHIMI (1977), and CARTMILL (1979) for the interpretation of their data. However, the factor K and the exponent n in the function have been assigned different quantitative values in the various experiments.

EL SHIMI (1977), who had used in his experiments stainless steel on the skin of the forearm, found values between 0.18 and 0.82 for the factor K, and for the exponent n

TABLE 9 - *Experiment 1, slide friction: Correlation coefficients r between normal force* $F_{(N)}$ *and friction resistance* $F_{(R)}$ *for push/pull (SZ) and for movements in radio-ulnar directions (RU).*

Material	r (SZ)	r (RU)
glass	0.81	0.85
wood	0.76	0.84
plastic	0.86	0.85
stone	0.73	0.79

TABLE 10 - *Experiment 2, static friction: Correlation coefficients r between normal force* $F_{(N)}$ *and friction resistance* $F_{(R)}$ *for push/pull (SZ) and for movements in radio-ulnar directions (RU).*

Material	r (SZ)	r (RU)
glass	0.68	0.67
wood	0.72	0.76
plastic	0.78	0.73
stone	0.74	0.76

values between 0.67 and 0.96. CARTMILL (1979) reports similar values for the exponent n. He has used wood as solid material and as visco-elastic material the entire palmar and plantar surfaces of hands and feet of primates. CARTMILL's experiment has been designed in such a way that the primates stood on their hands and feet on a wooden plate. The gradient of this plate was altered until an angle was reached where subjects began to slide down from the plate. K-values discovered in these designs were always bigger than one.

The differences in K- and n-values in the experiments reported so far are probably caused by the different experimental conditions. For theoretical reasons a «range of tolerance» from 0.67 to 1.0 can be deduced for the exponent n. The n-values obtained experimentally should fall within this range. A value of n = 0.67 should be found, if friction causes a purely «elastic» or reversible deformation. A value of n = 1.0 should be observed, if a «plastic» or permanent deformation is caused by friction. In this case the exponential approximation is transformed into a linear relation, which is also known as COULOMB's law of friction. The values of the exponent n we found in experiments 1 and 2 for different materials and directions of movement vary from 0.7 to 1.0. Therefore, the n-values fall in the range of tolerance mentioned above. This result means, that the skin possesses «plastic» (= permanent) as well as «elastic» (= reversible) characteristics of deformation.

The n-values in experiment 2 are higher than those in experiment 1. The normal forces are increased by the factor 2.5 to 3 in experiment 2 as compared to experiment 1. A possible reason for this could be that with an increase of the normal force applied the «plastic» characteristics of the skin become more important than its «elastic» characteristics. This may be the result of the compression of the skin between the bony terminal phalanx of the finger and the solid surface.

The K-values discovered in our experiments ranged from 0.4 to 1.2. As can be seen from *Tables 11* and *12*, the K-values for both directions, SZ and RU, increase from about 0.4 (wood), through plastic and stone to 1.2 for glass. The mean normal force in experiment 1 is smaller than 10 N. In contrast, the K-values in experiment 2 vary only

TABLE 11 - *Experiment 1, slide friction: Values of the factors K and n in the function* $F_{(R)} = K \cdot F_{(N)}^n$, *or* $\log F_{(R)} = \log K + n \cdot \log F_{(N)}$

Material	Direction	K	n
glass	RU	1.2	0.7
glass	SZ	1.2	0.7
wood	RU	0.4	1.0
wood	SZ	0.6	0.8
plastic	RU	0.7	0.8
plastic	SZ	0.9	0.7
stone	RU	0.7	0.8
stone	SZ	1.0	0.7

TABLE 12 - *Experiment 2, static friction: Values of the factors K and n in the function* $F_{(R)} = K \cdot F_{(N)}^n$.

Material	Direction	K	n
glass	RU	0.5	0.9
glass	SZ	0.5	0.9
wood	RU	0.5	0.9
wood	SZ	0.6	0.8
plastic	RU	0.6	0.8
plastic	SZ	0.5	0.9
stone	RU	0.6	0.9
stone	SZ	0.6	0.9

from 0.5 to 0.6. Here the mean normal force is higher than 20 N. Therefore, we can state that the K-values are dependent on the kind of material used, if low normal force (less than 10 N) is exerted. This dependency disappears with higher forces.

As *Tables 11* and *12* show, differences in the K-values are found also with respect to the different directions of movement. So, the values for the direction SZ are higher than those for the direction RU for wood, plastic, and stone. For glass values are alike. We therefore attribute the differences in the K-values for the different substrates to the differences in the structure of the surface of those substrates.

The differences in the friction evoked by movements in different directions which we found on 3 out of 4 substrates are probably caused by the different orientation of the dermatoglyphic ridges. The effect of the dermal ridges seems to decrease with increasing normal force, which is also plausible. For the exponent n, no differences depending on either material or normal force were found.

As already stated, the most important difference between experiment 1 and experiment 2 is the amount of normal force exerted. While in experiment 1, subjects had been asked to slide their fingers over the test substrate «*ad libitum*» — with the consequence that the normal forces measured were relatively low —, in experiment 2 subjects were instructed to press their fingers strongly on the substrates and then to try to push or pull them away. This was done to induce static friction. With these big differences in the forces acting (in experiment 2 the normal forces are 2.5 to 3 times higher than in experiment 1) it seemed inappropriate to compare the respective mean coefficients of friction over the whole range of forces. Rather, the comparison was carried out separately for the two possible directions of movements SZ and RU on the different materials for a mean range of normal force between 10 N and 20 N, where the μ-values overlap.

In the literature (GERTHSEN, 1982) on coefficients of static and slide friction, the following values of μ are documented for materials which resemble those used in our experiments.

system	teflon/teflon	static friction: 0.04
		slide friction: 0.02 - 0.04
system	rubber/asphalt	static friction: 0.8 - 1.1
		slide friction: 0.7 - 0.9

From these data it follows that the coefficients of static friction generally are higher than those of slide friction, while differences between the coefficients vary from 0 to 0.2.

The coefficients of friction calculated from our data are listed in *Table 13.* As can be seen, the coefficients for the materials stone and wood for the direction SZ are higher in experiment 2 than in experiment 1, if static friction is assumed. In the direction RU the value for μ for the material «stone» is higher as well, and for wood the μ-value is nearly the same as in experiment 1.

For the materials «glass» and «plastic» the coefficients of slide friction (experiment 1) are increased by 0.1 or 0.2, respectively, if compared to experiment 2.

For the explanation of this discrepancy a result obtained by ADELMAN (1975) seems to be important. The authors found it when investigating the sweat secretion at the soles during locomotion. Dogs have been tested on a tread mill with changeable slope. ADELMAN found an increase of the coefficient of (static) friction with increasing sweat secretion. ADAMS & HUNTER (1969) have investigated the paws of cats under different conditions and achieved a result directly opposite to the one of ADELMAN. In fact, they found a decrease of the coefficient of friction with increasing sweat secretion. According to ADELMAN the difference results from the fact that the plastic plate used for the experiment was in permanent contact with the paw. Beneath the paw, larger amounts of sweat could accumulate, since it was held in place passively for extended time. In the experiments of ADELMAN, however, during walking on the treadmill only a limited amount of sweat was accumulated and that remained on the surface of the treadmill after lift off of the paws. Our experiment 2 resembled the experiment of ADAMS & HUNTER in so far as there was a prolonged contact between the fingertip and surface of the material.

To summarize, the difference between the μ-coefficients in slide friction (exp. 1) and static friction (exp. 2) is in the predicted direction for materials «stone» and «wood», i. e. the materials with more structured surfaces produce greater μ-values when static friction is induced. However, the differences for the smooth materials «glass» and »plastic» are the other way round: higher μ-values are found in slide friction. We assume that sweat secretion (which was pronounced in numerous subjects) is the cause for the differences of the coefficients of friction shown in *Table 13.* The differences between experiment 1 (slide friction) and experiment 2 (static friction) are in the direction to be expected from a physical point of view: in the materials with more structured surfaces, like wood and stone, friction is greater than in the less structured materials «glass» and «plastic».

We conclude that the results of our *in vivo* experiments correspond excellently to those which have been obtained by other authors who investigated friction between skin or other visco-elastic materials and different solid materials, like steel, wood, plastic or stone. With our results we feel on firm ground.

TABLE 13 - *Comparison of coefficients of friction μ under normal forces between 10 N and 20 N.*

Material	Direction	Exp. 1 (slide)	Exp. 2 (static)
glass	RU	0.58	0.42
glass	SZ	0.60	0.44
wood	RU	0.41	0.40
wood	SZ	0.38	0.43
plastic	RU	0.42	0.36
plastic	SZ	0.46	0.38
stone	RU	0.42	0.48
stone	SZ	0.46	0.50

RU: movements in radio-ulnar directions,
SZ: push/pull

Effect of the dermal ridges on friction

We have also tried to compare the coefficients $\mu_{(R)}$ and $\mu_{(U)}$ and found no difference in both directions. As can be seen from *Tables 1* to *4* the coefficients μ for the direction SZ are always higher than those for RU. This is true for the measurements of each finger, each subject, and all substrates tested (data pool: 1) and 2)).

In experiment 1, the greatest differences between $\mu_{(SZ)}$ and $\mu_{(RU)}$ have been found for the two materials with more structured surfaces, «wood» and »stone», while the differences are smaller for the smooth material »plastic». For «glass» the difference is least. With the exception of «glass» all differences are highly significant ($p = 1\%$).

In experiment 2 the differences $\mu_{(SZ)} - \mu_{(RU)}$ are smaller, though more uniform. Nevertheless, all differences are statistically significant (glass, plastic, and wood: $p = 1$ %; stone: $p = 5$ %).

μ-values of each finger show the same pattern of results as those of all fingers taken together.

We conclude that the higher coefficients of friction obtained in movements in proximo-distal direction are caused by the fact that in these cases the vector of friction acts perpendicularly to the dermal ridges.

In the radio-ulnar direction RU, however, the vector acts in parallel to the ridges. Thus, in the pulling or pushing movement perpendicular to the dermal ridges additional power is needed to overcome the «obstacle» dermal ridge. This power is not needed, if the finger is moved in parallel to the ridges. In other words: in gripping, climbing, or walking the dermal ridges produce a greater frictional force, if the movement is carried out in proximo-distal, rather than radio-ulnar direction.

The quantitative relations between the forces and the coefficients of friction are displayed in *Figures 3* and *9*. In these histograms the arithmetic means of the coefficients of friction are plotted against the means of the respective normal forces. *Figures 16* and *17* are graphic representations of the μ-values obtained for identical normal forces on each kind of substrate. These μ-values were calculated from regressions based on the μ-values for the directions SZ and RU. From these figures can be seen that in experiment 1 friction on the material «glass» increases by 17%, if the same normal force is acting in the direction SZ, as compared to the direction RU. For the material «stone» this increase amounts to 31 %. For wood an increase of 22%, and for plastic an increase of 21% is observed.

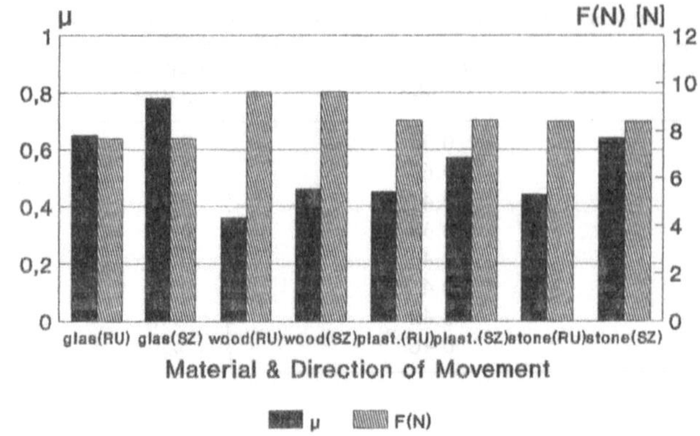

Figure 16 - Experiment 1: Mean coefficients of slide friction against push/pull $\mu_{(SZ)}$ and against movements in radio-ulnar directions $\mu_{(RU)}$ as percentages of identical normal forces $F_{(N)}$.

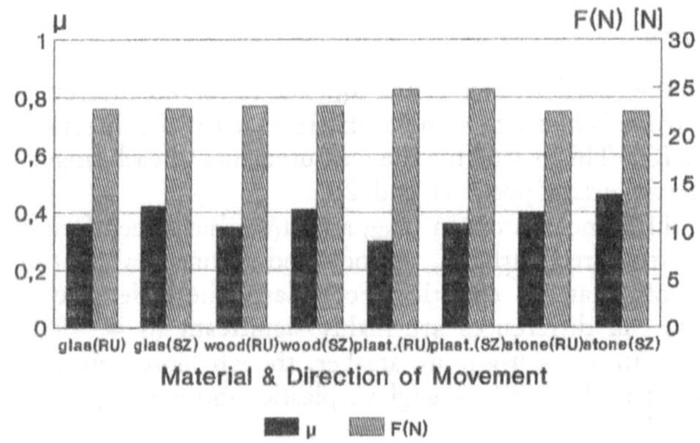

Figure 17 - Experiment 2: Mean coefficients of static friction against push/pull $\mu_{(SZ)}$ and against movements in radio-ulnar directions $\mu_{(RU)}$ as percentages of the means of identical normal forces $F_{(N)}$.

In experiment 2 the coefficients of friction increase for 14% with glass, for 15% with wood, for 17% with plastic, and for 13 % with stone.

Because of the notorious variability in these measurements, the above percentages should be understood as indicating the range of magnitude of the effect the dermatoglyphic ridges exert on friction. Under the relatively low normal force in experiment 1 the structure of the surface of a given material seems to be more important than in experiment 2, where the normal forces exerted were 2.5 to 3 times greater.

Because subjects spontaneously judged pulling (Z) the finger as more pleasant than pushing (S), we wanted to know, if the subjects' feelings were caused by some property of the dermal ridges. To compare the mean coefficients of Z and S, the total of the coefficients yielded for «SZ» was split into subsamples «S» and «Z». This comparison is displayed in *Tables 5* and *7*, and in *figures 14* and *15*. As expected, there are indeed differences between the means of the coefficients $\mu_{(S)}$ and $\mu_{(Z)}$. However, these differences are confounded with differences in the respective normal forces. In the sense of:

$$\mu = K \cdot F_{(N)}^{\,n-1} \qquad \text{(equ. 4)}$$

a higher value of $\mu_{(Z)}$ is combined with a lower value of $F_{(N)}$ in the direction «pull» (Z), and vice versa in the direction «push» (S). That means that the differences of the mean values $\mu_{(Z)}$ and $\mu_{(S)}$ — although significant — cannot be caused by the dermal ridges. If an effect of the ridges would exist, the relations between the coefficients of friction and the respective normal forces should be exactly the other way round like in the comparison between the values $\mu_{(SZ)}$ and $\mu_{(RU)}$.

The fact that the pulling movement of the finger is perceived as more pleasant, can be explained by the fact that in pushing the finger the pulp of the fingertip is torn away from the nail, which causes a feeling of pain. For pulling this is not the case.

As indicated in the «Results», for all test materials the smallest differences in the μ-values were found in finger I (thumb). This is probably due to the fact that the dermal ridges at the tip of the thumb are not orientated exactly from the radial to the ulnar side, as in all the other fingers, but instead they are bent conspicuously. Therefore, the thumb was the only finger, on which the ridges have not been exactly in the presumed orientation relative to the force exerted.

Experimental results and morphological correlates

The pattern of the ridges on the thumb leads to an interesting effect: in the precision grip, i. e. gripping with the tips of thumb and index or any other finger, ridges of the thumb are always orientated more or less perpendicularly to the ridges of the opposing finger, thereby achieving an optimization of the friction propensities.

Whenever the human hand is used for gripping or other manipulations, the primary areas of contact between skin and material are the fingers and the fingertips, which are equipped with patterns of dermatoglyphic ridges. Our results suggest that an arrangement of the ridges in patterns of concentric circles would lead to ideal adhesion in all directions. In such a pattern the vectors of the resulting frictional forces act always perpendicular to a number of ridges.

Investigating human embryos BONNEVIE (1929), however, showed that during the development of the pads at the fingers — covered with papillar ridges — a development of strictly concentric dermal ridges is impossible. Instead patterns of whirls and loops can be found, and whirls approach the ideal of a concentric pattern.

NIEMITZ (1977) observed in the hand of *Tarsius bancanus* the characteristic double rows of dermal papillae corresponding to the epidermal ridges. The latter are deflected towards the middle of the hand by an angle of ca. 45°. Furthermore, the fibrocytes and the collagenous fibres running from the depth of the pads towards their surfaces are arranged perpendiculary to the orientation of the dermal ridges. Consequently, if the vector of the most important frictional force acts perpendiculary to the ridges, the fibres are arranged parallel to those forces.

NIEMITZ stresses that the way in which the epidermis is tied to the corium is adapted optimally to the occurrence of forces acting perpendiculary on the dermal ridges. In his view the development of patterns of dermal ridges reflects a — possibly mechanical — function of the ridges. The findings of NIEMITZ (1977) are supported by stereoscan investigations of the connection between epidermis and corium by MISUMI & AKIYOSHI (1984).

These results show that by means of their ties to the dermal structures the dermatoglyphic ridges are adapted better to a frictional force acting perpendiculary, than to a force exerted parallel to their orientation.

From the results of our own experiments we conclude that the dermatoglyphic ridges in primates serve a biomechanical function. This conclusion does by no means preclude any other function of the dermatoglyphics, such as inforcement of tactile perception.

Although our experiments were confined to the finger tips there is no reason to believe that our results do not hold true for the ridge patterns at other parts of the palms or soles.

ACKNOWLEDGEMENT — The idea to investigate the problem was proposed by Prof. Preuschoft. We would like to thank him and the other members of the subdepartment for the stimulation and constructive criticism given during the study. Our thanks go also to Prof. Muhr for permission to use the working facilities in the biomechanic laboratory of the Surgical Clinic of the «Berufsgenossenschaftliches Krankenhaus Bergmannsheil». Dr. K. H. Buck has supported our work by providing a computer and the computer programs, and Mrs. M. Wischniwetzki has drawn or finished the figures.

References

ADAMS T. & HUNTER W.S., 1969. *Modification of skin mechanical properties by eccrine sweat gland activity.* J. Appl. Physiol. 26: 417-419.

ADELMAN S., 1975. *Sweating on paws and palms: what is its function.* Am. J. Physiol. 229: (5): 1400-1402.

ANDRES K.H., 1973. *Morphological criteria for the differentiation of mechanoreceptors in vertebrates.* Rheinisch-Westfälische Akad. Wiss., Band 53, Westdeutscher Verlag: 135-152.

BOWDEN F.P. & TABOR D., 1964. *The friction and lubrication of solids, Part II.* Oxford Univ. Press, London.

CARTMILL M., 1974. *Pads and claws in arboreal locomotion.* In: F. A. Jenkins, Primate Locomotion. Academic Press, New York.

CARTMILL M., 1979. *The volar skin of primates: Its frictional characteristics and their functional significance.* Am. J. Phys. Anthrop. 50: 497-510.

CHRISTEL, this volume.

COMAISH S. & BOTTOMS E., 1971. *The skin and friction: deviations from Amonton's laws, and the effects of hydration and lubrication.* Brit. J. Dermatol. 84: 37-43.

EL-SHIMI A.F., 1977. *In vivo skin friction measurements.* J. Soc. Cosmetic Chem. 28: 37-51.

GERTHSEN C., 1982. *Physik.* Springer Verlag Berlin, Heidelberg, New York: 34-36.

MISUMI Y. & AKIOSHI T., 1984. *Scanning electron microscopic structure of the fingerprint as related to the dermal surface.* The Anatomical Record 208: 49-55.

NAYLOR P.F.D., 1955. *The skin surface and friction.* Brit. J. Dermatol. 67: 239-248.

NIEMITZ C., 1977. *Zur funktionellen Anatomie der Papillarleisten und ihrer Muster bei Tarsius bancanus borneanus Horsfield, 1821.* Z. Säugetierk. 42: 321-346.

NIEMITZ C., 1990. *The evolution of primate skin structures in relation to gravity and locomotor patterns.* In: Jouffroy, F. K., Stack, M. H. and Niemitz, C. (eds.): Gravity, Posture and Locomotion in Primates. Editrice «Il Sedicesimo» Firenze: 129-156.

SACHS L., 1968. *Statistische Auswertungsmethoden.* Springer-Verlag, Berlin.

SCHLAGINHAUFEN O., 1905. Das Hautleistensystem der Primatenplanta unter Mitberücksichtigung der Palma. Gegenbaurs Morph. Jb. 34: 577-671.

SULZBERGER M.B., CORTESE T.A., FISHMAN L. AND WILEY H.S., 1966. *Studies on blisters produced by friction. I. Results of linear rubbing and twisting technics.* J. Invest. Derm., 47: 456.

WHIPPLE I.L., 1904. *The ventral surface of the mammalian chiridium with special reference to the conditions found in man.* Z. Morphol. Anthropol. 7: 261-368.

III

HAND DEVELOPMENT

M.F. Gutmann

The constructional preconditions of the basic organization of the tetrapod limb

In the phylogenetic model presented it is shown and explained that the basic anatomical pattern of the human hand and of the whole tetrapod extremity is determined in the early stages of tetrapod evolution. The interaction of bending movements of the body and supporting function of the paired fins during ambulation enforces the subdivision of the former paired fins by joints into a proximal part with one bone, a medial portion with crossable paired structures, and an autopodial region with a multitude of skeletal structures which become distally differentiated into the digits. The separately movable digits allow the adjustement of the distal part of the autopodium to the irregular contours of the substrate.

Only the existent configuration garantees the unimpaired combination of body flexures and ambulatory movements of the limbs while the autopodium can remain fixed on the ground. The observable configuration can be conclusively shown to represent the most economical solution possible.

The specific properties of primate arms and hands can only be conceived as phylogenetically late modifications of the invariant basic pattern of the tetrapod limb.

Key words: autopodium, evolution of tetrapods, metapodium, paired fins, phylogenetic transformation, pronation, stylopodium, zeugopodium.

Introduction

The aims of comparative morphology lie in the elucidation of configurational similarities between different organismic types. In the context of such studies organisms are usually not conceived as living mechanical systems, as «constructions». Consequently comparison of form yields only sequences of morphological configurations. If the methodological maxims of morphology are applied, unpolarized morphoclines can be set up. Morphoclines are very often interpreted by implicit and clandestine functional interpretations which suggest a polarity in the unpolarized morphological series. The great majority of authors seems to be satisfied with these morphological procedures, the unreflected suggestions of polarity, and reject the postulate that phylogenetic sequences might be directly set up by phylogenetic reconstructions explaining the necessary transformation steps (OSCHE, 1973; REMANE, 1985).

The organization of the paired fins of fishes and tetrapod limbs provide a good example for morphological similarities found by applying the criteria of homology research. The polarity of the morphocline, which is based on the supposition that piscine characters are ancestral and tetrapodal features derived, is universally accepted although a gradualistic interpretation of the transformation series is still lacking. The basic pattern of the skeletal elements in the limbs remains rather unchanged from the amphibians to the mammals and especially the primates. All deviations from the canonical pattern occuring in several mammal groups are considered to be secondary and irreversible transformations. In such a perspective the pattern of the human hand and the whole extremity are described as primitive because they display the same basic pattern of the limb as in primitive amphibians. In contrast to that, the use of the primate hand must be interpreted as extremely specialized and adapted to very specific environmental conditions. It is obvious that further comparison of morphological limb structures and additional studies of the usage of the extremities and the more specific adaptations of primates will not provide a better understanding of the basic limb construction and an insight into the selective constraints responsible for its arisal and persistence in the long phylogenetic history.

Sometimes a solution of this problem is sought by historical constructional models. The one set up to explain the skeletal elements in the limbs and the position of the joints suggests a lateral position of the primitive limb which was folded and twisted in a specific way to support the body in a mammal construction and to allow pronation and supination of the lower leg (*Figure 1*; after KUMMER, from STARCK, 1979). As this model tries to explain the configuration by pointing to the requirements in the late mammal stage of evolution it is teleological and therefore without explanatory value. The use of this model should be restricted to didactic purposes. Since the tetrapod limb appears in its perfected shape in primitive Devonian amphibians, constructional preconditions and the decisive selective constraints must have existed in the preceding piscine stages of phylogeny.

Therefore there can be little doubt that the solution must be worked out by reconstructing the transitory stages between fish constructions and the perfected tetrapod organization (GUTMANN, 1988; PETERS & GUTMANN, 1971; PETERS, 1985).

Attempts to arrive at such solutions by comparisons of the shapes of pretetrapod fish fins in crossopterygians were futile, because the pattern of bones in these fishes is very variable and in no way prospectively illuminating for the tetrapod situation. As in all cases of phylogentic transition, not the multiplication of comparative work is needed to corroborate suspected homologies, but the formulation of reasonable reconstructions of the transformation steps in order to establish the validity of suspected homologies. If homologies are defined as structures of common descent the only way of putting them on a firm foundation is by reconstruction of the transformation process leading to the observable organization (PETERS & GUTMANN, 1971; GUTMANN, 1977).

Steering in fish constructions

As the limbs of tetrapods show the same configuration as the paired fins of fishes, the constructional preconditions must be found on the fish-level of evolution. In additon to propulsion by lateral undulation almost every fish construction can perform up and down bending of the body for steering purposes. Normally the paired fins are used for steering upward and downward. In fully-developed fish constructions with two pairs of fins the energy consuming mode of steering by bending the body is only needed for sharper ascents or descents for which the paired fins are insufficient or while velocity is low and the fins cannot work effectively (GUTMANN & BONIK, 1978).

This is highly indicative of the selective advantage obtained in the phylogenetic development of paired fins. More precise information was elucidated by experimental work (BAINDRIDGE, 1961). It is well known that in a typical non-anguilliform fishes two pairs of fins represent the optimal piscine situation. Both pairs of fins work together to control the position of the body in the water. The anterior and posterior parts of the body are steered by separate finpairs in a very well-controlled way. If one or the other pair is amputated instability of the position of the body in water will result.

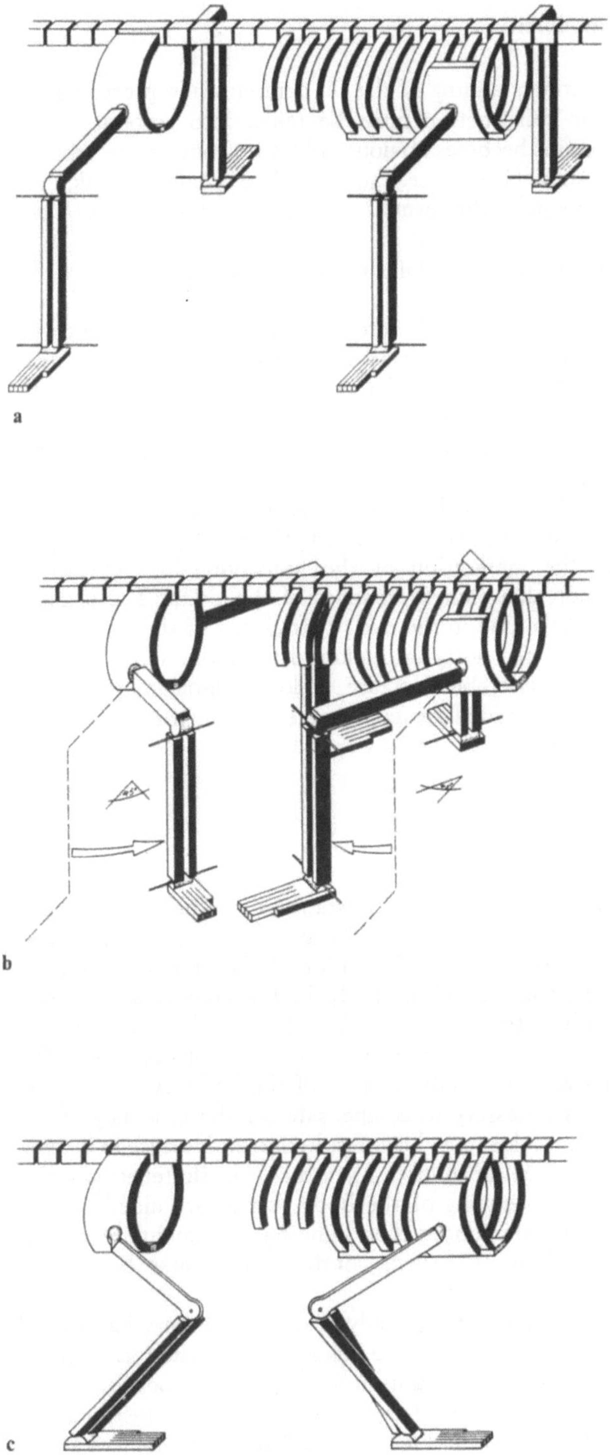

Figure 1 - Models to demonstrate the posture of the limbs in tetrapods. This sequence cannot represent the phylogenetic history because in the early stage (a) rotation of the lower legs and lower arms is not taken account of. The second stage (b) is entirely arbitrary and does not reflect the problems of the gliding movement of the autopodium on the ground. (c) The last stage suggests that pronation and supination should be conceived as a late stage of the transformation sequence. The derivation of pronation and supination from internal rotatory movements needed in lower tetrapods in which the limbs are laterally extended cannot be demonstrated. The figure also shows the basic components of the tetrapod limb: proximal stylopodium, middle zeugopodium, and distal autopodium. From STARCK after KUMMER.

The basic body construction of craniotes

In order to obtain an adequate understanding of the fin configuration preceding the paired limbs in tetrapods, the basic construction must be taken into account.

The selective value of the paired fins becomes obvious only if we compare vertebrates to other coelomates and lower chordates. Coelomates are worm-like organisms which use the whole body for propulsive movements in the form of undulations or peristalsis (*Figure 2*).

Such modes of propulsion will not automatically ensure the correct course to be steered. Thus, the direction of movement is determined by bending of the body in the required direction. Such steering movements are therefore continually superimposed on the nervously coordinated pattern of propulsive muscular activity. This means that locomotion and steering depend on muscle-generated deformations of the whole body. By involving the whole body construction such movements are energetically expensive.

In the early stages of chordate evolution the notochord appeared as an organ ensuring constancy of length. Such a restraint in an originally totally deformable coelomate worm has to be seen in connection with undulatory swimming as the sole mode of propulsion, which requires sideways bendings to travel down the body length. The high efficiency of swimming in chordates is due to the integration of the notochord into the worm construction. The notochord does not allow length changes of the body. The primary effect of the suppression of length changes lay in the rise of efficiency of the unimpaired bending movements. The appearance of the notochard allowed the reduction of circular and dissepimental muscles and the considerable increase of longitudinal muscles. The latter formed the somitic packages subdivided by connective tissue structures which interconnect muscles and the body axis.

The development of the paired fins

The incorporation of an organ ensuring length constancy into the chordate and vertebrate body was decisive for the development of tetrapod limbs in later stages. However, with regard to steering requirements, the formation of the notochord did not immediately affect the necessity to bend the whole body in the chosen directions.

Only flattening of the ventral side in the anterior section of the body could generate lift. Such a kind of lift generation became necessary in the course of cephalogenesis. The head of vertebrates established itself as a rigid anterior part of the body. As it could no longer contribute to propulsion it was necessary to compensate for the tendency of the head region to sink. Formation of lateral folds must have resulted in an increase in lift and an improvement in up and down steering capabilities for which the flattened anterior body region could be used. Steering movements of the body, which are aided by lift-generating horizontal folds, can be less pronounced than bending movements of a body lacking such structures. However, this kind of steering continued to require bending of the whole body.

A new level of efficiency in steering was attained only after the craniotes had passed the agnathan stage of evolution. The head became more voluminous by the formation of the jaw apparatus. This necessitated more lift generation in the anterior portion of the body and a more effective mode of steering. The solution opening up the transformation to the higher gnathostomes consisted in the formation of paired fins. The incipient stages

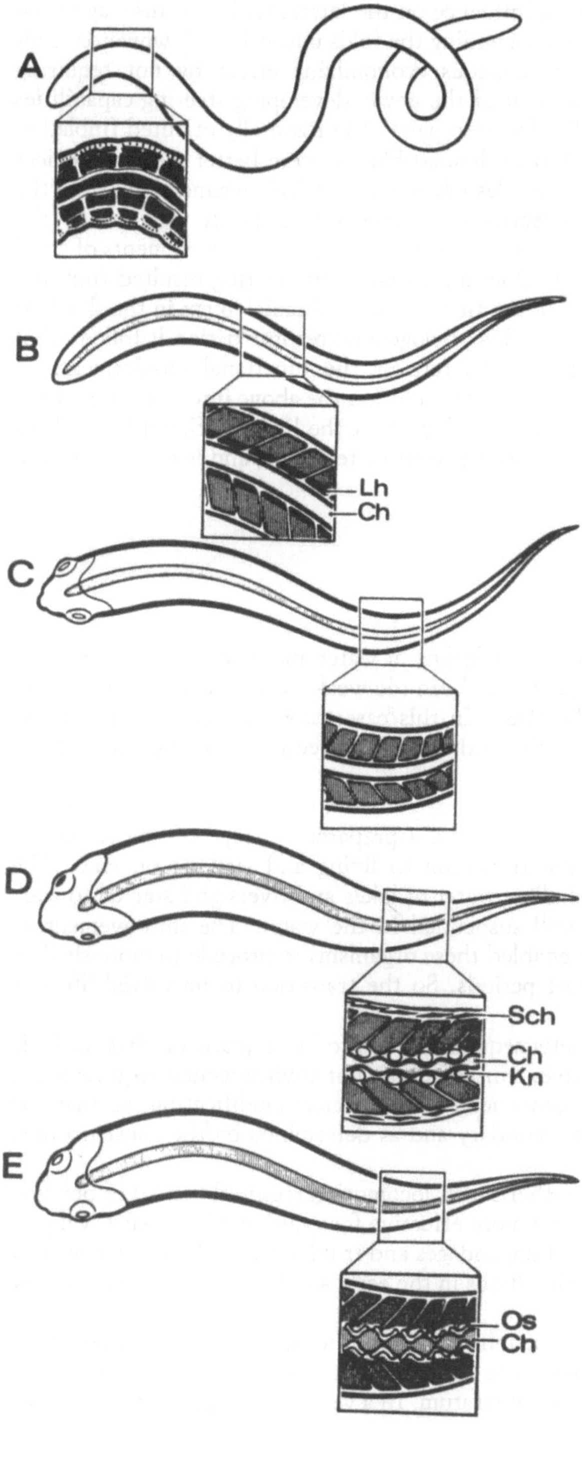

Figure 2 — Derivation of the trunk of craniotes from metameric worms.
A: Worm-like coelomate with highly deformable und fluid-filled body construction.
B: Formation of the notochord (Ch) ensuring constancy of length. In the branchiostoma stage, fluid filled coelomic spaces are present (Lh).
C: Primitive craniote.
D: Development of stabilizing skeletal structures (Sch - scales, Kn - cartilage around the notochord).
E: Prevalence of internal osseous skeletal structures (Os) as support of the body constrution. This allowed primitive tetrapods to tolerate the influence of gravity.

The existence of a length-constant body construction provides an essential constructional precondition for the development of the limbs and their subdivision by joints.

of the fin development may have consisted in the following steps: Parts of the somitic musculature could exert some deforming influence on the lateral folds. In this way minor steering movements could be generated by bending the folds upwards or downwards. This must have provided a selectively advantageous economizing effect by not requiring bending of the massive body. Improvement of the newly developing steering capabilities of the lateral folds enforced the subdivision into pairs of horizontally oriented finblades. Fins with narrow proximal portions, but undivided blades, were better steering devices and could be moved independently by muscles which were entirely emancipated from the somites of the trunk. The primary selective advantage was certainly attained by the substitution of energy consuming steering movements of the trunk by movements of small steering organs. Furthermore, a considerable improvement of steering resulted from the formation of narrow proximal portions of the fins. So the multiaxial joints in the shoulder and the hip established themselves. From this phylogenetic reconstruction it follows that the tetrapod configuration of fins and limbs reflects the functional requirements of ancestral swimming vertebrates. For this stage of evolution the above interpretation of the fin configuration in typical fishes is valid. It follows that the limb configuration reflects the steering requirements of the fish-like predecessors of tetrapods and is not determined by propulsive requirements.

The transition to terrestrial life

If fish constructions had crossed the borderline of water and land the outcome could not have been a tetrapod. The fins would have been too weak to support the trunk and to overcome the effect of gravity on the body. In this case the only mode of locomotion would have been snakelike bending of the body and total reduction of the fins (Bonik, 1978).

So the transformation process would not have ended in tetrapods. From this follows that support of the body on the paired fins and a preparatory stage of ambulation in shallow water must have preceded the transition to living and walking on land. The pretetrapods have certainly lived in shallow water of lakes and rivers and started to walk on the ground while the body was still suspended in the water. The improvement of movements acquired in this situation enabled these organisms to procede to more shallow water and to leave the water for short periods. So the transition to terrestrial life was achieved gradually.

In continuation of this phylogenetic sequence it has to be emphasized that intrinsic constructional conditions in relation to extrinsic factors, not environmental requirements, can provide the constraints for the subsequent morphological modification, so that the change of the habitat must appear as secondary and as determined by the constructional capabilities.

As the fins in pretetrapods were rather short, locomotion required the aid of bending movements of the body, which allowed a more effective foreward and backward swing of the paired appendages. This interplay of appendages and trunk reflects the involvement of the whole body frame and the developing limbs in the early and later stages of ambulatory locomotion.

The trunk movements were of course inherited from the bending movement used for swimming. In the tetrapods the motive force of the body-trunk with its massive muscles can be transmitted over the limbs to the substratum. In a very early stage undulation must have been replaced by standing waves which were not useful for propulsion in the water (EDWARDS, 1977). Nevertheless, the alternating trunk movements are responsible for the specific differentiation of the developing limbs.

As the body was rather heavy, and the fins soft, the distal parts of the fins lie flat on the ground to support the body (*Figure 3*). Because of their considerable body weight neither tetetrapods nor the early amphibians were capable of tiptoeing.

When, however, the distal portions of the fins rested on the ground, any improvement of locomotion could only come about by specific changes in the morphology of the developing limbs. In order to allow unimpaired bending movements of the trunk, the fin had to bend in three joints while its distal portions kept firm contact with the ground. One joint developed proximal at the body, one in the middle, forming the knee or elbow joints, and the third presumptive wrist and ankle regions.

The fins of the pretetrapods must have been organized according to the general pattern found in crossopterygians. Every fin was connected to the skeletal elements in the trunk on a narrow base forming a multiaxial joint. The fin blade had a skeletal support in the form of multiple osseus elements, which guaranteed reasonable flexibility and active deformability of the whole steering organ. The muscles extended over the fin blade and allowed controlled movements not only of the proximal portion of the fins, but also of the distal steering plate.

This morphological pattern became conducive of the subsequent alterations in the developing limb. In the course of the formation of the basic limb pattern all sections of the limbs, the stylopodia, zeugopodia, and the autopodia (*Figure 1*) could be moved relative to each other and in relation to the body by muscles bridging the newly forming joints between the major skeletal structures.

The dominating selective pressure leading to the formation of the limb skeleton and the joints becomes apparent if we realise the severe hindrance and disadvantageous situation that would arise if a fish with undivided fins tried to support its body on the ground and move head on. Such a movement would have caused gliding of the fin plane over the ground. Sideways movements caused by lateral bending of the trunk would have resulted in a lateral shift of the fin over the ground or a shift of the whole body in the opposite direction, thereby pushing the opposite fin over the substrate. A power stroke of the fin directed backwards to propel the body head on would require the rotation of the fin on its supporting plane. In this case locomotion would not only have been extremely clumsy but also energy-consuming, because much of the muscle force would have been expended on friction between the fin and the ground. Avoidance of such deleterious gliding movements was crucial for the development of the tetrapod limb.

Lateral dislocation of the fin plane in contact with the substrate could be reduced and eventually entirely eliminated by the subdivision of the limb into the segments mentioned above (*Figure 1, 3*). Only the autopodium remained in contact with the ground. The limb configuration just described allowed lifting the body from the ground and the independent movement of the trunk while the autopodium — hand or foot — rested firmly on the substrate.

However, the solution so far offered is still insufficient as it regards only the sideways movements of the trunk, but not the propulsion and forward movement of the body by the limbs. In order to propel the body the limbs have to swing backwards. The angular movements of the joints would allow the necessary bending of the body but could not prevent energy-consuming rotation of the autopodium over the ground. It is obvious, furthermore, that such a rotation cannot be eliminated because there is no alternative to swinging the whole fin backwards while the body moves foreward.

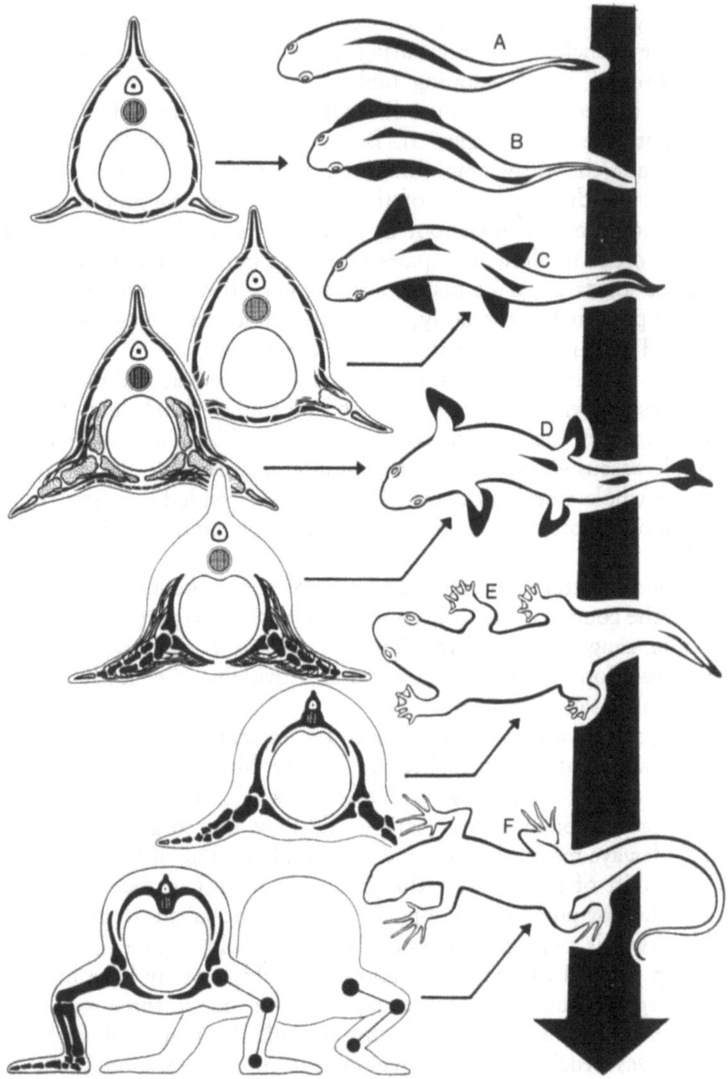

Figure 3 — Transformation of fins into limbs subdivided by joints.

A: Ancestral craniote performing propulsive movements with the whole body and steering by bending movements superimposed on the propulsive activity of the trunk.

B: Formation of lateral folds which aid steering up and down.

C: Subdivision of lateral folds into separately movable fins. Up and down steering does no longer require movements of the trunk.

D: Usage of paired fins as support of the body in shallow water. Flexibility of the fin is constituted by multiple osseous elements, and distally extending muscles provide preconditions for usage of fins for ambulation.

E: In the primitive amphibian stage subdivision of the fin by joints allows a combination of body flexures and firm support of the autopodium on the substrate.
In advanced tetrapods the body is elevated from the ground by elongated limbs.

F: The cross sections to the left show the organization of the developing limbs and the necessity of the usual subdivision of the bony bars and joints.

If we accept that the best solution is that the foot remains stationary on the ground we can infer the necessary consequences. Forward movement of the body inevitably requires rotation within the limb. This movement can occur in joints which are characterized by low frictional resistence. The demand of rotation within the limb can be met by two possible solutions: Either a second multiaxial joint is required, or two skeletal rods in one portion of the extremity are needed to comply to rotation by crossing over.

It is well known that the second solution became manifest in tetrapods. The zeugopodia in both, the fore- and hind-limbs contain two parallel and crossable skeletal rods. The parallel skeletal rods in the lower arms and lower legs are connected to the proximal portion of the appendage by an angular joint, the elbow and the knee joints respectively. Rotation within the limb is effected by crossing of the parallel bones. This movement is strictly guided in joints with low friction as compared to gliding of the autopodium on the substrate. So firm support on the sbustrate, energetically advantageous torsion within the limb, and strict guidance of the movements in joints are integrated in the limb. This solution is easily comprehended as the optimal one, if compared to the alternatives. The angular joints of elbows and knees are stabilized and controled by a simpler and more effective muscle system as compared to a ball-and-socket joint in the middle of the limb, which would have required a great number of muscles for stabilization of the joint. However, the narrow handles of the appendages did not offer enough insertion space for the voluminous muscular material for such a joint.

Some difficulties exist at the connection between zeugopodia and autopodia. Here an angular joint cannot allow the irregular movements required by the multiple motions of the limbs. This explains what can be observed. A large number of skeletal elements is incorporated into a connective tissue mat that permits a great variety of deformations while the deformed part of the limb remains sufficiently strong and resistant to non-guided movements.

Completion of the limb and increase of efficiency of its movements opened up the transformation path to terrestrial life. This change of environments has certainly taken place gradually and allowed members of the population to proceed further and further into the new habitat. As soon as swimming in a fish-like manner lost its selective influence, further changes could take place. The fading of steering and paddling requirements were prerequisites for the formation of the digits.

As shown earlier, differentiation and articulation of the limbs was very useful for moving on and over the irregularities of the ground, because the multiple joints allow the adjustment to varying levels of support for the fins. Further improvement of a firm stance was achieved by the subdivision of the autopodium into digits, which resulted in the capability to gain contact on a substrate that was rough and presented irregularities of minor dimensions.

From primitive tetrapods to primates

The reconstruction so far offered renders the basic pattern of the tetrapod extremity comprehensible. The limb construction is not explicable without the constraining mechanical influence of the length constant vertebrate trunk; in addition to that the limb constructions reflect the contradictory demands of a firm support on the ground, the bending requirements of the body and the unavoidable rotation of the distal portion of the limbs in relation to the forward moving body. As long as tetrapods move by bending the

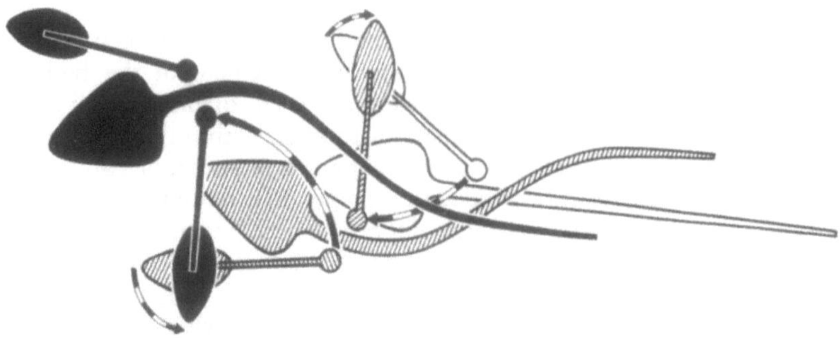

Figure 4 — Problems arising when ambulation is tried with laterally extended stiff limbs. During propulsion the body is forced to bend and the opposite limb to slide over the surface. Moreover, rotation of the limbs on the ground is unavoidable (after PETERS, 1985).

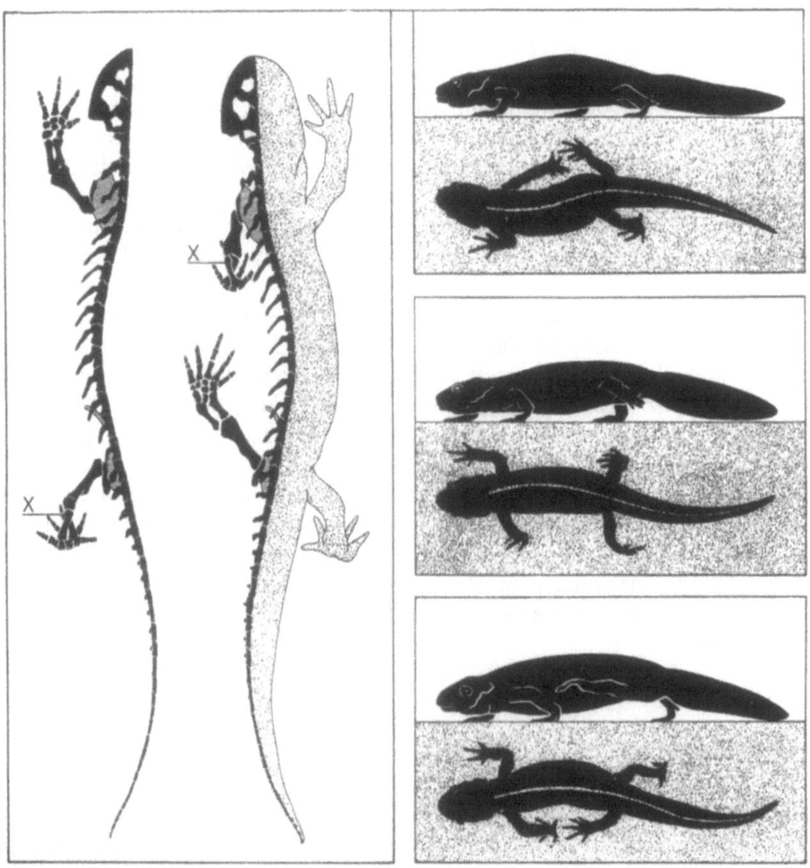

Figure 5 — Two stages in the ambulation process of a newt to show the interdependence of body movements, head on progress in locomotion, and internal rotatory movements in the limbs, X points to internal crossing of the osseous bars in the lower arms and lower legs. (Modified after SCHÄFER, 1941).

body in support of the gaits the general morphological pattern remains unchanged.

In conclusion, the reason for the persistance of the basic pattern of the limb skeleton in higher tetrapods has to be indicated. Additional constructional studies and reconstructions will have to elucidate much information about the locomotor mechanisms in the diverging tetrapod lineages. However, one conclusion can be drawn in advance. In all stages connecting the primitive tetrapod stage to the primate constructions all joints must have continually been used and in no intermediate stage can the torsion within the limb have ceased. Permanent selective confirmation must have operated to keep the pattern of skeletal elements and joints in existence. As long as the typical reptilian mode of locomotion was used the combination of lateral bending and limb movements enforced internal rotation and the use of all major joints. Transition to the early mammalian constructions brought the typical elevation of the body and the repositioning of the limbs. Although the planes of excursion of the limbs changed when they moved mainly in a parasagittal plane, rotational capabilities within the fore-arm allowed better adjustment to substrate irregularities and provided a greater versatility of the hand. By pronation and supination the grip could be employed in variable directions. Such capabilities became crucial when the primates turned to climbing in trees and using their hands in a primate-like manner. From this follows that pronation and supination represent late stages of the use of the phylogenetically old rotating capabilities of the early tetrapods.

Confirming negative evidence for the correlation of torsional movements and the persistence of the parallel and crossable skeletal rods in the lower portions of the extremities is found in forms which use their limbs by purely parasagittal movements. In this case crossing of the bones is no longer used and a single skeletal element is formed. This is observable in a paradigmatic way in many artiodactyls and perissodactyls.

Conclusion

The phylogenetic explanation for the so-called primitive features of the human hand and the whole extremity can be summed up as follows:

The tetrapod configuration of limbs was determined by the steering requirements of a typical fusiform fish-construction, which, after dissolving the primitive lateral folds into two pairs of fins, needed separate horizontal steering devices for the anterior and the posterior parts of the body. By the formation of the paired fins the movements of the body were freed of up and down steering functions, which in earlier stages had required energy-expensive bending of the whole body.

Transition to terrestrial life had to pass through intermediate stages of living in shallow water where the fins were used for ambulation while they still could serve for steering up- and down-wards in swimming.

The persistence of the basic pattern of skeletal elements in primates is highly indicative of the fact that some basic motor requirements of the primitive limbs have remained identical up to the primate level. The use of the main joints and use of the rotational capabilities of the forearm during pronation and supination represent a late stage of development in an uninterrupted sequence of transformation steps which have nevertheless left the basic pattern of the limb virtually unchanged.

From this it follows that the phylogenetic explanation of the basic properties of the limb cannot the elucidated by observations of primates or adaptational interpretations obtained by regarding the climbing behaviour in trees and other environmental conditions. Specific «adaptive» changes in the soft parts and the proportions of the skeletal structures can only have modified and modulated the basic construction or have become superimposed on the basic constructional pattern. Furthermore, the mode of use of the extremities and especially the hand is certainly highly dependent on the development of the brain and its delicate controlling influence.

Transition of effective walking by swinging the paired fins forwards and backwards depended on bending movements of the trunk which were derived from swimming with whole body movements. Energy-consuming gliding of the supporting plane of the developing limbs on the ground was avoided by three major flexures of the fin. These flexures were transformed into the typical jonts of the tetrapod extremity. The joints are organized in a way to allow free bending of the trunk while the limbs are firmly positioned on the ground.

Forward movement of the body while it is supported on the limbs would cause a backward swinging of the autopodium and its rotation on the ground if there were not an internal rotation within the limbs. Rotation is effected by the crossing of the double rods in the mid-limb section. So the necessity of the parallel and crossable skeletal rods for avoiding gliding of the autopodium on the substrate is explained. The alternative constructional solution in the form of a ball-and-socket joint in the knee and elbow regions would require a voluminous musculature for which insertion space on the humeri and femora is lacking.

The subdivision of the distal plate of the fin, the autopodium, into several digits arose when steering in water became obsolete. The flexible digits enable the limb to gain better contact with the irregular surface of the substrate.

While the subdivision of the distal part of the fin into rays is certainly a late modification that occurred after the tetrapods had turned to permanent living in a terrestrial habitat, the other properties have certainly preceded this step. As the basic pattern of skeletal structures and joints in the primate extremities are congruous to the construction of the limbs in lower tetrapods, the phylogenetic explanation can only be found by reconstructions reaching far back into vertebrate phylogeny.

The phylogenetic explanation of the typical limb pattern cannot be obtained simply by referring to environmental requirements because the position of the joints and the skeletal elements in the different portions of the limb are determined by the interaction of the already existing limbs with the whole body construction in locomotor movements.

Acknowledgements — I am grateful to D. Chivers for the aid in manuscript preparation and the stringent formulation of the main tenets. I have paricularly appreciated the invitation by Prof. Preuschoft to the inspiring symposium which was the incentive of our efforts, and the help for the publication of the paper.

References

Baindridge R., 1961. *Problems of fish locomotion.* In: Harris J.E. (ed.): Vertebrate Locomotion. Symp. Zool. Soc. London, 5: 13-22.

Bonik K., 1978. *Die Evolution der Tetrapoden als Problemlage. Zur Kritik an einem Modell.* Natur u. Museum, 108 (5): 133-136.

Bonik K & Gutmann W.F., 1978. *Die wichtigsten Beziehungen zwischen Fischgestalt und Hydrodynamik.* In Schäfer W. (Hrsg.): Evoluierende Systeme III. Aufsätze u. Reden senckenb. naturf. Ges., 29: 58-86; (W. Kramer) Frankfurt am Main.

BONIK K., GUTMAN W.F. & PETERS D.S., 1977. *Optimierung und Ökonomisierung im Kontext der Evolutionstheorie und phylogenetischer Rekonstruktionen.* Acta Biotheoretica, 26 (2): 75-119.

EDWARD J.L., 1977. *The evolution of terrestrial locomotion.* In: Hecht M.K., Goody P. & Hecht B. (eds.): Major Patterns in Vertebrate Evolution. Nato Advanced Study Inst. Series, Series A: Life Sciences (Plenum Press) New York, London, 14: 553-578.

GUTMANN W.F. 1977. *Phylogenetic reconstruction: Theory, methodology, and application to chordate evolution.* In Hecht M.K., Goody P.C. & Hecht B.M. (eds.): Major Patterns in Vertebrate Evolution. Nato Advanced Study Inst. Series, Series Q: Life Sciences (Plenum Press) New York, London, 14: 645-669.

GUTMANN W.F., 1985. *Anmerkungen zur Insuffizienz der Homologienforschung.* In: Ziegler W. (Hrsg.): Organismus and Selektion - Probleme der Evolutionsbiologie. Aufsätze u. Reden Senckenb. Naturf. Ges. (W. Kramer) Frankfurt am Main, 35: 201-215.

GUTMANN W.F., 1989. *Die Evolution hydraulischer Konstruktionen; organismische Wandlung statt altdarwinistischer Anpassung.* Senckenberg-Buch, 65, 200 S.; (W. Kramer) Frankfurt am Main.

GUTMANN W.F. & PETERS D.S., 1973 *Konstruktion und Selektion: Argumente gegen einen morphologisch verkürzten Selektionismus.* Acta Biotheoretica, 22 (4): 151-180.

OSCHE G. 1973. *Das Homologisieren als eine grundlegende Methode der Phylogenetik.* In: Schäfer W. (Hrsg.): Phylogenetische Rekonstruktionen - Theorie und Praxis. Aufsätze u. Reden Senckenb. Naturf. Ges. (W. Kramer) Frankfurt am Main, 24: 155-165.

PETERS D.S. & GUTMANN W.F., 1971. *Über die Lesrichtung von Merkmals- und Konstruktions-Reihen.* - Z. Zool. Syst. Evolutionsf., 9 (4): 237-263.

PETERS D.S. & GUTMANN W.F., 1978. Ausgangsform und Entwicklungszwänge der Gliedmaßen landlebender Wirbeltiere. Natur u. Museum, 108 (1): 16-21.

PETERS D.S., (1985. *Mechanical constraints canalizing the evolutionary transformation of tetrapod limbs.* Acta Biotheoretica, 34: 157-164.

REMANE J., 1985. *Beruht die phylogenetisch orientierte Homologienforschung auf einem Zirkelschluß?* In: Ziegler W. (Hrsg.): Organismus und Selektion - Probleme der Evolutionsbiologie. Aufsätze u. Reden Senckenb. Naturf. Ges., 35: 195-200.

SCHAEFFER B., 1941. *The morphological and functional evolution of the tarsus in Amphibians and Reptiles.* Bull. Amer. Mus. Nat. Hist., 78: 395-472.

STARCK D., 1979. *Vergleichende Anatomie der Wirbeltiere auf evolutionsbiologischer Grundlage.* vol. 2; (Springer) Berlin, Heidelberg, New York.

J.M.F. Landsmeer

Evolution and the hand

A variety of hands have been studied, including human hands, with particular emphasis on the extensor assembly, the system of extensor tendons and the flexor profundus. Data on the morphology of the thumb are included.
An attempt has been made to explain the relevant data in terms pertaining to the level of functional qualities. The terms functional domain and functional options have been introduced so as to designate a functional state.
In monitors the level of function of the hand und upper limb does not seem to surpass a certain stereotyped locomotion.
In opossums two functional domains prevail, locomotion and food collecting. Specialised structures in the hand will restrict the functional domain.
Squirrel monkeys keep to two basic patterns, but the domain of using the hands other than in locomotion is enlarged by grooming. Monkeys of the Old World still feature two major functional domains. The structure of the extensor assembly is such that the domain of tactile exploration is enlarged. In addition to this, monkeys have developed a certain quality of pinch, based on individualisation of the deep index flexor and on an appropriate positioning of the thumb.
In apes one functional domain prevails. Within this one and only domain, locomotion, food collection and grooming are possible options, though not the only ones. However, structurally this domain is skewed by the morphology and function of the thumb.
In man, with a full-fledged thumb, a functional universe or continuum has emerged, not impaired by any specific functional demand. The thumb as a most versatile component provides an opposing force to the fingers which are paired II — III and IV — V. Varying magnitude and direction of the opposing force yields specific data on qualities of objects held. Because of the setting of metacarpals II — V and the presence of a full-fledged thumb, the human hand is capable of a continuous perception of the universe it can hold, and of a unique encoding of each discrete situation. This quality of the hand finds its corollary in development of cortical structures and areas involved in processing this information.

Key words: extensor assembly, extensor tendons, flexor tendons, hand use, perception, release mechanism, thumb.

Introduction

Hands are not man's unique acquisition. Hand, and for that matter feet also, occur throughout the vertebrate realm, and even hands with impressive functional versatility are not man's only heritage. Still, upright posture and bipedal gait, and versatile hands figure as hallmarks of human existence. However, again bipedal gait as such is not strictly confined to man, rather the presence of an arched or vaulted foot, a pelvis of unique morphology and a specific spine. As for the latter point a recent experimental study on the lumbar spine by Preuschoft, Hayama & Günther (1988) should be given due attention.

A comparison of hands may shed some light on specific human qualities and it may also stir some thoughts on phylogenetic aspects of the hand and the importance of the hand for human evolution.

In the present paper, I will present observations made on various species of often remote relationship: *Varanus*, *Didelphis*, *Saimiri*, *Macaca mulatta*, *Cercopithecus*, *Presbytis*, *Pan* and *Homo sapiens*. Data from literature will be included.

Observations

It would be too much to present the relevant data *in extenso* in this paper; anyway a substantial part of it has been published (LANDSMEER, 1979, 1981, 1983, 1984, 1986, 1987).

The morphological evidence will be presented in such a way that its relevance for the major issues of this study and for the lines of reasoning followed is as clear as possible.

The major parts of the morphological evidence are:

- the extensor assembly,
- extensors of the fingers,
- flexor tendons of fingers and thumb.

In addition to this, other points will be brought into the picture, but basically the arguments of the issue hinge on these principal structures.

The extensor assembly

The extensor assembly constitutes the structural basis — in man — for the so called release of the terminal phalanx. This functional phenomenon is of basic importance for the pattern of flexion and extension of the interphalangeal joints, but also for a versatile use of the terminal phalanx, both in pinch and in active touch or tactile exploration. It permits the terminal phalanx to change from touch with the surface of the pulp to touch with the tip of the phalanx.

This specific structure is found in humans, and is quite similar in chimpanzees, gorillas (PREUSCHOFT, 1965), and monkeys of the Old World *(Figure 1)*, whereas the structure in *Cebus*, according to Van Zwieten (1980) gets very close to it. In *Saimiri* the structure is different in so far that the mechanism of release of the third phalanx is absent because in this species the arrangement of tendon fascicles at proximal interphalangeal level — the key to the release mechanism (LANDSMEER, 1949) — is not such that it could provide for the lateral bands to glide or shift along the dorso-lateral aspect of the trochlea of the first phalanx. Hence, the third phalanx in this species stays extended and even hyperextended with flexion of the proximal interphalangeal joint. After completed flexion, the third can be flexed as well. The extended position is most favourable for the «adhesive grip» this species is endowed with. In the opossum the situation is again quite different, as in this species the extensor assembly is provided with elastic structures and with an oblique ligament of conspicuous length. Together with the so-called accessory lumbricals, which occur in these species, these structures subserve the specific function of retraction of the claw. In the finger of the monitor lizard *Varanus* there is no such structure as an extensor assembly. A flat extensor tendon bridges all the joints of the

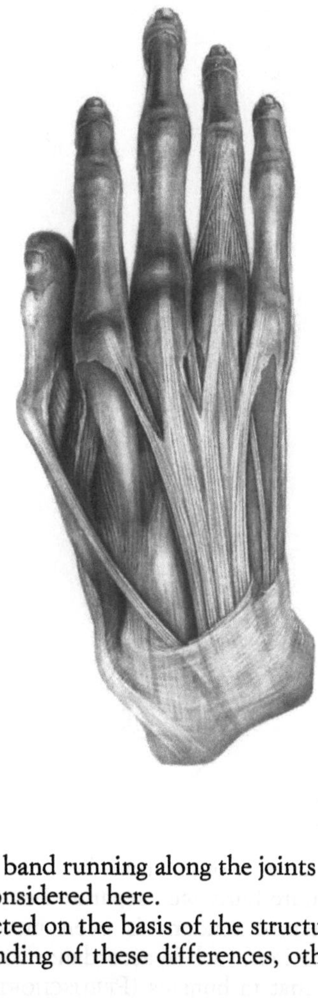

Figure 1 - The dorsal aspect of the hand in an Old World monkey (*Macaca mulatta*). In the fourth finger the structure of the extensor assembly has been exposed (Van Zwieten, 1980).

finger chain in succession *(Figure 2)*. There exists a lateral band running along the joints of the chain (LANDSMEER, 1981), but this will not be considered here.

Difference in finger and hand function can be expected on the basis of the structure of the extensor assembly only, but for a better understanding of these differences, other structures will be considered first.

Extensores digitorum

In humans and chimpanzees the fingers are under control of an extensor communis which takes origin from the extensor compartment in the forearm. The deep extensor and the extensor digiti quarti et quinti are suppressed but for an extensor indicis proprius. However, the tendon of this latter muscle is basically in an identical position to that of extensor communis, viz. on the dorsal side of the metacarpophalangeal joint, not on the ulno-dorsal side of the joint, the latter position taken in by the earlier mentioned extensor profundus and extensor digiti IV and V.

The two — actually three — sets of extensors prevail both in monkeys *(Figure 1)* and in the opossum.

Besides the number of muscular bodies involved, and in addition to the position of tendons at the level of the metacarpophalangeal joint, a conspicuous feature concerns the fasciculation and regrouping of tendons on the dorsum of the midhand and also the

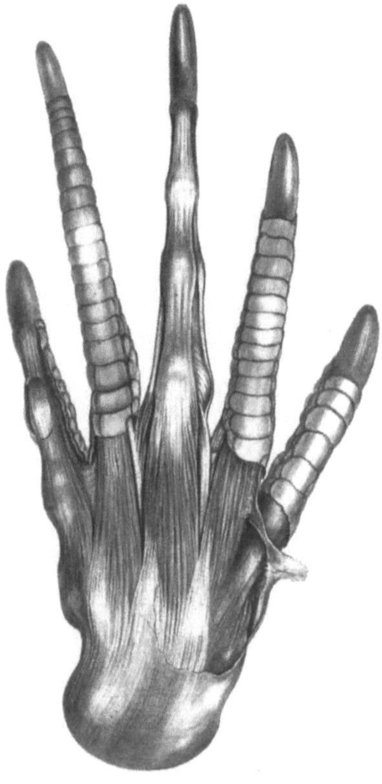

Figure 2 - The dorsal aspect of the hand in *Varanus*. Left side. Notice the asymmetry of the metacarpo-digital bands as they are exposed in the third finger. On the radial side the band runs obliquely towards the ulnar side of metacarpal II, on the ulnar side the digital band takes origin from an interosseus bundle.

occurrence of junctions between tendons. The latter are quite well known in humans, who feature fairly well separated extensor tendons. In chimpanzees, the separation of digital extensor tendons is even more complete than in humans, with few intertendinous junctions, such in accordance with STRAUS (1941). In gorillas the situation is quite similar to that in humans (PREUSCHOFT, 1965). Monkeys of the old world feature two sets of extensor tendons with fasciculation over the dorsum of the midhand of the tendons of the extensor digitorum longus (communis), and intertendinous bridges particularly between tendons III and IV, tendon IV contributing substantially to III.

In *Saimiri* there is a fasciculated tendon plate with major radiations to digits III and IV *(Figure 3)*. In the opossum a complex picture prevails with fasciculated tendons and junctions between fingers (LANDSMEER, 1979).

Flexor profundus, flexor pollicis longus

Man is the only species with a full-fledged flexor pollicis longus. Although in the gibbon the thumb is provided with a flexor longus, which arises from a separate body in the forearm, the pollex tendon and the tendon of digit II in the midhand are intertwined.

In chimpanzees the situation is such that separation in the muscular body of flexor profundus has progressed to the extent that separate bellies exist for the individual fingers *(Figure 4)*, confirming STRAUS (1942). However, tendons of digits III and IV are in-

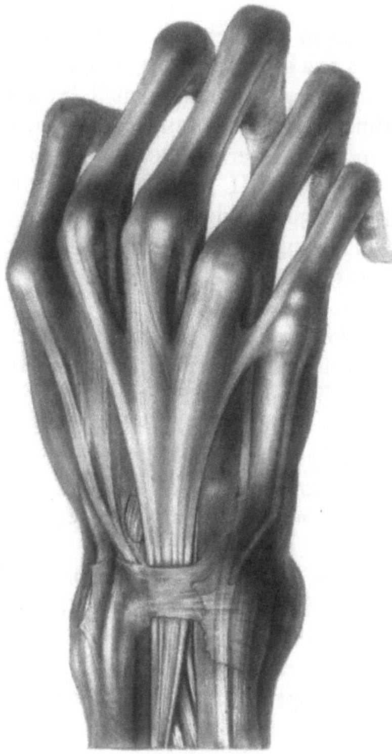

Figure 3 - Dorsal aspect of the right hand in *Saimiri sciureus*. Notice the two extensor tendons at each finger and the strong fasciculated «communis» tendons at the IIIrd and IVth fingers.

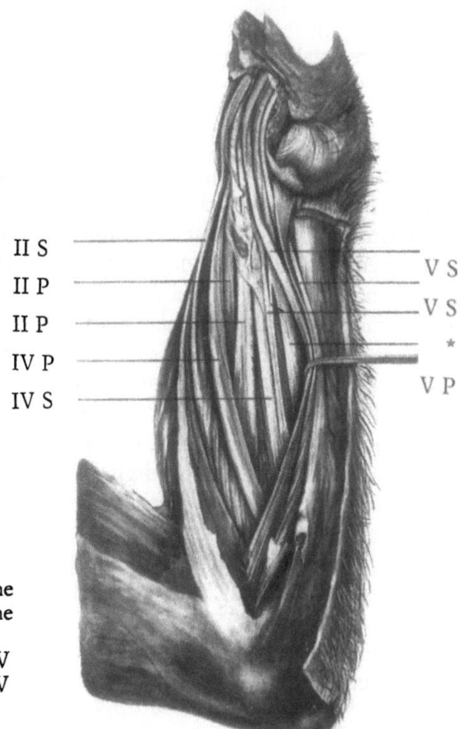

Figure 4 - The flexor digitorum profundus in the chimpanzee (*Pan troglodytes*). Left hand. Notice the outspoken digital division in this muscle.
II — V S = superficialis tendons to digits II — V
II — V P = profundus tendons to digits II — V

tertwined to some extent at the level of the midhand. The muscle belonging to digit II features as a separate entity, like flexor pollicis longus of man. A long flexor of the thumb in chimpanzees is just a small tendon, which splits up into a few nondescript cords proximally, losing themselves in subcutaneous structures of the thenar eminence, in the peritendinous layers of the long finger flexors, one of the strands following the tract of the tendon to digit III. Gorillas vary in this respect. Some individuals possess a deep flexor pollicis longus tendon, in others this tendon is absent (PREUSCHOFT, 1965).

It is remarkable that, in monkeys of the Old World, a long flexor of the thumb is not a separate muscle, but an offshoot of the tendon track to digit III which runs freely over the tendon to digit II. More or less in accordance with this feature is the independence of the tendon of digit II, at least in some species (a. o. rhesus). The tendon to digit V is also fairly well individualised, but intertwining at carpometacarpal level may reduce the individuality of the respective tendons. In *Saimiri* there is no separation in the flexor profundus whatsoever. The finger tendons spring from a parallel structured tendon plate with hardly any fasciculation present in it. The flexor pollicis springs from this plate and does so from the track of the tendon of the third digit. From here it crosses the track of digit II, though it remains adherent to it until it leaves the tendon plate, heading for the pollex.

In the opossum the flexor profundus features a solid tendon plate with no fasciculation in it whatsoever. The pollex tendon does not extend in a proximal direction into the radial border of this tendon plate, but rather can be followed as a tendinous strand over the palmar surface of this plate. Also in *Varanus* the individual digital tendons of flexor profundus spring from a solid tendon plate which shows no fasciculation whatsoever.

Discussion

The observations presented here constitute the main body of evidence pertaining to a concept regarding the significance of the primate hand for the evolution of humans. There are a few observations to be added on contrahens muscles and also on the extensor brevis of the thumb.

Digital morphology in *Varanus* proves the existence of the asymmetry of each digit as reflected in the proximal roots of the lateral stabilising system. Its relation to the mechanism of pronation and supination and of the latter to sprawling gait has been discussed earlier (LANDSMEER, 1983). In short, the horizontal component of humeral sway in sprawling gait is conveyed to the forearm, through a flexed elbow joint, as a twist of ulna and radius. (In the propulsive stroke, with retraction of the humerus, the twist of the two bones in the forearm is designated appropriately as pronation). The rotation of the forearm bones tends to become conveyed to the hand, resulting in an external rotation force on the substructure and a reaction force in opposite direction from the substructure to the hand. This reaction force loads the fingers in an asymmetrical way, viz. to the radial side. The bilateral stabilizing digital band is arranged in such way that it can resist the radially-directed loading by activation of muscle bundles which are attached to the ulnar band. Muscular attachments of this kind do not occur on the radial side. The arrangement of muscles in the distal part of the limb and of the shoulder joint with its ligamentous and muscular assets is such that basically one functional pattern, related to locomotion, prevails.

In the opossum, at the dawn of mammalian eruption, the hand is provided with two sets of extensors, with an extensor assembly of particular composition, featuring specific elastic structures and a conspicuously long oblique ligament. The presence of a set of small muscles, the accessory lumbricals, should not be ignored. The fingers are spanned by deep flexor tendons which spring from a solid plate. This means that digital flexion is carried out in all fingers simultaneously. However, two sets of extensors could be an indication of the existence of two functional optima. That these optima can be identified with locomotion on the one hand and catching food on the other is not implied here, but it seems likely to be so. The development of finger pads is indicative for a big step on the road to processing afferent information. It seems well warranted, however, to assign to the hand of the opossum specific functional patterns, so the hand may be said to be «pattern-controlled».

In *Saimiri*, basically, an identical situation prevails, but for the fact that claws are replaced by nails and for concomitant «loss» of the specialised system subserving claw retraction. The extensor assembly of *Saimiri* is arranged in such a way that adhesive grip is preferentially sustained by this structure. For strepsirhine primates this is shown in detail by NIESCHALK & DEMES (this volume). There is no doubt that the replacement of claws by nails and the loss of «elastic» structures increases the demands of active control, as it enhances at the same time possibilites in motor and perceptive behaviour. As a result, the functional potential of the hand in *Saimiri* seems to be considerably enlarged as compared to the hand of reptiles and of the opossum. In this respect it is significant to mention that in strepsirhines already, and definitely in *Saimiri*, grooming emerges as a functional acquisition (BISHOP, 1964). As functions like food catching, grooming and touch are not sharply, structurally engraved in the hand, it makes sense to distinguish many functional options. However, such options may very well be confined within a functional domain, the latter term related to the peripheral equipment of the hand. In this regard, the existence of two sets of extensors and the presence of a muscular group known as contrahentes is indicative of two major functional domains, one of them being locomotion, the other one comprising «handlike» functions. As regards these handlike, manipulatory, and finger functions, some well-defined structure qualities are of fundamental importance. These qualities are related in particular to the extensor assembly, flexor profundus and not the least to the structural setup of the thumb.

In monkeys of the Old World, apes and humans the extensor assembly is instrumental for the release of the terminal phalanx, a mechanism which in its turn is the keystone for active touch as it operates in man. Touch as an active functional quality is not confined to the groups mentioned, but is also present in monkeys of the New World (BISHOP, 1964). Release of the terminal phalanx, together with strong oblique retinacular ligaments is essential also for coupling the flexion of the interphalangeal joints of one finger and their decoupling. In the development of manipulatory functions it is important to be aware of a significant change in the setup of the flexor profundus. Fasciculation in the tendon plate becomes apparent, which leads at least to a separation of the tendons of the index and fifth digits in some monkey species (MAIER, 1971). The pollex flexor remains an offshoot of the flexor digitorum profundus, and more particularly of the tendinous track of the third ray. Hence the pollex tendon crosses on the palmar side the tendon of the index without mutual adherence. Functionally this arrangement, which is more conspicuous in one species than in the other, has to be integrated with the well known dexterous pinch of monkeys and the cortical control of this pinch. The functional morphology of the monkey thumb has to be assessed in this context, taking into account also the lack of a short extensor of the thumb and the abductor pollicis longus which attaches to the prepollex with an offshoot to the first metacarpal.

The function of the thumb in monkeys cannot be identical or even similar to the function of the thumb in humans, the former tending towards a stereotyped use, the latter towards a universal versatility.

In this respect it is extremely intriguing that a more discriminate use of components of the hand requires direct cortico-motoneural control, whereas the «whole hand patterns» can be controlled by recruitment at the spinal level of interneuronal systems. In apes and humans there is a loss of the second digital extensor and a reduction of the contrahens. This could mean that the hand is now entirely freed of locomotor duties, in such sense that a locomotor pattern is no longer engraved structurally in the hand. However, it is most remarkable that in the apes, except hylobatids, a substantial reduction of the tendon to the thumb has taken place.

In chimpanzees and gorillas (PREUSCHOFT, 1965), both flexor pollicis brevis and adductor brace the metacarpophalangeal and interphalangeal joints, establishing a kind of tenodesis between these joints.

On the basis of the morphology the function of the thumb in chimpanzees seems to be reduced to a stereotyped role. Pinch in chimpanzees is basically reduced to a pulp to side pinch of thumb and index, the thumb being moved as TUTTLE (1969) states: «..... as a unit at the carpometacarpal joint.» (p. 350 l.c.). This very point is documented in detail by CHRISTEL (this volume).

In chimpanzees the individualisation of tendons in the flexor profundus has proceeded to the extent that an individual recruitment of even one finger would be possible, although junctions at the level of the midhand may divert activation of a tendon to one finger slightly to the next one. It seems that the chimpanzee profits from the entire flexor mass to hold its heavy body weight suspended from one hand or even one finger. So, the hand of the chimpanzee is most remarkable. The digits have acquired a structural individuality not seen in humans, and from its structural morphology the chimpanzee seems to be endowed with an astonishing interdigital versatility (BOESCH & BOESCH, this volume), the thumb being recruited only to produce a stereotyped pinch or to close the grip on the lateral side. The intriguing point, however, is that a full-fledged thumb as seen in humans, and in humans only, requires for its optimal functioning four fingers of characteristic relative sizes and a specific setting. In apes the individualisation of the digits has progressed more than in any other primates. A pairing of digits on metrical data leads to III — IV and II — V pairs (SUSMAN, 1979). In this respect it is remarkable that deep flexor tendons to digits II and V are the most separated ones, tendons to III and IV showing intercrossing of fibres at the carpal region.

Summarising the situation in the hand of apes, the conclusion that they lack the characteristic human outfit can hardly be emphasised enough. The hand of apes is a hand in its own right, a hand endowed with conspicuous manipulative dexterity, but also able to meet demands issuing from locomotion, knuckle walking, including support of body weight, either in terrestrial or in brachiating situation.

The individualisation of the fingers could be helpful and instrumental for dealing with the complicated task of collecting food within a certain range of size and in all kinds of angles from the body, and preventing the body from falling from the tree with the other hand and the feet. Development of this manual and specifically interdigital versatility could have elicited the chimpanzee's capability of tool use.

It is most remarkable that humans have a full-fledged thumb, a separate flexor pollicis longus, a separate extensor pollicis brevis and an abductor pollicis longus which is inserted into the base of metacarpal I. A fairly completely separated index flexor prevails in humans, in addition limited separation of the flexor profundus to digits III, IV and V. In this regard, incomplete separation of the index tendon as reported by PREUSCHOFT

(1965), should not be ignored. In humans the fourth and fifth metacarpals have acquired a considerable mobility in the carpometacarpal joints (MARZKE, 1983). In addition, the metacarpophalangeal joints display a most conspicuous asymmetry (LANDSMEER, 1955). These features and other metrical data allow a pairing of digits in II — III and IV — V pairs. The possibilities that emerge from this morphological picture, to which other structural features can be added easily, are as astounding as they are logical.

Firstly, in humans no allowance is made any more for a specific function of direct mechanical support of body weight. This makes it possible that the thumb is also provided with a true and strongly developed flexor pollicis longus. It means that in the human hand there is one and only one functional domain, with no specific compartment for any other function whatsoever. Hence in man we face one functional continuum, a universe which requires basically a functional shaping for each particular event in which the hands are involved. Now it may be argued that such a domain exists in the chimpanzee as well, except for the slightly off-side position of the thumb. This point cannot be denied, but it should be observed as well that the chimpanzee's functional domain is digital versatility or rather interdigital versatility and, from what can be learned from field studies (BOESCH & BOESCH, this volume), it may very well be that chimpanzees surpass man in interdigital dexterity. However, the human hand can encompass any object within the apparent limits of size, by its hand and fingers, to obtain specific sets of data regarding its shape.

The pairing of fingers becomes easily apparent in the so-called cylindrical grip, in which digits II and III oppose the tip of the thumb, IV and V the thenar eminence. The thumb represents a counter-digital component which, by varying its position, at the carpometacarpal joint or at either the metacarpophalangeal joint or at both the metacarpophalangeal and interphalangeal joints, or by varying the force applied to an object, will provide many sets of data, which after due processing will give information on shape, consistency and surface qualities.

The afferent, perceptor functions of the hand, already alluded to in the context of opossum, gain a wider dimension in the capacity of perception of a specific form by a descrete set of data. Perception can now also be obtained by shaping of the hand of non-existing forms, which means that a new quality emerged which can best be qualified as abstraction and design. However, if by virtue of its capability of continuous shaping, the hand is introduced as a creational instrument, it should be understood that shaping, abstraction and design can be accomplished now also by-passing the hand, making use of neuronal circuitry, which is actually developed in order to process the specific afferent information and to program the required efferent sequences. Continuous shaping of the hand means that the hand can pass through an infinite continuous series of different shapes, as the result of anatomical properties explained above. The shaping of the hand is a motor function, but its direct implication is afferent informaton, which is a continuous faculty also with discrete stages of high resolution. That the hand is not the only structure that forced the central nervous system to hominization is self-evident. Development of the viscerocranium must have an equally decisive impact on development of the human central nervous system, and so of humans.

The point made here is that evolution of the human hand into an instrument endowed with a functional continuum must have had decisive influence on cerebral development as the result of the development of an afferent continuum, which is a necessary complement of the former.

It is obvious that the human hand in its function, both efferent and afferent, poses specific demands on neuronal control mechanisms (TEN DONKELAAR, 1983; KUYPERS, 1982; LEMON *et al.*, 1983; PHILLIPS, 1983). Pattern control does not seem to be adequate anymore, as patterns of motion no longer reside structurally in the hand. Therefore, other control mechanisms had to be developed which cannot better be described — on the basis of the peripheral situation — than as performance control. It means that performance itself which, as a design, by creation, invention and learning engraved in neuronal circuitry, constitutes the basis of control of movements in the hand. The opening of the area of cortical processing implies that patterns non-existent in the environment can be processed as well. This is basically the first step in abstraction or design. It also provides for a testing system, as qualities or properties of an environment can be tested with the quality *in abstracto*. Humans' specific quality is not to choose the sharpest flint out of a set offered, but to test the sharpness of a flint and to reject or use it. It is not likely that performance control is limited to the hands, considering the very important role of the other joints of the upper limb, to produce a successful performance. It is not likely either that in this respect the lower limb differs from the upper limb, and for that matter face and larynx have to be considered as well. The all-important point issuing from a structural functional study of the hand seems to be that human qualities as freedom, abstraction, invention, creation and responsibility are inherent in man's structure shaped by phylogeny and ontogenesis. If freedom is so much as the human state or the human condition, the «I» or subject state of humans seems to represent the momentous state of communication at either external or internal levels.

References

Bishop A., 1964. *Use of the hand in lower primates.* In Buettner-Janusch, J. (ed.), Evolutionary and Genetic Biology of the Primates, New York, Academic Press.

Donkelaar H.H. Ten, 1983. *A comparative approach to the central control of movement in vertebrates.* J Anat. 137: 407/408.

Kuypers, H.G.J.M., 1983. *A new look at the organization of the motor system.* Progress Brain Res. 57: 381-404.

Landsmeer J.M.F., 1949. *The anatomy of the dorsal aponeurosis of the human finger and its functional significance.* Anat. Rec. 140: 31-44.

Landsmeer J.M.F., 1955. *Anatomical and functional investigations on the articulation of the human fingers.* Acta Nat. (Suppl. 24): 2511-2569.

Landsmeer J.M.F., 1979. *The extensor assembly in two species of Opossum, Philander opossum and Didelphis marsupialis.* J. Morph. 161: 337-346.

Landsmeer J.M.F., 1981. *Digital morphology in Varanus and Iguana.* J. Morph. 168: 289-295.

Landsmeer J.M.F., 1983. *The mechanism of forearm rotation in Varanus exanthematicus.* J. Morphol. 175: 119-130.

Landsmeer J.M.F., 1984. *The human hand in phylogenetic perspective.* Bulletin of the Hospital for Joint Diseases Orthopaedic Institute, vol. 44, no. 7.

Landsmeer J.M.F., 1986. *A comparison of fingers and hand in Varanus, Opossum and Primates.* Acta Morphol. Neerl.-Scand. 24: 193-221.

Landsmeer J.M.F., 1987. *The hand and hominisation.* Acta Morphol. Neerl.-Scand. 25: 83-93.

Lemon R.N., R.B. Muir & G. Mantel, 1983. *The control of hand and finger movements in the macaque monkey.* J. Anat. 137: 410/411.

Marzke M.W., 1983. *Joint functions and grips of the Australopithecus afarensis hand, with special reference to the region of the capitate.* J. Human Evolution 12: 197-211.

Maier W., 1971. *Vergleichend- und funktionell-anatomische Untersuchungen an der Vorderextremität von Theropithecus gelada.* (Ruppell 1835). Abhandl. d. Senckenbergischen Naturforschenden Gesellschaft, Nr. 527. Kramer, Frankfurt am Main, 1971.

Phillips C.G., 1983. *Comparative aspects of the pyramidal tract and its role in motor control.* J. Anat. 137: 410.

Preuschoft H., 1965. *Muskeln und Gelenke der Vorderextremität des Gorillas (Gorilla gorilla Savage et Wyman, 1847).* Morphologisches Jahrbuch, Band 107, Heft 2: 99-183.

PREUSCHOFT H., S. HAYAMA & M.M. GÜNTHER, 1988. *Curvature of the lumbar spine as a consequence of mechanical necessities in Japanese macaques trained for bipedalism.* Folia Primatol 50: 42-58.
STRAUS W.L. jr., 1942. *The phylogeny of the human forearm extensors.* Hum. Biol. 13: 23-50 and 203-238.
STRAUS W.L. jr., 1942. *Rudimentary digits in primates.* Quart. Rev. Biol. 17: 228-243.
SUSMAN R.L., 1979. *Comparative and functional morphology of hominoid fingers.* Am. J. Phys. Anthrop. 50: 215-236.
TUTTLE R.H., 1969. *Quantitative and functional studies on the hands of the Anthropoidea.* J. Morph. 128: 309-364.
VAN ZWIETEN K.J., 1980. *The extensor assembly of the finger in man and non human primates.* Ph. D. Thesis, Leiden, 1980.

M. Godinot
K.C. Beard

A survey of fossil primate hands

The paleontological evidence pertaining to the evolution of the modern diversity in structure and function of primate hands is reviewed. The euprimate morphotype hand was elongated, had a short carpus and limited mobility, but the corresponding locomotor mode remains speculative. Hand remains of *Smilodectes*, *Adapis*, a Messel adapiform and *Pronycticebus* reveal a remarkable diversity in carpal structure achieved in these Eocene adapiforms, due to differing locomotor evolutionary pathways. The subfossil lemuriforms *Megaladapis* and *Palaeopropithecus* both show stereotyped (but different) grasping capabilities. We describe the strepsirhine morphotype hand, characterized by a relatively high degree of pollical divergence, features of the ulnocarpal articulation that imply an enhanced capacity for ulnar deviation, and relatively long digits; this hand is specialized for grasping.
The simiiform morphotype hand combines a relatively low degree of pollical divergence, features of the ulnocarpal articulation that imply a limited capacity for ulnar deviation, and relatively long metacarpals and short digits. This type of hand anatomy is mechanically well suited to arboreal palmigrade quadrupedalism. The hands of *Pliopithecus* and *Mesopithecus* are generally monkey-like. *Oreopithecus*' hand fits with its presumed suspensory habits. The hand of *Proconsul* suggests palmigrade quadrupedalism and climbing. *Australopithecus afarensis*' hand remains primarily a branch-grasping organ, with indications of enhanced manipulatory abilities. *Homo habilis* and *Paranthropus robustus* illustrate two lines of increased tool-use abilities.

Key words: Adapiformes, early primates, fossil hands, hand evolution, hand function, simiiform morphotype, strepsirhine morphotype.

Introduction

This chapter is a revised and updated version of a review of fossil primate hands, which was initiated after the Reisensburg 1985 Symposium and completed in 1989 (Godinot & Beard, 1991). The study of both fossil primate hands and euprimate origins has continuously advanced during this time, rendering it necessary to update this review. The initial version included an inquiry about early forms, with a detailed discussion of polarities and trends in euprimate hand characters that will not be repeated here. However, the results of this more detailed study are integrated into this survey, allowing what we know about fossil primate hands to be placed in an evolutionary perspective. Thanks to the generosity of our colleagues, this survey also includes more illustrations of the relevant fossils than did our previous paper.

Relatively complete and articulated fossil primate hand skeletons are exceptional indeed. Isolated hand bones are less rare, if not common. For example, metacarpals and phalanges have been identified for a variety of families of fossil primates. Isolated carpals have received much less attention, probably because they are very difficult to identify in mixed fossil assemblages. Yet even in this case, it has occasionally been possible to attribute isolated carpals to specific taxa (e.g., a hamate to *Megaladapis*, Lorenz von Liburnau, 1905; a capitate and a hamate to *Sivapithecus*, Rose, 1984; Spoor *et al.*, 1991). Despite their relative rarity, fossil primate hand remains have received much attention.

Metacarpals and phalanges have often provided critical information, especially for functional studies of subfossil Malagasy lemurs and fossil hominids. In addition, carpal remains of Eocene euprimates have recently begun to provide characters of significance for deciphering the phylogenetic interrelationships of primate higher taxa.

Historically, the first fossil primate hand to be described was that of *Mesopithecus pentelici* (GAUDRY, 1862). GAUDRY figured two partial, semi-articulated hands of *Mesopithecus* from the Greek Miocene locality of Pikermi. In his description of its anatomy, GAUDRY made several comparisons between *Mesopithecus* and living monkeys, especially *Presbytis* («les Semnopithèques»). SCHLOSSER (1888) described and figured, along with other limb bones, metacarpals and phalanges of *Adapis* and *Leptadapis*.

During the first half of this century, fossil primate hand bones were described as parts of the landmark early monographic studies of fossil primates: *Megaladapis* by LORENZ VON LIBURNAU (1905), *Notharctus* by GREGORY (1920), *Australopithecus* by BROOM (1946), and *Palaeopropithecus* by LAMBERTON (1947). Later, three important studies describing, at least in part, fossil hand remains of specific catarrhines were published at almost the same time. These classic works include the description by ZAPFE (1960) of very complete postcranial remains of the Miocene *Pliopithecus*, the description by NAPIER & DAVIS (1959) primarily of forelimb remains of the Miocene hominoid *Proconsul*, and the description by NAPIER (1962) of the hand remains of early *Homo* from Olduvai Gorge, Tanzania. These discoveries were followed by several detailed studies of the functional anatomy of the hands of living catarrhines, especially hominoids (e.g. NAPIER, 1961; LEWIS, 1965, 1969, 1972a, 1974; TUTTLE, 1967; PREUSCHOFT, 1965, 1970, 1973a,b; JENKINS & FLEAGLE, 1975; O'CONNOR, 1975). As a result of these and other seminal studies and because of its obvious implications for understanding human phylogeny, the evolution of hominoid wrists and hands has been the subject of increasing attention.

The study of the structure and function of the hands of mammals other than hominoids, including those of «primitive» mammalian groups and of prosimian primates, has also progressed, largely following the tradition of comparative anatomy (e.g. HAINES, 1958; JOUFFROY & LESSERTISSEUR, 1959; ETTER, 1974). However, an integration of fossil data with comparative anatomical evidence with respect to primate hand evolution has rarely been attempted, primarily because until very recently Paleogene primate hand remains have been inadequately represented in paleontological collections and even less adequately studied, with very few exceptions (GREGORY, 1920). In the most recent studies of early primate locomotion as related to hand evolution, the comparative anatomy of the hands of living forms remains prevalent (CARTMILL, 1974).

Our goal for this report is to review the fossil evidence concerning primate hand evolution, providing in so far as is possible functional interpretations of the various forms that are fairly well known. We have been able to include some previously undescribed fossils in this context. We will argue that, despite their relative rarity, these fossils provide critical information not only for understanding primate hand evolution, but also for locomotor and phylogenetic studies of several fossil primates, and even for understanding the phylogenetic interrelationships among primate higher taxa.

Early primate hand material is mainly represented by carpals and metacarpals, and almost never by complete digits. The functional interpretation of these fossils is hampered by a general lack of knowledge concerning the osteological adaptations of the wrists and hands of extant small quadrupedal primates as well as small arboreal and scansorial mammals (but see PREUSCHOFT *et al.* in this volume). Hence, the functional assessments we make for some of these fossil hand remains are tentative, and certainly will be enhanced in the future.

In this report, we give estimates of the proportions of the hand in the forelimb, and of the different parts of the hand, when this appears possible. These same proportions were used in the complementary osteometric study of JOUFFROY, GODINOT and NAKANO (1991 and this volume) and were completed with new fossil data (GODINOT, 1991a, 1992). The conclusions of these studies are integrated into our discussions. However, allometric effects on primate hand characters have yet to be studied in detail. The possible effects of body size on some of the characters we analyse are therefore generally unknown (but see ETTER, 1974) and may be significant.

We have been limited by several factors in our attempt to provide an overview of fossil primate hands: 1) the worldwide distribution of the pertinent material has forced us to rely partially on casts and published photographs and descriptions; 2) several of the most important specimens are currently under study by other workers; 3) our own interests strongly bias our coverage toward Paleogene primates. Nevertheless, we have studied original material of several of the forms treated below, including *Notharctus*, *Smilodectes*, *Adapis*, *Proconsul*, and *Mesopithecus*.

We use the taxon Simiiformes (= Anthropoidea) as proposed by HOFFSTETTER (1982), and Lemuriformes to include Lemuroidea and Lorisoidea. We use the family Proconsulidae, following HARRISON (1987).

Origin of the Euprimate Hand

The evolutionary transformation from the hand of non-primates to the hand of euprimates remains unclear. Obviously, the origin of the euprimate hand is intimately tied to the origins of euprimates themselves. For almost 20 years the taxon Plesiadapiformes has been used to include several families of supposedly «archaic primates». These primarily Paleocene families traditionally included Plesiadapidae, Paromomyidae, Carpolestidae, Saxonellidae, and Picrodontidae (e.g., SZALAY & DELSON, 1979), and sometimes included the more controversial Microsyopidae (e.g., BOWN & ROSE, 1976). More recent systematic studies recognize as additional families of plesiadiapiforms the Palaechthonidae, Micromomyidae and Purgatoriidae (GUNNELL, 1989; BEARD, 1989). In contrast to these primitive families lacking an opposable hallux and postorbital bar were the euprimates or «primates of modern aspect» (SIMONS, 1972) which possess these characters. However, over the past few years evidence has accumulated that plesiadapiforms should be excluded from the order Primates, and that such a broad and heterogeneous grouping of families has very low systematic value. More importantly, recent work has shown that the family Paromomyidae is probably the sister group of living Dermoptera (BEARD, 1989, 1990; KAY *et al.*, 1990). Most authors have considered Paromomyidae to be closely related to Plesiadapoidea (= Plesiadapidae, Carpolestidae and Saxonellidae), implying that this superfamily is also more closely related to Dermoptera than to Primates (BEARD, 1989). Hence, Plesiadapiformes in the traditional, broad sense is a gradistic, heterogeneous and provisional taxon.

Studies of postcranial anatomy support the monophyly of a clade including Dermoptera and Primates (= Primatomorpha), which would also include most or all of the former plesiadapiforms (BEARD, 1989, 1991, in press). This clade is a radiation of arboreal mammals characterized by extensive use of vertical postures and climbing and a series of correlative morphological characters (BEARD, 1989, 1991). Broader phylogenetic groupings between Primates and other mammals, such as the Archonta (=primates, dermopterans, tree shrews and bats), also imply that primates originated within an arboreal radiation of mammals (SZALAY & DRAWHORN, 1980). Clearly, the stem group of primates were arboreal mammals with claw-bearing hands.

One possible exception to arboreality being ubiquitous among close relatives of primates can be found in one of the early interpretations of the locomotion of *Plesiadapis tricuspidens*, then the best known plesiadapid. GINGERICH'S (1976) analysis of limb proportions showed a close similarity between *Plesiadapis tricuspidens* and the living *Marmota*, suggesting a terrestrial, or even fossorial, adaptation for *P. tricuspidens*. Other studies of the postcranial anatomy of *P. tricuspidens* (especially the anatomy of the elbow and ankle; SZALAY & DECKER, 1974; SZALAY *et al.*, 1975; SZALAY & DAGOSTO, 1980) maintained a higly arboreal habitus for this species. In our earlier paper we reconstructed a manual digit of *P. tricuspidens* (*Figure 1*), detailed study of which confirmed an arboreal lifestyle (GODINOT & BEARD 1991).

Hand remains of the smaller plesiadapid *Nannodectes intermedius* show that this animal possessed a divergent thumb and a reduced second metacarpal, suggesting a reduced second digit (*Beard*, 1989, 1990). A divergent thumb is also present in the living dermopteran *Cynocephalus*, suggesting that a divergent thumb also occurred in the dermopteran morphotype. This degree of thumb divergence may be homologous between

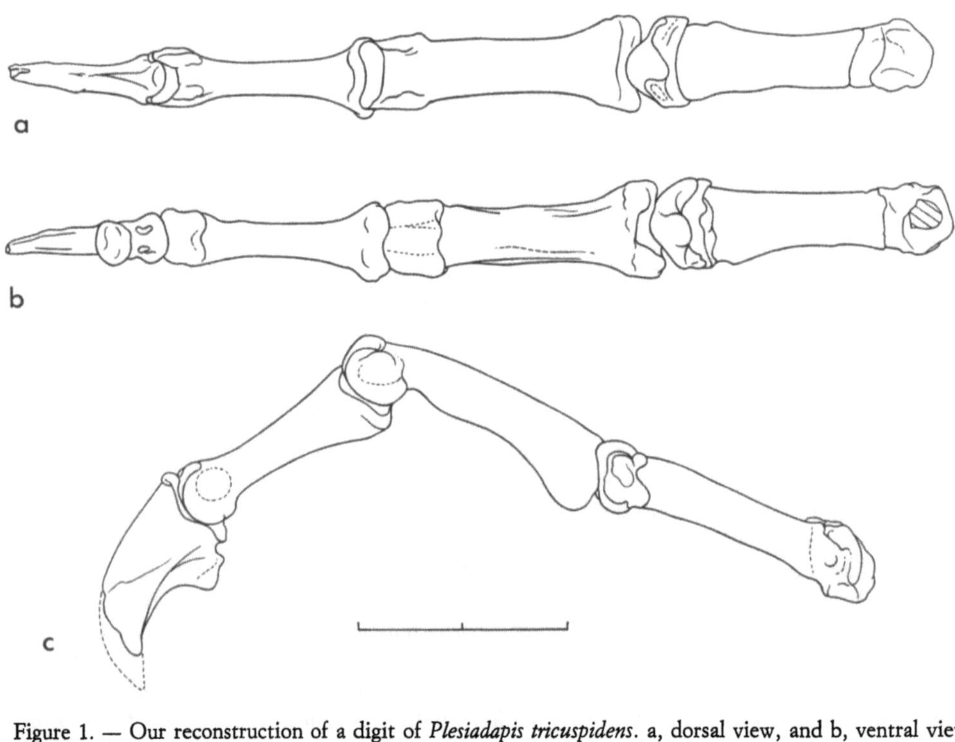

Figure 1. — Our reconstruction of a digit of *Plesiadapis tricuspidens*. a, dorsal view, and b, ventral view with the bones in the position of an extended digit; c, lateral view of the bones with the metacarpo-phalangeal joint in a position of high congruence, and the two inter-phalangeal joints in mid course. Scale = 1 cm.

dermopterans and primates (a view favored by C.B.) or it may have been convergently attained from a more primitive condition, because in dermopterans it is associated with specialized features in the carpus.

Historically, there has been considerable debate over the nature of the terminal phalanges of the earliest true primates. The presence of claws in *Daubentonia* and callitrichids suggested to several workers (e.g., LE GROS CLARK, 1936; CARTMILL, 1974; HERSHKOVITZ, 1977) that primitive primates bore claws that were replaced by nails in numerous primate lineages. More recent analyses imply that the claws of *Daubentonia* and callitrichids are secondarily derived (rather than being primitive retentions), but the exact nature of the terminal phalanges in their ancestors has never been discussed in detail. Recent analyses of terminal phalanges in Eocene adapiforms have demonstrated an intermediate structure (*Figure 2, 3*), in which the proximal part of the terminal phalanx is dorsopalmarly high and «claw-like», while the distal part is mediolaterally broadened and was probably nail-bearing (GODINOT, 1991a, 1992). This latter type of structure may more closely approximate the euprimate morphotype, and provides a plausible antecedent for the re-acquisition of claws in *Daubentonia* and callitrichids.

Analysis of the palmarly flattened distal phalanges of early euprimates suggests that stabilization of the tips of the digits may have been their most important function. Previous analyses of hand proportions have suggested that early euprimates possessed relatively long hands within the forearm and very long digits within the hand (JOUFFROY et al., 1991 and this volume; GODINOT & BEARD, 1991; GODINOT, 1991a, 1992). The biological role of this complex of hand characters may have been to seize and retain live insect prey while at the same time stabilizing the other hand on the support (GODINOT, 1991a). Such a scheme is consistent with CARTMILL's (1972) visual predation model of primate origins.

Adapiformes

Among the taxonomically diverse euprimate faunas known from the Eocene, hand remains are known mainly for various North American and European adapiforms. Omomyid hand remains are limited to some isolated distal phalanges (DAGOSTO, 1988). Adapiforms are usually considered to be primitive strepsirhines, closely related to, but outside the radiation of extant and fossil African, Asian, and Malagasy lemuriforms (including Lorisoidea). Postcranial characters have recently added support to this hypothesis (GEBO, 1986a; BEARD *et al.*, 1988; BEARD & GODINOT, 1988; DAGOSTO, 1988; COVERT, 1988; see below). Alternatively, some workers continue to follow GINGERICH (1980, and references therein) in advocating the likelihood of adapiform-simiiform affinities (e.g., RASMUSSEN, 1986; FRANZEN, 1988; RASMUSSEN & SIMONS, 1988; SIMONS, 1989).

Notharctinae

The skeletal anatomy of notharctines is perhaps the best known of any Paleogene primate taxon (e.g. GREGORY, 1920; COVERT, 1985; ROSE & WALKER, 1985). However, hand remains are not very common. Here, we will summarize what is known of the hand anatomy of notharctines, based primarily on the wrist of middle Eocene *Smilodectes* and the metacarpals and phalanges of middle Eocene *Notharctus* (GREGORY, 1920; BEARD & GODINOT, 1988). Phalanges of early Eocene *Cantius* are also known (COVERT, 1988).

Of the ten separate osseous elements that comprise the carpus of *Smilodectes*, all but

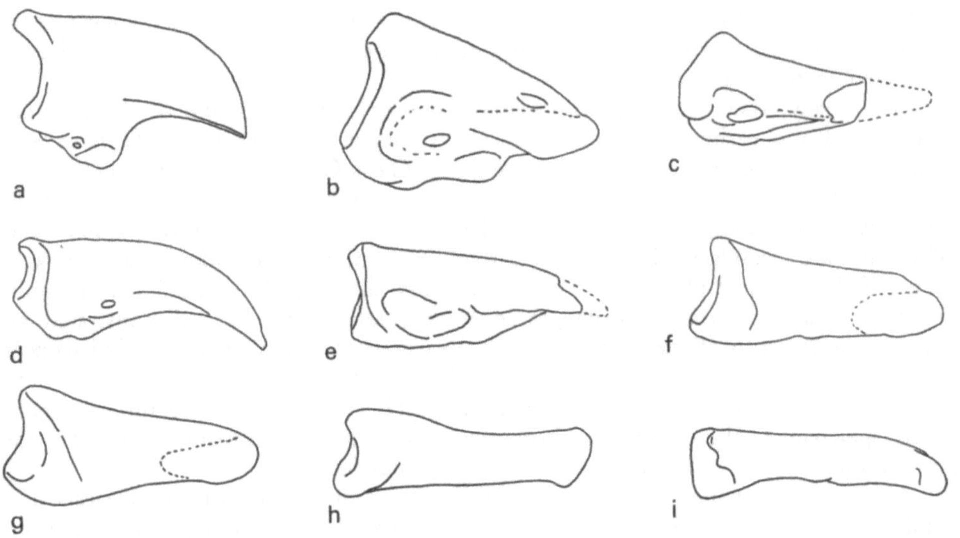

Figure 2. — A series of distal phalanges shown in profile, all drawn to the same proximo-distal length (completed from GODINOT, 1991a). a, *Plesiadapis tricuspidens* from Berru, Paleocene of the Paris Basin (MNHN R 5361); b, *Adapis* aff *betillei* from Escamps, late Eocene of the Quercy (ECA 1399); c, *Smilodectes gracilis* from the Bridger Basin, Middle Eocene of Wyoming (USNM 21815); d, *Daubentonia*; e, *Notharctus tenebrosus* from Grizzly Buttes, Bridger Basin, Middle Eocene of Wyoming (from a cast of AMNH 11473, probably slightly deformed); f, *Macaca*; g, *Hapalemur*; h, *Perodicticus*; i, *Cebus*.

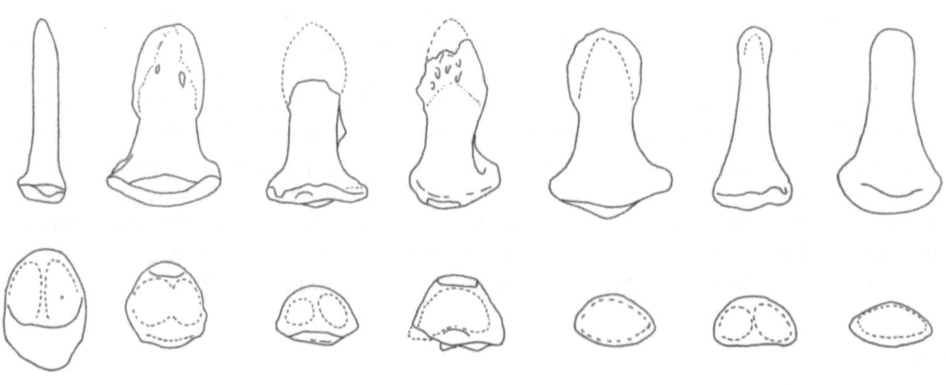

Figure 3. — Series of distal phalanges in dorsal (top row) and proximal (bottom row) views, all drawn to the same proximo-distal length (top row) and to the same radio-ulnar width (bottom row). Same individual specimens as in Figure 2, less *Plesiadapis* and *Macaca* (modified and completed from GODINOT, 1991a). a, *Daubentonia*; b, *Adapis*; c, *Smilodectes*; d, *Notharctus* (from a cast, probably slightly deformed); e, *Hapalemur*; f, *Cebus*; g, *Perodicticus*.

the trapezoid and prepollex are known (BEARD & GODINOT, 1988). In the proximal carpal row, the scaphoid possesses a proximal facet for the radius that is higher in the dorso-palmar dimension than it is radio-ulnarly. The lunate is very reduced radio-ulnarly as it is in many extant lemurids (e.g., *Lemur* and *Varecia*). The triquetrum bears articular facets proximally for the ulnar styloid process and for the pisiform. The facet for the ulnar styloid process is much larger than that for the pisiform and also differs from the latter in being highly concave dorso-palmarly. The dorsal limit of the facet for the ulnar styloid process is defined by a well-developed bony process, which must have served to stabilize the ulnocarpal joint at the expense of limiting ulnocarpal dorsiflexion. A second, nonarticular bony process of unknown function protrudes from the ulnar side of the bone. In dorsal view the triquetrum as a whole differs from those of *Adapis* and most or all lemuriforms in being quadrangular, rather than becoming increasingly narrow on the ulnar side of the bone (which yields a more triangular shape for this bone when it is viewed in dorsal perspective). In general, the ulnocarpal joint is typical of those of haplorhines and many other eutherians in possessing a facet for the ulnar styloid process on the pisiform that is neither as radio-ulnarly extensive nor as deeply excavated as it is in extant lemuriforms and late Eocene *Adapis*.

In the distal carpal row, the trapezium of *Smilodectes* is notable in that it is triangular in dorsal view as are those of many extant «prosimians» (e.g., ETTER, 1978; ROBERTSON, 1984), an indication of a highly divergent pollex (*Figure 4*). A very robust palmar tubercle occurs on the trapezium of *Smilodectes*, medial to which the tendon of the flexor carpi radialis passes on its way to insert into the base of the second metacarpal in extant

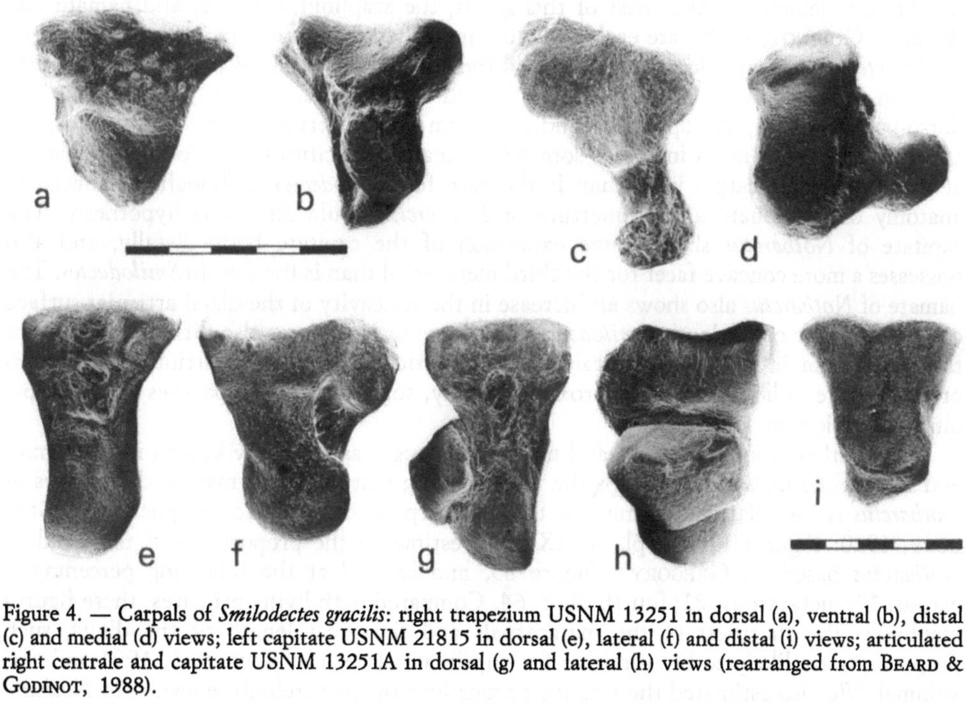

Figure 4. — Carpals of *Smilodectes gracilis*: right trapezium USNM 13251 in dorsal (a), ventral (b), distal (c) and medial (d) views; left capitate USNM 21815 in dorsal (e), lateral (f) and distal (i) views; articulated right centrale and capitate USNM 13251A in dorsal (g) and lateral (h) views (rearranged from BEARD & GODINOT, 1988).

lemuriforms (JOUFFROY, 1962). The extreme length of the trapezial tubercle might be related to the complete absence of a hamulus on the hamate, inasmuch as some critical depth of the carpal flexor tunnel is probably necessary in primates to allow passage of the extrinsic flexor musculature of the hand and digits. The articular facet for the first metacarpal on the trapezium in *Smilodectes* is sellar-shaped (BEARD & GODINOT, 1988; *Figure 4*), a notable difference from the condition observed in *Adapis* and possibly the Messel adapiform (see below). The relatively high degree of pollical divergence suggested by the form of the trapezium and the orientation of its distal facet for the first metacarpal suggest at least some capacity for pollical prehension in *Smilodectes*, although further analysis of the metacarpus and manual phalanges of this form are required before more definitive assessments in this regard can be made.

The anatomy of the rest of the carpal elements comprising the distal carpal row can be summarized as being very generalized. The centrale of *Smilodectes* (*Figure 4, g-h*), unlike those of all extant lemuriforms, is radio-ulnarly narrow, lacks articular contact with the hamate, and possesses a facet for articulation with the trapezoid that faces distally rather than disto-radially. Likewise, the capitate of *Smilodectes* (*Figure 4*) seems to be very generalized in that it is proximo-distally relatively short and lacks radio-ulnar expansion either of the capitate head proximally or of the capitate body distally. The hamate of *Smilodectes* shows contiguous articular facets for the bases of the fourth and fifth metacarpals, as do those of all extant «prosimians». The «spiral» facet on the hamate for articulation with the triquetrum is oriented relatively proximo-distally rather than radio-ulnarly, indicating a moderate degree of ulnar deviational capability at the midcarpal joint (O'CONNOR, 1975).

Although the wrist of *Notharctus* is much more poorly known than is that of *Smilodectes*, it is interesting and probably functionally significant that the three adequately known elements of the wrist of this genus, the scaphoid, capitate, and hamate (see BEARD & GODINOT, 1988), are each easily distinguished from the homologous elements of *Smilodectes*. The scaphoid of *Notharctus* differs from that of *Smilodectes*, among other ways, in possessing a proximal facet for the radius that is radio-ulnarly wider than it is dorso-palmarly high, the opposite condition from that observed in *Smilodectes*. Possibly this feature in *Notharctus* indicates somewhat greater capacities for radio-ulnar deviation at the antebrachio-carpal joint than is the case for *Smilodectes*, although the unknown anatomy of the lunate and triquetrum of *Notharctus* could alter this hypothesis. The capitate of *Notharctus* shows some expansion of the capitate body distally, and also possesses a more concave facet for the third metacarpal than is the case in *Smilodectes*. The hamate of *Notharctus* also shows an increase in the concavity of the distal articular surface compared to the condition in *Smilodectes*; this bone in *Notharctus* also differs from that of the latter form in bearing a «spiral» facet for articulation with the triquetrum that is oriented more radio-ulnarly than proximo-distally, suggesting less capacities of midcarpal ulnar deviation in *Notharctus*.

The notharctine metacarpus and manual phalanges are partially known in *Notharctus* and *Smilodectes* (*Figure 5*). Perhaps the most striking feature of the manual digital rays of *Notharctus* is the relative shortness of the metacarpals compared to the phalanges (GREGORY, 1920: Figures 11, 83; pl. XXIX). We estimated the proportions of the hand of *Notharctus* based on GREGORY's Figure 83, and arrived at the following percentages: carpus, 15; metacarpus, 21; fourth digit, 64. Compared with living primates, these figures emphasize the shortness of the metacarpals and the great length of the digits; such proportions are likely to be primitive for euprimates (see JOUFFROY *et al.*, 1991 and this volume). We also estimated the relative proportions of the forelimb segments of *Notharc-*

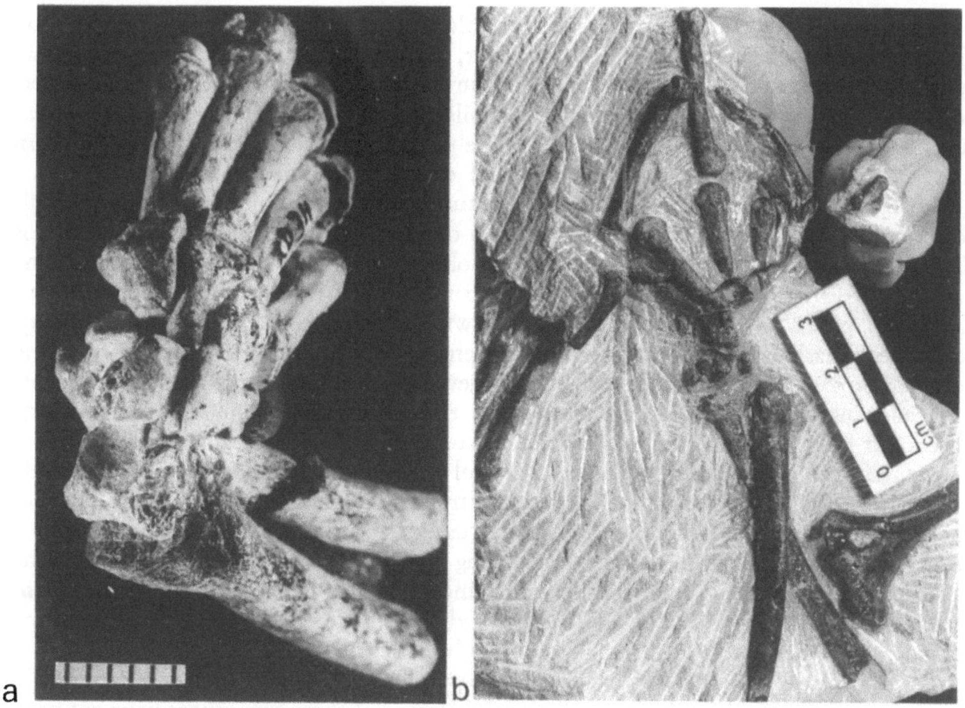

Figure 5. — Two Eocene adapiform hands showing the initial connections preserved by fossilization. a is a left hand with complete carpus and metacarpus, and extremities of radius and ulna, of *Adapis* cf *parisiensis* from Rosières 2, Quercy Phosporites, after cleaning of the specimen found in a clay nodule (scale in mm); b is a left articulated forearm of *Notharctus tenebrosus* from Grizzly Buttes, Bridger Basin, Wyoming, in the process of cleaning; further preparation revealed an almost complete hand with 9 of the 10 carpal elements, now under study (AMNH 127167; photograph courtesy and copyright of J.P. ALEXANDER; scale in cm.).

tus based on measurements and photographs in GREGORY (1920): humerus length, 32-33%; radius length, 31-33%; and hand length, 34-37%. These estimates differ considerably from those in NAPIER & DAVIS (1959: Table XVI), who possibly reversed two estimates. In comparison with extant primates (JOUFFROY *et al.*, this volume), it appears that *Notharctus* possesses a relatively short humerus and radius and a proportionately large hand.

Three metacarpals of *Smilodectes* (representing digits one to three) show proportions very similar to those of *Daubentonia*, with the second metacarpal being reduced. This suggests reduction of the second digit as occurs in living ectaxonic forms (GODINOT, 1992). Functional explanations for this reduction in the hand of lorisines (NIESCHALK & DEMES, this volume) probably also apply to very long ectaxonic hands. The morphology of the second metacarpal of *Smilodectes* is also similar to that of *Daubentonia* in the dorsal extension of its distal articular facet. This suggests frequent use of hand postures with a high degree of dorsiflexion of the proximal phalanges on the metacarpus in *Smilodectes*, as is well documented in *Daubentonia* (PETTER & PEYRIERAS, 1970; PETTER *et al.*, 1977)

In our previous paper we discussed the sole *Notharctus* distal phalanx described by GREGORY (GODINOT & BEARD 1991). However, subsequent examination of the original specimen has revealed that it is broken, which invalidates our previous conclusion. Other, complete *Notharctus* distal phalanges are similar to those of *Smilodectes* and *Cantius* (COVERT, 1988; GODINOT, 1991a, 1992). These phalanges show a distal broadening as in nail-bearing forms, and a relatively high and claw-like proximal part (*Figures 2, 3*).

A coherent hypothesis relating hand structure to function in notharctines is difficult to formulate despite the relative completeness of the fossil hand remains. The wrist of *Smilodectes* is most notable for its lack of the obvious carpal specializations for extreme radio-ulnar deviational capabilities such as occur in extant lorisids and hominoids (e.g., CARTMILL & MILTON, 1977; ETTER, 1978; LEWIS, 1985), and for its absence of lemuriform specializations relating to the development of a functional axis passing from the scaphoid proximally to the hamate distally, together with modifications of the ulnocarpal joint allowing enhanced ulnar deviation (e.g., triquetrum quadrangular in dorsal view, absence of highly concave and radio-ulnarly extensive facet for the ulnar styloid process on the pisiform). Yet the development of pollical divergence seems to have been relatively great in *Smilodectes* (judged by the form of the trapezium and the orientation of the facet for the first metacarpal), as is the case in most extant «prosimians». *Notharctus* exhibits a relatively short metacarpus and relatively long phalanges, also possibly indicating the development of prehensile capabilities. The inadequately known carpus of *Notharctus* differs surprisingly from that of *Smilodectes*, and these differences have been interpreted to indicate a somewhat greater emphasis on quadrupedalism in *Notharctus* compared to the latter form (BEARD & GODINOT, 1988). We are perplexed as to why similar differences in other anatomical complexes of the postcranial skeletons of these two genera have not been elucidated as of yet; possibly, a more detailed analysis of the postcranial anatomy of these well known animals would yield such differences (especially in light of the inferred phylogenetic distance between these genera; see BEARD, 1988). The notharctine postcranial skeleton as a whole shows traits that indicate the importance of leaping in a broad locomotor repertoire also including some quadrupedalism (e.g., GREGORY, 1920; ROSE & WALKER, 1985; COVERT, 1985; DAGOSTO, 1988), and possibly involving relatively infrequent vertical postures (GEBO, 1985). However, an influence of leaping on notharctine hand anatomy is not at all apparent.

Adapinae

Until recently, the hand of adapines was not very well known (SCHLOSSER, 1888; DAGOSTO, 1983). This situation is fortunately no longer the case as a nearly complete, articulated left hand skeleton of *Adapis parisiensis* has recently been discovered and described (GODINOT & JOUFFROY, 1984). The radius and ulna of this specimen are stout, fairly robust bones. The distal articular surface of the radius is dorso-palmarly high. The ulnar styloid process is high and round, radio-ulnarly broad, and dorso-palmarly compressed (*Figures 5, 6, 11, 12*).

The scaphoid is relatively thin, with a well defined palmar tubercle which projects palmarly away from the plane of the facet for the radius on this bone. The neck of the tubercle is slightly constricted. The ulnar surface of the scaphoid for articulation with the lunate is flat. The prepollex is very small and has two articular facets: a large and slightly concave one which most likely articulated with the scaphoid tubercle, and a smaller one which appears to have been for the trapezium. The centrale possesses a rectangular dorsal surface which seems to have been more extensive than that of *Smilodectes gracilis*. Its distal articulation with the trapezoid is flat, and appears to have been oriented much less disto-radially than is the case among lemuriforms. Contact between the centrale and the hamate, such as occurs ubiquitously in living strepsirhines, did not occur in *Adapis parisiensis*. The lunate has a convex, dorsally oriented articular facet for the radius which is only slightly smaller than the corresponding facet on the scaphoid. The palmar part of this bone is extensive, triangular, and apparently nonarticular. Well-marked facets occur radially for the scaphoid and ulnarly for the triquetrum. The distal articular surface for the midcarpal articulation (formed by the proximal ends of the capitate and hamate) is concave. The triquetrum has a dorsal surface which is similar to that of most lemuriforms in possessing a proximal process which defines the dorsal limit of the ulnocarpal articulation. The distal articular surface for the hamate is smooth and concavo-convex. Proximally, there are two confluent facets for the ulnar styloid process dorsally and the pisiform palmarly. The radial side of the triquetrum possesses a well-marked articular facet for the lunate. The pisiform is quite large and possesses the typical facets: a radio-ulnarly extensive, very concave facet for the ulnar styloid process (this feature is also typical of lemuriforms; neither haplorhines nor most other eutherians possess such a marked concavity), and a smaller, convex facet for the triquetrum.

The distal carpal row is composed of four bones: the trapezium, trapezoid, capitate, and hamate. The trapezium of *Adapis* is a very unusual bone, and is unlike those of either extant prosimians or *Smilodectes* (the Messel adapiform does resemble *Adapis* in this respect, however; see below). The trapezium is a relatively thin bone dorso-palmarly, with a convex, triangular dorsal surface. Ulnarly there is a thin, concave facet for articulation with the trapezoid. The distal facet for the first metacarpal is highly unusual in that it is almost completely flat. A small palmar tubercle is present at the radial extremity of the trapezium. The largest side of the trapezoid is the distal articular surface for the second metacarpal, which is dorso-palmarly concave and slightly convex radio-ulnarly. The proximal facet for the centrale is flat, the radial facet for the trapezium is curved and convex, and the ulnar facet for the capitate is very small. The capitate is proximo-distally elongated, and very narrow radio-ulnarly in relation to the hamate, a feature which is even more exaggerated in living lemuriforms. Distally the body of the capitate is extended radially; there is no radial torsion of the body with respect to the long axis of the bone such as occurs in some lemuriforms. The head of the capitate is similar to that of *Smilodectes* in that it is evenly rounded radially, but is very attenuated dorso-palmarly. The distal articular facet of the capitate was solely for the third metacarpal; dorso-palmarly, the facet is concave, and it becomes notably narrower palmarly. The hamate is large, and is particularly extensive proximo-distally and radio-ulnarly. The proximal articular facet for the triquetrum is broad and concavo-convex; it is oriented more radio-ulnarly than is typical of most strepsirhines. The distal articular facets for the fourth and fifth metacarpals are somewhat concave and poorly delimited from each other. A very moderate hamulus is present palmarly and ulnarly.

The metacarpals are stout and bear inflated distal heads (*Figure 6*). A palmar keel is present on the first metacarpal, and a less extensive one is present on the fifth metacarpal. On the original specimen, a sesamoid was preserved along the heads of two of the metacarpals. The lengths of the metacarpals decrease in the following order: third, fourth, second, fifth, first. The thickness of the metacarpal shafts in dorsal aspect decreases in the order: first, third, second, fourth, fifth. The proximal part of the first metacarpal possesses a small bony apophysis ventrally which has a round area for the insertion of M. abductor pollicis longus. The proximal surface of the first metacarpal is larger than the articular surface of the trapezium, but there is no clear demarcation between the articular

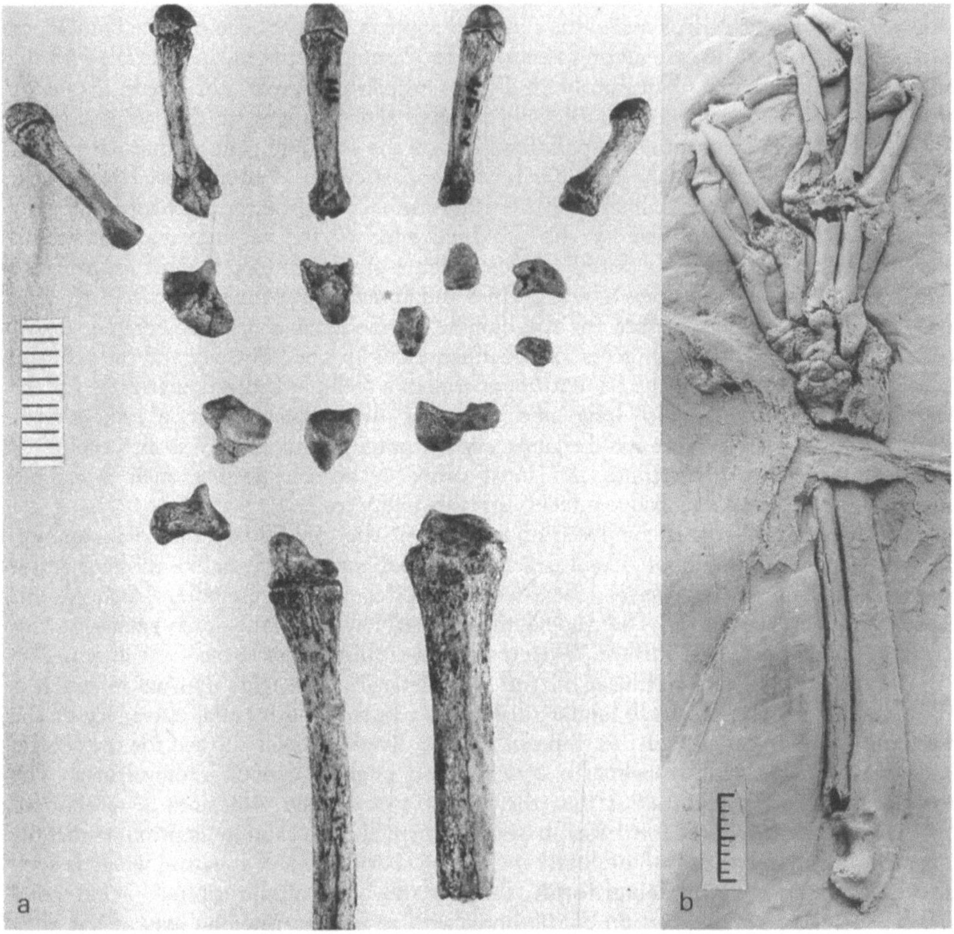

Figure 6. — Two very complete Eocene adapiform hands after preparation. a is the *Adapis* cf *parisiensis* hand from Rosières after separation of all isolated elements, metacarpals, two carpal rows, extremities of ulna and radius; b is the «Messel adapiform» showing radius and ulna, the complete carpus, metacarpus, and many attached phalanges, with the typical preservation of Messel fossils (Middle Eocene, Germany; photograph by E. Pantak-Wein, courtesy of Dr. J.L. Franzen, copyright Natur-Museum Senckenberg). Both scales are 1 cm.

surface, which was certainly present dorso-radially, and the evidently nonarticular surface on the palmar and ulnar margin of the bone. The proximal articular surfaces of the other metacarpals become increasingly more extensive dorso-palmarly from the second to the fifth metacarpals, and their dorso-palmar convexities increase correspondingly. Only on the fifth metacarpal does the proximal articular surface of the metacarpal extend onto the dorsal part of the bone.

Manual phalanges of *Adapis* were recently identified. The proximal phalanges are shorter than those of *Smilodectes*, slightly curved, and have distal articular surfaces palmarly extended (GODINOT, 1991a). They suggest frequent hand postures with a higher degree of digital flexion than in *Smilodectes*. Distal phalanges attributed to *Adapis* are very short, and consequently appear very high in lateral view (*Figure 2*). These phalanges allowed a provisional estimation of hand proportions in *Adapis*. Relative to forearm length, *Adapis* appears to have the shortest hand of all Eocene adapiforms, a character placing this genus within living generalized arboreal quadrupeds (GODINOT, 1991a). Concerning proportions within the hand, *Adapis* also appears to be distinct from other Eocene adapiforms, because of its relatively shorter digits and longer metacarpals. These proportions again place *Adapis* near living unspecialized arboreal quadrupeds (in spite of the fact that these proportions better reflect systematic than locomotor categories; JOUFFROY *et al.*, 1991, this volume).

While it is fairly certain that *Adapis parisiensis* was not specialized for vertical clinging and leaping, it is still unclear whether this species may have been specialized for lorisid-style slow climbing and grasp walking, as has been suggested by DAGOSTO (1983). Compared to lorisids, the wrist of *Adapis* is not very specialized (*Figures 11, 12*). The facet for the radius on the scaphoid is not radio-ulnarly expanded to the extent seen in lorisids. The corresponding facet on the lunate is expanded compared to that of *Smilodectes*, but the degree of expansion is similar to that in *Saimiri* and other ceboids, and is qualitatively different from the much more extreme radio-ulnar expansion of this facet characteristic of lorisids. Furthermore, the radial facets of the lunate and scaphoid are oriented dorsally in *Adapis*, whereas these facets are proximally oriented in lorisids. The pisiform and triquetrum of *Adapis* are neither reduced nor distally retracted as they are in lorisids. The «spiral» facet of the hamate in *Adapis* is oriented much more radio-ulnarly than in lorisids, limiting ulnar deviation at the midcarpal joint. The midcarpal articulation is not arcuate in proximal view as it is in lorisids; in fact, the head of the capitate is very reduced dorso-palmarly, suggesting little pronation-supination mobility at the midcarpal joint (see JENKINS, 1981). Finally, the trapezium of *Adapis* lacks any disto-ulnar expansion, and therefore the facet for the first metacarpal on this bone does not face so radially as it does in lorisids. The reduction of the second metacarpal characteristic of lorisids (especially *Perodicticus* and *Arctocebus*) is not present in either *Adapis* or *Leptadapis* (DAGOSTO, 1983).

Based on the evidence presented above, we conclude that the hand skeleton of *Adapis* shows no detailed convergences with lorisids which might further corroborate DAGOSTO's findings based primarily on the hindlimb. Alternatively, GODINOT & JOUFFROY (1984) suggested several similarities in wrist anatomy between *Adapis* and platyrrhines: the similar expansion of the lunate, the orientation and articular relationships of the centrale, and the large pisiform (*Figures 6, 12*). The dorsal orientation of the radial facets of the scaphoid and lunate are further similarities between *Adapis* and many ceboids. Together with the observations of GODINOT & JOUFFROY (1984), this feature suggests that the locomotion of *Adapis* may have involved a significant amount of palmigrade arboreal quadrupedalism as is the case for many living ceboids, because it indicates that the antebrachio-carpal joint was habitually dorsiflexed. We have seen that *Adapis* hand proportions place it within living generalized arboreal quadrupeds. It should be noted that our reconstruction of the locomotor adaptations of *Adapis* and the one previously proposed by DAGOSTO (1983) are not mutually exclusive. DAGOSTO's (1983) hypothesis emphasizes climbing as a central part of the locomotion of *Adapis*, while our hypothesis emphasizes palmigrade quadrupedalism.

Further study of two different *Adapis* species, from two different and contemporaneous localities, showed one of them to be more of a climber, and the other, from the same locality in which the hand was found, to be a more branch walking and running form

(GODINOT, 1991b). Hence the locomotor evolution of *Adapis* was complex. *Adapis* includes species variously adapted to climbing and rapid above-branch quadrupedalism, which explains previous differing hypotheses.

Cercamoniinae

We use this subfamily named by GINGERICH (1975) to include the European Eocene adapiforms which radiated before the arrival of the Adapinae at the beginning of the upper Eocene (FRANZEN, 1987). It comprises the two tribes Protoadapini and Anchomomyini of SZALAY & DELSON (1979). This group remained postcranially poorly known until specimens were found in the German middle Eocene locality of Messel. The first findings were articulated posterior parts of individuals (VON KOENIGSWALD, 1979, 1985). A more complete skeleton, including forelimb elements, of *Europolemur koenigswaldi*, was described by FRANZEN (1987), and another beautiful isolated forearm has been described by FRANZEN in 1988. As they are older than the remains of Adapinae, it would have been more logical to describe them first. However, their discovery is so recent that their study remains preliminary and mainly based on comparisons with the previously discovered forms.

The forelimb of *Europolemur koenigswaldi* is not very well preserved, but it nevertheless shows several interesting characters: a large pisiform, a relatively short first metacarpal, a longer basal phalanx of the first digit, and several other phalanges including two distal ones, which are very broad and short, clearly nail-bearing (FRANZEN, 1987). The position of the thumb on the palmar side of the hand points toward a high degree of thumb divergence and prehensile capabilities. *Europolemur* was interpreted as engaging in some degree of vertical clinging and leaping (FRANZEN, 1987), but corrected limb indices show more similarity to arboreal quadrupeds, and especially callitrichids (FRANZEN, 1988).

The second Messel specimen with forelimb parts is an isolated forearm. It is 50% larger than *E. koenigswaldi* in comparable parts (FRANZEN, 1988). Because it was not associated with craniodental remains, its precise taxonomic allocation is unknown, and it will be referred to here as the Messel adapiform. However, its size, morphology and provenance suggest that it too belongs to the Cercamoniinae. This articulated specimen shows radius, ulna, carpus, metacarpus and parts of the digits (*Figure 6*). The hand was on the whole longer than the radius. Among the most prominent characters of the carpus are the very large size of the scaphoid tubercle, the small size of the lunate, and the radio-ulnar narrowness of the capitate. The centrale is also fairly large, and is partially covered dorsally by the scaphoid. There was no articular contact between the centrale and the hamate. The distal articular facet for the trapezoid appears to be extensive and to have a transverse (radio-ulnar) orientation. A more detailed description is given by FRANZEN (1988 and in this volume).

An exhaustive, comparative and functional analysis of this carpus has not yet been accomplished. Similarities with the carpus of *Adapis* include the shape of the hamate, the shape of the trapezium in dorsal view and what can be seen of the trapezio-metacarpal joint (FRANZEN, 1988, infers the presence of a dish-shaped joint; this may require more scrutiny). These characters suggest a similar type of grasping in the Messel adapiform and *Adapis*, which was achieved with little movement at this joint. Differences with *Adapis*, which are also similarities with *Smilodectes*, include the radio-ulnar narrowness of the capitate and lunate. The capitate is probably primitive in this respect (GODINOT & BEARD, 1991), but the morphology of these two bones is at the same time suggestive of less emphasis on quadrupedalism in the Messel adapiform compared to *Adapis*.

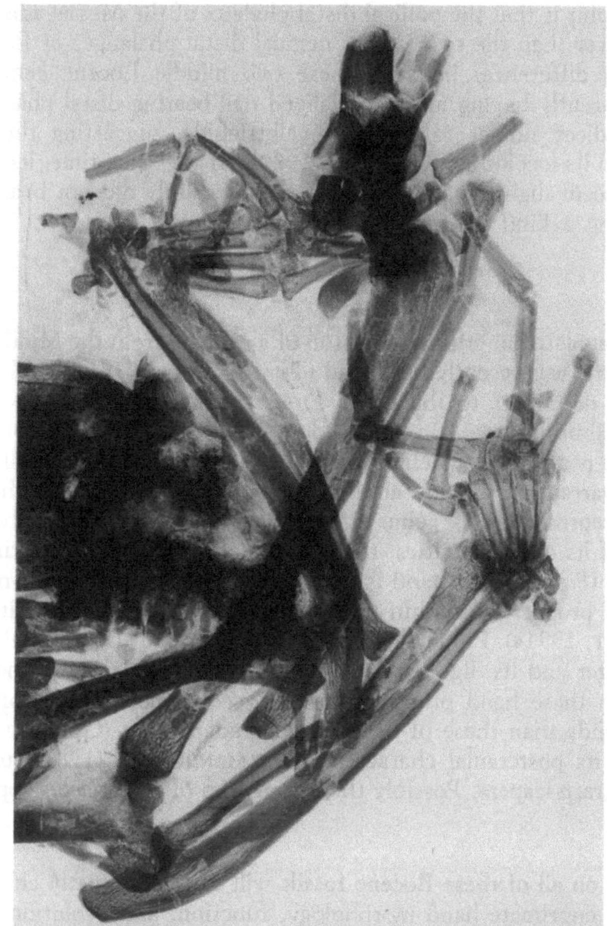

Figure 7. — Radiograph showing the forelimbs, traversed by a tibia and femur, of *Pronycticebus neglectus* from the Geiseltal Middle Eocene (Germany; radiograph courtesy of U. THALMANN).

The Messel adapiform is exceptional in showing articulated metacarpals and phalanges in an Eocene primate. Their proportions will have to be studied further. A first point is the great length of the basal phalanges in comparison with the metacarpals, a probably primitive character. FRANZEN (1988) gives estimated lengths for the metacarpals and preserved phalanges. Interestingly, the third metacarpal is the longest one, but the fourth basal phalanx is longer than that of the third digit. In adding the lengths of the metacarpals and basal phalanges, one obtains similar values for digits three and four. This indicates that the fourth digit was either equal in length with the third (paraxony), or longer, as in many living strepsirhines (ectaxony; see JOUFFROY *et al.*, this volume). This specimen also preserves two distal phalanges. The distal phalanx of the thumb is broad at its base, moderately elongated, and presents a distal broadening. The distal phalanx of the second digit is partly hidden and damaged, but seems to have been less distally broadened.

The interesting point is that the pollical distal phalanx of the Messel adapiform is clearly longer and narrower than the two known manual distal phalanges of *Europolemur*. This shows interesting differences between these two middle Eocene genera, the smaller *Europolemur* apparently having more specialized nail-bearing distal phalanges. *Europolemur* has limb indices similar to those of callitrichids, suggesting the importance of quadrupedalism in its locomotion (FRANZEN, 1988). At the same time, its broad and short distal phalanges seem slightly paradoxical, as they certainly did not bear claw-like nails, possibly suggesting a kind of grasping quadrupedalism.

Pronycticebus

An almost complete but crushed skeleton of a primate from the Middle Eocene of the Geiseltal (Germany) was recently described (THALMANN *et al.*, 1989). This specimen was attributed by its describers to the genus *Pronycticebus*, as a new species *P. neglectus*. However this attribution, based on limited dental evidence, has to be confirmed. Indeed, *Pronycticebus* is a peculiar genus characterized by very remarkable skull characters that may eventually warrant distinction at the subfamilial level. Whether the new Geiseltal species actually represents this genus is possible, but not demonstrated.

Regardless of its exact affinities, the new specimen reveals interesting characters in its hand anatomy (*Figure* 7). Its hand is relatively shorter than those of notharctines and *Europolemur*, and proportions within the forearm do not clearly place it in a locomotor category (GODINOT, 1991a). Proportions within the hand are also unusual: its metacarpals are relatively longer and its digits are relatively shorter than those of nothartctines and cercamoniines. In these hand proportions, *P. neglectus* more closely approximates the condition in indriids than those of other Eocene adapiforms (GODINOT, 1992). From a consideration of its postcranial characters, THALMANN *et al.* (1989) suggested that *P. neglectus* was a «grasp-leaper». Possibly the locomotion of *P. neglectus* approached that of cheirogaleids.

Future work on all of these Eocene fossils will certainly greatly enhance our understanding of early euprimate hand morphology, function, and evolution. The locomotor adaptations of these fossils is not easy to assess, as they differ so markedly from all living forms. However, as their number increases, more differences between them become apparent, and can be functionally interpreted. Their locomotion in this way becomes better and better circumscribed.

Lemuriformes

All living and subfossil lemuriforms are distinguished from other primates by derived characters of the midcarpal and ulnocarpal joints (JOUFFROY, 1975; GODINOT and JOUFFROY, 1984; BEARD *et al.*, 1988; BEARD and GODINOT, 1988). No primates other than lemuriformes have developed the distinctive midcarpal condyle, in which the centrale contacts the hamate, thereby affording greater deviational and rotational capability. As is well known, lorisines have greatly exaggerated their grasping and deviational capabilities.

The subfossil lemurs of Madagascar help document the true breadth of the extensive, endemic radiation of lemuriforms on that island. Some of these recently extinct forms were larger species of genera which are still extant (e.g., *Daubentonia robusta*, *Varecia insignis*), but others were uniquely adapted animals belonging to families or subfamilies that are now totally extinct (see TATTERSALL, 1982, and references therein). Postcranial remains of the more unusual species were not always correctly identified when they were initially discovered, and the peculiar morphology of some of these forms has yielded widely disparate functional interpretations in the past (see review by WALKER, 1974). Many limb bones of subfossil lemurs have been described and studied, but as yet no relatively complete hand material has been published. Nevertheless, from the very beginning of our knowledge of subfossil lemurs, metacarpals and phalanges have played an important role in reconstructing the locomotor adaptations of these animals. Hand remains have been reported for three of the large subfossil genera: *Megaladapis*, *Palaeopropithecus*, and *Babakotia*.

Megaladapis

The giant «lemuroid» *Megaladapis*, which belongs in its own family Megaladapidae, has been known since the end of the last century from skulls and important skeletal remains (FORSYTH-MAJOR, 1894; LORENZ VON LIBURNAU, 1905; GRANDIDIER, 1905). However, the known hand remains are limited to one carpal, four metacarpals, and some proximal and middle phalanges. LORENZ VON LIBURNAU (1905) described a hamate, three metacarpals, and some phalanges and GRANDIDIER (1905) described one phalanx. Later, LAMBERTON (1948) figured and described additional hand material of *Megaladapis* in greater detail.

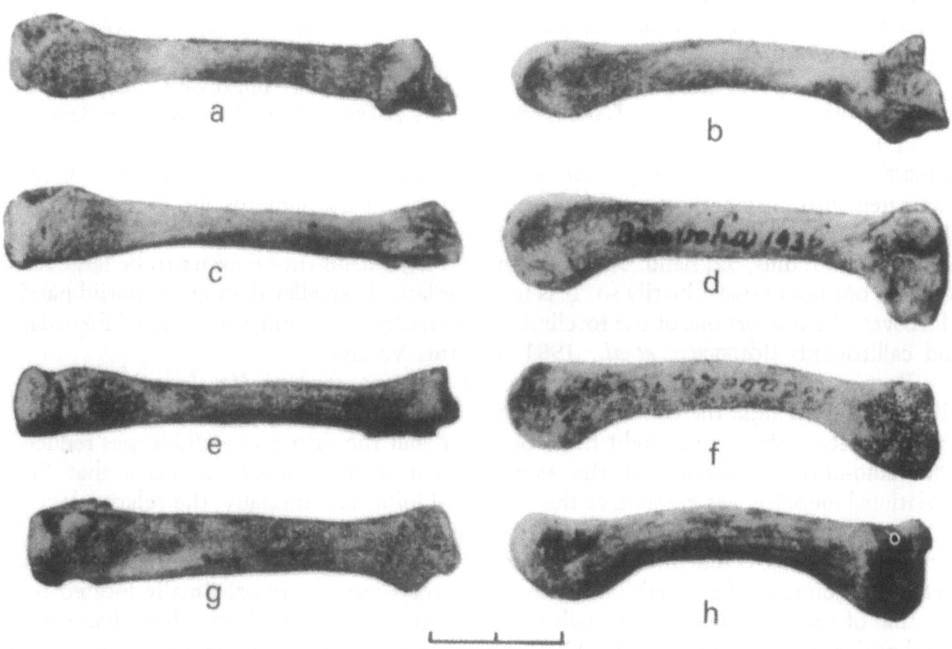

Figure 8. — Left metacarpals of *Megaladapis edwardsi*, from the plates of LAMBERTON (1946). Left, dorsal views; right, lateral views; a-b, second metacarpal; c-d, third metacarpal; e-f, fourth metacarpal; g-h, fifth metacarpal (scale is 2 cm)

In dorsal aspect, the hamate is proximo-distally short. Its shape is grossly similar to that of a human hamate, but the proximal articular facet extends onto the dorsal side of the bone, as in *Lemur*. This facet almost certainly represents articular contact between the hamate and the centrale. There is no separation between the distal facets for the fourth and fifth metacarpals. The entire distal surface of the hamate is convex, and the proximal facets of the metacarpals form a concave surface (LAMBERTON, 1948); this is the reverse of the curvature usually found in primates. The hamulus is relatively short, thick and rounded.

The metacarpals of *Megaladapis* are radio-ulnarly narrow and dorso-palmarly thick (*Figure 8*). Their proximal extremities are slightly enlarged compared to the shafts in dorsal aspect, and they are very thick dorso-palmarly in lateral view. The distal heads are also broader than the shafts. Their articular surfaces are roughly cylindrical and disto-palmarly oriented. The third metacarpal of *M. edwardsi* figured by LORENZ VON LIBURNAU is much thicker dorso-palmarly than the one of the same species figured by LAMBERTON. On both of these figures, the shaft of the third metacarpal is straight. LAMBERTON's figures show the convexity of the other metacarpals in the dorso-palmar plane: slight on the second metacarpal, more accentuated on the fourth and fifth metacarpals (*Figure 8*). LAMBERTON gives the following mean lengths (in mm): second metacarpal, 66.2; third metacarpal, 62.7; fourth metacarpal, 64.5; fifth metacarpal, 62. 0. He also gives robusticity indices, which increase from the second to the fifth metacarpals. See LAMBERTON (1948) for additional details.

The proximal and middle phalanges are dorso-palmarly flattened. The proximal phalanx has a strong curvature and its distal articular facet faces disto-palmarly. The middle phalanx is much less curved. The third digit of *Megaladapis* is reconstructed in lateral view in figure 9, based on the drawings of each bone given by LORENZ VON LIBURNAU (1905).

The large size of the hands and feet of *Megaladapis* has impressed many authors (LORENZ VON LIBURNAU, 1905; LAMBERTON, 1934; ZAPFE, 1963), but this size does not appear to us to be particularly unusual for a primate. From the dimensions given by LORENZ VON LIBURNAU, it is possible to extrapolate a hand length of about 230 mm. Compared with the lengths given by the same author for the humerus and radius, one may obtain the approximate length percentages of the various segments of the forelimb: humerus, 35; radius, 32; hand, 33. The hand of *Megaladapis* thus appears to be large for a primate, but not extraordinarily so. It is in fact relatively smaller than most indriid hands. Moreover, the proportions of the forelimb of *Megaladapis* are similar to those of *Plesiadapis* and callitrichids (JOUFFROY *et al.*, 1991 and this volume).

Because of the incomplete material thus far recovered of *Megaladapis*, it is very difficult to determine the overall adaptations of the hand of this animal. The hamate is proximo-distally short; one might therefore infer that the carpus as a whole was reduced. The radio-ulnar orientation of the «spiral» facet of the hamate suggests that little deviational mobility was possible at the midcarpal joint. Additionally, the relative lengths of the metacarpals are surprising in that the second metacarpal is the longest, followed in decreasing order by fourth, third and fifth. However, it is possible that the carpo-metacarpal joints of the fourth and fifth metacarpals were more proximally located than was that of the third metacarpal, such that these three metacarpals may have had similar distal projections (LAMBERTON, 1948). Likewise, the relatively great length of the second metacarpal may be partially explained by the palmar process (possibly for insertion of the flexor carpi radialis) of that bone, which extends proximally (LAMBERTON, 1948). Hence, we would like to have exactly comparable measurements before inferring a possible convergence with hominoids in second metacarpal lengthening. More significant aspects of metacarpal morphology may be the increasing robusticity observed from the second to the fifth metacarpals, and the increasing curvature observed from the third to the fifth metacarpals. These features suggest that the ulnar side of the metacarpus was subject to greater bending stresses during locomotor grasping (PREUSCHOFT, 1971). This would provide support for habitually vertical clinging postures for *Megaladapis*, as WALKER (1974) has pointed out that vertical clinging is often associated with ulnar-side dominance in the hand. However, an unusually high robusticity of the fifth metacarpal (equal to that of the third) has also been reported for the living *Pongo* (SIEGEL & PERNOTTO, 1975), in which it is certainly more related to climbing than to vertical clinging. Finally, the degree of curvature of the middle phalanges is intermediate between that of *Pan* and that of *Hylobates* (*Figure 9*), suggesting less suspension than in gibbons, but possibly more branch grasping than in *Pan*. These characters, associated with limited carpal mobility, indicate some kind of stereotyped grasping of slender substrates, for climbing or any kind of clinging (see also NIESCHALK & DEMES and LANDSMEER, both in this volume).

The most commonly proposed locomotor hypothesis for *Megaladapis* is that the animal must have been adapted for climbing and clinging to vertical supports, with the extant koala, *Phascolarctos*, perhaps being the best living analogue (WALKER, 1967, 1974; JUNGERS, 1977, 1978). The evidence provided by the known aspects of hand anatomy in *Megaladapis* suggests good but stereotyped grasping ability, which is consistent with this interpretation.

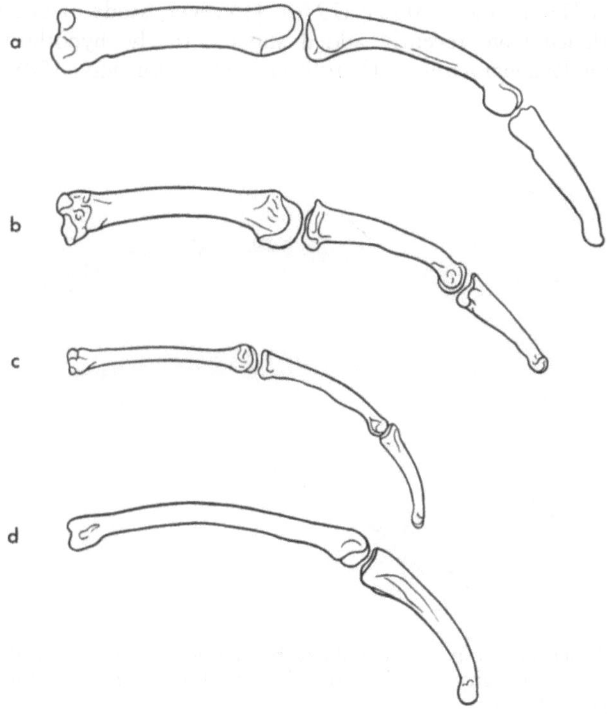

Figure 9. — Comparison of third metacarpal and phalanges of the third digit of two subfossil Malagasy lemurs and two living apes, in lateral view. a, *Megaladapis*, after LORENZ VON LIBURNAU (1905); b, *Pan*, and c, *Hylobates*, both after ETTER (1974); e, *Palaeopropithecus*, after LAMBERTON (1947). Not to scale.

Palaeopropithecus

The postcranials of *Palaeopropithecus* are perhaps the most uniquely derived of any primate. When first discovered, its femora were originally attributed to a sloth of the family Bradypodidae (GRANDIDIER, 1901). LAMBERTON (1947) described much of the skeleton in detail, and elucidated characters which he felt were similarities to both *Pongo* and sloths.

The metacarpals of *Palaeopropithecus* are very narrow and elongated, and are strongly curved in the dorso-palmar plane, probably more so than in any other primate (*Figure 10*). Their distal articular surfaces are noteworthy, as they possess trochleas rather than the globular heads typical of most primate metacarpals. This character is a clear convergence with sloths. The metacarpals of *Palaeopropithecus* were of similar length. The proximal phalanges are strongly curved and their dorso-palmar height decreases rather markedly distally. The proximal ends of the proximal phalanges fit into the trochleas of the distal metacarpals, rendering radio-ulnar movement at the metacarpo-phalangeal joints negligible. *Figure 9* is our reconstruction of the third metacarpal and proximal phalanx of *Palaeopropithecus*, based on the figures of Lamberton (the phalanx is referred by him to the second digit). *Figure 9* shows that the hand of *Palaeopropithecus* was thinner and more elongated than that of *Megaladapis*, and that the bones were also more curved.

The metacarpals and phalanges of *Palaeopropithecus* show that the hand was habitually flexed, that the digits possessed reduced radio-ulnar mobility, and that the digits of this creature therefore probably flexed simultaneously to form a hook-like grasping organ. The strong curvature of the bones indicates high bending stresses suggestive of suspension (PREUSCHOFT, 1971). These hand characters do not allow a clear choice between the sloth-like locomotion favored by THENIUS (1969) and JUNGERS (1980) and the more *Pongo*-like pattern advocated by LAMBERTON (1947) and WALKER (1974). However, study of other postcranials and new fossil evidence have recently added support to the hypothesis favoring a sloth-like adaptation for *Palaeopropithecus* (MACPHEE, 1986; GODFREY, 1988).

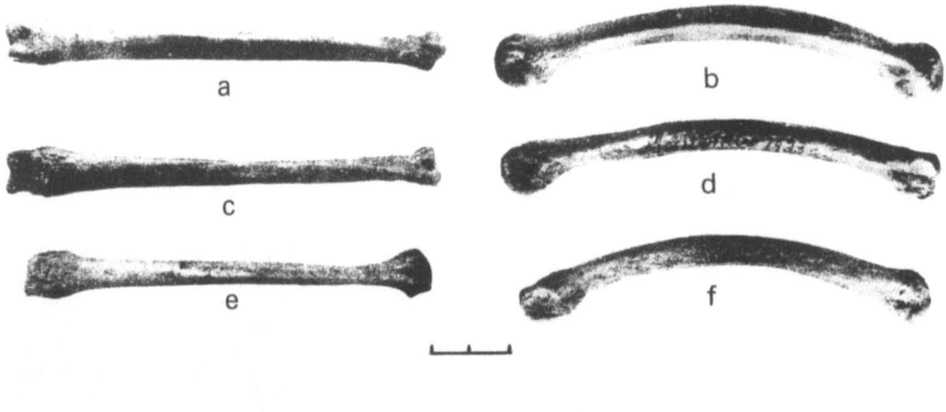

Figure 10. — Three metacarpals of *Palaeopropithecus ingens* from Manombo, Madagascar, from the plates of LAMBERTON (1947). Left, dorsal views, and right, medial views; a-b, second metacarpal; c-d, third metacarpal; e-f, fifth metacarpal (scale is 2 cm).

Babakotia

Babakotia is the most recently discovered of the subfossil lemurs of Madagascar (GODFREY *et al.*, 1990). Its dental and postcranial anatomy identify it as a member of the extinct indrioid family Palaeopropithecidae. The hand of this subfossil lemur is known from much of the carpus, a metacarpal, a left pollical proximal phalanx and some proximal phalangeal fragments (JUNGERS *et al.*, 1991). The scaphoid and centrale are fused as in *Indri*. The metacarpal and pollical phalanx are also similar to those of *Indri*, but are more robust. The proximal phalangeal fragments are curved like those of other palaeopropithecids. The metacarpophalangeal joints are less specialized than those of *Palaeopropithecus*, but seem to have emphasized flexed postures. *Babakotia* was apparently a climbing and suspensory palaeopropithecid that was less specialized than the more sloth-like *Palaeopropithecus* (JUNGERS *et al.*, 1991). It represents a plausible structural antecedent to the more derived but contemporary *Palaepropithecus*.

Early strepsirhine and early simiiform hands

The strepsirhine hand morphotype is much better known than the simiiform one, because of the characters known in the Eocene adapiforms. As these fossils show a number of primitive characters in comparison with lemuriforms, all of the latter traits have been removed from the strepsirhine morphotype (GODINOT & BEARD, 1991), making it much more accurate, but at the same time more difficult to interpret functionally. The characters that we have listed for the strepsirhine morphotype are: relatively short metacarpals and relatively long digits (compared with simiiforms), a very divergent first metacarpal and associated trapezial features. The relatively long digits may be a retention from the euprimate morphotype. The very divergent thumb shows an emphasis on grasping, which may possibly have been associated with an ulnar shift in hand posture with respect to the axis of the forearm during locomotion. Such a shift is clearly seen in *Adapis*, in which it is reflected on the triquetrum and pisiform (*Figures 6, 11, 12*). The relatively great length of the fourth proximal phalanx in the Messel adapiform also suggests an ectaxonic hand correlated with more ulnarly deviated hand postures. Such an ulnar shift is not apparent on the carpus of *Smilodectes*, but is suggested by its metacarpals. In any case, a strong thumb divergence and emphasis on grasping are clearly derived compared to the primitive mammalian condition.

We have also noted a number of characteristics comprising part of the simiiform hand morphotype: a subquadrangular triquetrum in dorsal view, an emphasis in size on lunate and capitate, a more quadrangular trapezium in dorsal view correlated with a more distally oriented first metacarpal, and relatively shorter digits and longer metacarpals than in early strepsirhines (*Figures 11, 12*). The less divergent thumb of simiiforms is associated with a different type of prehension. This type of hand is in general mesaxonic, functioning with its main axis (digit three) broadly parallel to the support. Relatively longer metacarpals and shorter digits compared to the strepsirhine morphotype suggest a distant analogy with digitigrade runners. On the whole, presuming that these early simiiforms were arboreal, this hand seems best suited for palmigrade quadrupedalism on large horizontal supports (see also PREUSCHOFT *et al.* in this volume). In its lesser degree of thumb divergence and general mesaxony, the early simiiform hand is more similar to the primitive mammalian condition than is the early strepsirhine hand.

In attempting to place the differences between the two morphotypes within an

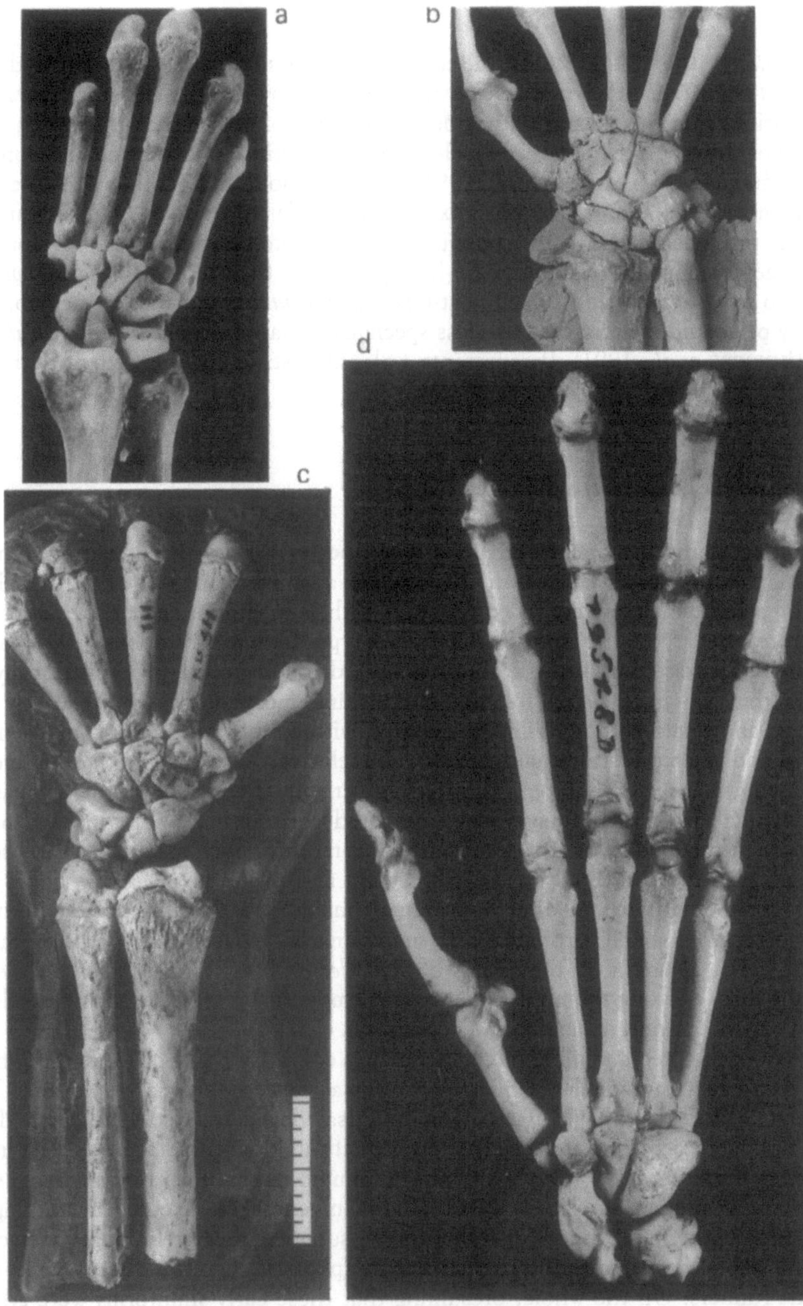

Figure 11. — Three living primate hands compared to a reconstruction of *Adapis* hand, all in dorsal view. a, *Aotus*, South-American simiiform; b, *Lemur mongoz* (photograph courtesy of Dr. F.K. JOUFFROY); c, *Adapis*; d, *Lemur catta*; b and d show the midcarpal condyle typical of living lemuriforms; a shows a low degree of first metacarpal divergence, possibly exaggerated on this dry specimen (same scale, 2 cm, for a, c, and d).

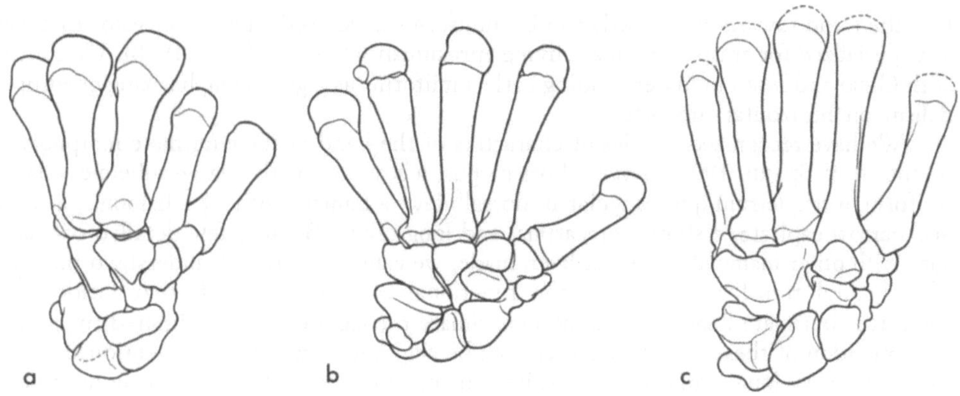

Figure 12. — Comparison between the carpus and metacarpus (in dorsal view) of the left hand of: a, *Nycticebus* (articulated specimen); b, *Adapis* (reconstructed); and c, *Cebus* (articulated). Not to scale.

evolutionary framework, three alternatives are possible: 1) the early simiiform hand is primitively less specialized, and therefore approximates the euprimate morphotype; 2) the early simiiform hand has secondarily evolved toward a less grasping type, and is therefore derived in comparison with the strepsirhine morphotype. This second hypothesis is clearly influenced by the scala naturae of living primates. It was, for example, advocated by GREGORY (1920); 3) the simiiform and the strepsirhine types of hands are two different lines of specialization derived from a structure more primitive than each of them. It seems unparsimonious to hypothesize the secondary decrease of a primitively strong thumb divergence in early simiiforms because this would involve the reversal of a complex character. It would also seem unlikely insofar as a decrease in the functional role of the thumb would lead to its reduction more easily than to a structural reorganization of the carpals. These considerations would imply that early simiiforms retained a primitively low degree of pollical divergence from the euprimate morphotype, an hypothesis that is favored by one of us (M.G.; see similar conclusions, based on other aspects of postcranial anatomy, in FORD, 1988). At the same time, their metacarpals and digits and their emphasis on the median line of the carpus show an early specialization toward horizontal quadrupedalism, which was probably not attained by the euprimate morphotype. On the whole, the most likely hypothesis is that the hands of early simiiforms and early strepsirhines represent two divergent lines of specialization.

This hypothesis is easy to accommodate with recent studies on the foot. GEBO (1986) argued for a de-emphasis on vertical climbing and an increased emphasis on quadrupedalism on horizontal supports to explain the acquisition of simiiform foot characters from a presumed «prosimian» stage. However, is it necessary to hypothesize such a preceding prosimian stage, which is necessarily correlated with the loss by early simiiforms of a number of derived characters, including «de-emphasis of the strong grasping hallux, decrease in length of the fourth digit» (GEBO, 1986a)? As several of the simiiform characters, especially in the astragalus, are more primitive than the alternative character states found in «prosimians» (GEBO, 1986a), the most parsimonious hypothesis could be that the trend toward quadrupedalism in simiiforms developed from an ancestor that was less specialized for grasping than are living «prosimians» (see also FORD, 1988). We concur with GEBO and FORD in understanding early simiiforms as a group emphasizing quadrupedalism on horizontal supports.

We have recognized a series of characters of the hand of the euprimate morphotype (GODINOT & BEARD, 1991): a very short carpus, a scaphoid with a large tubercle, a large pisiform, a triquetrum quadrangular in dorsal view, a hamate without a hamulus, a short and narrow capitate, a short metacarpus, and long digits. Because few detailed comparisons with other mammals have yet been made, we cannot at present understand the type of structural transformation, and the functional correlates thereof, that yielded the primitive euprimate hand from its non-euprimate predecessor. Only a limited functional interpretation of these carpal characters can be achieved now. Still, we mentioned that a hand with very long digits is poorly suited for manipulation, although it may have been capable of more generalized behaviors and postures such as clinging and grasping. Also, several of the above mentioned characters suggest a low degree of mobility for the carpus. These characters are not sufficient to delimit any specific locomotor mode: they seem compatible with vertical clinging and leaping, «grasp-leaping» (SZALAY & DAGOSTO, 1980), climbing, and other kinds of arboreal activities. In contrast to the hand, the foot shows a high degree of specialization in the hallux in all euprimates, hence providing more clues about their ancestral morphotype. The degree of hallux divergence that simiiforms and strepsirhines have in common has been assumed by almost all recent authors to be homologous in both groups, inherited from the euprimate morphotype (e.g. GEBO, 1986a; SZALAY & DAGOSTO, 1988; but see FORD, 1988). We will also assume that this is parsimonious, but one of us (M.G.) also notes that the structural differences between the feet of the two groups could lead to question this hypothesis, possibly indicating their eventual divergence from an animal with less hallux opposition (in the same way that the differing morphologies of the hands of the two groups can be interpreted as two different specializations from a less prehensile form). Assuming that at least some degree of hallucal prehension is homologous in euprimates, this character could indicate differentiation of the ancestral euprimate from a plesiadapiform-grade predecessor by more emphasis on vertical climbing, some leaping, a shift toward a fine branch milieu, or a combination of these, probably involving some hindlimb domination (see e.g. NAPIER & WALKER, 1967; MARTIN, 1972; CARTMILL, 1974; SZALAY & DAGOSTO, 1980, 1988; GEBO, 1986a; DAGOSTO, 1988; FORD, 1988). It is very hard to reconstruct the early phases of such an ancient (at least Paleocene) history without direct fossil evidence, and thus hypotheses concerning the euprimate morphotype locomotor mode remain very speculative.

In this context, the morphology of *Tarsius* should shed light on a haplorhine, or «Simiiformes plus *Tarsius*» morphotype, but the authors disagree on its significance. The senior author favors the view that *Tarsius* (or its ancestors) has secondarily developed a relatively high degree of thumb divergence related to its vertical clinging and leaping adaptation; this character appears then convergent with that of the strepsirhine morphotype, and not primitively retained. The peculiar foot of *Tarsius* is known to function with much more parasagittal movements than that of galagos (JOUFFROY *et al.*, 1984; GEBO, 1987). The peculiarities of *Tarsius* suggest that its foot evolved from a structure different from that of strepsirhines. It seems to have evolved from an early simiiform type of foot, which utilized more strictly parasagittal movements (GEBO, 1986a). Hence, *Tarsius* could be a vertical clinger and leaper that evolved from an animal similar to the simiiform morphotype, in accordance with haplorhine monophyly.

The second author favors the view that *Tarsius*, with its triangular trapezium and relatively high degree of thumb divergence, more likely approximates the primitive condition for Haplorhini than does the simiiform morphotype. In this view, the degree of pollical divergence found in the euprimate, strepsirhine, and haplorhine morphotypes would all be roughly equivalent, and there would be no need to invoke the convergent acquisition of a relatively divergent pollex in the strepsirhine morphotype and in *Tarsius*. This schema implies that early simiiforms have de-emphasized their pollical prehensile capacities compared to their «prosimian» predecessors (e.g. GREGORY, 1920), as they are assumed to have done for the prehensile capacities of their hallux (GEBO, 1986a; SZALAY & DAGOSTO, 1988).

A more detailed analysis of the carpus of *Tarsius* in comparison with other primitive forms should help to choose between these hypotheses, or to generate new ones. With luck, the fossil record of early simiiform and early haplorhine hands will also be enhanced in the relatively near future, in order to provide more precise answers to some of the problems outlined above.

Simiiformes: non-hominoid Catarrhini

Pliopithecidae

Several phalanges of one of the Oligocene Fayum propliopithecids, *Aegyptopithecus zeuxis*, are known, but they have never been illustrated. According to FLEAGLE (1983): «They are curved, indicating grasping digits, but lack either the elongation of living apes or the robustness seen in terrestrial monkeys». They suggest arboreal habits (PREUSCHOFT, 1975).

ZAPFE (1960) described much of the postcranium of *Pliopithecus vindobonensis* from the early Miocene of Czechoslovakia. Since that time, several later workers have reviewed the postcranial anatomy of this species; most notable among these studies are those of SIMONS & FLEAGLE (1973) and FLEAGLE (1983). These studies have emphasized the overall similarities in postcranial anatomy between *Pliopithecus* and large extant platyrrhines. Morphometric studies by CORRUCCINI *et al.* (1975, 1976) have generally agreed with these conclusions.

Among the postcrania of *Pliopithecus* treated by ZAPFE (1960) were four carpals and most of the metacarpus and phalanges of three different individuals (*Figure 13*). The scaphoid of *Pliopithecus* figured by ZAPFE (1960) reveals that *Pliopithecus* had an unfused centrale. In the lack of palmar deflection of the tubercle, *Pliopithecus* is similar to *Proconsul*, and different from most extant monkeys (*Figures 13, b-d*). Cercopithecoid monkeys all possess a beaklike process immediately radial to the facet for the centrale, and *Pliopithecus* lacks this feature as does *Proconsul* (NAPIER & DAVIS, 1959). The lunate of *Pliopithecus* was also relatively unspecialized, with minimal expansion of the radial facet characteristic of living hominoids. Moreover, the radial facet of the lunate lies nearly in the same plane as the articular region for the scaphoid, rather than being nearly orthogonal to this facet as in extant hominoids. Overall, the lunate of *Pliopithecus* does not appear to be very different from those of either *Proconsul* or *Alouatta*. The trapezium of *Pliopithecus* possesses a facet for the first metacarpal which is relatively palmarly oriented, similar to *Alouatta*; this facet does not extend onto the dorsal surface of the bone as it does in *Proconsul*. Dorsally, a tubercle for the posterior oblique ligament of the carpometacarpal joint of the thumb, a putative hominoid synapomorphy (LEWIS, 1977), is not

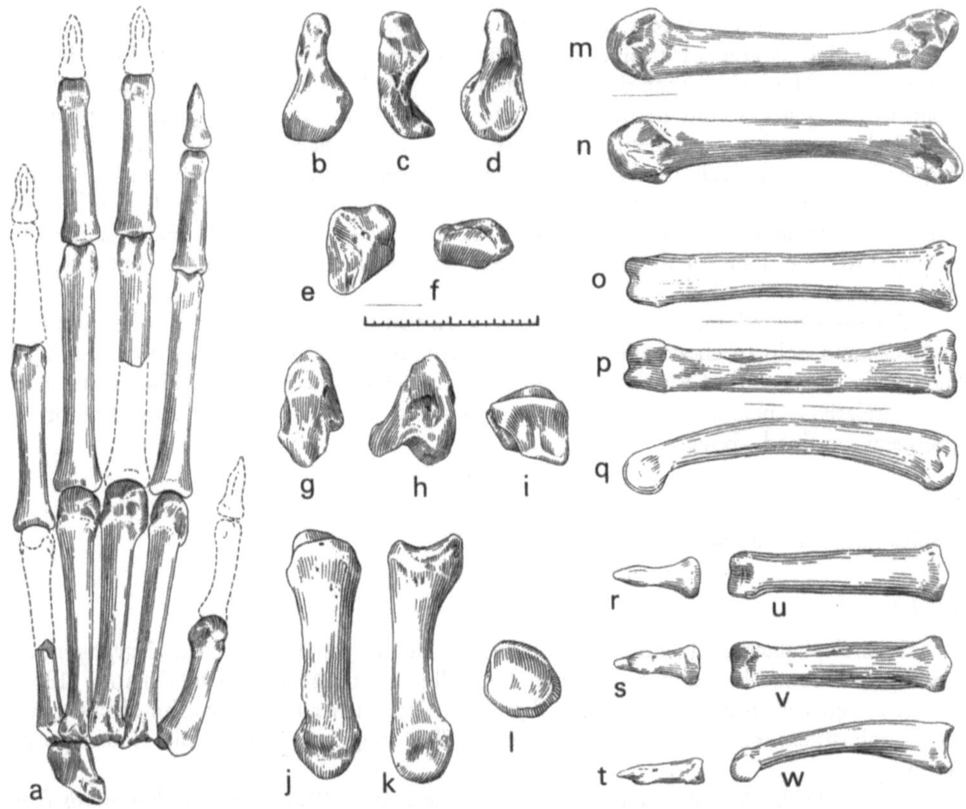

Figure 13. — Hand bones of *Pliopithecus vindobonensis* from Neudorf an der March, Miocene of Czechoslovakia (drawings from ZAPFE, 1960). a, left hand of holotype with first metacarpal from another specimen, in dorsal view; b-d, left scaphoid in dorsal (b), profile (c) and palmar (d) views; e-f, left trapezium in palmar (e) and distal (f) views; g-i, left hamate in palmar (g), medial (h), and distal (i) views; j-l, right first metacarpal in dorsal (j), medial (k), and proximal (l) views; m-n, left fourth metacarpal in medial (m), and lateral (n) views; o-q, left fourth proximal phalanx in dorsal (o), palmar (p), and lateral (q) views; r-t, distal phalanx in dorsal (r), palmar (s), and lateral (t) views; u-w, middle fourth phalanx in dorsal (u), palmar (v), and lateral (w) views. All drawings from b to w at the same scale, scale bar is 2 cm; a at smaller scale.

apparent. The hamate of *Pliopithecus* is similar to those of platyrrhines such as *Ateles*, but also to *Proconsul*. The hamate is narrow proximally, the «spiral» facet is oriented rather proximo-distally, and there is only a moderately developed hamulus present (*Figure 13, g-i*). The distal articular facets for the fourth and fifth metacarpals form a concavity, rather than being slightly convex as in living pongids. The preserved portions of the metacarpus and phalanges show that the hand of *Pliopithecus* was relatively shorter than those of most extant hominoids and was generally monkey-like (FLEAGLE, 1983). Phalanges that are most likely attributable to *Pliopithecus* from Rudabanya, Hungary, are notable for their curvature in the dorso-palmar plane, well-developed palmar ridges for fibrous flexor sheaths, and deeply-grooved distal articular trochleas (BEGUN, 1988). These features are interpreted as indicative of suspensory adaptations (BEGUN, 1988).

The late Miocene Chinese genus *Laccopithecus* is dentally very similar to *Pliopithecus*. A manual proximal phalanx attributed to this genus differs, however, from those of *Pliopithecus*, and resembles more closely those of the living Siamang, suggesting grasping and perhaps manual suspension in this genus (MELDRUM & PAN, 1988).

Cercopithecidae

The anatomy of the hands of fossil cercopithecids has been poorly studied, partly because of the paucity of good material. As noted in our introductory remarks above, one exception to this is the Miocene colobine *Mesopithecus pentelici*. The hand of *Mesopithecus* was originally described by GAUDRY (1862), who advocated that *Mesopithecus* must have been somewhat terrestrially adapted, based mainly on the robusticity of the forelimb and metacarpus. Such an interpretation was supported by SZALAY & DELSON (1979).

The hand of *Mesopithecus* is characterized by short, robust metacarpals, and a relatively long pollex which was also more gracile than the other digits. One of the specimens figured by GAUDRY, MNHN PIK 294, is under further technical preparation and study (*Figure 14*). Compared with the living *Presbytis entellus* it shows several

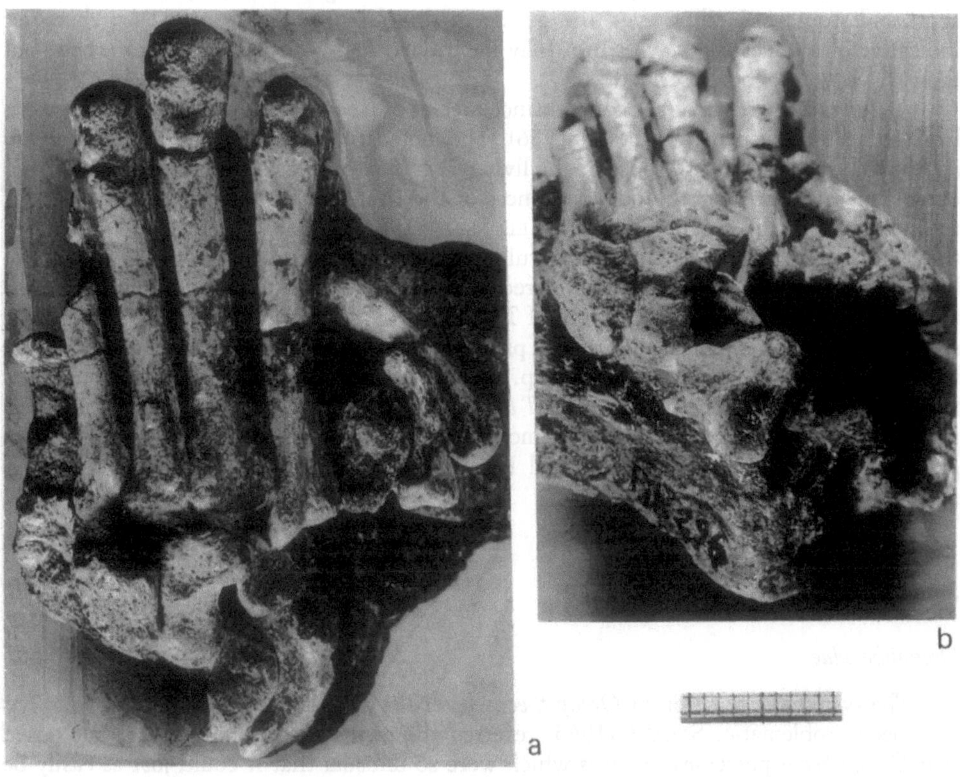

Figure 14. — A partial hand of *Mesopithecus pentelici* from Pikermi, Miocene of Greece (MNHN PIK 296) in dorsal (a) and proximo-dorsal (b) views. Scale is 1 cm.

differences in the carpals: scaphoid tubercle slightly less elongated, capitate body more radio-ulnarly extended, and line of contact between capitate and hamate less proximo-ulnarly deflected. These differences are possibly the result of retention of primitive characters in *Mesopithecus*, but their functional significance remains difficult to interpret. The hamate of *Mesopithecus* seems proximo-distally short, but it is eroded in its distal and ulnar parts; only the proximal part of the «spiral» facet is preserved (*Figure 14, b*). Judging by GAUDRY's published illustrations (GAUDRY, 1862: pl. IV, Figures 4-5), *Mesopithecus* also possessed a relatively large, robust pisiform. These bones will be more completely prepared and described elsewhere. Possibly the anatomy of the hands of early colobines such as *Mesopithecus* may provide evidence to test the hypothesis (e.g., STRASSER, 1988) that colobine monkeys are secondarily adapted to arboreal life, and that the basal cercopithecid was a terrestrially adapted form. Preliminary study of the anatomy of the hand and other postcranial elements of two of the earliest fossil colobines, *Mesopithecus* and *Dolichopithecus*, is certainly consistent with such a view (see discussion of these genera in SZALAY & DELSON, 1979, for additional details and references). Phalanges from the pivotal Miocene cercopithecid *Victoriapithecus* are also known (SZALAY & DELSON, 1979); further study of these specimens should also help to resolve the debate concerning the locomotor adaptations of early cercopithecids.

Well-preserved hand remains of fossil cercopithecines are known for two extinct species of the genus *Theropithecus*. The proximal phalanges (manual and pedal) of the large Plio-Pleistocene baboon *Theropithecus oswaldi* from Olorgesailie, Kenya, are extremely short and stout «exceeding even the gelada in this feature» (JOLLY, 1972), a character indicative of a terrestrial way of life. Jolly mentioned many carpals and metacarpals, but these have yet to be described.

A study of the hand of a second extinct species of *Theropithecus*, Pliocene *T. brumpti*, has also been published (JABLONSKI, 1986). The metacarpals and phalanges of *T. brumpti* appear to be very similar to those of the living *T. gelada*. Similarity between the fossil and extant species in the length and slenderness of the first metacarpal and the shortness of the phalanges of the second digit led JABLONSKLI (1986, p. 180) to conclude: «Although the hand of *T. brumpti* was more powerfully built than is that of the gelada, it appears to have been capable of the same high degree of precise opposition of the thumb and index finger». The derived traits of the hand of *Theropithecus* are shared by two distantly related species within the genus. They therefore probably represent a specialization «primitive for the genus as a whole» (JABLONSKI, 1986, p. 181). A difference between these two species is the slight curvature of the phalanges of *T brumpti*, by contrast with the absence thereof in the gelada. JABLONSKI (1986) takes this morphological evidence, together with paleoecological data, to indicate a more arboreal habitus for *T. brumpti* compared to its living relative.

Simiiformes: Hominoidea

Oreopithecidae

The systematic position of *Oreopithecus bambolii* from the late Miocene of Italy has long been problematic. STRAUS (1963) referred the genus to the Hominidae, but noted that *Oreopithecus* possesses features which were so unusual that it could just as easily be classified in its own family Oreopithecidae, a procedure followed here. SZALAY & DELSON (1979) referred this family to the Cercopithecoidea. HARRISON (1986a, b) and SARMIENTO

(1987) have recently provided convincing evidence which supports the hominoid status of the family, and their taxonomy is followed here.

We were not able to study the original material of *Oreopithecus*, so we will only briefly summarize the work of other authors here. HÜRZELER (1958) gave a first description of the postcranials of *Oreopithecus* found before 1957. Concerning the hand, he wrote, «Contrary to conditions in pongids, the metacarpalia are comparatively short, and the first metacarpal is rather stout». He also described the distal phalanges, which have a shaft that is circular in cross section and a robust apical tuft for supporting the nail. A new skeleton found in 1958 was described by STRAUS (1963), who added: «the metacarpals and phalanges are slightly (but distinctly) curved, and the phalanges (especially the basals) give the appearance of being slightly scooped out ventrally». SZALAY & DELSON (1979: Figure 220a) provided the best illustration published to date of the hand remains of *Oreopithecus*. SUSMAN (1985) studied this material and described similarities in the scaphoid between *Oreopithecus* and *Pongo*, noted the great length of the hand, and suggested that the phalangeal curvature exhibited by *Oreopithecus* was similar to that of extant African apes and less than that of *Pongo*. Based on hand morphology and the relatively high intermembral index of *Oreopithecus*, SUSMAN suggested suspensory habits for this genus intermediate between those of *Pongo* and *Pan*. In a study that was primarily aimed at establishing the phylogenetic relationships of the genus, HARRISON (1986b) noted that the preserved lunate of *Oreopithecus* is distorted but probably would have resembled Asian apes and humans in its proportions. The scaphoid of *Oreopithecus* demonstrates that the centrale was not fused in this genus; it also lacks the apparently apomorphous palmar beak-like process characteristic of cercopithecids (NAPIER & DAVIS, 1959; HARRISON, 1986b). These characters are consistent with the climbing and suspensory habits that have been inferred for this primate. The locomotor repertoire of *Oreopithecus* probably approached that of female orang-utans (ANDREWS & GROVES, 1976; JUNGERS, 1987).

Proconsulidae

The original hand material of *Proconsul africanus* described by NAPIER & DAVIS (1959) has been among the most important fossil evidence for hypotheses concerning the early evolution of hominoid locomotor adaptations (*Figure 15*). Two schools of thought have formed with respect to the locomotor adaptations of the wrist and hand in *Proconsul*: 1) *Proconsul* possessed the (apparently apomorphous) wrist adaptations characteristic of extant hominoids, and therefore probably practiced some suspensory behaviors («brachiation» *sensu lato*); or 2) *Proconsul* lacked these derived wrist adaptations of extant hominoids and was therefore most likely an arboreal quadruped, similar to extant cebids and colobines. The former view has been advocated almost solely by LEWIS (1971, 1972a, 1972b); the latter view has been adopted by most other workers (NAPIER & DAVIS, 1959; PREUSCHOFT, 1973b; MORBECK, 1975, 1977; CORRUCCINI *et al.*, 1975; O'CONNOR, 1976; MCHENRY & CORRUCCINI, 1983).

We corrected the forelimb and hand proportions estimated by NAPIER & DAVIS (1959), using the new information from recently discovered parts of the same individual (WALKER & PICKFORD, 1983). The length of the proximal phalanx of the third digit is 31 mm, slightly less than estimated by NAPIER & DAVIS (33.5 mm from their Appendix IV). This gives new proportions for the hand: carpus, 18.5; metacarpus, 33.5; and third digit, 48. Using the slightly different hand length and the new humeral length of 167 mm (WALKER & PICKFORD, 1983), the percentages of forelimb segments become: humerus, 38.3; radius, 35.7; and hand, 26.0. Comparison with living primates confirms the

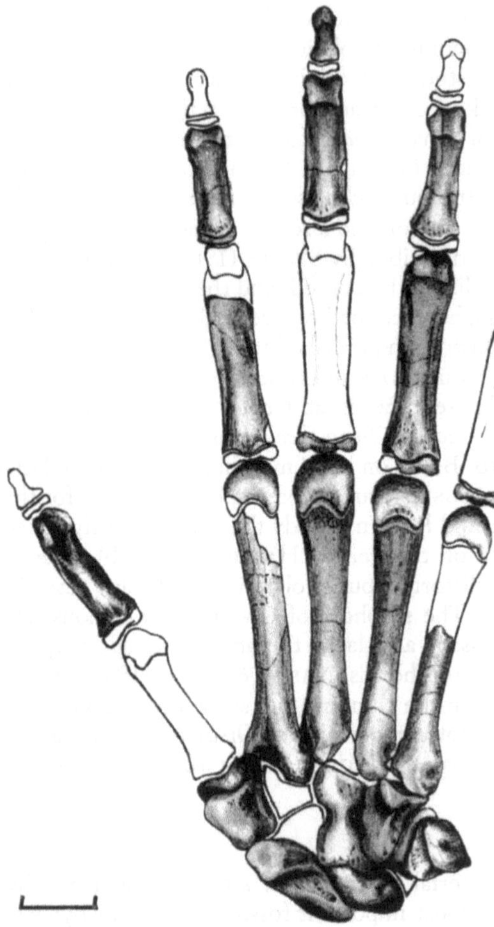

Figure 15. — The partial left hand of *Proconsul africanus* from the Miocene of Rusinga Island, Kenya, as reconstructed and figured by NAPIER & DAVIS (1959); palmar view. Scale is 1 cm.

relatively high humeral length (probably primitive) and the relatively short hand length found by NAPIER & DAVIS (1959). *Proconsul* appears very close to *Cercopithecus* for these proportions (JOUFFROY *et al.*, 1991, and this volume; *Figure 16*).

Proportions within the hand place *Proconsul* near *Macaca*. This seems to corroborate the findings of MCHENRY & CORRUCCINI (1983) of morphometric similarity between the hands of these two genera. However, their morphometric study of the carpus failed to reveal several functionally significant characters in which these genera differ (see below). The osteometric study of JOUFFROY *et al.* (this volume) shows that forelimb proportions reflect locomotor categories better than do the intrinsic proportions of the hand. The functionally significant metric affinities of *Proconsul* therefore appear to lie with the highly arboreal *Cercopithecus*, rather than with the more terrestrial *Macaca*.

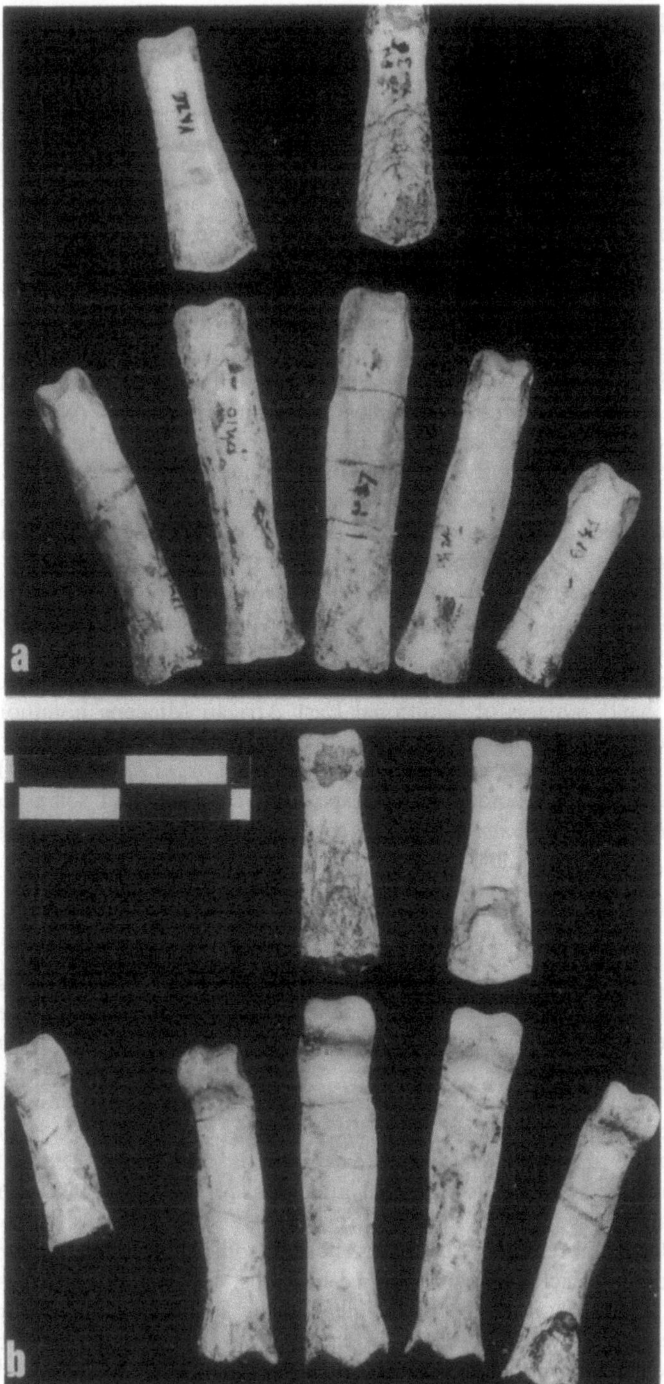

Figure 16. — Middle and distal phalanges of the hand of a sub-adult male of *Proconsul africanus* from the Miocene Kaswanga Primate Site, Rusinga Island, Kenya (photograph courtesy of Dr. D. BEGUN; scale units are 1 cm).

Newly-collected wrist and hand fossils of *Proconsul africanus* and *P. nyanzae* (Beard *et al.*, 1986, this volume) underscore the apparently unique anatomy of the wrist in this fossil primate. In particular, the ulnocarpal joint in *Proconsul* is modified away from the presumably primitive simiiform morphology found in platyrrhines and cercopithecids in two important ways: (1) the articular facets on the triquetrum and pisiform for the ulnar styloid process are more extensive in a proximoradial-distoulnar direction than is the case in extant monkeys; these facets also lack the bony stops on the ulnar side of the pisiform and on the dorsal side of the triquetrum that are found in monkeys, and (2) the ulnar tapering of the triquetrum as a whole favors enhanced ulnar deviation in *Proconsul* compared to extant monkeys, in which this bone is much more quadrangular in dorsal view (i.e., there is more intervening nonarticular bone between the facets for the ulnar styloid process and the hamate on the triquetrum of monkeys; see Beard *et al.*, 1986). Greater capacities of ulnar deviation are one of the hallmark adaptations of the wrists of extant hominoids compared to those of most other simiiforms (O'Connor, 1975; O'Connor & Rarey, 1979), and extant hominoids lack direct ulnocarpal contact partly as a structural adaptation to allow this greater range of movement (O'Connor, 1975; O'Connor & Rarey, 1979; Lewis, 1972, 1985; Cartmill & Milton, 1977). Although *Proconsul* clearly retained direct ulnocarpal contact as do most simiiforms, the anatomy of the ulnar side of the wrist in *Proconsul* appears to provide insight into the means by which the structural/functional transformation of the ulnocarpal joint of extant hominoids initially took place (Beard *et al.*, 1986, this volume). These features are consistent with a locomotor repertoire that included palmigrade quadrupedalism, but with a much greater emphasis on climbing than is typical of simiiforms other than hominoids.

Recently, hand remains of additional East African Miocene hominoids have been recovered, although these have yet to be analyzed in great detail. A first metacarpal fragment and some phalanges are known for *Turkanapithecus kalakolensis* (Leakey *et al.*, 1988a), and a partial scaphoid, lunate, trapezium, capitate, and some metacarpals are known for *Afropithecus turkanensis* (Leakey *et al.*, 1988b). The latter specimens seem to be closely comparable to the homologous elements in *Proconsul nyanzae* (Leakey *et al.*, 1988b).

Pongidae

The fossil record of the paraphyletic family Pongidae (the extant great apes and fossil forms that are more closely allied to them than is Hylobatidae) is poor, even in terms of craniodental remains (see review by Andrews & Martin, 1987). Hand remains of fossil pongids are limited to a hamate, a capitate, a partial first metacarpal, and three manual phalanges, all from the Miocene Siwalik series of Pakistan. These fossils all seem to be attributable to *Sivapithecus* (Rose, 1984, 1986; Kelley, 1988; Spoor *et al.*, 1991).

The curvature of the radial side of the capitate head of *Sivapithecus* differs from that of *Pongo* in such a way as to indicate that midcarpal pronation-supination mobility must have been relatively reduced in the fossil form (Rose, 1984). The shape of the capitate head in *Sivapithecus* more closely resembles those of some ceboids, particularly *Ateles*. The hamate is robust and lacks a well-developed hamulus. Its triquetral facet is dissimilar to that in extant great apes: it is proximodistally shorter; its dorsal half is markedly concave, providing a very stable weight-bearing surface in the extended position of the wrist. The anatomy of the proximal articular surface of the proximal phalanges is «markedly cercopithecid-like,» attesting to the lack of predominance of either knuckle-walking or suspensory behaviors in the positional repertoire of *Sivapithecus* (Rose, 1986, p. 366). The curvature and robusticity of the shafts argue for some limited capacities for manual suspension, however. The general pattern that emerges from consideration of all of the available postcranial fossils is one of an animal that used both quadrupedalism and climbing. It is likely that *Sivapithecus* also engaged in some rather limited suspensory activities (ROSE, 1986), but the hamate suggests a predominance of quadrupedal locomotion. In fact the two neighbouring carpals, hamate and capitate, indicate differing ranges of extension at the mid-carpal joint, suggesting locomotor variations in *Sivapithecus* (SPOOR *et al.*, 1991).

Hominidae

Because so much has been (and undoubtedly will continue to be) written on the subject of fossil hominid hands, we wish only to provide a very general overview of what we consider to be some of the most important fossil hominid hand remains discovered and treated to date (for more details see, for example, NAPIER, 1959, 1962; LEWIS, 1973, 1977; SUSMAN, 1979, 1988; SUSMAN & CREEL, 1979; BUSH *et al.*, 1982; JOHANSON *et al.*, 1982; SUSMAN & STERN, 1982; RIGHTMIRE, 1982; STERN & SUSMAN, 1983; MARZKE, 1983; MCHENRY, 1983; SUSMAN *et al.*, 1984; MARZKE & SHACKLEY, 1986; RICKLAN, 1987). We will limit our discussion of fossil hominid hand remains to three important, extinct species of hominids: *Australopithecus afarensis*, *Paranthropus robustus*, and *Homo habilis*.

The hand material of *A. afarensis* comes from the Pliocene of Hadar, Ethiopia (BUSH *et al.*, 1982; JOHANSON *et al.*, 1982). This material is more complete than that from any other African Plio-pleistocene hominid locality; it is also the oldest, dated at more than 2.8 million years. It contributes to the controversy surrounding the locomotion of this species (e.g. SUSMAN *et al.*, 1984). From its hand characters, TUTTLE (1981) concluded that this species probably used its hands «primarily as manipulatory and climbing organs». The main documented hand features of *A. afarensis* include: 1) an «elongate and rod-shaped» (BUSH *et al.*, 1982) pisiform, implying powerful development of the flexor carpi ulnaris by analogy with *Pan*; 2) a robust hook on the hamate (TUTTLE, 1981); 3) a capitate with dorsal trapezoid facet and lateral waisting of the bone, as in *Pan*; 4) trapezial carpometacarpal joint more concavoconvex than in *Homo* (including *H. habilis*) and similar again to the condition observed in *Pan*; 5) metacarpals with large heads and bases and curved shafts (like *Pan*); unlike African apes, the metacarpals lack transverse dorsal ridges, and thus do not appear to have been adapted for knuckle-walking (TUTTLE, 1967); 6) the pollical metacarpal and proximal phalanx indicate that the thumb of *A. afarensis* was gracile, but longer relative to the index finger than it is in the chimpanzee (MARZKE, 1983); 7) the proximal phalanges of digits two to five are curved and possess marked bony ridges for digital flexor sheaths; 8) the middle phalanges possess well-marked areas for insertion of the flexor digitorum superficialis; and 9) the distal phalanges exhibit relatively reduced apical tufts.

Many of these characters «recall the bony apparatus which accomodates the well developed flexor musculature in living apes and positions it for efficient hook-like grip of the branches by the flexed fingertips during arboreal climbing and feeding» (MARZKE, 1983, p. 198). This agrees with other anatomical complexes of *A. afarensis* which suggest a considerable amount of arboreal activity for this species (SUSMAN *et al.*, 1984). Neverthe-less, several other hand features are either unique or similar to humans. The relatively long thumb and the pronatory ability of the second metacarpal facilitated two kinds of precision grips, especially the pad-to-side pinch of objects (MARZKE, 1983). This suggests better manipulatory capabilities in *A. afarensis* than in chimpanzees. However, the thumb was not human-like, and a saddle joint between the hamate and the fifth metacarpal was lacking; MARZKE (1983) deduced that power grips other than the hook-like power grip of apes were limited, implying that tool-use still was far from the human condition. Locomotor branch grasping probably was the primary function of the power grip in the hand of early *Australopithecus*.

The hand remains of *Homo habilis* were recovered from Bed 1, Olduvai Gorge, Tanzania (NAPIER, 1962); they are dated 1.76 million years. The O.H.7 hand has 13 bones. The proximal phalanges have thick cortices and marked curvatures; the middle phalanges are robust and curved, with well-marked flexor digitorum superficialis insertions. As in *A. afarensis*, these characters suggest powerful grasping capabilities, such as those of living apes (SUSMAN & CREEL, 1979; SUSMAN & STERN, 1979). However, the *H. habilis* material differs from that of *A. afarensis*, chiefly in the more human-like trapeziometacarpal joint (STERN & SUSMAN, 1983) and in the greater expansion of the apical tufts of the distal phalanges (SUSMAN & CREEL, 1979). These characters suggest, despite the fact that the length of the thumb is unknown, enhanced «ability to maneuver objects by the thumb and fingertips in varied orientations and to also maintain the grip of these objects against external forces» (MARZKE & SHACKLEY, 1986, p. 454). On the whole, «changes in the human direction are well advanced in the thumb and fingertips of *Homo habilis* but power-grasping capabilities are retained in the more proximal segment of the fingertips» (SUSMAN & STERN, 1982). H. *habilis* presumably retained a significant amount of climbing behavior, but had enhanced tool-use abilities relative to *A. afarensis*. In fact, stone tools also occur in Bed 1, Olduvai.

The hand material of *Paranthropus robustus* (sometimes referred to as *Australopithecus robustus*) is derived from Member 1, Swartkrans, South Africa (SUSMAN, 1988 a, b; 1989). It is dated at about 1.8 million years. *P. robustus* shares a number of traits of hand anatomy in common with living humans, including: a distal pollical phalanx with an expanded apical tuft and a well-marked area for insertion of M. flexor pollicis longus (*Figure 17*); a pollical metacarpal with a relatively thick shaft and broad base that also bears a distinct lateral crest for insertion of the opponens pollicis; and proximal manual phalanges of digits two to five that are relatively shorter and straighter than those of extant great apes and *A. afarensis* (*Figure 18*). These characters may have some phylogenetic significance, but this will not be considered here (see SKELTON *et al.*, 1986; TOBIAS, 1988). The enlarged apical tuft of the pollical distal phalanx suggests a «... well-innervated and well-vascularized, fleshy fingertip» (SUSMAN, 1988a); the thick, broad-based pollical metacarpal shaft implies an adaptation for transmission of relatively high compressive loads through the pollical carpometacarpal joint, as occurs in a variety of prehensile activities; the marked areas for insertion of the flexor pollicis longus and opponens pollicis indicate a human-like emphasis on these muscles; and the relatively short, straight manual

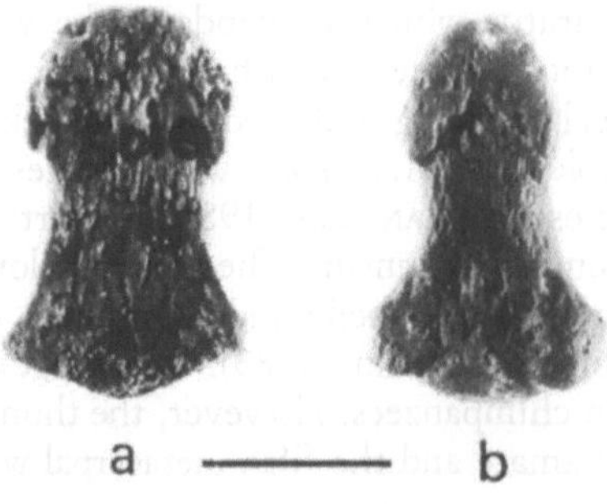

Figure 17. — Pollical distal phalanx SKX 5016 from Member 1, Swartkrans, attributed to *Paranthropus robustus* (a), and of a modern human (b) in palmar views. The apical tuft is prominent and the excavation for the insertion of the flexor pollicis longus tendon is large in *Paranthropus* (photographs courtesy of Dr. R.S. SUSMAN; scale bar is 1 cm).

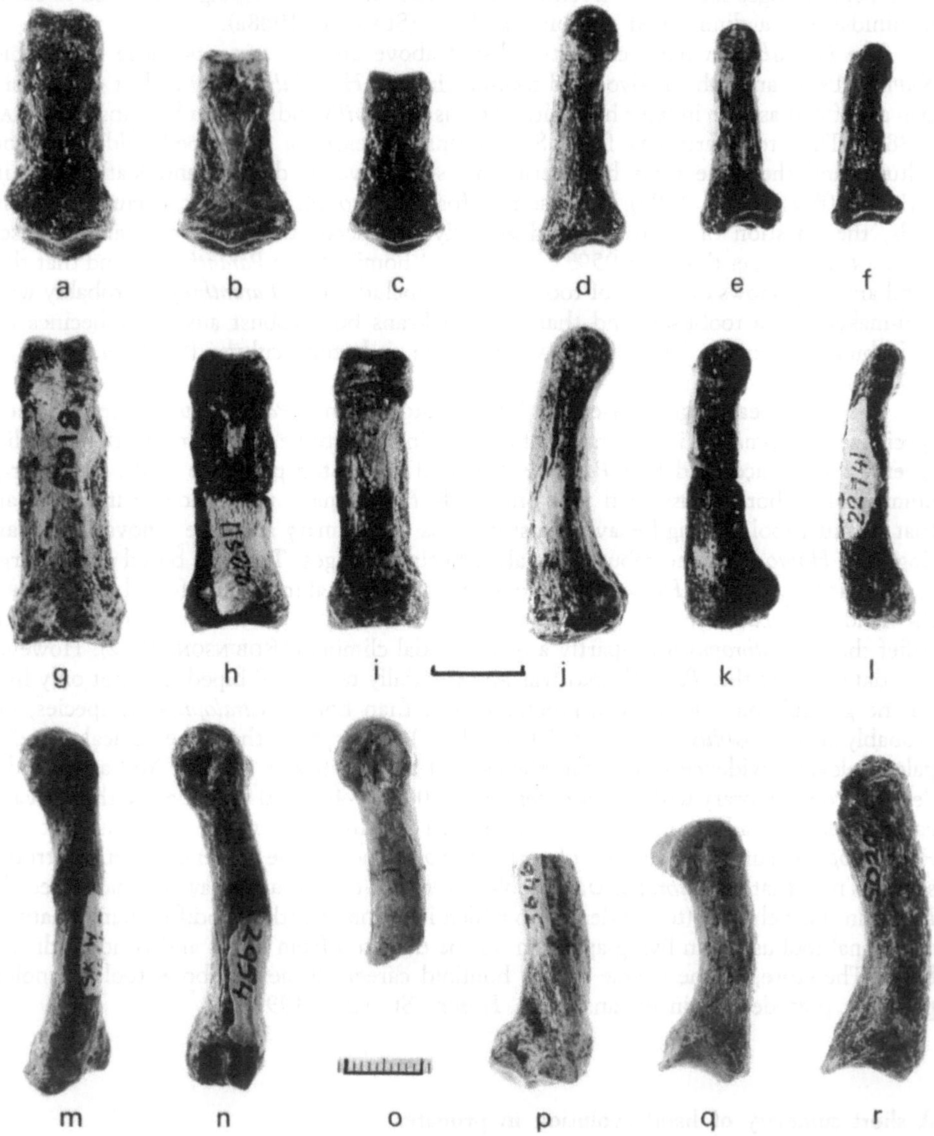

Figure 18. — Isolated hand bones: middle phalanges (a-f), proximal phalanges (g-l), and metacarpals (m-r) from Swartkrans, South Africa. Most of them can be attributed to *Paranthropus robustus*, but some others can be attributed to *Homo* cf *erectus*, e.g. the first metacarpal SK 84 (q), which is smaller and morphologically different from the other first metacarpal SKX 5020 (r; probably *Paranthropus*). a-c, palmar views, and d-f, profile views of middle phalanges SKX 5021 from Member 1 (a, d), SKX 36712 from Member 3 (b, e), and SKX 9449 from Member 1 (c, f); g-i, palmar views, and j-l , profile views of proximal phalanges SKX 5018 from Member 1 (g, j), SKX 22511-30220 from Member 3 (h, k), and SKX 22741 from Member 3 (i,l); m-r, profile views of metacarpals SK 14147 from Member 1 (m), SKW 2954 from Member 2 (n), SK 85 from Member 2 (o), SKW 3646 from Member 2 (p), SK 84 from Member 2 (q), SKX 5020 from Member 1 (r). Both scale bars are 1 cm; the upper one is for the two top rows (all photographs courtesy of Dr. R.L. SUSMAN).

proximal phalanges suggest «... a reduced emphasis on power grasping (compared to earlier hominids) and a diminished climbing ability» (SUSMAN, 1988a).

The *Paranthropus* hand characters listed above are the same as those from which NAPIER (1962) and others advocated tool-making for *H. habilis*. They reflect «a precision grip as refined as that in later hominids such as *H. habilis* and modern humans» (SUSMAN, 1988b). The stone artifacts from Swartkrans represent a developed Oldowan stone culture, and the bone tools bear striations suggestive of digging and leather-working activities (BRAIN *et al.*, 1988). Because rare fossil *Homo*, cf. *H. erectus*, occur in the same beds, the question of what hominid actually produced these artifacts can be raised. SUSMAN emphasizes that over 95% of Member 1 hominids are *Paranthropus* and that their hand anatomy shows evidence of tool-use. He concludes that *Paranthropus* probably was a tool-maker and a tool-user, and that at Swartkrans both robust australopithecines and early humans seem to have used a simple bone and stone culture (SUSMAN, 1988a, b, 1991).

The new idea, largely based on hand anatomy, that *Paranthropus* probably was a specialized tool-maker is a considerable step in the studies of hominization. Earlier, several authors accepted that *Paranthropus* used associated primitive tools, but a larger number of authorities assumed that only early *Homo* made the Oldowan industry, and that habitual tool-making behavior possibly was the primary adaptive innovation separating early *Homo* from the robust australopithecine lineages. This was based on the larger cranial capacity of early *Homo*, on the evidence of tool-making ability found by NAPIER in the hand of *H. habilis* (in the quasi-absence of *Paranthropus* hand bones), and also on the belief that *Paranthropus* was «partly a quadrupedal climber» (ROBINSON, 1972). However, new data suggest that *Paranthropus* was an essentially terrestrial biped that not only lived on the ground, but did so to a greater extent than both *Australopithecus* species, and probably also *H. habilis* (SUSMAN & STERN, 1991). A survey of the archeological as well as paleontological evidence shows that the earliest stone artifacts (2.5-2.3 My) are found in deposits contemporary with *Paranthropus*, and 300,000-400,000 years before the arrival of early *Homo*. In addition, there are two later localities in which stone artifacts and *Paranthropus* occur in the absence of early *Homo*. Hence all the evidence together strongly suggests now that *Paranthropus* was a tool-maker, and even that it may well have preceded *Homo* in this behavior (to the degree to which it is linked with a modified hand anatomy; occasional tool-use as in living apes can not be detected from bones and is not addressed here). Therefore, in the course of the hominid career, stone and bone tool technology probably preceded brain expansion in *Homo* (SUSMAN, 1991).

A short summary of hand evolution in primates

Plesiadapiforms were early primate and dermopteran relatives, long-handed and arboreal, some of them gliding, possessing a short carpus, long mobile digits, and sharp claws (see BEARD, 1989, 1990; GODINOT & BEARD, 1991).

Early euprimates similarly had hands with a short carpus exhibiting little deviational capabilities, a short metacarpus and long digits. The hand retained some attributes of the primitive convergent hand (e.g. some degree of convergence between digits two to five). The degree of thumb divergence is controversial, low (M.G.) or high (C.B.).

Early strepsirhines (Eocene Adapiformes) emphasized thumb divergence and the grasping capability of the hand (for C.B., they retained these characters), as seen in *Smilodectes* and the Messel adapiform. The Messel adapiform had a certain degree of paraxony or ectaxony in its digits. *Adapis* shows several characters of the triquetrum and pisiform which also reflect an emphasis on ulnarly deviated hand postures relative to the forearm. These derived characters, shared with lemuriforms, suggest a close affinity between the two groups.

Early lemuriforms evolved their midcarpal condyle, which increased their carpal deviational capabilities. Several of the various groups of later lemuriforms increased their capacities for thumb prehension and enhanced the functional parallelism of the remaining digits. Thumb prehension was not emphasized, and was possibly diminished, in *Microcebus* and other cheirogaleids, perhaps in *Daubentonia* too. Some subfossil Malagasy lemuriforms evolved particularly specialized types of hands (e.g., the hook-like hand of *Palaeopropithecus*).

Early lorisids emphasized hand grasping. This hypothesis is suggested by the unusual degree of thumb divergence as well as osteometric characters of both galagines and lorisines (e.g. relative elongation of the first metacarpal). Galagines may therefore have secondarily evolved toward vertical clinging and leaping (but see interpretation of the foot evidence by GEBO, 1986b). Lorisines greatly enhanced their carpal deviational capabilities by modifying their lunate, ulnocarpal joint, and midcarpal condyle (CARTMILL & MILTON, 1977; LEWIS, 1985).

Early simiiforms (and early platyrrhines) retained a general hand mesaxony and acquired relatively shorter digits and longer metacarpals as a result of their emphasis on horizontal quadrupedalism. They had a very moderate degree of thumb divergence and prehensile capabilities. Grasping specialization in later simiiforms involves digits two to five, and is usually accompanied by thumb reduction (platyrrhine atelines and catarrhine colobines).

Early cercopithecids seem to have further emphasized horizontal quadrupedalism, or even semi-terrestriality, as is suggested by their relatively shorter digits and longer metacarpals compared to the simiiform morphotype. Early fossil cercopithecid hands (*Mesopithecus*) corroborate this view.

Early hominoids, as exemplified by *Proconsul*, enhanced ulnar deviational capabilities by modifying their ulno-carpal joint (BEARD *et al.*, 1986, this volume). Later hominoids also emphasized their second digit (and metacarpal) and secondarily acquired a relatively high degree of thumb divergence correlated with their prehensile specializations.

Early hominids retained numerous features of the hand that have been interpreted as adaptations for climbing (STERN & SUSMAN, 1983; SUSMAN *et al.*, 1984). Modifications of this morphology possibly occurred in parallel both in later *Homo* and in the robust australopithecines (SUSMAN, 1988 a, b). These modifications suggest a reduced emphasis on climbing and an increased emphasis on pollical dexterity in these more derived hominids.

The broad picture provided by the evolutionary history of primate hands underscores the repeated transformations in hand structure and function that have preceded the evolution of the hand of modern humans. We agree with LEWIS (1977) that the human hand can hardly be viewed as essentially «primitive», as its anatomy and proportions are the result of a complex history, including the acquisition of a number of recently derived characters. Increasing hand capability was traditionally considered an important factor in human brain evolution. However the new evidence seems to suggest that tool-making hands and brains evolved before major brain capacity increases took place in early *Homo*.

ACKNOWLEDGEMENTS — Dr. R.J. Emry and Dr. L. Ginsburg allowed us to study notharctine and *Mesopithecus* hand bones, respectively. Dr. R.W. Thorington, Jr. and Prof. P. Taquet provided access to comparative collections. Dr. M.J. Novacek and J.P. Alexander provided casts of an articulated hand specimen of *Notharctus*. A. Collier loaned the hand of *Adapis* for study. Dr. J. Franzen made available to us his manuscript and photographs of the new Messel adapiform. Other original photographs were given to us by J.P. Alexander, Dr. D. Begun, Dr. F. Jouffroy, Dr. R. Susman, U. Thalmann. To all of them we express our gratitude. Dr. A. Walker, Prof. H. Preuschoft and Dr. D.L. Gebo have provided much useful discussion on fossil primates and functional anatomy. Dr. D.E. Russell, Dr. B. Senut, Dr. M. Dagosto, Dr. D. Begun, C. Ward, R. Heinrich, J. Runestad, and Dr. W.P. Luckett provided useful criticism and editorial help. Technical help was provided by A. Beaux, J. Martin, and M. Pons. M. Godinot acknowledges the support from the CNRS, Montpellier, and from the Carnegie Museum of Natural History, Pittsburgh.

References

ANDREWS P. and GROVES C., 1976. *Gibbons and brachiation.* In: Gibbon and Siamang, (D.M. Rumbaugh, ed.), Karger, Basel, vol. *4*: 167-218.

ANDREWS P. and MARTIN L.,1987. *Cladistic relationships of extant and fossil hominoids.* J. Hum. Evol. *16*: 101-118.

BEARD K.C., 1988. *New notharctine primate fossils from the early Eocene of New Mexico and southern Wyoming and the phylogeny of Notharctinae.* Am. J. Phys. Anthrop. *75*: 439-469.

BEARD K.C., 1989. *Postcranial anatomy, locomotor adaptations, and paleoecology of early Cenozoic Plesiadapidae, Paromomyidae, and Micromomyidae (Eutheria, Dermoptera).* Ph. D. Dissertation, The Johns Hopkins University, Baltimore (Maryland).

BEARD K.C., 1990. *Gliding behaviour and palaeoecology of the alleged primate family Paromomyidae (Mammalia, Dermoptera).* Nature, *345*: 340-341.

BEARD K.C., 1991. *Vertical postures and climbing in the morphotype of Primatomorpha: implications for locomotor evolution in primate history.* In: Origine (s) de la Bipédie chez les Hominidés (Y. Coppens and B. Senut eds.), Editions du CNRS, Paris, pp. 79-87.

BEARD K.C., in press. *Phylogenetic systematics of the Primatomorpha, with special reference to Dermoptera.* In: Mammal Phylogeny (F.S. Szalay, M.J. Novacek and M.C. McKenna eds.). Springer, New York.

BEARD K.C. and GODINOT M., 1988. *Carpal anatomy of* Smilodectes gracilis *(Adapiformes, Notharctinae) and its significance for lemuriform phylogeny.* J. Hum. Evol. *17*: 71-92.

BEARD K.C., TEAFORD M.F. and WALKER A. 1986. *New wrist bones of* Proconsul africanus *and* P. nyanzae *from Rusinga Island, Kenya.* Folia primatol. *47*: 97-118.

BEARD K.C., TEAFORD M.F. and WALKER A. (this volume).

BEARD K.C., DAGOSTO M., GEBO D.L. and GODINOT M., 1988. *Interrelationships among primate higher taxa.* Nature *331*: 712-714.

BEGUN D.R., 1988. *Catarrhine phalanges from the late Miocene (Vallesian) of Rudabanya, Hungary.* J. Hum. Evol. *17*: 413-438

BOWN T.M. and ROSE K.D., 1976. *New early Tertiary primates and a reappraisal of some Plesiadapiformes.* Folia Primatol. *26*: 109-138.

BRAIN C.K., CHURCHER C.S., CLARK J.D., GRINE F.E., SHIPMAN P., SUSMAN R.L., TURNER A., and WATSON V., 1988. *New evidence of early hominids, their culture and environment from the Swartkrans cave, South Africa.* S. African J. Sci., *84*: 828-835.

BROOM R., 1946. *The occurrence and general structure of the South African Ape-Men.* Mem. Transvaal Mus. 2: 7-153.

BUSH M.E., LOVEJOY C.O., JOHANSON D.C. and COPPENS Y., 1982. *Hominid carpal, metacarpal, and phalangeal bones recovered from the Hadar Formation: 1974-1977 collections.* Am. J. Phys. Anthrop. *57*: 651-677.

CARTMILL M., 1972. *Arboreal adaptations and the origin of the order Primates.* In: Functional and Evolutionary Biology of Primates (R.H. Tuttle, ed.), Aldine-Atherton, Chicago, pp. 97-122.

CARTMILL M., 1974. *Pads and claws in arboreal locomotion.* In: Primate Locomotion (F.A. Jenkins, Jr., ed.), pp. 45-83, Academic Press, New York.

CARTMILL M. and MILTON K., 1977. *The lorisiform wrist joint and the evolution of "brachiating" adaptations in the Hominoidea.* Am. J. Phys. Anthrop. 47: 249-272.

CORRUCCINI R.S., CIOCHON R.L. and MCHENRY H.M., 1975. *Osteometric shape relationships in the wrist joint of some anthropoids*. Folia Primatol. *24*: 250-274.

CORRUCCINI R.S., CIOCHON R.L., and MCHENRY H.M., 1976. *Osteometric shape relationships in the wrist joint of some anthropoids*. Folia Primatol., *24*: 250-274.

COVERT H.H., 1985. *Adaptations and evolutionary relationships of the Eocene primate family Notharctidae*. Ph. D. Dissertation, Duke University (North Carolina).

COVERT H.H., 1988. *Ankle and foot morphology of* Cantius mckennai: *Adaptations and phylogenetic implications*. J. Hum. Evol. *17*: 57-70.

DAGOSTO M., 1983. *Postcranium of* Adapis parisiensis *and* Leptadapis magnus *(Adapiformes, Primates)*. Folia Primatol. *41*: 49-101.

DAGOSTO M., 1988. *Implications of postcranial evidence for the origin of euprimates*. J. Hum. Evol. *17*: 35-56.

ETTER H.-U.F., 1974. *Morphologisch- und metrisch-vergleichende Untersuchung am Handskelet rezenter Primaten*. Gegenb. Morph. Jahrb. *120*: (1), 1-21; (2), 153-171; (3), 299-322; (4), 457-484.

ETTER H.-U.F., 1978. *Lorisiform hands and their phylogenetic implications: A preliminary report*. In: Recent Advances in Primatology. Vol. 3. Evolution. (D.J. Chivers and K.A. Joysey, eds.), London, Academic Press, pp. 161-170.

FLEAGLE J.G., 1983. *Locomotor adaptations of Oligocene and Miocene hominoids and their phyletic implications*. In: New Interpretations of Ape and Human Ancestry. (R.L. Ciochon, R.S. Corruccini, eds.), Plenum Press, New York, pp. 301-324.

FORD S.M., 1988. *Postcranial adaptations of the earliest platyrrhine*. J. Hum. Evol. *17*: 155-192.

FORSYTH-MAJOR C.J., 1894. *On* Megaladapis madagascariensis, *an extinct gigantic lemuroid from Madagascar; with remarks on the associated fauna and on its geological age*. Phil. Trans. Roy. Soc. London (Series B). *185*: 15-38.

FRANZEN J.L., 1987. *Ein neuer Primate aus dem Mitteleozän der Grube Messel (Deutschland, S-Hessen)*. Cour. Forsch.-Inst. Senckenberg *91*: 151-187.

FRANZEN J.L., 1988. *Ein weiterer Primatenfund aus der Grube Messel bei Darmstadt*. In: Der eozäne Messelsee. Eocene Lake Messel. (J.L. Franzen and W. Michaelis, eds.), Cour. Forsch.-Inst. Senckenberg. *107*: 275-289.

GAUDRY A., 1862. *Animaux fossiles et géologie de l'Attique*, F. Savy, Paris.

GEBO D.L., 1985. *The nature of the primate grasping foot*. Am. J. Phys. Anthrop. *67*: 269-277.

GEBO D.L., 1986a. *Anthropoid origins - the foot evidence*. J. Hum. Evol. *15*: 421-430.

GEBO D.L., 1986b. *Miocene lorisids - The foot evidence*. Folia primatol *47*: 217-225.

GEBO D.L., 1987. *Functional anatomy of the tarsier foot*. Am J. Phys. Anthrop. *73*: 9-31.

GINGERICH P.D., 1975. *A new genus of Adapidae (Mammalia, Primates) from the late Eocene of Southern France, and its significance for the origin of higher primates*. Contr. Mus. Paleontol. Univ. Michigan *24*: 163-170.

GINGERICH P.D., 1976. *Cranial anatomy and evolution of early Tertiary Plesiadapidae (Mammalia, Primates)*. Univ. Michigan Papers on Paleontol. *15*: 1-141.

GINGERICH P.D., 1980. *Eocene Adapidae, paleobiogeography, and the origin of South American Platyrrhini*. In: Evolutionary biology of the New World Monkeys and Continental Drift. (R.L. Ciochon, A.B. Chiarelli, eds.), Plenum Press, New York, pp. 123-138.

GODFREY L.R., 1988. *Adaptive diversification of Malagasy strepsirhines*. J. Hum. Evol. *17*: 93-134.

GODFREY L.R., SIMONS E.L., CHATRATH P.S., and RAKOTOSAMIMANANA R., 1990. A new fossil lemur (*BabaKotia*, Primates) from Northern Madagaskar. C.R. Acad. Sci. Paris, *310*: 81-87.

GODINOT M., 1991a. *Approches fonctionnelles des mains de primates paléogènes*. Géobios, Mém. Spéc., *13*: 161-173.

GODINOT M., 1991b. *Toward the locomotion of two contemporaneous Adapis species*. Z. Morph. Anthrop., *78*: 387-405.

GODINOT M., 1992. *Early euprimate hands in evolutionary perspective*. J. Hum. Evol., *22*: 267-283.

GODINOT M. and BEARD K.C., 1991. *Fossil primate hands: a review and an evolutionary inquiry emphasizing early forms*. Hum. Evol., *6*: 307-351.

GODINOT M. and JOUFFROY F.K., 1984. *La main d'*Adapis (*Primate, Adapidé*). In: Actes du symposium paléontologique Georges Cuvier (E. Buffetaut, J.M. Mazin, E. Salmon, eds.). Le Serpentaire, Montbéliard, pp. 221-242.

GRANDIDIER G., 1901. *Un nouvel Edenté sub-fossile de Madagascar*. Bull. Mus. Hist. Nat., *7*: 54-56.

GRANDIDIER G., 1905. *Recherches sur les lémuriens disparus et en particulier sur ceux qui vivaient à Madagascar*. Nouv. Arch. Museum, 4eme serie *7*: 1-142.

GREGORY W.K., 1920. *On the structure and relations of* Notharctus, *an American Eocene primate.* Mem. Am. Mus. Nat. Hist. *3*: 49-243.

GUNNELL G.F., 1989. *Evolutionary history of Microsyopoidea (Mammalia, Primates) and the relationship between Plesiadapiformes and Primates.* Univ. Mich. Pap. Pal., *27*: 1-157.

HAINES R.W., 1958. *Arboreal or terrestrial ancestry of placental mammals.* Quart. Rev. Biol. *33*: 1-23.

HARRISON T., 1986a. *New fossil anthropoids from the middle Miocene of East Africa and their bearing on the origin of the Oreopithecidae.* Am. J. Phys. Anthrop. *71*: 265-284.

HARRISON T., 1986b. *A reassessment of the phylogenetic relationships of* Oreopithecus bambolii *Gervais.* J. Hum. Evol. *15*: 541-583.

HARRISON T., 1987. *The phylogenetic relationships of the early catarrhine primates: A review of the current evidence.* J. Hum. Evol. *16*: 41-80.

HERSHKOVITZ P., 1977. *Living New World Monkeys (Platyrrhini).* Univ. of Chicago Press, Chicago.

HOFFSTETTER R., 1982. *Les Primates Simiiformes (= Anthropoidea) (Compréhension, Phylogénie, Histoire Biogéographique).* Ann. Paléont. *68*: 241-290.

HÜRZELER J., 1958. Oreopithecus bambolii *Gervais: A preliminary report.* Verh. Naturf. Ges. Basel. *69*: 1-48.

JABLONSKI N.G., 1986: *The hand of* Theropithecus brumpti. In: Primate Evolution (J.G. Else and P.C. Lee, eds.), Cambridge Univ. Press, Cambridge, pp. 173-182.

JENKINS F.A. Jr., 1981. *Wrist rotation in primates: A critical adaptation for brachiators.* Symp. Zool. Soc. Lond. *48*: 429-451.

JENKINS F.A. Jr. and FLEAGLE J.G., 1975. *Knuckle-walking and the functional anatomy of the wrists in living apes.* In: Primate Functional Morphology and Evolution (R.H. Tuttle, ed.). Mouton, The Hague, pp. 213-227.

JOHANSON D.C., LOVEJOY C.O., KIMBEL W.H., WHITE T.D., WARD S.C., BUSH M.E., LATIMER B.M., COPPENS Y., 1982. *Morphology of the Pliocene partial hominid skeleton (A.L. 288-1) from the Hadar Formation, Ethiopia.* Am. J. Phys. Anthrop. *57*: 403-451.

JOLLY C.J., 1972. *The classification and natural history of* Theropithecus (Simopithecus) *(Andrews, 1916), Baboons of the African Plio-Pleistocene.* Bull. British Mus. (Nat. Hist), Geology. 22: 1-123.

JOUFFROY F.K., 1962. *La musculature des membres chez les lémuriens de Madagascar: Etude descriptive et comparative.* Mammalia *26*: 1-326.

JOUFFROY F.K., 1975. *Osteology and myology of the lemuriform postcranial skeleton.* In: Lemur Biology (I. Tattersall and R.W. Sussman, eds.), New York, Plenum Press., pp. 149-192.

JOUFFROY F.K. and LESSERTISSEUR J., 1959. *La main des lémuriens malgaches comparée à celle des autres primates.* Mém. Inst. Sci. Madagascar, Série A. *13*: 195-219.

JOUFFROY F.K., BERGE C. and NIEMITZ C., 1984. *Comparative study of the lower extremity in the genus* Tarsius. In: Biology of Tarsiers (C. Niemitz, ed.). Gustav Fischer, Berlin., pp. 167-190.

JOUFFROY F.K., GODINOT M. and NAKANO Y., 1991. *Biometrical characteristics of primate hands.* Hum. Evol., 6: 269-306.

JOUFFROY F.K., GODINOT M. and NAKANO Y. (this volume).

JUNGERS W.L., 1977. *Hindlimb and pelvic adaptations to vertical climbing and clinging in* Megaladapis, *a giant subfossil prosimian from Madagascar.* Yrbk. Phys. Anthropol. *20*: 508-524.

JUNGERS W.L., 1978. *The functional significance of skeletal allometry in* Megaladapis *in comparison to living prosimians.* Am. J. Phys. Anthrop. *49*: 303-314.

JUNGERS W.L., 1987. *Body size and morphometric affinities of the appendicular skeleton in Oreopithecus bambolii (IGF 11778).* J. Hum. Evol., *16*: 445-456.

JUNGERS W.L., GODFREY L.R., SIMONS E.L., CHATRATH P.S. and RAKOTOSAMIMANANA B., 1991. *Phylogenetic and functional affinities of Babakotia (Primates), a fossil lemur from northern Madagascar.* Proc. Natl. Acad. Sci. U.S.A., *88*: 9082-9086.

KAY R.F., THORINGTON R.W. and HOUDE P., 1990. *Eocene plesiadapiform shows affinities with flying lemurs not primates.* Nature, *345*: 342-344.

KELLEY J., 1988. *A new large species of* Sivapithecus *from the Siwaliks of Pakistan.* J. Hum. Evol., *17*: 305-324.

KOEBKE J. (this volume).

KOENIGSWALD W. von, 1979. *Ein Lemurenrest aus dem eozänen Ölschiefer der Grube Messel bei Darmstadt.* Paläontol. Z. *53*: 63-76.

KOENIGSWALD W. von, 1985. *Der dritte Lemurenrest aus dem mitteleozänen Ölschiefer der Grube Messel bei Darmstadt.* Carolinea *42*: 145-148.

LAMBERTON C., 1934. *Contribution à la connaissance de la faune subfossile de Madagascar: lémuriens et ratites.* Mém. Acad. Malgache. *17*: 1-168.

LAMBERTON C., 1947. *Contribution à la connaissance de la faune subfossile de Madagascar.* Note 16. Bull. Acad. Malgache. *26*: 1-52.

LAMBERTON C., 1948. *Contribution à la connaissance de la faune subfossile de Madagascar.* Note 19. Bull. Acad. Malgache. *27*: 19-23.

LEAKEY R.E., LEAKEY M.G. and WALKER A., 1988a. *Morphology of* Turkanapithecus kalakolensis *from Kenya.* Am. J. Phys. Anthrop. *76*: 277-288.

LEAKEY R.E., LEAKEY M.G. and WALKER A., 1988b. *Morphology of* Afropithecus turkanensis *from Kenya.* Am. J. Phys. Anthrop. *76*: 289-307.

LE GROS CLARK W.E., 1936. *The problem of the claw in primates.* Proc. Zool. Soc. London, 1-24.

LEWIS O.J., 1965. *Evolutionary change in the primate wrist and inferior radio-ulnar joints.* Anat. Rec. *151*: 275-286.

LEWIS O.J., 1969. *The hominoid wrist joint.* Am. J. Phys. Anthrop. *30*: 251-268.

LEWIS O.J., 1971. *Brachiation and the early evolution of the Hominoidea.* Nature *230*: 577-579.

LEWIS O.J., 1972a. *Evolution of the hominoid wrist.* In: The Functional and Evolutionary Biology of Primates (R. Tuttle, ed.). Aldine-Atherton, Chicago, pp. 207-222

LEWIS O.J., 1972b. *Osteological features characterizing the wrists of monkeys and apes, with a reconsideration of this region in* Dryopithecus (Proconsul) africanus. Am. J. Phys. Anthrop. *36*: 45-58.

LEWIS O.J., 1973. *The hominid os capitatum, with special reference to the fossil bones from Sterkfontein and Olduvai Gorge.* J. Hum. Evol. *2*: 1-12.

LEWIS O.J., 1974. *The wrist articulations of the Anthropoidea.* In: Primate Locomotion (F.A. Jenkins Jr., ed.), Academic Press, New York, pp. 143-169.

LEWIS O.J., 1977. *Joint remodelling and the evolution of the human hand.* J. Anat. *123*: 157-201.

LEWIS O.J., 1985. *Derived morphology of the wrist articulations and theories of hominoid evolution.* Part I. The lorisine joints. J. Anat. *140*: 447-460.

LORENZ VON LIBURNAU L., 1905. Megaladapis edwardsi *G. Grandidier.* Denkschr. Kais. Akad. Wiss. Wien. *77*: 451-490.

MACPHEE R.D.E., 1986. *Primates.* In: McGraw-Hill Yearbook of Science and Technology. McGraw-Hill, New York, pp. 362-364.

MARTIN R.D., 1972. *Adaptive radiation and behaviour of the Malagasy lemurs.* Phil. Trans. Roy. Soc. London. *264* (B): 295-352.

MARZKE M.W., 1983. *Joint functions and grips of the* Australopithecus afarensis *hand, with special reference to the region of the capitate.* J. Hum. Evol. *12*: 197-211.

MARZKE M.W. and SHACKLEY M.S., 1986. *Hominid hand use in the Pliocene and Pleistocene: Evidence from experimental archaeology and comparative morphology.* J. Hum. Evol. *15*: 439-460.

MCHENRY H.M., 1983. *The capitate of* Australopithecus afarensis *and* A. africanus. Am. J. Phys. Anthrop. *62*: 187-198.

MCHENRY H.M. and CORRUCCINI R.S., 1983. *The wrist of* Proconsul africanus *and the origin of hominoid postcranial adaptations.* In: New Interpretations of Ape and Human Ancestry (R.L. Ciochon, R.S. Corruccini, eds.). Plenum Press, New York., pp. 353-367.

MELDRUM D.J. and PAN Y., 1988. *Manual proximal phalanx of* Laccopithecus robustus *from the latest Miocene site of Lufeng.* J. Hum. Evol., *17*: 719-731.

MORBECK M.E., 1975. Dryopithecus africanus *forelimb.* J. Hum. Evol. 4: 39-46.

MORBECK M.E., 1977. *The use of casts and other problems in reconstructing the* Dryopithecus (Proconsul) africanus *wrist complex.* J. Hum. Evol. 6: 65-78.

NAPIER J.R., 1959. *Fossil metacarpals from Swartkrans.* Fossil Mammals of Africa (Br. Mus. Nat. Hist.) *17*: 1-18.

NAPIER J.R., 1961. *Prehensility and opposability in the hands of primates.* Symp. Zool. Soc. Lond. *5*: 115-132.

NAPIER J.R., 1962. *Fossil hand bones from Olduvai Gorge.* Nature. *196*: 409-411.

NAPIER J.R. and DAVIS P.R. 1959. *The fore-limb skeleton and associated remains of* Proconsul africanus. Br. Mus. (Nat. Hist.), London, Fossil Mammals of Africa. *16*: 1-69.

NAPIER J.R. and WALKER A.C. 1967. *Vertical clinging and leaping - a newly recognized category of locomotor behaviour of primates.* Folia Primatol. *6*: 204-219.

NIESCHALK U. and DEMES B. (this volume).

O'CONNOR B.L. 1975. *The functional morphology of the cercopithecoid wrist and inferior radioulnar joints, and their bearing on some problems in the evolution of the Hominoidea.* Am. J. Phys. Anthrop. *43*: 113-122.

O'Connor B.L. 1976. Dryopithecus (Proconsul) africanus: *quadruped or non-quadruped?* J. Hum. Evol. *5*: 279-283.

O'Connor B.L. and Rarey K.E., 1979. *Normal amplitudes of radioulnar pronation and supination in several genera of anthropoid primates.* Am. J. Phys. Anthrop *51*: 39-44.

Petter J.-J. and Peyrieras A., 1970. *Nouvelle contribution à l'étude d'un lémurien malgache, le Aye-aye* (Daubentonia madagascariensis *E. Geoffroy).* Mammalia, *34*: 167-193.

Petter J.-J., Albignac R. and Rumpler Y., 1977. *Faune de Madagascar, 44, Mammifères lémuriens (Primates, Prosimiens)*, Imprimerie Nouvelle, Paris.

Preuschoft H., 1965. *Muskeln und Gelenke der Vorderextremität des Gorilla.* Morph. Jb. *107*: 99-183.

Preuschoft H., 1970. *Functional anatomy of the lower extremity.* In: The Chimpanzee, (G.H. Bourne, ed.) Karger, Basel, Vol. 3, pp. 221-294.

Preuschoft H., 1971. *Mode of locomotion in subfossil giant lemuroids from Madagascar.* In: Proc. Third Int. Congr. Primatol. Karger, Basel, Vol. 1, pp. 79-90.

Preuschoft H., 1973a. *Functional anatomy of the upper extremity.* In: The Chimpanzee, (G.H. Bourne, ed.) Karger, Basel, Vol. 6, pp. 34-120.

Preuschoft H., 1973b. *Body posture and locomotion in some East African Miocene Dryopithecinae.* In: Human Evolution, (M.H. Day, ed.), Taylor & Francis, London, Vol. II, pp. 13-46.

Preuschoft H., 1975. *Body posture and mode of locomotion in fossil primates - Method and example:* Aegyptopithecus zeuxis. In: Proceedings from the Symposia of the Fifth Congress of the International Primatological Society 1974. Japan Science Press, Tokyo, pp. 345-359.

Rasmussen D.T., 1986. *Anthropoid origins: a possible solution to the Adapidae-Omomyidae paradox.* J. Hum. Evol. *15*: 1-12.

Rasmussen D.T. and Simons E.L., 1988. *New specimens of* Oligopithecus savagei, *early Oligocene primate from the Fayum, Egypt.* Folia Primatol., *51*: 182-208.

Ricklan D.E., 1987. *Functional anatomy of the hand of* Australopithecus africanus. J. Hum. Evol. *16*: 643-664.

Rightmire G.P., 1982. *Multivariate analysis of an early hominid metacarpal from Swartkrans.* Science *176*: 159-161.

Robertson M.L., 1984. *The carpus of* Proconsul africanus: *Functional analysis and comparison with selected nonhuman primates.* Ph. D. Dissertation, University of Michigan.

Robinson J.T., 1972. *Early Hominid Posture and Locomotion.* University of Chicago Press, Chicago.

Rose K.D. and Walker A., 1985. *The skeleton of early Eocene* Cantius, *oldest lemuriform primate.* Am. J. Phys. Anthrop. *66*: 73-89.

Rose M.D., 1984. *Hominoid postcranial specimens from the middle Miocene Chinji Formation, Pakistan.* J. Hum. Evol. *13*: 503-516.

Rose M.D., 1986. *Further hominoid postcranial specimens from the late Miocene Nagri Formation of Pakistan.* J. Hum. Evol. *15*: 333-367.

Sarmiento E.E., 1987. *The phylogenetic position of* Oreopithecus *and its significance in the origin of Hominoidea.* Am. Mus. Novitates *2881*: 1-44.

Schlosser M., 1888. *Die Affen, Lemuren, Chiropteren, Insectivoren, Marsupialer, Creodonten und Carnivoren des europäischen Tertiärs.* Beitr. Paläont. Öst.-Ung. *6*: 1-226.

Schmid P., 1982. *Die systematische Revision der europäischen Microchoeridae Lydekker, 1887 (Omomyiformes, Primates).* Juris Druck, Zürich.

Siegel M.I. and Pernotto B, 1975. *Hand use and metacarpal robusticity in Catharrhini.* Primates *16*: 371-377.

Simons E.L., 1972. *Primate Evolution: An Introduction to Man's Place in Nature.* MacMillan, New York.

Simons E.L., 1989. *Description of two genera and species of Late Eocene Anthropoidea from Egypt.* Proc. Natl. Acad. Sci. U.S.A., *86*: 9956-9960.

Simons E.L. and Fleagle J.G., 1973. *The history of extinct gibbon-like primates.* In: Gibbon and Siamang (D.M. Rumbaugh, ed.) Karger, Basel, Vol. 2: 121-148.

Skelton R.R., McHenry H.M. and Drawhorn G.M., 1986. *Phylogenetic analysis of early hominids.* Curr. Anthrop. 27: 21-43.

Stern J.T. and Susman R.L., 1983: *The locomotor anatomy of* Australopithecus afarensis. Am. J. Phys. Anthrop. *60*: 279-317.

STRASSER E., 1988: *Pedal evidence for the origin and diversification of cercopithecid clades.* J. Hum. Evol. *17*: 225-245.

STRAUS W.L. Jr., 1963. *The classification of* Oreopithecus. In: Classification and human evolution (S.L. Washburn, ed.). Aldine, Chicago, pp. 146-177.

SUSMAN R.L., 1979. *Comparative and functional morphology of hominoid fingers.* Am. J. Phys. Anthrop. *50*: 215-236.

SUSMAN R.L., 1985. *Functional morphology of the* Oreopithecus *hand.* Am. J. Phys. Anthrop. *66*: 235 (abstract).

SUSMAN R.L., 1988a. *Hand of* Paranthropus robustus *from Member* 1, *Swartkrans: fossil evidence of tool behavior.* Science *240*: 781-784.

SUSMAN R.L., 1988b. *New postcranial remains from Swartkrans and their bearing on the functional morphology and behavior of* Paranthropus robustus. In: Evolutionary History of the Robust Australopithecines (F.E. Grine ed.), Aldine de Gruyter, New York, pp. 149-172.

SUSMAN R.L., 1989. *New hominid fossils from the Swartkrans Formation (1979-1986 excavations): postcranial specimens.* Am. J. Phys. Anthrop. *79*: 451-474.

SUSMAN R.L., 1991. *Who made the Oldowan tools? Fossil evidence for tool behavior in Plio-Pleistocene hominids.* J. Anthrop. Res., *47*: 129-151.

SUSMAN R.L. and CREEL N., 1979: *Functional and morphological affinities of the subadult hand (O.H. 7) from Olduvai Gorge.* Am. J. Phys. Anthrop. *51*: 311-332.

SUSMAN R.L. and STERN J.T. Jr, 1979. *Telemetered electromyography of flexor digitorum profundus and flexor digitorum superficialis in* Pan troglodytes *and implications for interpretation of the O.H. 7 hand.* Am. J. Phys. Anthrop. *50*: 565-574.

SUSMAN R.L. and STERN J.T. Jr., 1982. *Functional morphology of* Homo habilis. Science. *217*: 931-934.

SUSMAN R.L. and STERN J.T., 1991. *Locomotor behavior of early hominids: epistemology and fossil evidence.* In: Origine (s) de la Bipédie chez les Hominidés (Y. Coppens and B. Senut, eds.), Editions du CNRS, Paris, pp. 121-131.

SUSMAN R.L., STERN J.T. Jr. and JUNGERS W.L., 1984. *Arboreality and bipedality in the Hadar hominids.* Folia Primatol. *43*: 113-156.

SZALAY F.S. and DAGOSTO M., 1980. *Locomotor adaptations as reflected on the humerus of Paleogene primates.* Folia primatol. *34*: 1-45.

SZALAY F.S. and DAGOSTO M., 1988. *Evolution of hallucial grasping in the primates.* J. Hum. Evol. *17*: 1-33.

SZALAY F.S. and DECKER R.L., 1974. *Origins, evolution and function of the tarsus in late Cretaceous eutheria and Palaeocene primates.* In: Primate Locomotion. (F.A. Jenkins, Jr., ed.). Academic Press, New York, pp. 223-259.

SZALAY F.S. and DELSON E., 1979. *Evolutionary History of the Primates*, Academic Press, New York.

SZALAY F.S. and DRAWHORN G., 1980. *Evolution and diversification of the Archonta in an arboreal milieu.* In: Comparative Biology and Evolutionary Relationships of Tree Shrews (W.P. Luckett, ed.). Plenum Press, New York, pp. 133-169.

SZALAY F.S., TATTERSALL I. and DECKER R.L., 1975. *Phylogenetic relationships of* Plesiadapis *- postcranial evidence.* In: Contributions to primatology. Approaches to Primate Paleobiology (F.S. Szalay, ed.), Karger, Basel, Vol. 5, pp. 136-166.

TATTERSALL I., 1982. *The Primates of Madagascar,* Columbia Univ. Press, New York.

THALMANN U., HAUBOLD H. and MARTIN R.D., 1989. Pronycticebus neglectus, *an almost complete adapid primate specimen from the Geiseltal (G.D.R.).* Palaeovertebrata, *19*: 115-130.

THENIUS E. 1969. *Stammesgeschichte der Säugetiere (einschliesslich der Hominiden).* In: Handbuch der Zoologie (J.-G. Helmcke, D. Starck, H. Wermuth, eds.), 8 Band, 2 Teil. Walter de Gruyter, Berlin, pp. 1-368.

TOBIAS P.V., 1988. *Numerous apparently synapomorphic features in* Australopithecus robustus, Australopithecus boisei, *and* Homo habilis: *support for the Skelton-McHenry-Drawhorn hypothesis.* In: Evolutionary History of the «Robust» Australopithecines (F.E. Grine, ed.). Aldine-de Gruyter, New York, pp. 203- 308.

TUTTLE R.H., 1967. *Knuckle-walking and the evolution of hominoid hands.* Am. J. Phys. Anthrop. *26*: 171-206.

TUTTLE R.H., 1981. *Evolution of hominid bipedalism and prehensile capabilities.* Phil. Trans. R. Soc. *292*: 89-94.

WALKER A., 1967. *Patterns of extinction among the subfossil Madagascan lemuroids.* In: Pleistocene

Extinctions: The Search for a Cause (P.S. Martin, H.E. Wright, Jr., eds.). Yale Univ. Press, New Haven, pp. 425-432.

WALKER A, 1974. *Locomotor adaptations in past and present prosimian primates.* In: Primate Locomotion (F.A. Jenkins, Jr., ed.). Academic Press, New York, pp. 349-381.

WALKER A.C. and PICKFORD M., 1983. *New postcranial fossils of* Proconsul africanus *and* Proconsul nyanzae. In: New Interpretations of Ape and Human Ancestry (R.L. Ciochon, R.S. Corruccini, eds.). Plenum Press, New York, pp. 325-351.

ZAPFE H., 1960: *Die Primatenfunde aus der Miozänen Spaltenfüllung von Neudorf an der March (Devinska Nova Ves), Tschechoslowakei. Mit Anhang: Der Primatenfund aus dem Miozän von Klein-Hadersdorf in Niederösterreich.* Schweiz. Palaeontol. Abh. *78*: 1-293.

ZAPFE H., 1963. *Lebensbild von* Megaladapis edwardsi *(Grandidier).* Folia Primatol. *1*: 178-187.

J.L. Franzen

The oldest primate hands: additional remarks and observations

Two specimens discovered in 1982 from the Middle Eocene lake deposits of Grube Messel near Darmstadt (Germany) comprise most of the hand skeleton of adapiform primates. Clearly belonging to different species they represent the oldest articulated primate hand skeletons up to now. The smaller one was described by FRANZEN (1987a, 1988) as *Europolemur koenigswaldi* n. sp. Except for its opposable thumb and scutiform nail-bearing end-phalanges it does not display many details. The larger one is not determinable in the absence of any pertinent dentition. Contrasting with all lemuriforms but corresponding with *Smilodectes, Adapis*, and the haplorhines its centrale is excluded from any contact with the hamate, while it differs from *Adapis* by its broad articulation with the lunate. Supplementary observations to FRANZEN (1988) are made through the transparent epoxy raisin from the palmar side. Functional and phylogenetic interpretations are discussed.

Key words: adapiform primates, Eocene, *Europolemur*, Grube Messel, hand skeleton.

Introduction

Two specimens discovered in 1982 by the team of the Senckenberg Research Institute at the famous Messel locality of fossil vertebrates in West Germany (FRANZEN, 1987a, 1988) comprise most of the hand skeleton of adapiform primates. Dating from the beginning of the Middle Eocene (Early Geiseltalian, MP unit 11; FRANZEN & HAUBOLD, 1986, FRANZEN, 1987b) they represent the oldest articulated primate hand skeletons known up to now. Each of them is clearly referring to a different species. While the smaller specimen (SMF-ME 1228B) belongs to half of a skeleton of *Europolemur koenigswaldi* FRANZEN, 1987, the other one is part of an isolated right forearm, which is about 50% larger (SMF-ME 1683). In the absence of any pertinent dentition, the second specimen is presently not determinable on species or genus level. Although the smaller hand derives from a juvenile, the difference of individual age is not sufficient to explain the morphological gap between the two specimens, which is not restricted to size only. Evidently due to the early stage of ossification, the smaller hand appears not so well preserved, particularly in the carpal region, whereas the larger one is very well presented. Both are now embedded in epoxy raisin from the dorsal or the palmar side respectively due to the applied transfer technique of preparation.

The purpose of this contribution is not only to draw attention to these discoveries, but also to add new observations, particularly in the light of the review of fossil primate hands published by GODINOT & BEARD (this volume).

The hand of *Europolemur koenigswaldi*

As indicated by the position of the ulna and the pisiform, the hand of the right forelimb is seen from its palmar side (*Figure 1*). The pisiform is the only carpal bone which

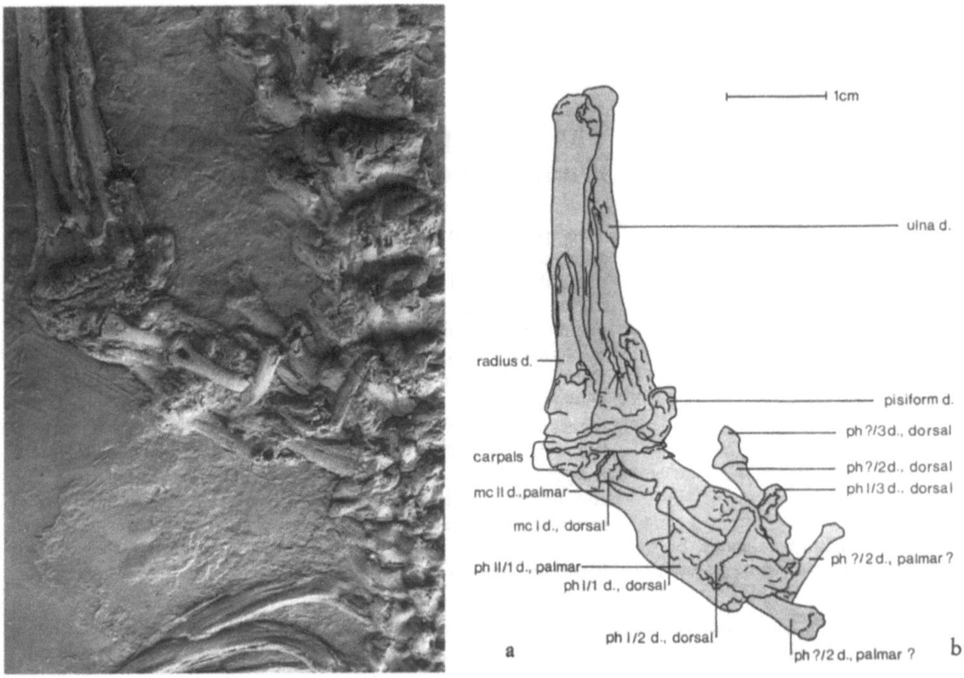

Figure 1. — *Europolemur koenigswaldi*, lower Middle Eocene of Messel, right anterior zeugo- and autopodium in palmar view (SMF-ME 1228 B); scale indicated. - a) Photo of the original, coated with ammoniumchloride. - b) Explanatory drawing. (a and b by Elke Pantak-Wein).

can be identified, being somewhat isolated and projecting ulnarly. Even by application of the stereo x-ray technique it is not possible, however, to trace any of the other carpal bones. Therefore nothing can be said about their outlines, their arrangement, and the articulations. There is some evidence, however, concerning the opposability of the pollex, because metacarpal I, together with the basal phalanx of that digit, is exposed from its dorsal side and crosses the palm diagonally from proximo-radially to disto-ulnarly, while all the other metacarpals are seen palmarly. How metacarpal I articulates with the trapezium is not discernible.

Interestingly, the metacarpals appear to be more slenderly built than in *Adapis parisiensis* (compare Godinot & Jouffroy, 1983), except for the first metacarpal, which is the shortest and the strongest one. The basal phalanges are much longer than the second ones; they are also longer than the metacarpals. The end-phalanges, two of which are clearly visible, contrast with *Notharctus* in their very short and flat proportions, which are comparable with those of *Lemur*. They are scutiform, and there can be no doubt that they were nail-bearing. The length/breadth-ratio of the end phalanx of presumably the fourth digit is 157.

The second hand of an adapiform primate from Messel

The second primate hand from Messel is also from the right side (*Figure 2*). Its excellently preserved dorsal aspect has been described in much detail by Franzen (1988).

There are, however, some additional remarks to be made particularly with respect to GODINOT & BEARD (this volume).

In the proximal carpal row, the lunate is very reduced while the scaphoid tubercle is large as it is in *Smilodectes* and in many extant lemurids (BEARD & GODINOT, 1988; GODINOT & BEARD, this volume). There is no prepollex identifiable. The dorsal surface of the triquetrum is damaged, so that nothing can be said concerning the presence of a proximal process. The triquetrum, however, differs from *Adapis parisiensis* and the lemuriforms in showing a trapezoidal outline tapering somewhat radially (*Figure 2, 4*).

Only recently a pisiform was discovered through the transparent epoxy plate on the palmar side of the hand (*Figures* 3-4). Its size and visible outline corresponds almost exactly with that of *Lemur rubriventer* and *Adapis parisiensis* (being only a little bit larger), which means that it is relatively large. In its present position it is horizontally attached to the other carpals. In this view a very deep concave facet can be seen proximo-ulnarly for the styloid process of the ulna and a convex one on the proximo-radial side for the triquetrum, as is the case with recent lemuriforms as well as fossil *Adapis parisiensis* (BEARD & GODINOT, 1988; BEARD *et al.*, 1988). The ulnar facet appears to be even deeper than in *Adapis* and *Lemur*, whereas the contour line of the radial side of the pisiform is less concave.

In the distal carpal row nothing can be said about the existence of a robust palmar tubercle on the trapezium, as is known of *Smilodectes* (BEARD & GODINOT, 1988; GODINOT & BEARD, this volume), because the bone is damaged at that spot. Concerning the distal articular facet of the trapezium for metacarpal I, it looks almost identical with *Adapis parisiensis* in its dorsal aspect. But in contrast to the statement of BEARD *et al.* (1988), it appears to me that it is only in the trapezium that this articulation is more or less flat in *Adapis parisiensis*, while the proximal facet of metacarpal I is slightly sellar-shaped. The same may be the case with the Messel adapiform of which this articulation is more or less hidden (*Figures 2-4*). Its orientation, however, is more distally like *Adapis parisiensis*, and not radially as it is the case with lorisids.

Considering the other carpo-metacarpal relationships, there is no contact between the hamate and the third metacarpal, hence a derived situation compared with many lemuriforms (LEWIS, 1977: 174). On the other side, a clearly developed articulation between the second metacarpal and the capitate excludes metacarpal III from any contact with the trapezoid (*Figure* 2). Being apparently dorso-palmarly continual, the articular facet between the second metacarpal and the capitate reflects a situation described by LEWIS (1977) in many platyrrhines. This is related to the loss of the corresponding carpometacarpal ligament, which is considered by GODINOT & BEARD (this volume) a derived character. Contrasting with recent lemuriforms, but corresponding with living haplorhines as well as *Adapis parisiensis*, the contact between centrale and trapezoid is transversely oriented. The capitate is proximo-distally long and radio-ulnarly narrow, as is the case with *Adapis parisiensis* and living lemuriforms. The hamate is the largest of all the carpal bones. As in *Smilodectes* and all extant prosimians, it displays almost confluent articular facets for metacarpals IV and V. The "spiral" facet between the triquetrum and the hamate is more proximo-distally directed as in *Smilodectes*, compared with its more transverse orientation in *Notharctus* and *Adapis*. Palmarly, the hamate is partly covered by the pisiform, but a well developed hamulus can be observed disto-ulnarly (*Figure 3*).

Very interesting are the articulations of the centrale (*Figures 2-4*). Corresponding with *Smilodectes, Adapis,* and the Haplorhini, but contrasting with all lemuriforms, the

a

b

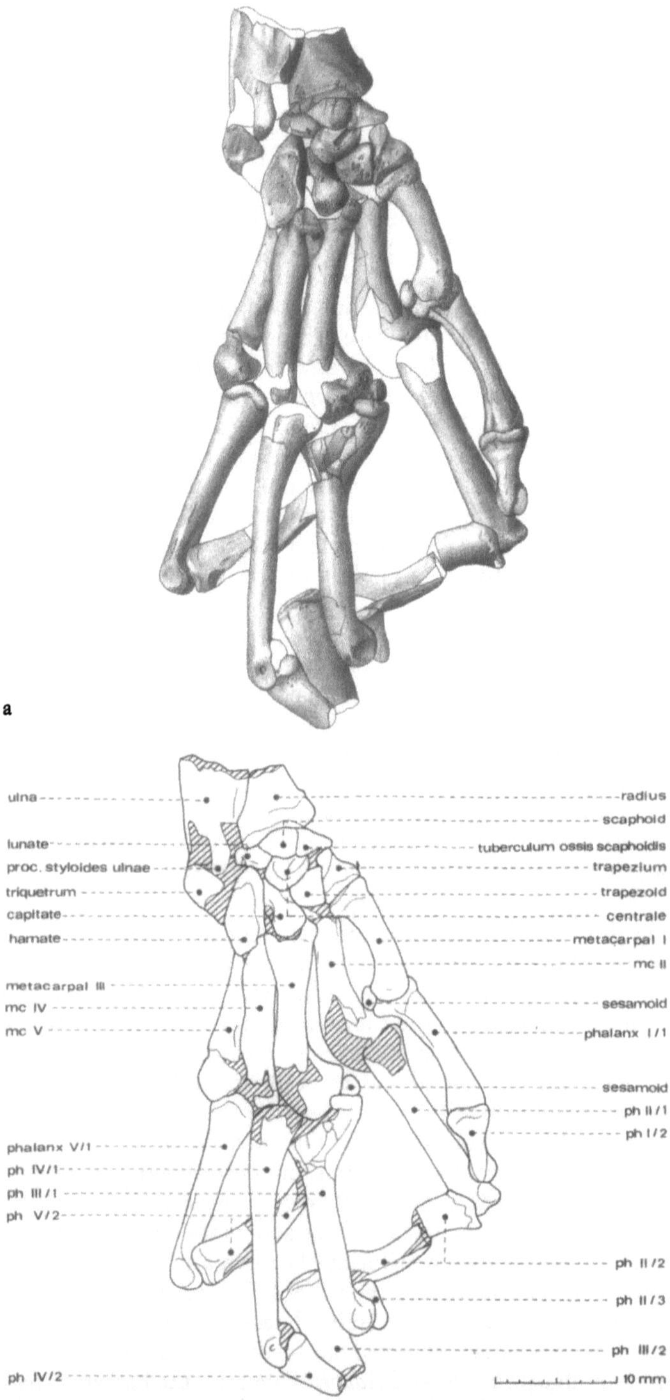

Fig. 2. Messel adapiform; lower Middle Eocene of Messel, right hand in dorsal view (SMF-ME 1683); scale indicated. - a) Drawing. - b) Explanatory drawing. Hatched = damaged bone. (a and b by Christine Hemm).

a

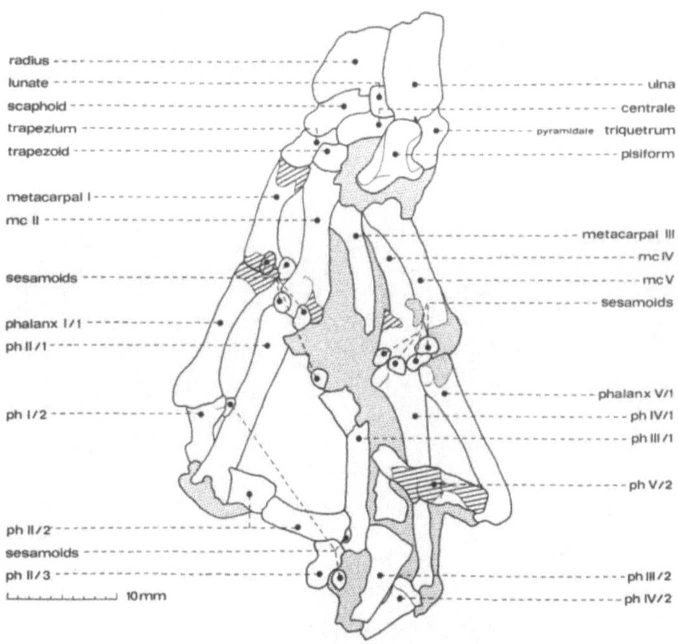

b

Fig. 3. Messel adapiform; lower Middle Eocene of Messel; right hand in palmar view (SMF-ME 1683), embedded in transparent epoxy raisin; scale indicated. - a) Photo (by Elke Pantak-Wein). - b) Explanatory drawing (by Christine Hemm). Hatched = damaged bone, dotted = covered by sediment.

TABLE 1 - *Proportions of the hand*

	carpus	metacarpus	fourth digit
Notharctus [1]	15,0%	21,0%	64,0%
Europolemur koenigswaldi	12,5%	24,0%	63,5%
Messel adapiform	13,0%	24,5%	62,5%

[1] after GODINOT & BEARD, this volume.

centrale is separated from any contact with the hamate. On the other side it differs from *Adapis* by its broad proximo-ulnarly directed articulation with the lunate.

Like in *Notharctus* the metarcarpals are short compared with the basal phalanges. They display about the same size and proportions as in *Adapis parisiensis*. The metacarpal formula is III.IV.V.II.I., as it is with *Notharctus*. Taken together with the basal phalanges, a paraxonic situation is indicated for the hand of the Messel adapiform (GODINOT & BEARD, this volume). While almost complete middle phalanges are only preserved with digits II and V, fragments of those occur also with digits III and IV. Metacarpals and phalanges are more or less curved palmarly. Their arrangement converges in a proximal direction as typical for a prehensile hand. In their original position of discovery metacarpal I and II enclose an angle of about 22°, but the possible maximum angle of excursion was evidently larger. The pollical distal phalanx is longer and narrower than the two manual ones known of *Europolemur koenigswaldi*. Its length/breadth-ratio is 168. Altogether the length proportions of the hand correspond almost exactly with those of *Europolemur koenigswaldi*, while in *Notharctus* the metacarpus appears to be relatively shorter and the carpus longer (*Table 1*).

Additional observations are possible from the palmar side considering the sesamoids (*Figure 3*). In digits I, II, IV and V two sesamoids can be observed in their original position at the metacarpo-phalangeal joints, while in digit III most of this articulation is covered by

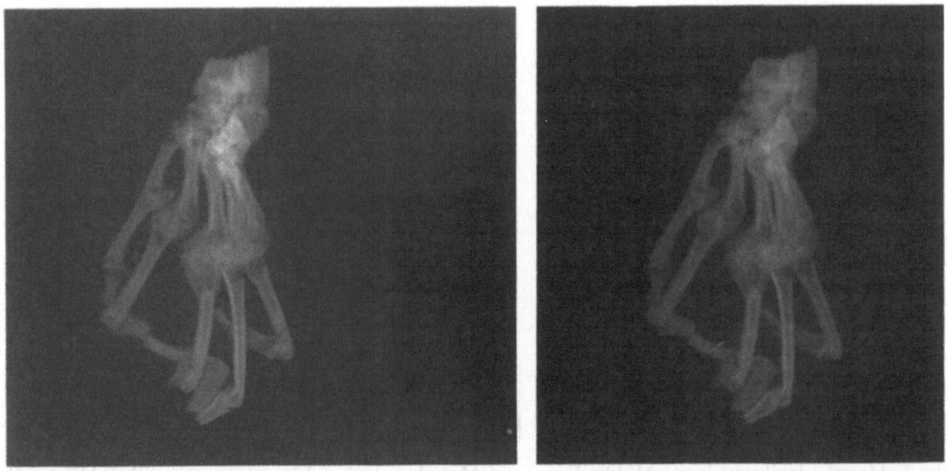

Fig. 4. Messel adapiform (SMF-ME 1683); stereo x-ray of carpal and metacarpal area in palmar view (by Dr. J. Habersetzer), primarily magnified ca. 2X; for osteological explanation compare figs. 2b and 3b.

sediment, so that only one sesamoid remains visible. Isolated sesamoids occur at the distal phalangeal articulations, which are preserved in digits I and II only.

Discussion

Except for their size the two adapiform hands from Messel are morphologically very similar. But the proportions of the distal phalanges are broader in the manual ones of *Europolemur koenigswaldi* than in the pollical end-phalanx of the large Messel adapiform.

Although very near in size and proportions, the hand of the Messel adapiform (SMF-ME 1683) is clearly distinguished from *Adapis parisiensis* by the following characters:

1. The lunate is relatively considerably smaller in the Messel adapiform.

2. While the centrale displays a broad articulation with the lunate in the Messel adapiform, both carpals are clearly separated from each other by a contact between the capitate and the scaphoid in *Adapis parisiensis.*

3. The capitate is proximo-distally relatively shorter in the Messel adapiform.

Similarities with living lemuriforms do not necessarily mean synapomorphies. They could also be explained as retained (symplesiomorphic) characters or by convergence. The only way to decide between these alternatives is to interpret the functional context of the respective characters in order to relate them to evolution which is understood as a process of optimization or "economization" (PETERS & GUTMANN, 1971). Functional analysis is also the only way to figure out what characters are functionally interconnected, i.e. correlated in their evolution (FRANZEN, 1985: 224).

Very important in this context appear to be the articulations of the centrale. BEARD *et al.* (1988) found that it is only in the lemuriforms that the centrale articulates with the hamate, while in all other primates it is excluded from that contact by the capitate. Thus BEARD et al. regarded the lemuriform situation as synapomorphic, and all the other primates as symplesiomorphic in this respect.

Another interpretation is also possible, however (FRANZEN, 1988: 283-284). While the centrale of the mammals is phylogenetically a relic of a whole complex of carpalia situated between the proximal and the distal row and extending in amphibians and reptiles from the radial to the ulnar side (ROMER & PARSONS, 1983: 213), its evolution within the mammals is characterized by reduction and by fusion (STARCK, 1979: 661-663). This would certainly improve the stability of the carpus. In this sense, it would not be the centrale but the hamate, which became enlarged within the lemuriforms. Thus the contact between these two bones, which was originally present, became secondarily enlarged. On the other hand, the development of the centrale within the other primates would not be symplesiomorphic, but synapomorphic in another direction. It was within this process of reduction of the centrale complex by which the original articulation with the hamate got lost.

While the first process, which involves an enlargement of the hamate, could functionally be dealing with the typical ectaxonic grip of lemuriforms walking on branches (GODINOT & BEARD, this volume), the alternative development could not only be related to an increase of the stability of the carpus but also to a further improvement of the opposability of the thumb (FRANZEN, 1988: 284).

It must be confessed, however, that (in spite of PREUSCHOFT *et al.* in this volume) the functional details of these developments are not yet really understood. Concerning the non-lemuriform primates, it could also be an enhanced transfer of weight to the third digit, which would favour a direct contact between the lunate and the capitate, excluding incidentally the centrale from its original articulation with the hamate. Correspondingly, the transverse articulation between the centrale and the trapezoid (as well as the direct contact between the capitate and metacarpal II) would improve the weight transfer to digit II in haplorhines and adapiforms, while in lemuriforms the position of the centrale would refer to ectaxony, transferring most of the weight to the ulnar side of the hand.

Whatever may be, the oldest primate hands should not be regarded as showing necessarily the most primitive state of character development within that order. Further discoveries and functional analyses are imperative in order to solve those problems.

ACKNOWLEDGEMENTS — The author thanks the team and particularly the technicians of the Senckenberg Research Institute, who unearthed and prepared the oldest hands of primates, as well as Professor Holger Preuschoft who invited and encouraged him to contribute to this volume. Very stimulating and elucidating were also discussions with Chris Beard and Marc Godinot during a recent stay at the Johns Hopkins University at Baltimore, thanks to an invitation from Kenneth D. Rose. Photographs and drawings are owing to the talent and the efforts of Elke Pantak-Wein and Christine Hemm.

References

BEARD K.C., DAGOSTO M., GEBO D.L. & M. GODINOT, 1988. *Interrelationships among primate higher taxa.* Nature, 331: 712-714.

BEARD K.C. & GODINOT M., 1988. *Carpal anatomy of Smilodectes gracilis (Adapiformes, Notharctinae) and its significance for lemuriform phylogeny.* J. Human Evolution, 17(1/2): 71-92.

FRANZEN J.L., 1985. *What is "Pithecanthropus dubius Koenigswald, 1950"?* In DELSON, E. (ed.): Ancestors: The Hard Evidence: New York (Alan R. Liss, Inc.); 221-226.

FRANZEN J.L., 1987a. *Ein neuer Primate aus dem Mitteleozän der Grube Messel (Deutschland, S-Hessen).* Cour. Forsch.-Inst. Senckenberg, 91: 151-187.

FRANZEN J.L., 1987b. *Mammalian reference levels MP 11-13.* Münchner Geowiss. Abh., A 10: 24-25.

FRANZEN J.L., (1988). *Ein weiterer Primatenfund aus der Grube Messel bei Darmstadt.* Cour. Forsch. - Inst. Senckenberg, 107: 275-289.

FRANZEN J.L. & HAUBOLD H., 1986. *The Middle Eocene of European mammalian stratigraphy. Definition of the Geiseltalian.* Modern Geology, 10: 159-170.

GODINOT M. & BEARD K.C. (this volume).

GODINOT M. & JOUFFROY F.K., 1983. *La main d'Adapis (Primate, Adapidé).* In E. BUFFETAUT, J.M. MAZIN & E. SALMON (eds.): Actes du Symposium Paléontologique G. CUVIER, Montbéliard - France 1982: Montbéliard, 221-242.

LEWIS O.J., 1977. *Joint remodelling and the evolution of the human hand.* J. Anat., 123(1): 157-201.

PETERS D.S. & GUTMANN W.F., 1971. *Über die Lesrichtung von Merkmals- und Konstruktions-Reihen.* Z. Zool. Syst. Evolutionsforsch., 9: 237-263.

ROMER A.S. & PARSONS T.S., 1983. *Vergleichende Anatomie der Wirbeltiere.* - 5. Ed.; Hamburg, Berlin (Parey).

STARCK D., 1979. *Vergleichende Anatomie der Wirbeltiere auf evolutionsbiologischer Grundlage, 2: Das Skeletsystem. Allgemeines, Skeletsubstanzen, Skelet der Wirbeltiere einschließlich Lokomotionstypen.* Berlin, Heidelberg, New York (Springer).

K.C. Beard
M.F. Teaford
A. Walker

New hand bones of the early Miocene hominoid *Proconsul* and their implications for the evolution of the hominoid wrist

Since the structure and function of the hands in early hominoids has doubtlessly placed constraints on the structure and function of the hands of their living descendents, it is only reasonable to investigate the known fossil material in the present context. Unfortunately, there is only one genus (*Proconsul*) of which relevant postcranial remains, and handbones in particular, are available. As a basis of a reliable interpretation of evolutionary processes, the geology of the known and new fossiliferous sites is described. Previous ideas about the locomotor adaptations of early hominoids are reviewed and reconsidered in the light of the new material. We conclude that the traditional hominoid status of *Proconsul* is well founded. The wrist morphology seems to indicate a very basal position within the superfamily – although the possibility cannot be ruled out that the seemingly primitive features are adaptations to the use made of the hands. No matter which reason holds true, the morphological evidence does not directly support suspensory habits in *Proconsul*, but its characters fit well to the assumption of a predominance of climbing and generalised arboreal quadrupedalism in its locomotor repertoire, without visible dependence on body size.

Key words: arboreal quadrupedalism, climbing, hand skeleton, locomotor adaptation, postcranial remains, Rusinga Island.

Introduction

The structure and function of the hands of early hominoids have, of course, placed constraints on the structure and function of the hands of their living descendants. Paleoprimatologists have long recognized that early hominoids were taxonomically diverse during the Miocene epoch, but such assessments have been based almost exclusively on analyses of fossil teeth and jaws (e.g., ANDREWS, 1978). Indeed, recent paleontologic fieldwork in the Miocene of East Africa has revealed a taxonomic diversity of early hominoids much greater than that which was envisioned only a short time ago (HARRISON, 1986; LEAKEY & LEAKEY, 1986a, 1986b, 1987; LEAKEY *et al.*, 1988a, 1988b): Despite these advances in our knowledge of early hominoid diversity, our understanding of the structure and function of early hominoid hands remains hampered by the fact that very few of these «dental taxa» are known from relevant postcranial remains (see also Godinot and Beard, this volume).

The most notable exception to this rule is the early Miocene hominoid genus *Proconsul*, originally named by HOPWOOD (1933). Species attributed to *Proconsul* apparently differ from each other primarily in body size, and together they span almost the same range of body size as do extant hominoids. Although there is some disagreement concerning the alpha-level taxonomy of the genus (see especially KELLEY, 1986), three species are generally recognized; from smallest to largest, they are: *Proconsul africanus*, *Proconsul nyanzae*, and *Proconsul major* [see MARTIN (1981) and LEAKEY *et al.* (1988b) for discussion of current concepts concerning the species *P. major*]. All of these species have been found at a variety of sites clustered around the northern and eastern shores of Lake Victoria in Kenya and in northeastern Uganda. The anatomy of the hand and wrist of *Proconsul* was first documented by NAPIER & DAVIS (1959) in their classic description of KNM-RU 2036, a specimen of *Proconsul africanus* from Rusinga Island, Kenya.

Three fossiliferous localities on Rusinga Island have recently yielded new *Proconsul* wrist material. We have described the anatomy of these fossils in detail elsewhere (BEARD *et al.*, 1986). In this paper we reassess previous ideas concerning the early evolution of the hominoid wrist in the light of this new material.

Geology of Sites

Rusinga Island has long been recognized as one of the most important sources of Miocene hominoid fossils (ANDREWS, 1981). The island itself lies at the mouth of the Nyanza Gulf in the northeast corner of Lake Victoria. It is an erosional remnant of the northern edge of a massive Miocene volcano known as Kisingiri (VAN COUVERING & VAN COUVERING, 1976). As such, it is made up of a complex series of alternating volcanic deposits and their weathering products. These deposits have been dated by the potassium-argon method and they span the period between 17.5 and 18.0 million years B.P. (DRAKE *et al.*, 1988). A number of the deposits are fossiliferous, but 80% of Rusinga's fossils have come from the Fossil Bed Member of the Hiwegi Formation. All of the three sites on Rusinga Island which have recently yielded new *Proconsul* wrist material lie within this Fossil Bed Member.

Site R114

This is the site of discovery of the most complete specimen of *Proconsul*, KNM-RU 2036. Discovered in 1951 by WHITWORTH (1953), it is located on the northwest slope of Kiakanga Hill at the western end of Rusinga Island. The deposit itself has proven to be very interesting, as it was probably formed in a large hollow tree trunk. The fossils in the deposit appear to be the remains of animals that either used the tree trunk during their lives or were carried there by carnivores or scavengers after death (WALKER *et al.*, 1986). Initially, an articulated left forelimb and associated cranial and foot remains of a single subadult *Proconsul africanus* were described from this site (NAPIER & DAVIS, 1959; DAVIS & NAPIER, 1963). Interpretations of this material have played a pivotal role in the development of ideas about the evolution of the hominoid skeleton.

Thirty years after the original discovery, more bones of the same individual were found at the National Museums of Kenya in Nairobi. These fossils included the proximal two-thirds of the left first metacarpal and the missing proximal phalanx of the left third finger (WALKER & PICKFORD, 1983). A re-examination of the R114 site in 1984 yielded yet more bones of this skeleton on the dumps of the original excavation. The 1984 material included bones of the right hand that were either missing or damaged in the left hand skeleton. As a result, we now know nearly all the elements of the wrist and hand of this individual from one side or the other (*Figure* 1). The additional material of the hand of KNM-RU 2036 confirms the hand proportions originally proposed by Napier and DAVIS (1959).

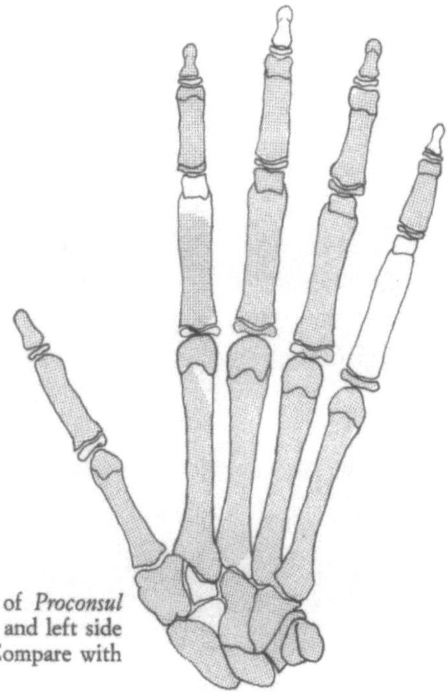

Figure 1 - Schematic illustration of reconstructed hand of *Proconsul africanus* (KNM-RU 2036), based on specimens from right and left side of this single individual. Stippled areas are now known. Compare with Napier and Davis (1959) Plate IX.

The Kaswanga Primate Site

At the end of the 1984 season at R114, Bwana Peter Nzube found a concentration of *Proconsul* bones and teeth washing out on the surface of another site in the eastern exposures at Kaswanga on Rusinga Island (*Figure* 2a,b; WALKER & TEAFORD, 1988). Surface collection and water washing in 1984, and detailed excavation in 1985, led to the recovery of parts of the skeletons of nine *Proconsul* individuals. They range in age from very young infants, through juveniles and subadults, to adults. It is not yet clear whether both *Proconsul africanus* and *P. nyanzae* are represented, but most of the specimens (and all of the adult specimens) clearly belong to *P. africanus*. Most of the bones were found broken up or whole on the surface, but about one quarter were found *in situ*. The only articulated bones were the left and right feet of one adult, an infant foot and lower leg, and a partial adult hand skeleton (*Figure* 3a,b). The hand material found at the Kaswanga Primate Site includes 49 carpals, with every element represented several times, 16 whole or partial metacarpals and several metacarpal epiphyses, and 219 phalanges (*Figure* 4, *Table* 1). Apart from the terminal phalanges of digit I in both hand and foot, we have not yet been able to sort these into hand and foot phalanges. Given the size differences between individuals and the different growth stages represented, we may never be able to do so. In any event, our knowledge of the hand of *Proconsul* has been increased by the finds at the Kaswanga Primate Site by at least an order of magnitude.

The excavation at the site shows it to be at the junction of the Kaswanga Point Grits and the overlying Fossil Bed Member of the Hiwegi Formation (WALKER & TEAFORD, 1988). The *Proconsul* bones were deposited in small channels which were cut into the uppermost flaggy beds of the Kaswanga Point Grits after minor earth movements. Our

Figure 2a - Kaswanga Primate Site, June 17, 1984, the day after the discovery of *Proconsul* fossils there. View is west-northwest.

2b - Kaswanga Primate Site later in 1984 field seasons. View is north-northwest, Wanyama Hill is in distance.

Figure 3a - Articulated right partial hand skeleton of *Proconsul africanus in situ* at Kaswanga Primate Site, Rusinga Island, Kenya.

TABLE 1 - *Preliminary list of new* Proconsul *wrist and hand material from Rusinga Island, Kenya.*

	Proconsul africanus (KNM-RU 2036) from R114	*Proconsul** from Kaswanga Primate Site	*Proconsul nyanzae* (KNM-RU 15100) from R5
Carpals			
Triquetrum	1	6	1
Pisiform	—	5	1
Lunate	—	6	1
Scaphoid	—	8	1
Hamate	—	4	—
Capitate	—	6	—
Trapezoid	—	5	1
Trapezium	-	4	—
Centrale	—	5	1
Metacarpals	2	14	2
Phalanges	4	(219) (not yet sorted into hands & feet)	7

* *Proconsul* from Kaswanga Primate Site possibly includes a mixture of *Proconsul africanus* & subadult *Proconsul nyanzae.*

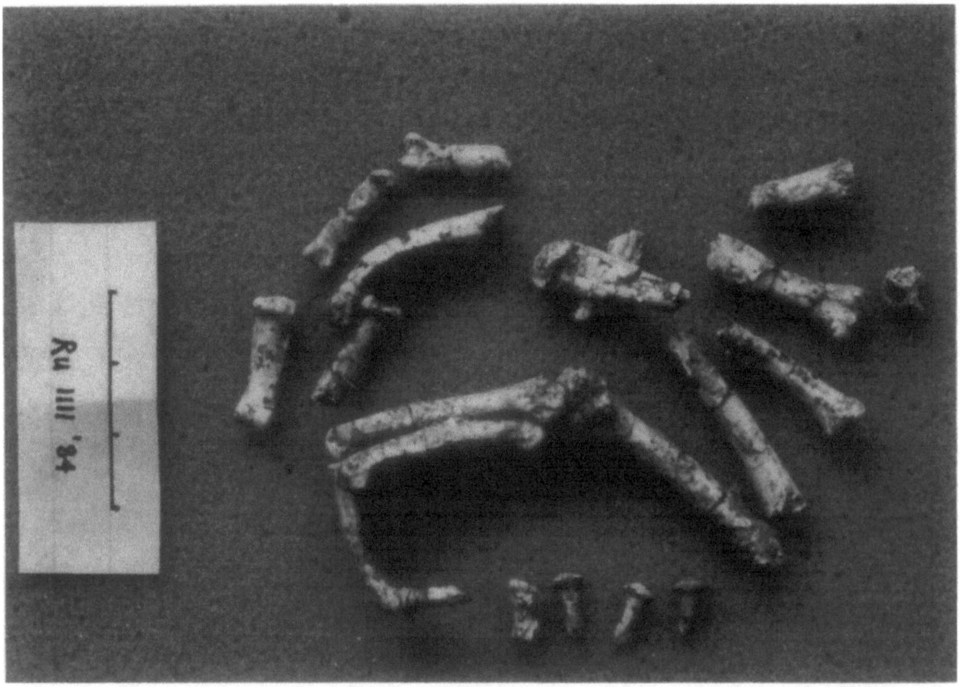

Figure 3b - Same specimen, prepared after excavation. Scale equals 3 cm.

excavations at Kaswanga support the view of NESBIT EVANS *et al.* (1981) that the Kaswanga fossil beds contain typical forest dwelling species which were buried in flood-plain sedimentary environments. Most of the individual skeletons appear to have been disarticulated before or during burial in the channels. The primate bones were associated with two lagomorph skeletons, fragments of tragulids and rodents, and many fossil leaves, seeds, and wood.

Site R5

In 1985, Alan Walker discovered a partial hand skeleton of an adult *Proconsul nyanzae*, KNM-RU 15100, at the R5 exposures on the western side of Kaswanga Point. The specimen had been washing out for some time and nothing further was recovered after extensive sieving and washing. It may be that only a partial hand skeleton was buried there. The specimen preserves a centrale, lunate, triquetrum, pisiform, trapezoid, partial scaphoid, two metacarpal bases and several phalanges, including terminals. As the first *Proconsul nyanzae* hand skeleton, it is very important. It is in the same Fossil Bed Member as the other Kaswanga primates, but a little higher in the section.

In summary, the hand skeleton of *Proconsul africanus* is considerably better known now than before. All bones of the wrist are now known several times over. In addition, for the first time hand remains are also known for the larger *Proconsul nyanzae*. We hope that

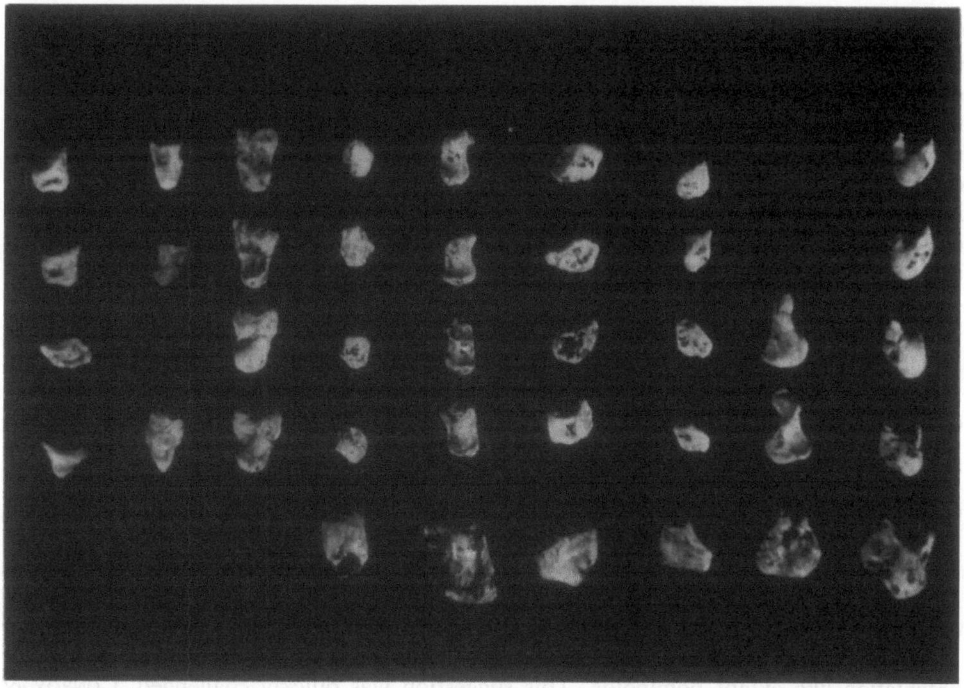

Figure 4 - New carpal bones of *Pronconsul africanus* (top four rows) and *Proconsul nyanzae* (bottom row). From left to right, columns are: trapezia, hamates, capitates, trapezoids, pisiforms, triquetra, centrales, scaphoids, and lunates.

future fieldwork will yield more complete material of this last species so that additional problems concerning its adaptations and evolution can be resolved. Paleoenvironmental information indicates that these species were contemporaneous, if not sympatric, and that they lived in forest or woodland near floodplains on Rusinga. NESBIT EVANS *et al.* (1981), on the basis of a taphonomic study, favor a forested environment for the early Miocene of Rusinga, while PICKFORD (1983), on the basis of gastropod species, favors a dry forest or woodland. In either case, it is clear that enough trees were present for *Proconsul* to have carried out an arboreal existence.

Previous ideas on the locomotor adaptations of *Proconsul* and early hominoids

When they first described the forelimb material of *Proconsul africanus*, NAPIER & DAVIS (1959) noted similarities to *Presbytis*, and inferred that *Proconsul* was most likely an arboreal quadruped, although features considered to be adaptations for suspension were also noted. The interpretations of NAPIER & DAVIS on the hand anatomy of *Proconsul* were limited by the lack of comparative anatomical studies of the hands of living primates and, as ours are today, by the lack of any detailed knowledge of primate hand use especially in the wild (but see BOESCH & BOESCH, CHRISTEL, MAIER, NIESCHALK & DEMES, RAWLINS, REDSHAW in this volume).

Our knowledge concerning the functional morphology of the hand of extant primates was greatly enhanced by the work of LEWIS (1969, 1972a, 1974, 1977), who documented several traits characteristic of the wrists of hominoids. Among these traits, perhaps the most important for later ideas on hominoid functional morphology and evolution is the proximal retraction of the ulnar styloid process away from articular contact with the carpus. Lewis suggested that this trait allowed increased abilities of supination. Because increased capacities for supination are required for suspensory progression and are characteristic of hominoids (LEWIS, 1972a; O'CONNOR & RAREY, 1979), Lewis further suggested that this retraction of the ulnar styloid process must be an adaptation for «brachiation».

The notion that brachiation might have been involved in the initial radiation of the Hominoidea can be traced to AVIS (1962), who proposed that this mode of locomotion may have allowed the exploitation of a terminal branch niche by the last common ancestor of living hominoids (see also GRAND, 1972; RIPLEY, 1979); LEWIS (1971, 1972a, 1972b), following Avis, has suggested that the derived features of the wrist of hominoids may have been similar in evolutionary importance to the reorganization of the tarsus observed in early artiodactyls, and commonly held to be a major adaptative breakthrough, allowing for the rapid radiation of that higher taxon (SCHAEFFER, 1948).

More specifically, LEWIS (1971, 1972a, 1972b) examined casts of the forelimb of *Proconsul africanus* and suggested that the derived features of the wrist of extant hominoids, most notably the lack of ulnocarpal articular contact, were also present in *Proconsul*. Moreover, Lewis suggested that these traits indicated that *Proconsul* was capable of brachiation like extant hominoids. This suggestion was quickly challenged. CONROY & FLEAGLE (1972) and ZWELL & CONROY (1973) argued that the evidence of the postcranium of *Proconsul* supported knuckle-walking rather than brachiation as the locomotor mode of this primate. Most other workers have argued that *Proconsul* was essentially monkey-like in its carpal adaptations and must therefore have been a palmigrade arboreal quadruped (SCHÖN & ZIEMER, 1973; MORBECK, 1975, 1977; CORRUCCINI *et al.*, 1975, 1976; O'CONNOR, 1976; MCHENRY & CORRUCCINI, 1983).

An alternative hypothesis of locomotor behavior for early hominoids has used prosimian models for the first time (CARTMILL & MILTON, 1977). Noting the retraction of the ulnar styloid process from articulation with the carpus in lorisids, as well as other similarities in carpal morphology (CARTMILL & MILTON, 1977: *Table 1*), CARTMILL & MILTON suggest that the carpal modifications observed by LEWIS in extant hominoids need not have been adaptations for bimanual suspension. Rather, these workers propose that lorisid-style slow climbing might better explain the origin of the extant hominoid carpus. Moreover, ROBERTSON (1984) has suggested that this model may be specifically applicable to *Proconsul*. Perhaps the most attractive aspect of CARTMILL & MILTON'S model is that it is more in line with what consensus there is concerning the adaptations of the wrist of Miocene hominoids. Given an arboreal quadrupedal ancestry (e.g., *Proconsul africanus*), an increase in body size in early hominoids should, according to CARTMILL & MILTON, «inescapably» lead to cautious arboreal locomotor habits because of the increased dangers associated with falling.

The slow-climbing hypothesis of early hominoid locomotor adaptations is also more consonant with field observations of hominoid locomotor behavior, and goes a long way toward answering CONROY & FLEAGLE'S (1972) major objection to Lewis' brachiating hypothesis: «If Lewis's suggestion is followed that the advanced hominoid wrist (as seen in *Pan* and *Gorilla*) is a brachiating adaptation, it would seem that the only primates which consistently brachiate are ill adapted for such activity and the knuckle-walking *Gorilla* possesses brachiating adaptations which it does not use». Implicit in such a view of hominoid locomotor evolution is the idea that hylobatid-style bimanual suspension is a uniquely derived locomotor adaptation, and that the ancestors of the extant great apes probably never engaged in such specialized locomotor activity (e.g., see ANDREWS & GROVES, 1976). Theoretical mechanical considerations (which logically have antedated experiments; see PREUSCHOFT, 1970, 1973a) as well as direct experimentation in the form of electromyographic studies and bone strain analyses by FLEAGLE *et al.* (1981) indicate that similar patterns of muscle recruitment and bone strain occur during climbing and brachiation behaviors among anthropoids. Therefore, FLEAGLE *et al.* (1981) suggest that specialized locomotor behaviors such as brachiation *per se* need not be invoked to explain the carpal adaptations of hominoids.

The degree of carpal convergence purported to exist between lorisids and hominoids (CARTMILL & MILTON, 1977) has been forcefully challenged by LEWIS (1985). Lewis notes several features of the lorisid wrist and hand which are very different from the condition in hominoids. These include: (1) the proximal position of the lunate and the great mediolateral expansion of the radial facet of that bone, (2) the lack of an intra-articular meniscus on the ulnar side of the wrist, (3) the articulation between the ulnar styloid process and the triquetrum (and variably between the ulnar styloid process and the pisiform) which occurs in ulnar deviation, (4) the unique «midcarpal condyle» which results in conjoint motions of rotation and ulnar deviation, and (5) the reduction of digit II and the pincer-like grasp of lorisines (for more detailed discussion of this last feature, see NIESCHALK and DEMES, this volume). These features, along with those described by ETTER (1978), clearly point to the extreme degree of structural specialization exhibited by the hands of lorisids. These differences in hand and wrist anatomy between lorisids and hominoids may indicate very different functional adaptations for the two groups (LEWIS, 1985). The functional meaning of the traits is investigated in this volume by PREUSCHOFT *et al.*

In summary, two main hypotheses have been put forward to explain the specializations of the hominoid wrist. LEWIS (1971, 1972a, 1972b) suggested that these specializations must have been adaptations for brachiation. CARTMILL & MILTON (1977) proposed that these same specializations could have been adaptations for lorisid-style slow climbing. Both models possess good points and bad points, and both can be interrelated by theoretical, biomechanical, and electromyographic studies (PREUSCHOFT, 1970, 1973a; FLEAGLE *et al.*, 1981). That is, since similar patterns of muscle recruitment and bone stress occur during climbing and brachiation, the two models are not necessarily mutually exclusive. Both locomotor modes, and others as well, have been proposed for *Proconsul* in the past. For example, several other analyses of various aspects of forelimb morphology in *Proconsul* have emphasized that the fossil form shows some features that are not closely matched in any extant anthropoid, but have certainly tended to be in more general agreement with conclusions supporting generalized arboreal quadrupedalism than more specialized modes of locomotion such as brachiation and knuckle-walking (e.g., PREUSCHOFT, 1973b; FLEAGLE, 1983; ROSE, 1983, 1988; BEARD *et al.*, 1986; see also GEBO *et al.*, 1988). Below, we will reassess these different hypotheses for the locomotion of *Proconsul* in light of newly collected hand material.

New hand bones of *Proconsul africanus* and *Proconsul nyanzae*

The new hand material of two species of *Proconsul* (see BEARD *et al.*, 1986: *Table 1*) allows a reassessment of the diverse locomotor inferences which were largely based on one specimen in the past. Elements of the *right* hand of the KNM-RU 2036 specimen of *Proconsul africanus* provide very good information on the overall proportions of the hand of one individual of this species (*Figure 1*). Also, several bones of the wrist of *Proconsul africanus* were previously either not known at all (i.e., the trapezoid and the centrale), or known only from specimens so poorly preserved (i.e., the triquetrum and the hamate) that even rational scholars could disagree about their anatomy. Each of these bones are now known from several specimens. More importantly, hand bones of *Proconsul nyanzae* are now known for the first time. We have presented a detailed account of the descriptive osteology of this new material elsewhere (BEARD *et al.*, 1986). Here we would like to reassess some of the major points of contention that have been argued about in the past. Also, we will draw some phylogenetic conclusions regarding the systematic position of *Proconsul* within the Catarrhini, and we will comment on the implications of this new material for previous notions of early hominoid locomotor evolution.

The idea that *Proconsul* was a brachiator (LEWIS, 1971, 1972a, 1972b) rests largely on the claim of a hominoid-like triquetrum and pisiform in *Proconsul,* a condition in which these carpal elements would not have articulated with the ulnar styloid process as they do in monkeys. Some of the disagreement on this point in the past can be attributed to the use of casts by workers doing original research on the fossils (MORBECK, 1977); however, the poor preservation of the partial left triquetrum of KNM-RU 2036 clearly has played an important role in fostering this controversy as well. In fact, this bone fragment is so equivocal that some workers (e.g., ROBERTSON, 1984) have not been able to assert with absolute confidence that the specimen represents part of the triquetrum rather than some other bone. However, it was in proper association in the original matrix (NAPIER & DAVIS, 1959: plate 8), and comparison with the more complete, newly-collected triquetra of this species confirms its original allocation.

Based on the newly collected material, the articulated pisiforms and triquetra of *P. africanus* and *P. nyanzae* are illustrated in *Figure 5.* Clear articular facets for the ulnar styloid process are present on both bones in both species. These facets are best preserved in *P. africanus,* in which they are most similar to the corresponding facts in *Colobus* among the comparative sample of extant primates we utilized. Nevertheless, functionally important differences in morphology between the ulnocarpal articulations of *P. africanus* and *Colobus* do exist (BEARD *et al.*, 1986). Although both the triquetrum and the pisiform of *P. nyanzae* are somewhat eroded and the facet on the pisiform for the ulnar styloid process is broken medially, the preserved portions of these bones are similar in every respect to the corresponding bones in *P. africanus,* and there can be no question that an ulnocarpal articulation occurred in this larger species as well.

Along with the observations of other workers (SCHÖN & ZIEMER, 1973; MORBECK, 1975, 1977; CORRUCCINI *et al.*, 1975, 1976; O'CONNOR, 1976; MCHENRY & CORRUCCINI, 1983), our findings should end the controversy concerning the ulnocarpal articulation in *Proconsul.* Together with functional analyses of other aspects of forelimb morphology in *Proconsul* (e.g., FLEAGLE, 1983; ROSE, 1983, 1988; see also GEBO *et al.*, 1988), the occurrence of an ulnocarpal articulation in *Proconsul* suggests that it is highly unlikely that brachiation *per se* was a significant component of *Proconsul's* locomotor repertoire. Such an interpretation is further corroborated by the independent analysis of other aspects of hand and carpal anatomy in *Proconsul* (PREUSCHOFT, 1973a,b; JENKINS, 1981).

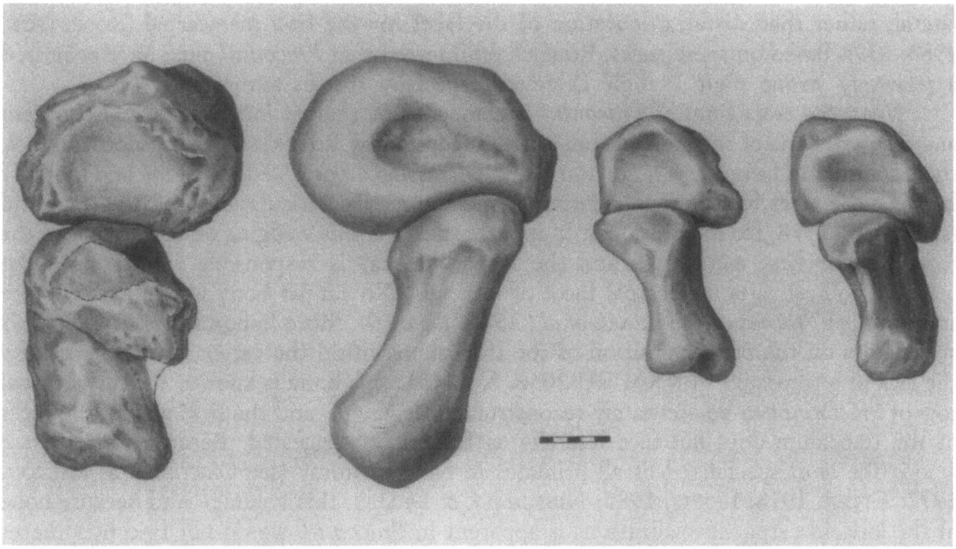

Figure 5 - Articulated pisiform and triquetra in proximomedial view. From left to right: *Proconsul nyanzae* (KNM-RU 15100, left; dotted lines outline missing portions); *Pan troglodytes* (right); *Colobus guereza* (right); *Proconsul africanus* (right). Scale equals 5 mm. Note facets for ulnar styloid process on all specimens except *Pan*.

As we noted above, three other types of locomotor adaptations have also been suggested for *Proconsul* in the past. These include knuckle-walking (CONROY & FLEAGLE, 1972; ZWELL & CONROY, 1973), palmigrade arboreal quadrupedalism (SCHÖN & ZIEMER, 1973; PREUSCHOFT, 1973b; MORBECK, 1975, 1977; CORRUCCINI *et al.*, 1975, 1976; O'CONNOR, 1976; MCHENRY & CORRUCCINI, 1983), and grasp-walking and climbing (ROBERTSON, 1984). In light of the new discoveries, we feel that the second of these hypotheses is still the most reasonable. We will briefly discuss the first and last of these suggested locomotor modes, to illustrate why we disagree with them.

The proposal that *Proconsul* was a knuckle-walker (CONROY & FLEAGLE, 1972; ZWELL & CONROY, 1973) was partly based on the putative resemblances of the carpus of *Proconsul* to extant hominoids. Inasmuch as these resemblances have been refuted (usually in answering Lewis), the argument for knuckle-walking in *Proconsul* suffers as much as, or even more than, that for brachiation. Moreover, SCHÖN & ZIEMER (1973) and PREUSCHOFT (1973b) presented convincing evidence that the hand of *Proconsul* was ill-adapted for knuckle-walking. Finally, MORBECK (1975) offered methodological criticisms of ZWELL & CONROY'S paper, and the possibility that *Proconsul* was a knuckle-walker has not really been seriously considered since that time.

A more interesting proposal for the locomotor mode of *Proconsul* is that offered by ROBERTSON (1984). She argues that similarities between the trapezium of *Proconsul* and those of lorisoids (especially lorisids) indicate that *Proconsul* may have employed «grasp-walking» and climbing as significant aspects of its locomotor repertoire. In effect, this represents an extension of CARTMILL & MILTON'S (1977) scenario of early hominoid locomotor adaptations to *Proconsul,* something CARTMILL & MILTON themselves did not do. The principal similarities that ROBERTSON found between the trapezium of *Proconsul* and those of lorisoids are the «flatness» and «gracility» of the bone and the relatively lateral, rather than distal, orientation of the facet for the first metacarpal (ROBERTSON, 1984: 239). Based on these traits, ROBERTSON inferred that *Proconsul* must have possessed a relatively strong digit I- digit II grasp, especially for an anthropoid.

When the trapezium of *Proconsul* is compared to that of lorisoids (e.g., *Nycticebus* and *Galago*), none of the general similarities suggested by ROBERTSON are obvious. In fact, the trapezium of lorisoids is quite unusual in possessing a bony wedge distally and medially between the facet for the first metacarpal and the medial articulation with the trapezoid (see ETTER, 1978, for illustrations). It appears that it is this wedging of bone between the facet for the first metacarpal and the trapezoid that is responsible for the laterally-oriented pollical carpometacarpal facet of lorisoids. No similar bony wedge exists in the trapezium of *Proconsul* (see BEARD *et al.*, 1986: *Figure 9*). More importantly, ROBERTSON's speculation on the proper position of the trapezium within the carpus was hampered by the lack of a trapezoid in KNM-RU 2036. Now that this bone is known, the distal carpal row of *Proconsul* can be accurately reconstructed (*Figure 6*), and the first metacarpal facet of the trapezium does *not* face laterally as Robertson suggested. Because lorisoids are among the *most* specialized of all primates in hand anatomy (see CARTMILL & MILTON, 1977; ETTER, 1978; LEWIS, 1985; NIESCHALK & DEMES, this volume), and because none of the lorisoid carpal specializations is apparent in *Proconsul,* we do not find hypotheses which emphasize lorisid-style grasp-walking and climbing for *Proconsul* especially convincing.

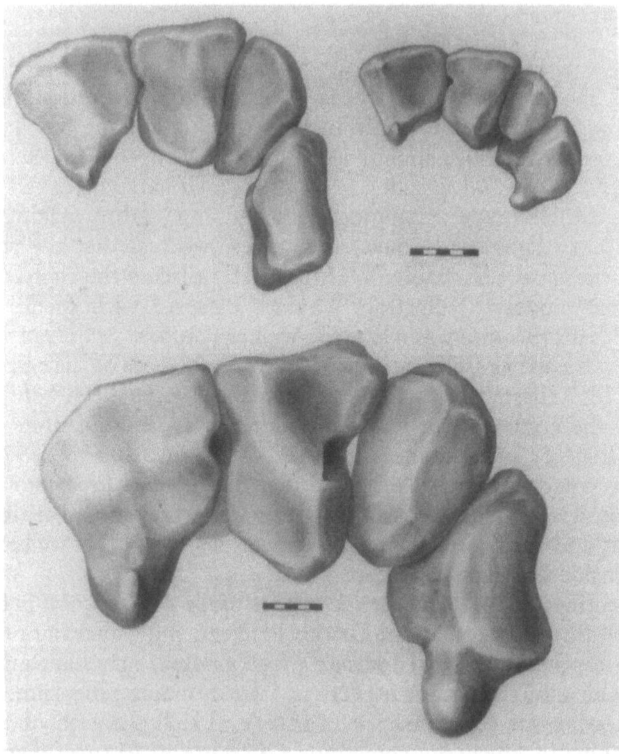

Figure 6 - Distal view of distal carpal row. Clockwise from upper left: *Proconsul africanus, Macaca fascicularis, Pan troglodytes.* Both scales equal 5 mm. Lower scale is for *Pan* only.

Evolutionary and Phylogenetic Implications

The phylogenetic position of Proconsul

As recently reviewed by ANDREWS (1985), *Proconsul* has traditionally been viewed as a hominoid. There has been little consensus on the phylogenetic position of the genus within the Hominoidea, however, and HARRISON (1987) has suggested *Proconsul* may be a basal catarrhine, less closely related to extant hominoids than extant cercopithecoids are. We feel that several shared derived features unite *Proconsul* with later hominoids, including the features indicated by ANDREWS (1985: Table 2). Other shared derived characters uniting *Proconsul* with hominoids include the relatively high encephalization quotient of the genus (WALKER *et al.*, 1983), the presence of frontal sinuses, and sacral morphology. Therefore, we feel that the traditional hominoid status of *Proconsul* is well-founded on both cladistic and phenetic bases.

While we firmly believe that *Proconsul* should be considered, as it has been since its original discovery, as a hominoid, evidence is accumulating which supports a very basal position for this genus within the superfamily (ANDREWS, 1985; ANDREWS & MARTIN, 1987; HARRISON, 1987). In particular, our confirmation of an ulnocarpal articulation in *Proconsul* takes away one of the most important postcranial synapomorphies previously suggested to link *Proconsul* with extant hominoids. Other aspects of carpal morphology corroborate the very primitive condition of this joint complex in *Proconsul:* (1) the radial facet of the lunate lacks any mediolateral expansion; (2) the radial facet of the lunate is oriented in nearly the same plane as the facet for the scaphoid on this bone, whereas in all extant hominoids these facets are basically orthogonal to one another; (3) the proximal articular facets on the trapezium (for articulation with the centrale and the scaphoid tubercle) are neither dorsopalmarly extensive nor confluent with the corresponding facet of the trapezoid; (4) the hamate possesses a very moderately developed hamulus (see O'CONNOR, 1975, 1976). While it is conceivable that these traits could have evolved in parallel among the various extant lineages of hominoids, this possibility appears to us to be highly unlikely. Therefore, the evidence from the anatomy of the carpal joints supports the placement of *Proconsul* as the sister taxon of extant Hominoidea, including Hylobatidae (ANDREWS, 1985; ANDREWS & MARTIN, 1987; see also GEBO *et al.*, 1988). Future study of other postcranial bones of *Proconsul* should provide a good test of this phylogenetic hypothesis.

Comments on the early evolution of hominoid postcranial adaptations

Based largely on LEWIS' scenario of early hominoid locomotor adaptations, several workers have suggested that brachation (or bimanual suspension in general) could have played an important role in the initial radiation of the Hominoidea (LEWIS, 1971, 1972a, 1972b; RIPLEY, 1979; TEMERIN & CANT, 1983). CARTMILL & MILTON (1977) suggest that adaptations for slow-climbing among early hominoids are more likely to have been involved, although they do not completely discount the possibility of some suspensory behaviors among early hominoids: «A large slow-climbing quadruped which had already developed long and mobile forelimbs... Would develop increasingly suspensory habits if it continued to increase in body size» (CARTMILL & MILTON, 1977: 269). However, since FLEAGLE *et al.* (1981) have demonstrated that similar biomechanical and musculoskeletal demands are placed on the forelimb in both climbing and brachiation (see also PREUSCHOFT, 1973a, b), and since the term «brachiation» means something much more specific today than it did a decade or so ago (LEWIS, 1985), many of the apparent differences in these two scenarios disappear. In either case, the uniquely hominoid features of the wrist are most logically explained as adaptations for more antipronograde (STERN, 1975) positional behaviors.

Antipronograde susensory behaviors are often felt to be imperative if large animals are to maintain an arboreal existence (STERN, 1975; CARTMILL & MILTON, 1977; RIPLEY, 1979). CARTMILL & MILTON (1977: 269) have stated this relationship most explicitly: «The importance of suspensory locomotion and postures in an arboreal mammal's behavioral repertoire should vary as a function of its body weight...». Obviously, living pongids possess larger body size than do other living anthropoids, and even hylobatids fall in the upper size range of living non-hominoid anthropoids. Therefore, one could argue that the derived hominoid carpal features allowed the radiation of hominoids into the larger size distribution among anthropoids that they presently enjoy. In fact, such an interpretation is implicit in many scenarios of early hominoid locomotor evolution.

To what extent is the fossil record of early hominoid evolution consistent with such a hypothesis? Basically, the new paleontological evidence argues that the opposite is true. During the early Miocene, species of *Proconsul* exhibited a range in size approaching that of living hominoids (GINGERICH *et al.*, 1982; WALKER & PICKFORD, 1983; FLEAGLE & KAY, 1985; CONROY, 1987; RUFF, in press). Yet, from the available evidence of the hand of *P. nyanzae*, the intermediate-sized species of the genus, there is no indication whatsoever of increased capacities of suspensory behavior compared to *P. africanus*. If this pattern holds for the rest of the postcranium of this species (and for the largely unknown *P. major*), it would appear that the acquisition of relatively large body size among early hominoids *preceded* the evolution of modern hominoid carpal specializations.

Clearly, there are at least three possible explanations for such a situation: 1) the larger species of *Proconsul* were more terrestrially adapted than *P. africanus* (SIMONS & PILBEAM, 1972; FLEAGLE, 1983; FLEAGLE & KAY, 1985); 2) *P. nyanzae* was capable of suspensory behaviors not revealed in its carpal morphology; 3) *P. nyanzae* had similar locomotor adaptations to *P. africanus*, and both were generalized arboreal quadrupeds. The first of these possibilities is conceivable, and perhaps even probable in the case of *P. major*, but little paleoenvironmental or anatomical evidence has been presented to support the hypothesis. The second possibility also exists, but is untestable given our present state of knowledge of the postcranium of *P. nyanzae*. At present we feel most comfortable with the third of the possibilities listed above. The known corresponding bones in *Proconsul nyanzae* and *P. africanus* are extremely similar to each other in all respects except for size (WALKER & PICKFORD, 1983; BEARD *et al.*, 1986). Furthermore, we do not think that the possibility of a 23-40 kg primate (GINGERICH *et al.*, 1982; WALKER & PICKFORD, 1983; CONROY, 1987; RUFF, in press) being an above-branch quadruped is so unlikely that one must resort to one of the other two possibilities on the basis of size alone. Nevertheless, as the recent advances in Miocene hominoid field paleontology remind us, much remains to be learned about these enigmatic animals. We look forward to future work, in the hope that more synthetic analyses of the locomotor and postural adaptations of the earliest members of our superfamily may yield further insights into a way of life that cannot be exactly matched among living primates.

ACKNOWLEDGEMENTS — We thank the government of Kenya and the trustees of the National Museums of Kenya for permission to carry out the research. We are also grateful to R.E.F. Leakey, and to Bwana Kamoya Kimeu and his team. For access to comparative collections, we thank R.W. Thorington, Jr. and Linda Gordon (National Museum of Natural History. Smithsonian Institution). Elaine Kasmer skillfully prepared figures 5 and 6. M.D. Rose kindly reviewed the manuscript and made helpful suggestions. Linda Szabelski typed the manuscript and provided logistic help while we were in the field, for which we are grateful. A.W. and M.F.T. were supported by NSF grant BNS 8418567. K.C.B. was supported by a predoctoral fellowship from NSF and by funds from the Lucille P. Markey Foundation. Finally, we are most grateful to Prof. Holger Preuschoft for inviting us to participate in the symposium, which stimulated this research work.

References

ANDREWS P., 1978. *A revision of the Miocene Hominoidea of East Africa.* Bull. Br. Mus. Nat. Hist. (Geol.), 30: 85-224.

ANDREWS P., 1981. *A short history of Miocene field paleontology in western Kenya.* J. Hum. Evol., 9: 3-9.

ANDREWS P., 1985. *Family group systematics and evolution among catarrhine primates.* In: Ancestors: the Hard Evidence. E. Delson, (ed.), Alan R. Liss, New York, pp. 14-22.

ANDREWS P. & Groves C.P., 1976. *Gibbons and brachiation.* In: D.M Rumbaugh (ed.) Gibbon & Siamang, 4: 167-218.

ANDREWS P. & MARTIN L., 1987. *Cladistic relationships of extant and fossil hominoids.* J. Hum. Evol., 16: 101-118.

AVIS V., 1962. *Brachiation: the crucial issue for man's ancestry.* Southwest J. Anthropol., 18: 119-148.

BEARD K.C., TEAFORD M.F. & WALKER A., 1986. *New wrist bones of Proconsul africanus and P. nyanzae from Rusinga Island, Kenya.* Folia Primatol., 47: 97-118.

CARTMILL M. & MILTON K., 1977. *The lorisiform wrist joint and the evolution of «brachiating» adaptations in the Hominoidea.* Am. J. Phys. Anthropol., 47: 249-272.

CONROY G.C., 1987. *Problems of body-weight estimation in fossil primates.* Int. J. Primatol., 8: 115-137.

CONROY G.C. & FLEAGLE J.G., 1972. *Locomotor bahaviour in living and fossil pongids.* Nature, 237: 103-104.

CORRUCCINI R.S., CIOCHON R.L. & MCHENRY H.M., 1975. *Osteometric shape relationships in the wrist joint of some anthropoids.* Folia Primatol., 24: 250-274.

CORRUCCINI R.S., CIOCHON R.L. & MCHENRY H.M., 1976. *The postcranium of Miocene hominoids: were dryopithecines merely «dental apes»?* Primates, 17: 205-223.

DAVIS P.R. & NAPIER J., 1963. *A reconstruction of the skull of Proconsul africanus.* Folia Primatol., 1: 20-28.

DRAKE R.L., VAN COUVERING J.A., PICKFORD M., CURTIS G.H. & HARRIS J.A., 1988. *New chronology for the early Miocene mammalian faunas of Kisingiri, western Kenya.* J. Geol. Soc. Lond., 145: 479-491.

ETTER H.-U.F., 1978. *Lorisiform hands and their phylogenetic impplications: a preliminary report.* In: Recent Advances in Primatology. vol 3. Evolution. D.J. Chivers and K.A. Joysey (eds.), Academic Press, London, pp. 161-170.

FLEAGLE J.G., 1983. *Locomotor adaptations of Oligocene and Miocene hominoids and their phyletic implications.* In: New interpretations of Ape and Human Ancestry. R.L. Ciochon and R.S. Corruccini (eds.), Plenum. New York, pp. 301-324.

FLEAGLE J.G. & KAY R.F., 1985. *The paleobiology of catarrhines.* In: Ancestors: The Hard Evidence. E. Delson (ed.), Alan R. Liss. New York, pp. 23-36.

FLEAGLE J.G., STERN J.T., JR., JUNGERS W.L., SUSMAN R.L., VANGOR A.K. & WELLS J.P., 1981. *Climbing: a biomechanical link with brachiation and with bipedalism.* Symp. Zool. Soc. Lond., 48: 359-375.

GEBO D.L., BEARD K.C., TEAFORD M.F., WALKER A., LARSON S.G., JUNGERS W.L. & FLEAGLE J.G., 1988. *A hominoid proximal humerus from the early Miocene of Rusinga Island, Kenya.* J. Hum. Evol., 17: 393-401.

GINGERICH P.D., SMITH B.H. & ROSENBERG K., 1982. *Allometric scaling in the dentition of primates and prediction of body weight from tooth size in fossils.* Am. J. Phys. Anthropol., 58: 81-100.

GODINOT & BEARD K.C., this volume.

GRAND T.I., 1972. *A mechanical interpretation of terminal branch feeding. J. Mamm., 53: 198-201.*

HARRISON T., 1986. *New fossil anthropoids from the middle Miocene of East Africa and their bearing on the origin of the Oreopithecidae* Am. J. Phys. Anthropol., 71: 265-284.

HARRISON T., 1987. *The phylogenetic relationships of the early catarrhine primates: a review of the current evidence.* J. Hum. Evol., 16: 41-80.

HOPWOOD A.T., 1933. *Miocene primates from East Africa.* Ann. mag. Nat. Hist. 11: 96-98.

JENKINS F.A., JR., 1981. *Wrist rotation in primates: a critical adaptation for brachiators.* Symp. Zool. Soc. Lond., 48: 429-451.

KELLEY J., 1986. *Species recognition and sexual dimorphism in Proconsul and Rangwapithecus.* J. Hum. Evol., 15: 461-495.

LEAKEY R.E. & LEAKEY M.G., 1986a. *A new Miocene hominoid from Kenya.* Nature, 324: 143-146.

LEAKEY R.E. & LEAKEY M.G., 1986b. *A second new Miocene hominoid from Kenya.* Nature, 324: 146-148.

LEAKEY R.E. & LEAKEY M.G., 1987. *A new Miocene small-bodied ape from Kenya.* J. Hum. Evol., 16: 369-387.

LEAKEY R.E., LEAKEY M.G. & WALKER A.C., 1988a. *Morphology of Turkanapithecus kalakolensis from Kenya.* Am. J. Phys. Anthropol., 76: 277-288.

LEAKEY R.E., LEAKEY M.G. & WALKER A.C., 1988b. *Morphology of Afropithecus turkanensis from Kenya.* Am. J. Phys. Anthropol., 76: 289-307.

LEWIS O.J., 1969. *The hominoid wrist joint.* Am. J. Phys. Anthropol., 30: 251-268.

LEWIS O.J., 1971. *Brachiation and the early evolution of the Hominoidea.* Nature, 230: 577-578.

LEWIS O.J., 1972a. *Evolution of the hominoid wrist.* In: The Functional and Evolutionary Biology of Primates. R.H. Tuttle (ed.), Aldine-Atherton. Chicago, pp. 207-222.

LEWIS O.J., 1972b. *Osteological features characterizing the wrist of monkeys and apes, with a reconsideration of this region in Dryopithecus (Proconsul) africanus.* Am. J. Phys. Anthropol., 36: 45-58.

LEWIS O.J., 1974. *The wrist articulations of the Anthropoidea.* In: Primate Locomotion. F.A. Jenkins, Jr, (ed.), Academic, New York, pp. 143-169.

LEWIS O.J., 1977. *Joint remodelling and the evolution of the human hand.* J. Anat., 123: 157-201.

LEWIS O.J., 1985. *Derived morphology of the wrist articulations and theories of hominoid evolution. Part I. The lorisine joints.* J. Anat., 140: 447-460.

MARTIN L., 1981. *New specimens of Proconsul from Koru, Kenya.* J. Hum. Evol., 10: 139-150.

MCHENRY H.M. & CORRUCCINI R.S., 1983. *The wrist of Proconsul africanus and the origin of hominoid postcranial adaptations.* In: New Interpretations of Ape and Human Ancestry. R.L. Ciochon and R.S. Corruccini (eds.), Plenum, New York, pp. 353-367.

MORBECK M.E., 1975. *Dryopithecus africanus forelimb.* J. Hum. Evol., 4: 39-46.

MORBECK M.E., 1977. *The use of casts and other problems in reconstructing the Dryopithecus (Proconsul) africanus wrist complex.* J. Hum. Evol., 6: 65-78.

NAPIER J.R. & DAVIS P.R., 1959. *The forelimb skeleton and associated remains of Proconsul africanus.* Fossil Mammals of Africa (Br. Mus. Nat. Hist.), 16: 1-69.

NESBIT EVANS E.M., VAN COUVERING J.A.H. & ANDREWS P., 1981. *Palaeoecology of Miocene sites in western Kenya.* J. Hum. Evol., 10: 99-116.

O'CONNOR B.L., 1975. *The functional morphology of the cercopithecoid wrist and inferior radioulnar joints, and their bearing on some problems in the evolution of the Hominoidea.* Am. J. Phys. Antropol., 43: 113-122.

O'CONNOR B.L., 1976. *Dryopithecus (Proconsul) africanus: quadruped or non-quadruped?* J. Hum. Evol., 5: 279-283.

O'CONNOR B.L. & RAREY K.E., 1979. *Normal amplitudes of radioulnar pronation and supination in several genera of anthropoid primates.* Am. J. Phys. Anthropol., 51: 39-44.

PICKFORD M., 1983. *Sequence and environments of the lower and middle Miocene hominoids of western Kenya.* In: New Interpretations of Ape and Human Ancestry. R.L. Ciochon and R.S. Corruccini (eds.), Plenum, New York. pp. 421-439.

PREUSCHOFT H., 1970. *Functional anatomy of the lower extremity.* In: The Chimpanzee, G.H. Bourne (ed.), vol. 3. Karger, Basel, pp. 221-294.

PREUSCHOFT H., 1973a. *Functional anatomy of the upper extremity.* In: The Chimpanzee, G.H. Bourne (ed.), vol. 6 Karger, Basel, pp. 34-120.

PREUSCHOFT H., 1973b. *Body posture and locomotion in some East African Miocene Dryopithecinae.* In: Human Evolution. M.H. Day (ed.), Taylor and Francis, London, pp. 13-46.

RIPLEY S., 1979. *Environmental grain, niche diversification, and positional behavior in Neogene primates: an evolutionary hypothesis.* In: Environment, Behavior, and Morphology: Dynamic Interactions in Primates. M.E. Morbeck, H. Preuschoft, and N. Gomberg (eds.), Gustav Fischer, New York. pp. 75-93.

ROBERTSON M.L., 1984. *The carpus of Proconsul africanus: functional analysis and comparison with selected nonhuman primates.* Ph.D. Dissertation, University of Michigan, Ann Arbor.

ROSE M.D., 1983. *Miocene hominoid postcranial morphology: monkey-like, apelike, neither, or both?* In:

New Interpretations of Ape and Human Ancestry. R.L. Ciochon and R.S. Corruccini (eds.), Plenum, New York, -pp. 405-417.

ROSE M.D., 1988. *Another look at the anthropoid elbow.* J. Hum. Evol., 17: 193-224.

RUFF C.B., 1990. *Body mass and hind limb bone cross-sectional and articular dimensions in anthropoid primates.* In: Body Size in Mammalian Paleobiology. J. Damuth and B.J. MacFadden (eds.), Cambridge University Press, Cambridge, pp. 119-149.

SCHAEFFER B., 1948. *The origin of a mammalian ordinal character.* Evolution, 2: 164-175.

SCHÖN M.A. & ZIEMER L.K., 1973. *Wrist mechanism and locomotor behavior of Dryopithecus (Proconsul) africanus.* Folia Primatol., 20: 1-11.

SIMONS E.L. & PILBEAM D.R., 1972. *Hominoid paleoprimatology.* In: The Functional and Evolutionary Biology of Primates. R.H. Tuttle (ed.), Aldine-Atherton, Chicago, pp. 36-62.

STERN J.T., JR., 1975. *Before bipedality.* Yrbk. Phys. Anthropol., 19: 59-68.

TEMERIN L.A. & CANT J.G.H., 1983. *The evolutionary divergence of Old World monkeys and apes.* Am. Nat., 122: 335-351.

VAN COUVERING J.H. & VAN COUVERING J.A., 1976. *Early Miocene mammal faunas from East Africa: aspects of geology, faunistics and paleoecology.* In: Human Origins. G.L. Isaac and E.R. McCown (eds.), W.A. Benjamin, Menlo Park. pp. 155-207.

WALKER A., FALK D., SMITH R. & PICKFORD M., 1983. *The skull of Proconsul africanus: reconstruction and cranial capacity.* Nature, 305: 525-527.

WALKER A.C. & PICKFORD M., 1983. *New postcranial fossils of Proconsul africanus and Proconsul nyanzae.* In: New Interpretations of Ape and Human Ancestry. R.L. Ciochon and R.S. Corruccini (eds.), Plenum. New York, pp. 325-351.

WALKER A. & TEAFORD M. 1988. *The Kaswanga primate site: An early Miocene hominoid site on Rusinga Island, Kenya.* J. Hum. Evol., 17: 539-544.

WALKER A.C., TEAFORD M.F. & LEAKEY R.E., 1986. *New information concerning the R114 Proconsul site, Rusinga Island, Kenya.* In: Proceedings of the Xth Congress of the International Primatological Society. vol 1. Primate Evolution. J.G. Else and P.C. Lee (eds.), Cambridge Univ. Press. Cambridge, pp. 143-149.

WHITWORTH T., 1953. *A contribution to the geology of Rusinga Island.* Quart. J. Geol. Soc. London., 109: 75-96.

ZWELL M. & CONROY G.C., 1973. *Multivariate analysis of the Dryopithecus africanus forelimb.* Nature, 244: 373-375.

B. Christ
H.J. Jacob
B. Brand-Saberi
M. Grim

On the development of the human hand *

Morphogenesis (shaping), cell differentiation and the spatial arrangement of different tissues (pattern formation) are found to be basic events of limb and hand development. Hand shaping is the result of AER-mesoderm interactions, regionally different cell proliferation and cell death. Cell diversification and pattern formation in the developing hand, on the one hand, occur due to the differentiation of pluripotent somatopleure-derived limb bud cells, which is controlled by positional information. On the other hand, these events are found to be the results of immigration and goal recognition of different, already determined cells. Further investigations, which must include the mechanisms at the molecular level, are to be done in order to find out how the different developmental events are connected and the genetic programme becomes realized.

Key words: cell differentiation, human hand, morphogenesis, myogenesis, pattern formation.

Introduction

In this book many facts have been collected that are concerned with the comparative anatomy and functions of the hand. Complex movements, such as grasping or piano-playing, may give an idea of the functional multiplicity of the human hand. The extraordinary sensibility and capability of different movements correspond to the extensive areas which represent the hand on the cerebral cortex. The functions of the human hand are the result of a multi- stage developmental process during ontogenesis, which can be characterized by different events.

The successive steps of hand development, which starts at the early limb bud and which results in an enormous complexity, have already been described in detail by Keibel & Elze (1908), Hochstetter (1955) and Blechschmidt (1951, 1956, 1961). Jacob & Hinrichsen (1984) presented a scanning electron microscopic documentation of the development of the human arm and leg.

However, these descriptive observations are not sufficient to clarify the basic problems of limb development. Thus, results have to be taken into account which were obtained experimentally from embryos of higher vertebrates. Christ & Jacob (1986) and Christ *et al.* (1986 and 1987) compared similar limb rudiments taken from human embryos and those of other vertebrate classes. According to these authors, shaping, cell differentiation, and spatial arrangement of different cells are the basic processes of limb development, which take place in limb buds of different types and different species in a quite similar way. Species-specific differences only concern the shape of limbs and the spatial arrangement of different limb tissues. It is the aim of this paper to review the

* This paper is dedicated to Professor Dr. med. E. Blechschmidt, our teacher of Human Embryology.

particular features of the development of the human hand and to link them with experimental results of general validity.

Results

In early human arm buds, which are first visible at Carnegie-stage 12 (4.2 mm), there is no border between the anlagen of hand and forearm. The bud, initially pointing laterally shows a rapid outgrowth and becomes increasingly adducted. *Figure 1* shows the arm bud of an 8 mm embryo. At this stage, first signs of a proximo-distal subdivision can be found. The hand plate is already flattened and can be delimited from the stalk-like forearm. It is characteristic of the human arm development that the hand turns more ulnarly, compared with the axis of the forearm. The mesodermal core of the hand rudiment consists of apparently undifferentiated, mesenchymally arranged cells and blood vessels of a capillary type (*Figure 2*).

The metacarpal and finger rays can be clearly identified at the distal edge of the hand plate in human embryos ranging from 12 -15 mm. *Figure 3* shows the hand anlage of a 17.5

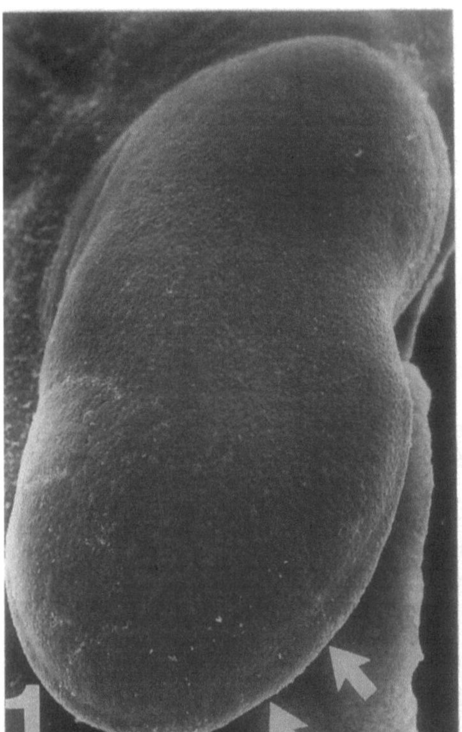

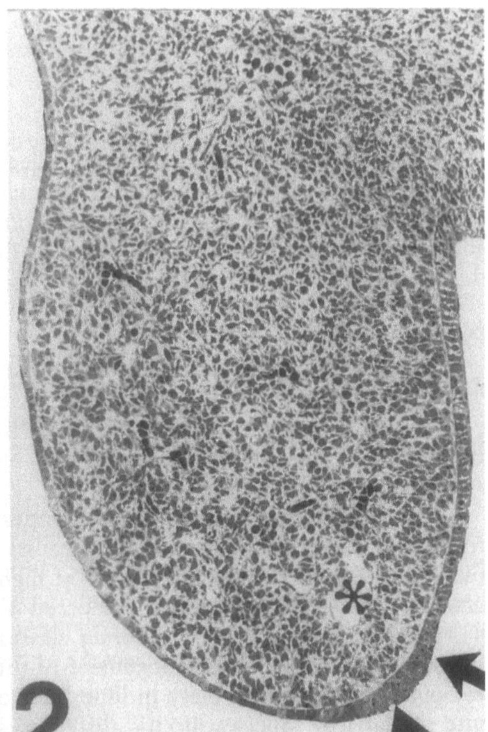

Figure 1. — Scanning electron micrograph (SEM) of the arm anlage, 8mm human embryo. The hand is demarcating from the forearm. Arrows: apical ectodermal ridge (AER).

Figure 2 — Section of the arm anlage, 8 mm human embryo. Asterisk: marginal sinus; arrows: AER.

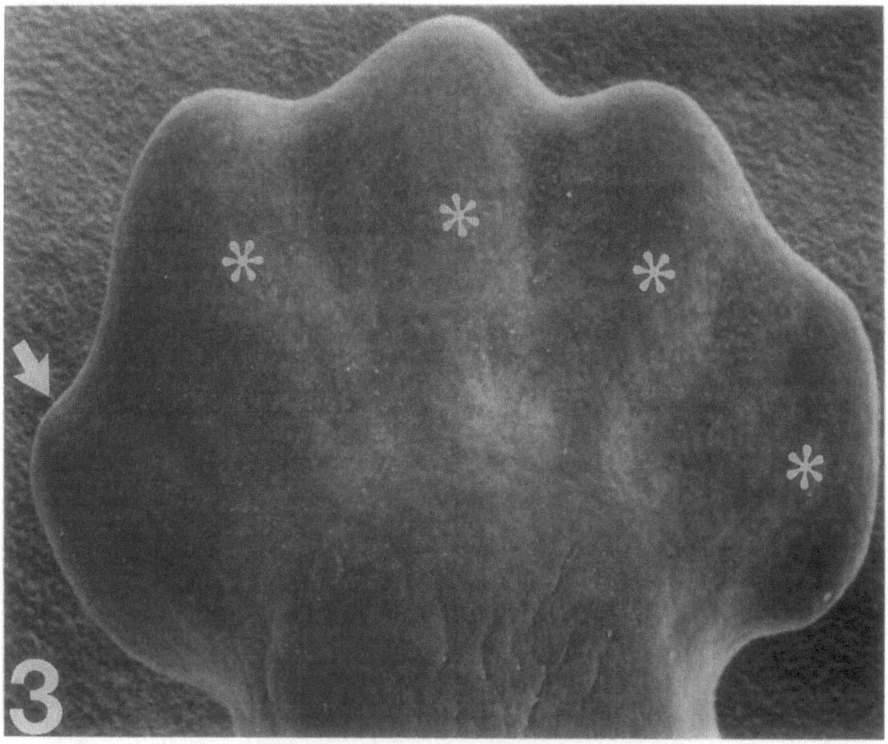

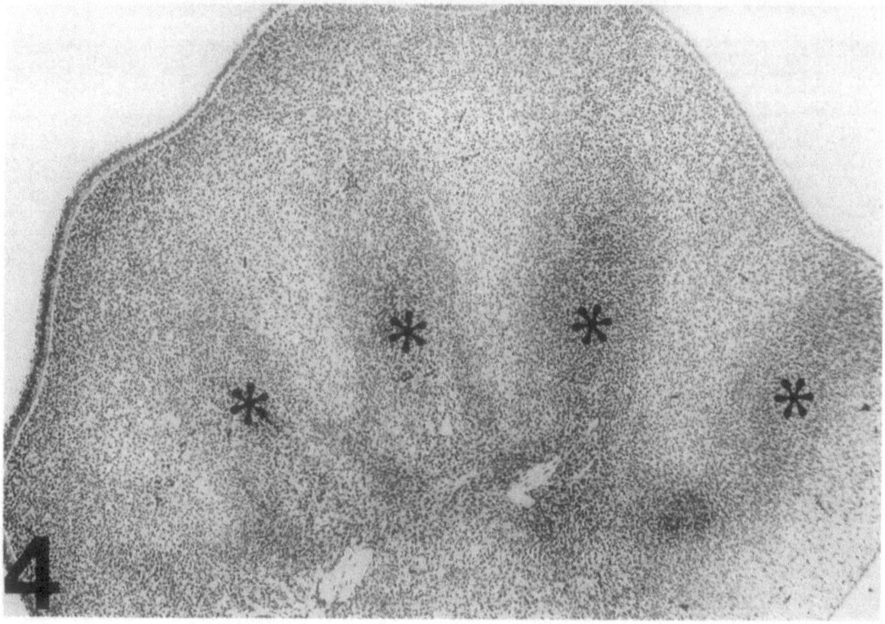

Figure 3 — SEM of the hand anlage, 17,5 mm human embryo. Arrow: thumb anlage; asterisks: finger rays.

Figure 4 — Section of the hand plate, 18 mm human embryo. Asterisks: cartilage rudiments of metacarpalia and fingers.

mm human embryo. In this developmental stage, the five finger rays are already clearly outlined and the thumb bud has reached its characteristic position of radial abduction. Inside the hand plate (*Figure 4*) different zones can be distinguished. Each finger ray is characterized by a tissue condensation representing precartilage and cartilage blastemas from which the skeletal elements of the metacarpus and the fingers develop.

During further development the hand plate changes its position in a way that it comes into pronation. The shaping and separation of the fingers are completed. The fingers get a position of flexion and the thumb shifts into an opposition to the fingers (*Figures 5-7*).

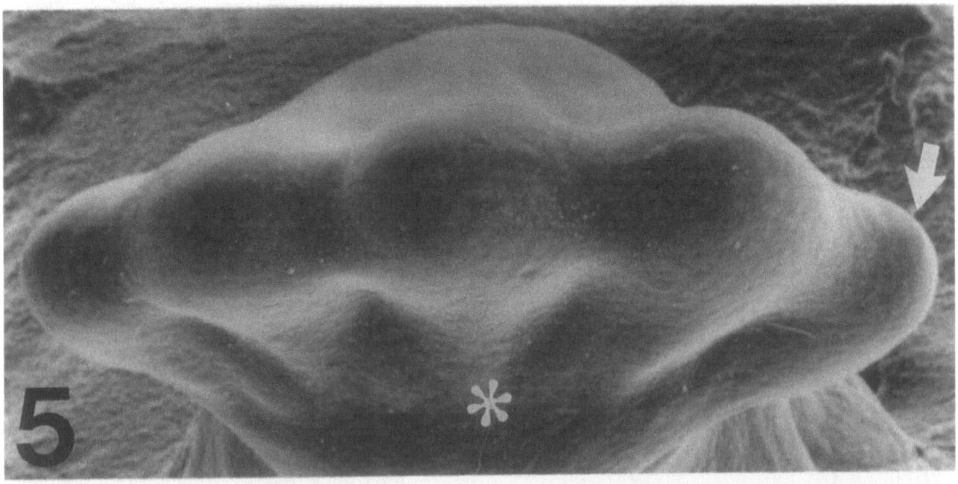

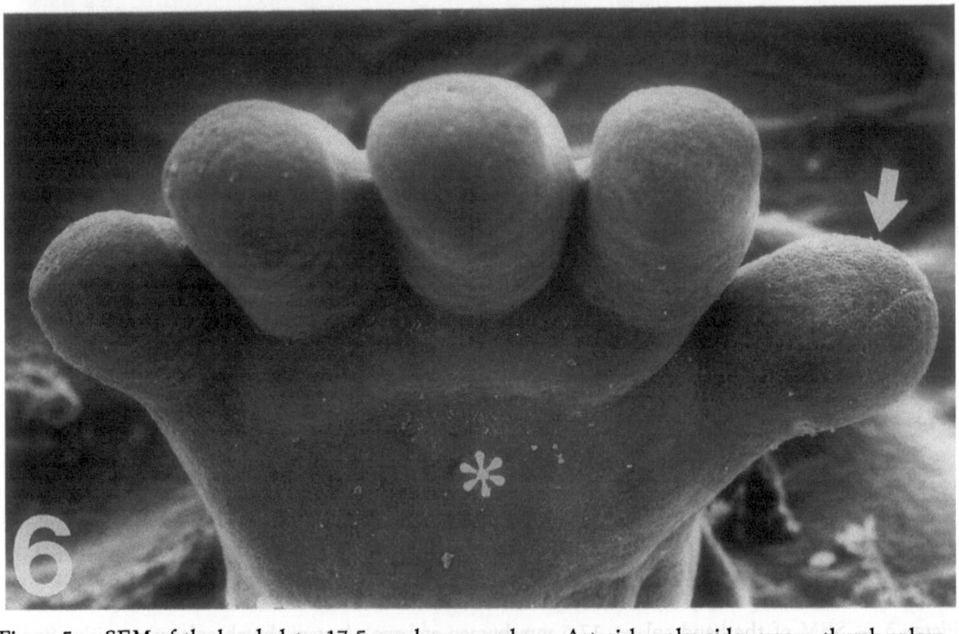

Figure 5 — SEM of the hand plate, 17,5 mm human embryo. Asterisk: volar side; arrow: thumb anlage.

Figure 6 — SEM of the hand anlage, 19 mm human embryo. Asterisk: volar side; arrow: thumb anlage.

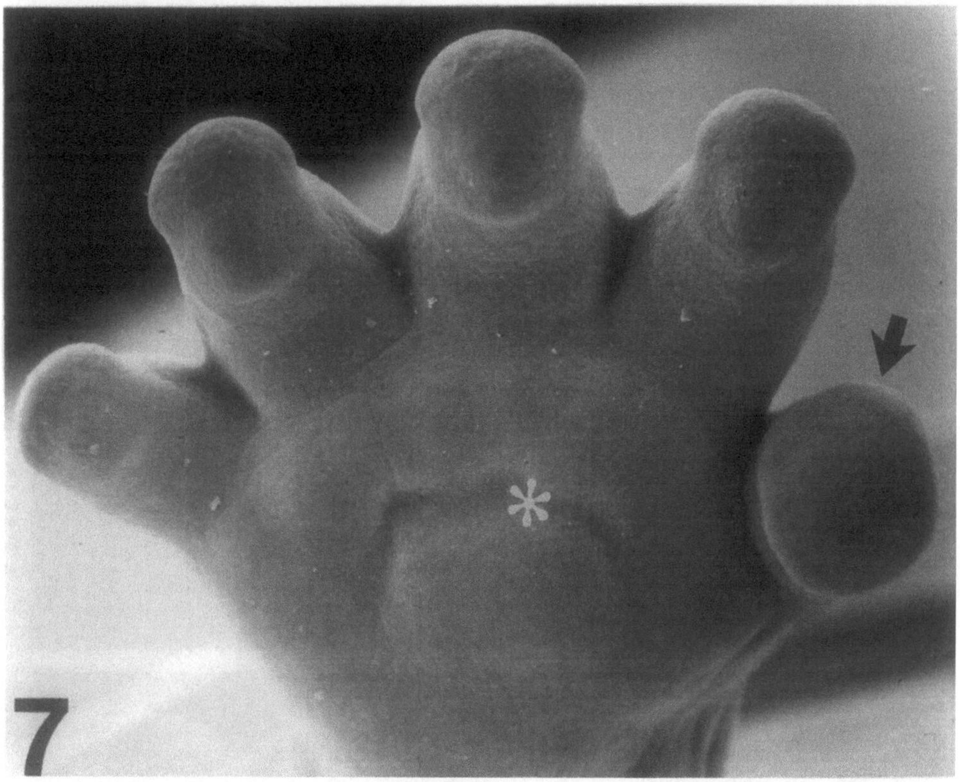

Figure 7 — SEM of the hand anlage, 28 mm human embryo. Note the opposition of the thumb (arrow); asterisk: volar side.

Due to the growth and the developmental movements the fetal hand has reached a position which is similar to that of a grasping hand. (*Figures 8 and 9*). This view is in line with the statement of Blechschmidt (1961) «that developmental movements anticipate future movements in a primitive way».

It is evident that human hand anlagen of different developmental stages exhibit characteristic changes in shape and structure. In addition to the increasing complexity of shape, various tissues such as cartilage, connective tissue, muscle, nerves, etc. are each characterized by specific spatial arrangements (*Figure 4*). The question arises as to how these different events, morphogenesis, cell differentiation and the spatial arrangement of different cells, are controlled. Looking at hand shaping we have to recall that the early arm anlage, including the hand rudiment, consists of the still epithelial somatopleure and its ectodermal covering. Harrison (1921), Murray (1926) and Detwiler (1933) have shown that the determination of the limb field is a very early event which is restricted to the somatopleure. The somatopleure itself becomes initially stabilized by influences of the paraxial mesoderm (Pinot, 1970, Kieny, 1969, Agnish & Kochhar, 1977, Christ & Jacob, 1986). The somatopleure gives rise to a mesenchyme and induces the formation of the apical ectodermal ridge (AER), which can be seen in the *Figures 10 - 12*. The AER is an important structure which induces and maintains limb bud outgrowth (Saunders, 1948,

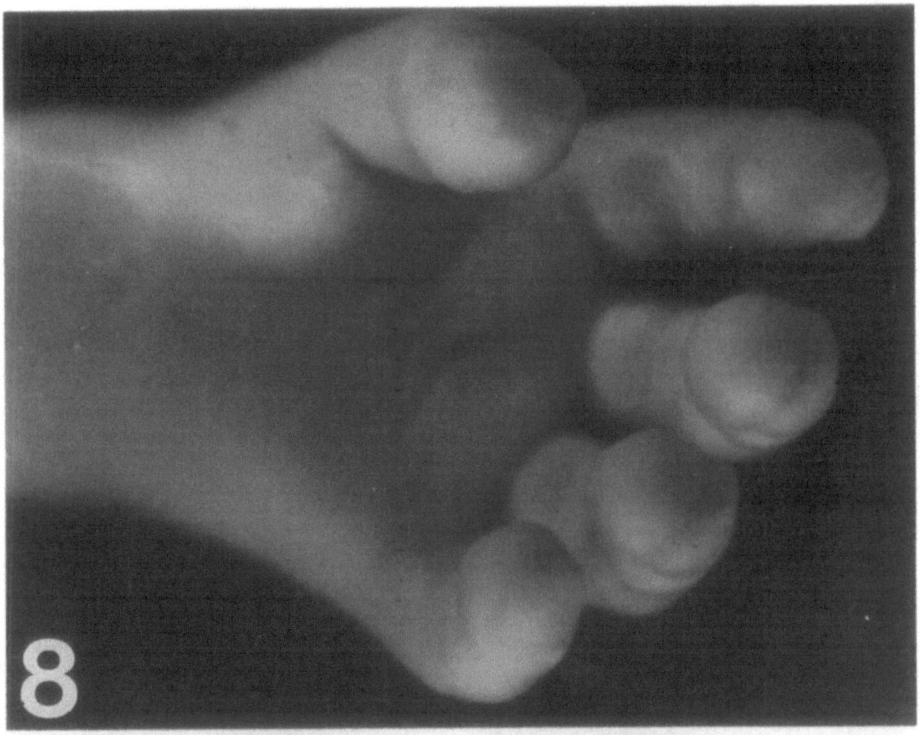

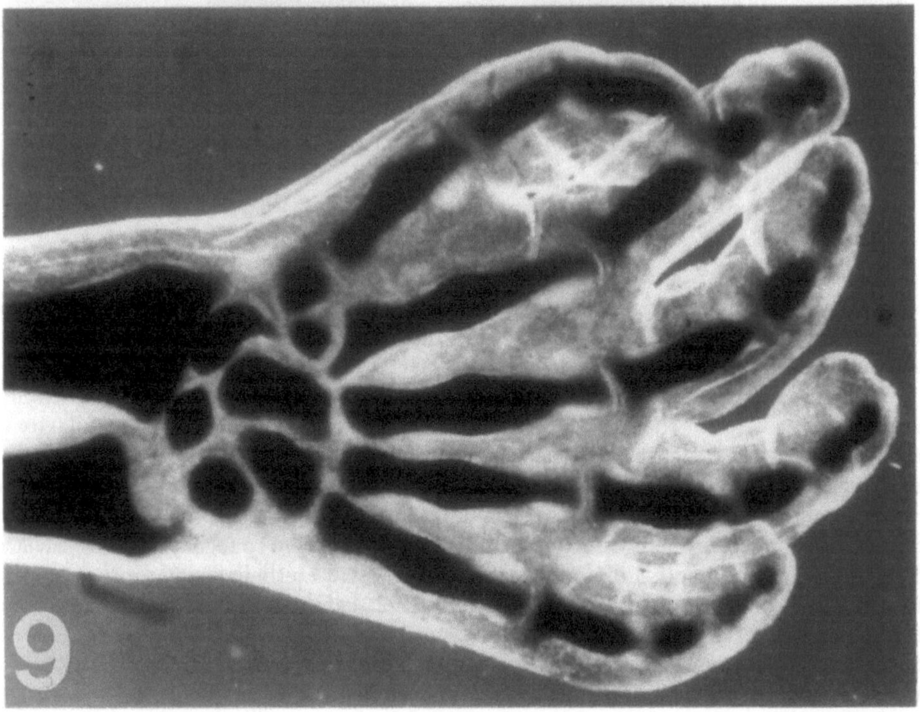

Figure 8 — Hand anlage of a 5 cm human fetus. Note the progressive flexion of the fingers and the opposition of the thumb.

Figure 9 — Hand anlage of a 5,8 cm human fetus. The anlagen of the cartilageneous skeletal elements are visible after staining according to the Lundvall-technique.

ZWILLING, 1955). Its function is to keep the underlying mesoderm (progress zone) in a labile state and at a high level of mitotic activity. Removal of the AER leads to deficiencies of those distal limb parts which have not yet been formed in the progress zone mesoderm (SAUNDERS, 1948, SUMMERBELL, 1974, 1977).

It is interesting to note that the inducing effect of the AER is not a specific one. As was shown by ZWILLING (1955), AERs could be exchanged between wing and leg bud of chicken embryos without affecting the typical development of the wing or the leg. Combinations of limb bud ectoderm from duck or rat embryos with limb bud mesoderm from chick embryos resulted in characteristic chick limbs covered by duck-like or rat-like skin (PAUTOU, 1968, JORQUERA & PUGIN, 1971). The structure and the growth-inducing influence of the AER depend on a factor called «apical ectodermal ridge maintenance factor» (AEMF), which seems to be produced by the mesoderm in the posterior part of the limb bud (SAUNDERS *et al.*, 1958).

The fully-developed human AER consists of 4 - 5 cell layers covered by an outer layer of periderm cells (*Figures 12*). The latter show tight junctions and separate the extracellular compartment of the AER from the amniotic cavity (KELLEY & FALLON, 1976, 1981).

According to these authors, other cells of the AER are connected by gap junctions, which have been found to be important for chemical and electrical coupling. The epithelium of the AER is limited by a continuous basal lamina. Beneath the AER a special composition of the extracellular matrix (ECM) and long processes of mesodermal cells can be seen which, however, do not make contact with AER cells (KELLEY & FALLON, 1976, 1981).

It can be emphasized that the axial elongation of the limb including that of the hand, is controlled by interactions between the AER and the underlying mesoderm. Moreover, CHRIST *et al.* (1986 a) assumed that the flexion of the fingers occurs due to a dorso-ventral shifting of the AER activity during hand development. Another factor which is suggested to help the hand to form is the occurrence of cell death in different zones. The reduction of the number of digits in birds might go back to the distinct anterior and posterior necrotic zones (ANZ, PNZ), which limit the amount of distal mesenchyme available for digit formation. According to MILAIRE (1977), the ANZ seen in rat, mouse and man is absent in the mole forelimb, which consequently forms an additional anterior digit. Mutants of the chicken, with different cell degeneration activities in both the ANZ and PNZ, show different numbers of digits (HINCHLIFFE, 1974). In the talpid mutant the ANZ and PNZ are lacking and the resulting limbs are polydactylous exhibiting up to 8 digits (HINCHLIFFE & EDE, 1967, CAIRNS, 1977). In the chicken mutant wingless, on the other hand, an increased size of the ANZ leads to deficiencies of the preaxial digits.

Figure 13 shows degenerating cells within the interdigital spaces of a rat embryo. These interdigital necrotic zones (INZ) have been found in all amniotes, including human embryos (MENKES, *et al.* 1965, KELLEY, 1970). Their morphogenetic role is to separate and shape the digits. An experimental inhibition of interdigital cell death, which, for example, can be achieved by the use of Janus Green leads to syndactylies (MENKES & DELEANU, 1964). *Figures 14 and 15* show interdigital areas of a developing human hand plate.

According to KELLEY & FALLON (1981), the AER is actively involved in finger separation: the basal cells contain bundles of microfilaments oriented in a radio-ulnar direction. The contraction of these microfilaments might support the initial separation of the fingers. This can be understood if we keep in mind that the interdigitally-situated collagen fibrils extend at right angles from the basal lamina of the ectoderm into the depth of the apical mesoderm, as was shown by KELLEY & FALLON (1981). In temporal

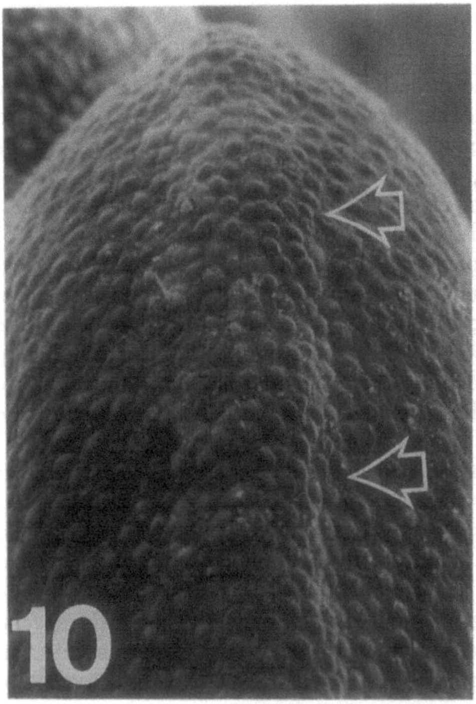

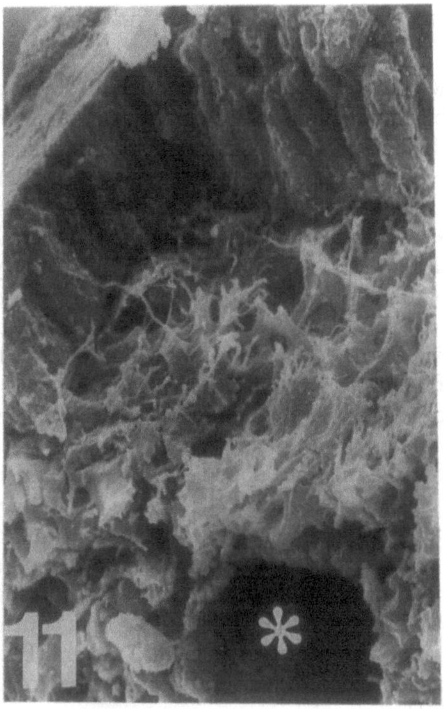

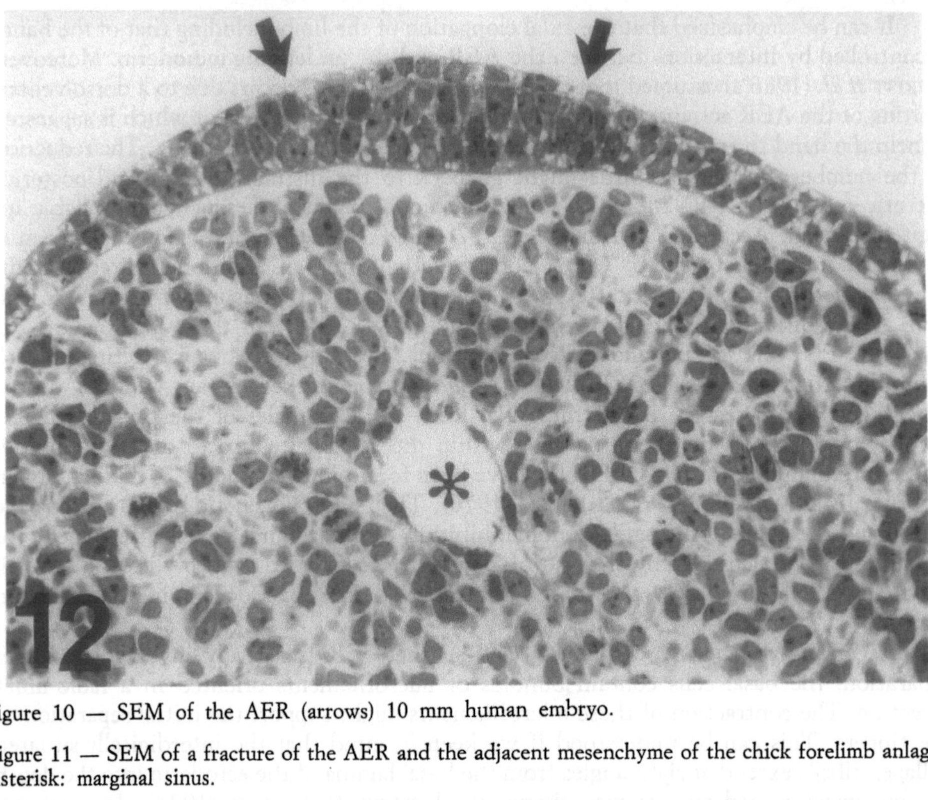

Figure 10 — SEM of the AER (arrows) 10 mm human embryo.

Figure 11 — SEM of a fracture of the AER and the adjacent mesenchyme of the chick forelimb anlage. Asterisk: marginal sinus.

Figure 12 — Transverse section of the hand plate, 10 mm human embryo. Arrows: multilayered AER; asterisk: marginal sinus. Note the mitotic figures in the periderm and the mesenchyme.

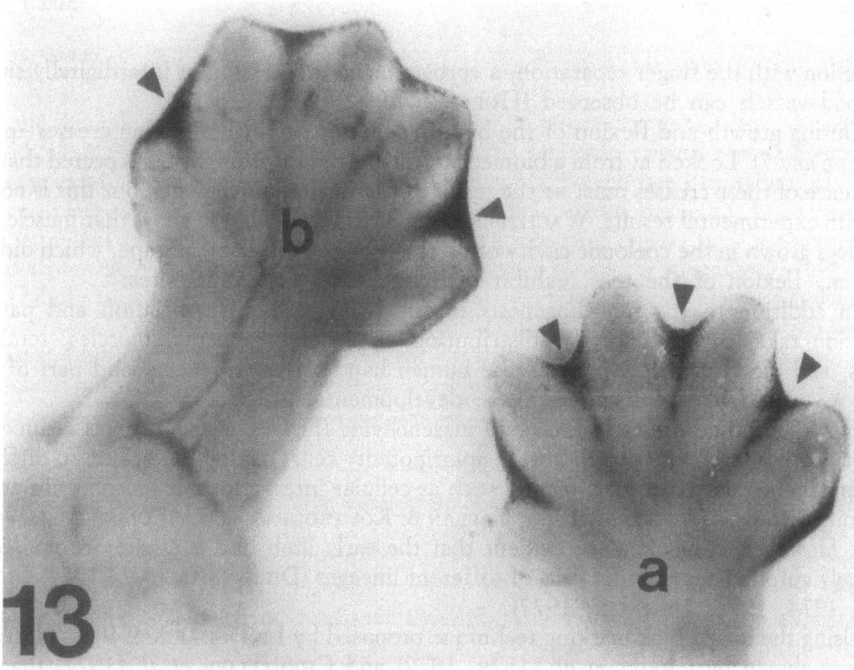

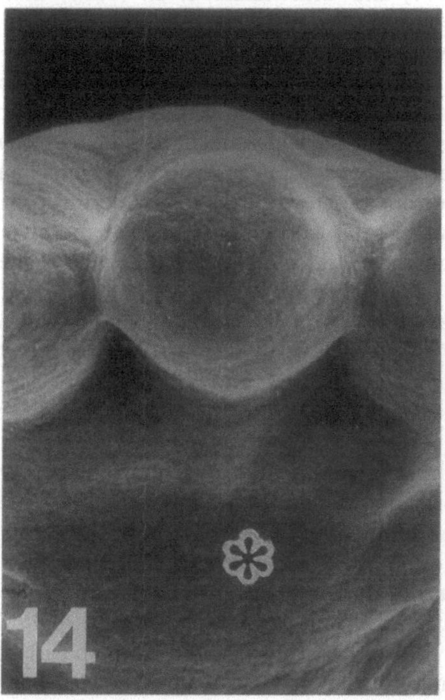

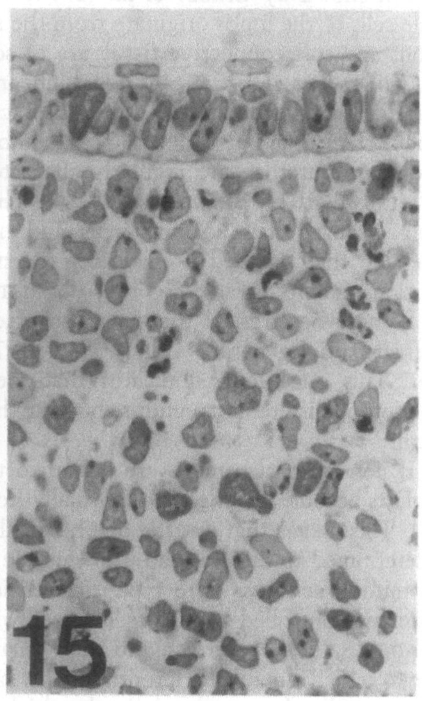

Figure 13 — Whole mount preparation of rat hand (a) and foot (b) plates after in vivo staining with Nile blue sulfate. Arrows: interdigital necrotic zones (INZ).

Figure 14 — Hand anlage of a 19 mm human embryo. Arrows: AER slightly visible. The finger anlagen are clearly separated. Asterisk: volar side.

Figure 15 — Micrograph of the interdigital zone, 18 mm human embryo. Note cell degeneration.

connection with the finger separation, a «programmed» regression of interdigitally situated blood vessels can be observed (HURLE *et al.*, 1985).

During growth and flexion of the hand plate and the fingers flexion creases appear (*Figures 6 and 7*). Looked at from a biomechanical view point it might be expected that the occurrence of these creases must be the result of the flexion movements, but this is not in line with experimental results. WACHTLER *et al.* (1981) were able to show that muscle-free avian legs grown in the coelomic cavity or on the chorioallantoic membrane, which did not show any flexion of the toes, exhibit distinct creases at normal sites.

In addition to the morphogenesis of the hand, cell differentiation and pattern formation take place. A specific distribution of skeletal elements, muscles, tendons, nerves, and blood vessels typical of the human hand arises. In the second part of this paper we would like to focus on these developmental processes.

For a long time the early limb bud mesenchyme (*Figures. 2 and 12*) was assumed to consist of a homogeneous population of «pluripotent» cells, having the option to differentiate under the control of local factors such as cellular interactions or morphogens acting over long distances (ZWILLING, 1966, CAPLAN & KOUTROUPAS, 1973, SUMMERBELL *et al.*, 1973). However, it has become evident that the early limb bud is already composed of relatively autonomous founder cells of different lineages (DIENSTMAN *et al.*, 1974, CHRIST *et al.*, 1974, 1977, NEWMAN, 1977).

Using the quail-chick marking technique proposed by LE DOUARIN & BARQ (1969), it has been shown by CHRIST *et al.* (1974, 1977) and CHEVALLIER *et al.* (1977) that the muscle cells of the limbs originate from the somites (*Figures 16 and 17*), whereas cartilage, tendons and soft connective tissue are of somatopleural origin. This is also true for the hand (CHRIST *et al.*, 1979). When limb anlagen, which had not yet been invaded by somitic cells, were isolated and heterotopically cultivated, they were often found to be normally shaped and articulated. These limbs exhibit normally arranged skeletal elements and dermis, but they are regularly characterized by a total absence of skeletal muscle (CHRIST *et al.*, 1977, BRAND *et al.*, 1985). The zone between cartilage and dermis contains loosely arranged mesenchymal cells and numerous blood vessels and might give the impression that the bed is made for the muscle cells which under these experimental conditions fail to invade the bud (CHRIST *et al.*, 1986 b).

It must be stressed that in muscle-free limbs normally arranged tendons develop which later on degenerate (CHRIST *et al.*, 1979, KIENY & CHEVALLIER, 1979, BRAND *et al.*, 1985). This once more indicates that mechanical factors are not necessary for the formation of different structures, but for the maintenance and the adaption of different structures.

If non-brachial somites are joined to the arm rudiment, the actually inappropriate myoblasts become included into normally- formed arm muscles (JACOB *et al.*, 1983, KIENY, 1983). This means that, on the one hand, the somite-derived myoblasts in the early limb are determined to differentiate into muscle cells, but, on the other hand, their distribution, mitotic activity, fibre type differentiation, and later arrangement into individual muscles are controlled by factors localized within the limb itself (GRIM *et al.*, 1989). GRIM & WACHTLER (1990) were able to show that the epiblast cells isolated from an early quail embryo, which do not differentiate into muscle cells, can make up a normally-shaped extensor medius longus muscle in the chick. This means that shaping and differentiation of muscle are different events.

The early limb bud grows by proliferation of the distal mesenchyme and the myogenic cells are forced to colonize the bud from proximal to distal. BRAND-SABERI *et al.* (1989) have offered experimental evidence that the differentiation state («juvenility») of

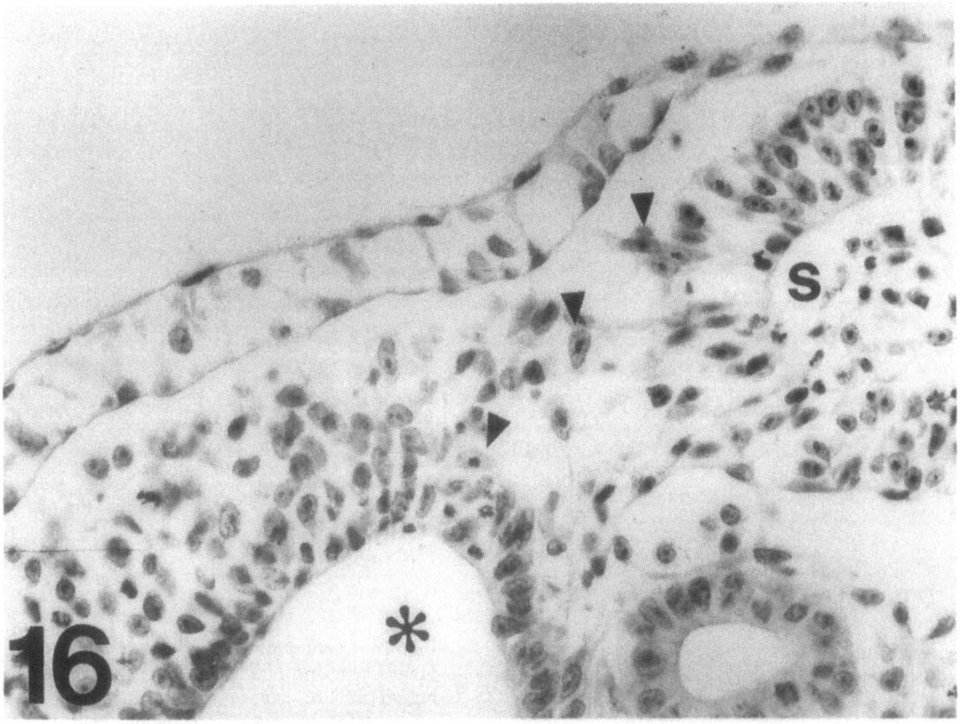

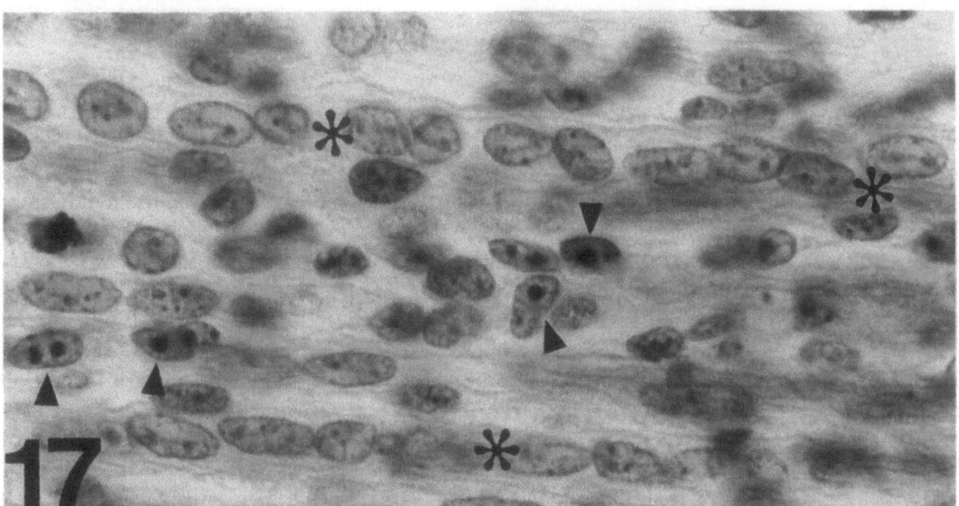

Figure 16 — Transverse section of a 3-day chick embryo. Somites (s) of the wing level have been replaced by corresponding quail somites. Arrows: migrating myogenic cells containing quail nuclei; asterisk: coelomic cavity.

Figure 17 — Longitudinal section of a limb muscle of a chick embryo 7 days after replacement of limb somatopleure by quail somatopleure. The myotubes (asterisks) contain chick nuclei, whereas the muscular connective tissue is made up of quail cells (arrows). Feulgen-Rossenbeck reaction.

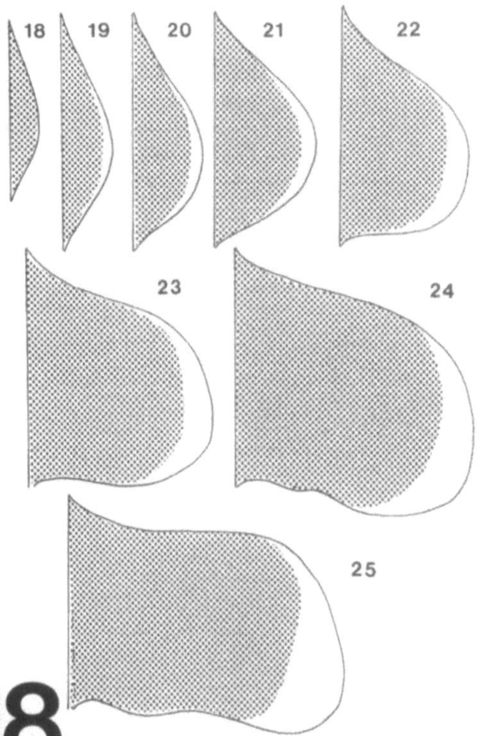

18

Fig. 18. Quail wing buds of the stages 18 - 25 Hamburger and Hamilton (HH) that show the boundaries of myogenic cell invasion. The white areas are free of myogenic cells.

the limb bud mesenchyme controls the directed migration of myogenic cells. *Figure 18* shows the distal boundary of myoblast colonization in limb buds of different stages according to BRAND *et al.* (1985).

In addition to the somite-derived myoblasts, neural crest cells invade the avian limb. They differentiate into Schwann cells, epidermal melanocytes and branched melanocytes situated within the dermis and the muscular connective tissue (CHRIST *et al.*, 1983, WACHTLER, 1984). Endothelial cells and axons complete the limb bud tissues. Experimental studies with avian chimeras have given evidence that the endothelial lining of the blood vessels is constituted by immigrated angioblasts (CHRIST *et al.*, 1979, WILSON, 1983, BRAND *et al.*, 1985). The modalities of angioblast distribution and vessel pattern formation are still under investigation. SEICHERT & RYCHTER (1971), as well as CAPLAN & KOUTROUPAS (1973), have shown that the subectodermal mesenchyme of the limb bud is devoid of blood vessels (*Figure 19*). According to FEINBERG & BEEBE (1983), this lack of vessels might be due to the hyaluronate produced by the ectoderm. However, this subectodermal mesenchyme contains single angioblastic cells which are not yet included in vessel linings (*Figure 20*).

Axons grow on stereotyped highways into the limb bud and the relation between motoneurons and muscle cells displays a high specificity (HOLLYDAY, 1980, LANDMESSER, 1980). According to GRIM *et al.* (1989), a sequence of hierarchically-acting factors controls motoneuron innervation. There are two identical programmes of which the one under neuronal control has priority over that one localized in the muscles. Under normal

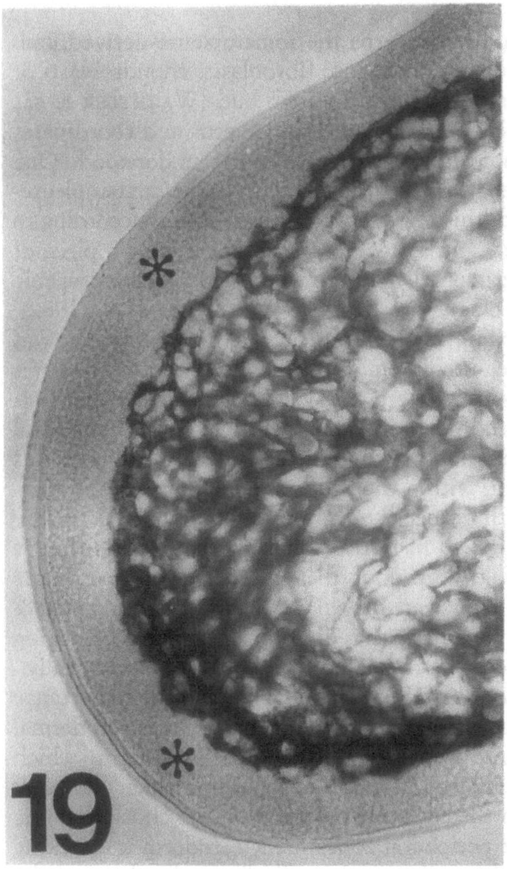

Figure 19 — Wing anlage of a 4-day chick embryo. The capillary network of the growing vessels is marked by injection with Indian ink. Note the zone just below the ectoderm which is devoid of capillaries (asterisks).

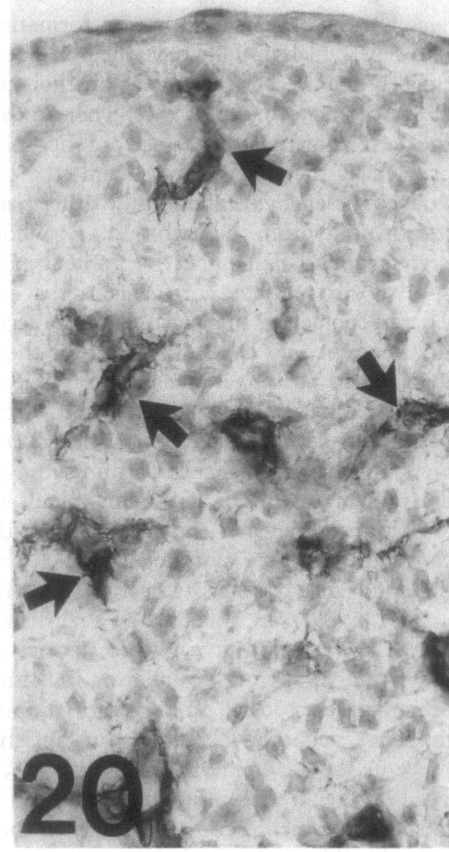

Figure 20 — Section of the avascular zone of a 4-day quail embryos. Angiogenic cells (arrows). Stained with MB1-antibody (kindly provided by B. Péault, Nogent-sur-Marne).

conditions both programmes do not differ and might generate the high neuro-muscular specifity.

Later on, blood-derived cells contribute to the development of the skeleton. It has been proved that the chondroclasts and osteoclasts are of haematogenous origin (JOTEREAU & LE DOUARIN, 1978, JACOB *et al.*, 1986).

Summing up, it can be stated that the limb bud mesenchyme is already composed of cells of different origin, which are committed to different lineages. The somatopleure-, somite-, neural crest-, and blood-derived cells cannot replace one another during development.

Regarding the mechanisms leading to the characteristic spatial arrangement of different cells, at first the migration, distribution and goal recognition modalities of already-committed cells must be taken into consideration. Regarding the somite-derived myogenic cells as an example, these events have been discussed previously.

On the other hand, pattern formation also concerns the somatopleure-derived limb bud cells, which initially have the option to differentiate into fibroblasts, chondroblasts or smooth muscle cells dependent on their position within the limb bud (WACHTLER *et al.*, 1982, CHRIST & JACOB, 1986). There must be a framework that governs and coordinates the developmental processes, but up to now this is far from being understood. One problem is that, on the one hand, the prepattern is situated within the somatopleure-derived mesenchyme and, on the other hand, this mesenchyme differentiates according to its position within the limb bud as a whole. If in a stage 24 (HH) chick wing a piece of dermis or muscular connective tissue is experimentally shifted to the central chondrogenic zone, it differentiates «ortsgemäß», that is to say, into cartilage (WACHTLER *et al.*, 1982). This phenomenon that parts develop according to their position within the whole, was first described by DRIESCH (1892) for the sea-urchin embryo. The molecular mechanisms involved in positional signalling of limb bud cells are not yet known. The hypothesis postulated by SUMMERBELL *et al.* (1973) is an interesting approach to explain the skeletal pattern formation along the posterior-anterior and proximo-distal axes. According to this view, the mesodermal cells born in succession within the progress zone beneath the AER are thought to obtain different values with respect to various limb parts. Along the anterior-posterior axis the cells are considered to have the ability to recognize different concentrations of a morphogen, which should be produced in the posterior part of the limb bud (zone of polarizing activity, ZPA). However, BRAND (1987) and BRAND-SABERI & CHRIST (1989) were able to show that distal wing parts that did not contain cells belonging to the ZPA region give rise to normal feet when isolated and cultivated ectopically.

Another approach came from BLECHSCHMIDT (1963) who emphasized the importance of non-ridge ectoderm on the stratification of the limb bud mesenchyme. This ectodermal influence might be based on an «antichondrogenic» activity of the ectoderm, which prevents the differentiation of cartilage within the dermis and myogenic zone. BRAND *et al.* (1985) reported on limb buds cultured in the coelomic cavity or on the chorioallantoic membrane, which used to chondrify in their peripheral zones if the ectoderm was lacking or replaced by another kind of epithelium. According to these findings, the shape of the early chondrogenic zone can be understood to reflect the shape of the whole limb bud. This might also be the case in the developing hand. It is not yet clear how the shaping of single skeletal elements takes place. Cell death might play a role in the separation of skeletal elements. FELL & CANTI (1935) described an area of cell death in the chick limb («opaque patch») which can be found between radius and ulna. Experimental results have given evidence that the early shaping and articulations of the skeletal elements do not depend on muscle functions (JACOB *et al.*, 1986).On the other hand, MÜLLER (1986) has shown that changes in the skeletal pattern leads to alterations in muscle pattern.

ACKNOWLEDGEMENTS — The studies presented in this paper were supported by a grant Ch 44/6-1 from the Deutsche Forschungsgemeinschaft. The authors are much obliged to Mrs. I. Schaeben-Hamm, Mrs. U. Ritenberg and Mr. H. Hake for expert technical assistance and to Mrs. B. Scharf for typing the manuscript.The authors wish to thank Prof. Dr. med. K. Hinrichsen, Ruhr-University Bochum, Dept. of Anatomy and Embryology, for the permission to use the Figures 1, 3, 5, 7, 8, 10.

References

AGNISH N.D., KOCHHAR D.M., 1977. *The role of somites in the growth and early development of mouse limb buds.* Develop Biol, 56: 174-183.

BLECHSCHMIDT E., 1951. *Die frühembryonale Lageentwicklung der Gliedmaßen.* (Entwicklung der Extremitäten beim Menschen, Teil I). *Die frühembryonale Formentwicklung der Gliedmaßen.* (Entwicklung der Extremitäten beim Menschen, Teil II). *Die frühembryonale Strukturentwicklung der Gliedmaßen.* (Entwicklung der Extremitäten beim Menschen, Teil III). Z Anat Entw-Gesch, 115: 529-540; 597-616; 617-657.

BLECHSCHMIDT E., 1956. *Entwicklungsfunktionelle Untersuchungen am Bewegungsapparat.* Acta Anat, 27: 62-68.

BLECHSCHMIDT E., 1961. *Die vorgeburtlichen Entwicklungsstadien des Menschen.* S Karger, Basel.

BLECHSCHMIDT E., 1963. *Die Gestaltungsfunktionen der menschlichen Oberhaut. I. Mitteilung. Die Schichtenbildung des Integuments.* Arch f klin exp Derm, 215: 567-578.

BRAND B. 1987. *Zur Morphogenese und Musterbildung der Vogelextremität unter besonderer Berücksichtigung der Migration der myogenen Zellen. Eine experimentelle, ultrastrukturelle und vergleichende Untersuchung.* Dissertation, Bochum.

BRAND-SABERI B., CHRIST B., 1989. *Zum Einfluß der zone of polarising activity (ZPA) auf die Musterbildung von Vogelextremitäten.* Verh Anat Ges, 84: (im Druck).

BRAND B., CHRIST B., JACOB H.J., 1985. *An experimental analysis of the developmental capacities of distal parts of avian leg buds.* Am J Anat, 173: 321-340.

CAIRNS J.M., 1977. *Growth of normal and talpid 2 chick wing buds: an experimental analysis.* In: Vertebrate Limb and Somite Morphogenesis (DA Ede, JR Hinchliffe, M Balls, eds). Cambridge University Press, pp 123- 137.

CAPLAN A.J., KOUTROUPAS S., 1973. *The control of muscle and cartilage development in the chick limb: the role of differential vascularization.* J Embryol Exp Morph, 29: 571-583.

CHEVALLIER A., KIENY M., MAUGER A., 1977. *Limb-somite relationship: origin of the limb musculature.* J Embryol Exp Morph, 41: 245-258.

CHRIST B., JACOB H.J., 1986. *Morphogenesis, Musterbildung und Zell- migration als Teilprozesse der Extremitätenentwicklung bei Amniotenembryonen. Eine Synopsis experimenteller Befunde.* Verh Anat Ges 80: 115-126.

CHRIST B., JACOB H.J., JACOB M., 1974. *Über den Ursprung der Flügelmuskulatur. Experimentelle Untersuchungen mit Wachtel- und Hühnerembryonen.* Experientia, 30: 1446-1448.

CHRIST B., JACOB H.J., JACOB M. 1977. *Experimental analysis of the origin of the wing musculature in avian embryos.* Anat Embryol, 150: 171-186.

CHRIST B., JACOB H.J., JACOB M., 1979. *Differentiating abilities of avian somatopleural mesoderm.* Experientia, 35: 1376-1378.

CHRIST B., HINRICHSEN K., JACOB H.J., 1987. *Morphologische Aspekte der Handentwicklung (Referat).* Verh Anat Ges, 81: 201-210.

CHRIST B., JACOB H.J., JACOB M., WACHTLER F., 1983. *On the origin, distribution and determination of avian limb mesenchymal cells.* In: Limb Development and Regeneration (RO Kelley, PF Goetinck, JA MacCabe, eds), Part B. Alan R Liss, New York, pp. 281-291.

CHRIST B., JACOB H.J., JACOB M., BRAND B. 1986a. *Principles of hand ontogenesis in man.* Acta Morphol Neerl-Scand, 24: 249-268.

CHRIST B., JACOB M., JACOB H.J., BRAND B., WACHTLER F., 1986b. *Myogenesis: a problem of cell distribution and cell interactions.* In: Somites in Developing Embryos (R Bellairs, DA Ede, JW Lash, eds) NATO ASI Series Vol 119. Plenum Press, New York, London, pp 269-275.

DETWILER S.R., 1933. *On the time of determination of the antero-posterior axis of the fore-limb in Amblystoma.* J Exp Zool, 64: 405-414.

DIENSTMAN S.R., BIEHL J., HOLTZER S., HOLTZER H. 1974. *Myogenic and chondrogenic lineages in developing limb buds grown in vitro.* Develop Biol 39: 83-95.

DRIESCH H., 1892. *Entwicklungsmechanische Studien. III. Die Verminderung des Furchungsmaterials und ihre Folgen (Weiteres über Theilbildungen). IV. Experimentelle Veränderungen des Typus der Furchung und ihrer Folgen (Wirkungen von Wärmezufuhr und von Druck). V. Von der Furchung doppeltbefruchteter Eier. VI. Über einige allgemeine Fragen der theoretischen Morphologie.* Zeitschr f Wiss Zool 55: 1-62.

FEINBERG R.N., BEEBE D.C. 1983. *Hyaluronate in vasculogenesis.* Science, 220: 1177-1179.

FELL H.R. & CANTI R.G., 1935. *Experiments on the development in vitro of the avian knee joint.* Proc R Soc London, B 116: 316-351.

GRIM M., WACHTLER F., 1990. *Muskelmorphogenese ohne Muskelzellen.* Verh Anat Ges 85.

GRIM M., NENSA K., CHRIST B., JACOB H.J., TOSNEY K.W., 1989. *A hierarchy of determining factors controls motoneuron innervation. Experimental studies on the development of the plantaris muscle (PL) in avian chimeras.* Anat Embryol 180: 179-189.

HARRISON R.G., 1921. *On relations of symmetry in transplanted limbs.* J Exp Zool 32: 1-136.

HINCHLIFFE J.R. 1974. *The pattern of cell death in chick limb morphogenesis.* Libyan J Sci, 4A: 23-32.

HINCHLIFFE J.R., EDE D.A., 1967. *Limb development in the polydactylons talpid 2 mutant of the fowl.* J Embryol Exp Morph, 17: 385-404.

HOCHSTETTER F. 1955. *Über die Entwicklung der Form der menschlichen Gliedmaßen.* Akad Wiss Wien, math-naturwiss Kl 109: 3. Abhdl.

HOLLYDAY M., 1980. *Motoneuron histogenesis and the development of limb innervation.* Curr Top Dev Biol, 15: 181-215.

HURLE J.M., COLVEE E., FERNANDEZ-TERAN M.A. 1985. *Vascular regression during the formation of the free digits in the avian limb bud: a comparative study in chick and duck embryos.* J Embryol Exp Morph, 85: 239-250.

JACOB H.J., HINRICHSEN K., 1984. *Rasterelektronenmikroskopische Dokumentation zur Entwicklung von Hand und Fuß bei menschlichen Embryonen.* Verh Anat Ges 78: 285-287.

JACOB H.J., CHRIST B., GRIM M. 1983. *Problems of muscle pattern formation and of neuromuscular relations in avian limb development.* In: Limb Development and Regeneration (RO Kelley, PF Goetinck, JA MacCabe, eds), Part B. Alan R Liss, New York, pp. 333-341.

JACOB H.J., CHRIST B., JACOB M. 1986. *Origin and development of avian limb muscles.* Acta Histochem Suppl-Bd XXXII: 145-150.

JORQUERA B., PUGIN E. 1971. *Sur la compartement du mésoderme et de l'ectoderme du bourgeon de membre dans les échanges entre le poulet et le rat.* CR Hebd Séanc Acad Sci Paris D 272: 1522- 1525.

JOTEREAU F, LE DOUARIN N., 1978. *The developmental relationship between osteocytes and osteoclasts. A study using quail-chick nuclear markers in endochondral ossification.* Develop Biol 63: 253-265.

KEIBEL F., ELZE C. 1908. *Normentafel zur Entwicklungsgeschichte des Menschen.* Fischer, Jena.

KELLEY R.O. 1970. *An electron microscopic study of mesenchyme during development of interdigital spaces in man.* Anat Rec, 168: 43-54.

KELLEY R.O., FALLON J.F. 1976. *Ultrastructural analysis of the apical ectodermal ridge during vertebrate limb morphogenesis. I. The human forelimb with special reference to gap junctions.* Develop Biol, 51: 241-256.

KELLEY R.O., FALLON J.F. 1981. *The developing limb: an analysis of interacting cells and tissues in a model morphogenetic system.* In: Morphogenesis and Pattern Formation (TG Connelly, LL Brinkley BM Carlson, eds) Raven Press, New York, pp 49-85.

KIENY M. 1969. *Sur les relations entre le mésoderme somitique et le mésoderme somatopleurale avant et en cours de l'induction primaire des membres de l'embryon du poulet.* CR Hebd Séanc Acad Sci Paris D 268: 3183-3186.

KIENY M. 1983. *Cell and tissue interactions in the organogenesis of the avian limb musculature.* In: Limb Development and Regenetation (RO Kelley, PF Goetinck, JA MacCabe, eds), Part B. Alan R Liss, New York, pp 293-302.

KIENY M. & CHEVALLIER A. 1979. *Autonomy of tendon development in the embryonic chick wing.* J Embryol Exp Morph, 49: 153-165.

LANDMESSER L. 1980. *The generation of neuromuscular specifity.* Ann Rev Neurosci, 3: 279-302.

LE DOUARIN N., BARQ G. 1969. *Sur l'utilisation des cellules de la caille japonaise comme «marqueurs biologiques» en embryologie expérimentale.* CR Hebd Séanc Acad Sci Paris, D 269: 1543- 1546.

MENKES B., DELEANU M., 1964. *Leg differentiation and experimental syndactyly in chick embryo.* Rev Roumaine Embryol Cytol, 1: 69-77.

MENKES B., DELEANU M., Ilies A., 1965. *Comparative study of some areas of physiological necrosis of the embryo of man, laboratory mammalians and fowl.* Rev Roumaine Embryol Cytol, 2:161- 172.

MILAIRE J., 1977. *Rudimentation digital au cours du dévelopment normale de l'autopode chez les mammifères.* In: Mécanismes de la Rudimentation des Organes chez les Embryons de Vertébrés, Edition du CNRS 266, pp 221-233, Paris.

MÜLLER G. 1986. *Effects of skeletal change on muscle pattern formation.* Biblthca Anat, Karger, Basel, 29: 91-108.

MURRAY P.D.D.F., 1926. *An experimental study of the development of the limbs of the chick.* Proc Linn Soc NSW, 51: 187-263.

NEWMAN S.A. 1977. *Lineage and pattern in the developing wing bud.* In: Vertebrate Limb and Somite Morphogenesis (DA Ede, JR Hinchliffe, M Balls, eds). Cambridge University Press, London, p 181.

PAUTOU M.P., 1968. *Rôle déterminante du mésoderme dans la différenciation spécifique de la patte de l'oiseau.* Arch Anat Micr Morph Exp, 48:311-328.

PINOT M. 1970. *Le rôle du mésoderme somitique dans le morphogenèse précoce des membres de l'embryon de poulet.* J Embryol Exp Morph, 23: 109-151.

SAUNDERS J.W. 1948. *The proximo-distal sequence of the origin of the parts of the wing and the role of the ectoderm.* J Exp Zool, 108: 363-404.

SAUNDERS J.W., GASSELING M.T. & GFELLER, SR M.D. 1958. *Interactions of ectoderm and mesoderm in the origin of axial relationship in the wing of the fowl.* J Exp Zool, 137: 39-74.

SEICHERT V., RYCHTER Z. 1971. *Vascularization of the developing anterior limb of the chick embryo.* Folia Morphol XIX: 367-377.

SUMMERBELL D. 1974. *A quantitative analysis of the effect of the excision of the AER from the chick limb bud.* J Embryol Exp Morph, 32: 227-237.

SUMMERBELL D. 1977. *Reduction of the rate of growth, cell density and cell division following removal of the apical ectodermal ridge of the chick limb bud.* J Embryol Exp Morph, 40: 1-21.

SUMMERBELL D., LEWIS J.H., WOLPERT L., 1973. *Positional information in chick limb morphogenesis.* Nature 224: 492-496.

WACHTLER F. 1984. *On the differentiation and migration of some non-neuronal neural crest derived cell types.* Anat Embryol 170: 161-168.

WACHTLER F., CHRIST B., JACOB H.J. 1981. *Ein experimenteller Beitrag zur Entstehung digitaler Beugefurchen.* Acta Anat, 110: 103-107.

WACHTLER F., CHRIST B., JACOB H.J. 1982. *Grafting experiments on determination and migratory behaviour of presomitic, somitic and somatopleural cells in avian embryos.* Anat Embryol, 164: 369-378.

WILSON D. 1983. *The origin of the endothelium in the developing marginal vein of the chick wing bud.* Cell Diff., 13: 63-67.

ZWILLING E. 1955. *Ectoderm-mesoderm relationship in the development of the chick embryo limb bud.* J Exp Zool, 128: 423-441.

ZWILLING E. 1966. *Cartilage formation from so-called myogenic tissue of chick embryo limb buds.* Ann Med Exp Biol Fenn, 44: 134-139.

MIX
Papier aus verantwortungsvollen Quellen
Paper from responsible sources
FSC® C105338

If you have any concerns about our products,
you can contact us on
ProductSafety@springernature.com

In case Publisher is established outside the EU,
the EU authorized representative is:
Springer Nature Customer Service Center GmbH
Europaplatz 3, 69115 Heidelberg, Germany

Printed by Libri Plureos GmbH
in Hamburg, Germany